MC68HC11

An Introduction

Software and Hardware Interfacing

MC68HC11

An Introduction

Software and Hardware Interfacing

2nd Edition

Han-Way Huang
Minnesota State University • Mankato

Africa • Australia • Canada • Denmark • Japan • Mexico • New Zealand • Philippines
Puerto Rico • Singapore • Spain • United Kingdom • United States

NOTICE TO THE READER

Delmar Staff
Publisher: Alar Elken
Executive Editor: Sandy Clark
Acquisitions Editor: Gregory L. Clayton
Developmental Editor: Michelle Ruelos Cannistraci
Editorial Assistant: Jennifer Thompson
Executive Marketing Manager: Maura Theriault

Channel Manager: Mona Caron
Executive Production Manager: Mary Ellen Black
Production Manager: Larry Main
Senior Project Editor: Christopher Chien
Art/Design Director: David Arsenault
Marketing Coordinator: Paula Collins
Technology Project Manager: Tom Smith

Asia
Thomson Learning
60 Albert Street, #15-01
Albert Complex
Singapore 189969

Japan
Thomson Learning
Palaceside Building 5F
1-1-1 Hitotsubashi, Chiyoda-ku
Tokyo 100 0003 Japan

Australia/New Zealand:
Nelson/Thomson Learning
102 Dodds Street
South Melbourne, Victoria 3205
Australia

UK/Europe/Middle East
Thomson Learning
Berkshire House
168-173 High Holborn
London WC1V 7AA United Kingdom

Thomas Nelson & Sons LTD
Nelson House
Mayfield Road
Walton-on-Thames
KT 125 PL United Kingdom

Latin America
Thomson Learning
Seneca, 53
Colonia Polanco
11560 Mexico D.F. Mexico

Canada
Nelson/Thomson Learning
1120 Birchmount Road
Scarborough, Ontario
Canada M1K 5G4

Spain
Thomson Learning
Calle Magallanes, 25
28015-MADRID
ESPANA

International Headquarters
Thomson Learning
International Division
290 Harbor Drive, 2nd Floor
Stamford, CT 06902-7477

Library of Congress Cataloging-in-Publication Data
ISBN# 0-7668-1600-1

Huang, Han-Way.
    MC 68HC11, an introduction : software and hardware interfacing / Han-Way Huang.—
    2nd ed.
        p.   cm.
    Includes bibliographical references and index.
    ISBN 0-7668-1600-1 (alk. paper)
    1. Digital control systems. 2. Electronic controllers. I. Title.

TJ223.M53 H95 2000
629.89—dc21

00-023001

Contents

Chapter 3   Data Structures and Subroutine Calls   113

Chapter 4   C Language Programming   163

Chapter 5   Operation Modes and Memory Expansion   199

Chapter 6   Interrupts and Resets   241

Chapter 9   68HC11 Serial Communication Interface   401

Appendixes

Glossary   783

Index   791

Preface

Intended Audience

This book is intended to serve two groups of readers:

1. Students in Electrical and Computer Engineering, Electrical, Electronics, and Computer Engineering Technology Programs who would like an introduction to microprocessor and microcontroller applications. For those who are exposed to the microprocessor for the first time, this book provides a broad and systematic introduction to microprocessor interfacing using the Motorola 68HC11 as an example. This book can also benefit senior year college students if they want to learn the 68HC11 microcontroller and use it in design projects.

2. Working technicians and engineers who are familiar with other microcontrollers and would like to learn specifically about the Motorola 68HC11 microcontroller. Comprehensive examples are provided to explore every function available in the 68HC11. Fundamental topics can be skipped for this group of readers.

Prerequisites

This book is designed for a one-semester or two-quarter course sequence at the introductory level with or without programming background and for practicing engineers who want to learn the 68HC11. Basic knowledge about number systems and digital logic design capability are required to read this book. An extensive appendix on number systems is provided for those who are not familiar with this subject. Readers who are not familiar with logic design should refer to a book on digital logic design. Many good books can be found by searching the Internet, for example, at www.amazon.com and www.barnesandnoble.com.

Approach

During the past ten years, we have seen two trends in microprocessor education:

1. Using the microcontroller as a vehicle for teaching microprocessor interfacing.
2. Using high-level languages in microprocessor programming.

This text follows these trends. The Motorola 68HC11 is one of the most popular 8-bit microcontrollers. People who are familiar with the 68HC11 can learn other microcontrollers easily. Microprocessor evaluation boards are essential for learning microprocessor programming and I/O interfacing. This text covers the Motorola EVB and EVBU because they have been around since the 68HC11 was introduced. The CMD-11A8 is also used because it is less expensive and it also performs several I/O experiments easily. Several high-level languages have been used in microprocessor programming. This text uses C language because of its popularity.

Advantages of Microcontrollers

Recent advances in electronic semiconductor technology have resulted in the development of highly integrated microprocessors and microcontrollers. Microprocessors today are used not only in many desktop personal computers but also in many more embedded applications such as fax and copying machines, laser printers, and communication controllers. However, a microprocessor needs external memories, peripheral interface chips and other logic circuits to build a complete system, and these requirements complicate the design of microprocessor-based products. A microcontroller incorporates a small amount of memory, timers, A/D and D/A converters, and other peripheral functions in one chip (along with the CPU), and these extra resources are adequate for many embedded applications. Microcontrollers not only simplify the design of many embedded products but also have the advantage of smaller size and lower power consumption.

Assembly Language and C Language

An embedded product consists of both hardware and software. In the early days of microprocessor applications, most design engineers programmed in assembly language. However, with the sophistication of application problems at hand, design engineers quickly discovered the low productivity provided by the assembly language. Therefore, using high-level languages to program microprocessors and microcontrollers has quickly become the common practice in developing embedded systems. This second edition reflects this trend. Although many high-level languages are being used in embedded programming, we chose C language for its popularity. Chapter 4 of this edition serves as a tutorial to the C language for those who are not familiar with C. The ICC11 C compiler from ImageCraft has been chosen as the compiler to be used in testing all of the C programs in this edition. ICC11 is inexpensive and easy to use. Some minor tunings need to be done in order to make it work with the demo boards that are used in this book. Chapter 4 and several later chapters explain them.

Most example programs in chapter 6 to chapter 11 are written in both assembly and C languages. Assembly language is covered for several reasons:

- Assembly language is very useful for understanding the architecture of microprocessors and microcontrollers.

- Programs written in assembly language are normally much smaller than those written in high-level languages. This is especially important for those applications that can afford only a very small amount of memory and in which running time is critical.

- Assembly language is still used in the industry. Many companies mix the use of both languages.

Hardware Evaluation Boards

Hardware evaluation boards allow us to execute programs translated into the 68HC11 machine language. They enable us to practice the basic assembly languange and C language programming, and I/O interfacing for the 68HC11. This book introduces the Motorola EVB and EVBU demo boards and Axiom Manufacturing's CMD-11A8 board. These three demo boards all include the Buffalo monitor. The Buffalo monitor allows us to set and display the contents of memory locations and registers, set breakpoints, trace the execution of instructions, and download programs onto the board for execution. A tutorial on using these boards is provided in Chapter 2.

The CMD-11A8 was designed and fabricated by Axiom Manufacturing Inc. The CMD-11A8 uses the same Buffalo monitor as EVB and EVBU. People who have been familiar with EVB and EVBU should feel at home with this board. The CMD-11A8 provides easy interface to LCD and keypads and is sold with a serial cable and power adapter.

Textbook Organization

Both programming and hardware interfacing issues are explored in detail in this book. Software and hardware development goes hand-in-hand throughout. The text begins with the programming model of the 68HC11. Chapter 1 gives a brief overview of the hardware and software of a computer system, introduces different types of memory technologies, and presents the programming model and the instruction execution cycles of the 68HC11 microcontroller. Chapter 2 introduces assembly language syntax and assembler directives, explores the implementation of single- and multi-precision arithmetic, logic operations, delay times, and program loops using appropriate 68HC11 instructions. Software and hardware development tools are also reviewed in this chapter. Chapter 3 describes different types of data structures and their operations, details the issues related to subroutine calls including parameter passing, results returning, and local variables allocation. Many examples on the use of the 68HC11 EVB I/O library routines are also included at the end of this chapter.

Chapter 4 is a tutorial to C language programming. All C language constructs that will be used in this book are covered. A tutorial on how to use the ICC11 Compiler to develop C programs and download to the demo board for execution is also provided. Most of the examples from chapter 6 to 11 include both assembly and C programs.

Chapter 5 introduces the setup of the 68HC11 operation mode and the DRAM and SRAM technologies, explores memory system design issues including memory space assignment, decoding methods, and the decoder design. It presents the conventions used in timing diagrams and the 68HC11 read and write bus cycle timing diagrams. An extensive example explains the design process of adding an external SRAM chip to the 68HC11. Many diagrams are used to explain the timing verification process.

Chapter 6 introduces the concepts and handling of interrupts and resets, and explores the details of the 68HC11 interrupt and reset mechanisms. Chapter 7 presents basic input/output concepts and explores the 68HC11 I/O ports. Examples are given to demonstrate the interfacing of input/output devices including DIP switches, LEDs, seven-segment displays, LCDs, D/A converters, keyboards and keypads, and printers to the 68HC11. The Intel Parallel Peripheral Interface (PPI) chip i8255 is also discussed. This demonstrates that we can mix the use of chips from different vendors.

Chapter 8 discusses the functions and applications of the 68HC11 timer system including the input-capture, output-compare, real-time interrupt, and pulse accumulator functions. Numerous examples are used to demonstrate the measurement of frequencies and pulse widths, generation of delays and digital waveforms, generation of periodic interrupts, the capture of arrival times, and so on.

Chapter 9 discusses asynchronous serial communication and the serial communication interface (SCI) of the 68HC11. Examples are given to illustrate hardware interfacing, timing analysis, and programming of the SCI and also of the Motorola 6850 chip. Chapter 10 presents the 68HC11 Serial Peripheral Interface (SPI). This versatile interface implements a protocol that enables any peripheral device that conforms to this protocol to interface to the 68HC11. Chapter 11 deals with the 68HC11 A/D converter. This chapter not only explains the operation of the 68HC11 A/D converter but also explains analog signal conditioning, result interpretation, and result processing. Complete examples on interfacing the temperature sensor and the humidity sensor are included.

At the end of the book, you will find useful information included as appendices. Appendix A is an extensive treatment of number systems. Appendix B explores the process of developing a microcontroller-based product and also suggests several workable design projects. Appendix C is a summary of I/O register contents for the 68HC11A8. Appendix D lists all the 68HC11 instructions in alphabetical order. Appendix E is a table of 68HC11 interrupt vectors. Appendix F is a partial list of SPI-compatible devices. Development tools vendors are listed in Appendix G. Standard resistors are listed in Appendix H. All the terms used in this text are listed in the Glossary at the end of this book for easy search.

Retained Features

Special attention is paid to the pedagogy. Each chapter opens with a list of Objectives. Every subject is presented in a step-by-step manner. Background issues are presented before the specifics related to each 68HC11 function are discussed. Numerous examples are then presented to demonstrate the use of each 68HC11 I/O function. Procedural steps and flowcharts are used to help the reader understand the program logic in most examples. Each chapter concludes with a Summary and numerous exercise problems and lab assignments. A comprehensive Glossary is placed at the end of the book.

New Features

- Integrated coverage of both assembly language and C language.
- New chapter on C Language, which includes a tutorial on using a C compiler.
- Added C program to all of the examples.
- New coverage on the CMD-11A8 Axiom board in addition to the EVB and EVBU boards.
- Expanded Appendices include reference material that should be useful in looking up technical information and gathering data that is not covered in this book.
- New Design Projects provides design examples and discusses the design analysis process.
- CD in back of the book will help readers learn the 68HC11. Program examples, data sheets and several freeware programs — including as11, small C compiler, and sim68 — are included. The demo version of the ICC11 is also included. Data books and reference manuals about the A and E series of the 68HC11 microcontroller are also included in this CD.
- Program examples available on the CD are indicated in the textbook margin by a CD icon.
- Additional chapter problems.
- Additional lab exercises.

Supplements

Instructor's Resource CD. This CD includes solutions to all exercise problems and more than 700 Power Point Presentation Slides created by the author. ISBN: 0-7668-1601-X

Delmar's Electronics Technology Website. Visit *MC68HC11: An Introduction, 2e's* Online Companion at www.electronictech.com for additional programming examples, projects, and text updates.

Acknowledgments

This book would not have been possible without the help of a number of people, and I would like to express my gratitude to all of them. I would like to thank Jay Farnam, Chad Friedler, Tony Plutino, and other people at Motorola who supported my teaching over the years. I would like to thank my EE and EET students who gave me the motivation to write this book. I would also like to thank Greg Clayton, Acquisitions Editor of Delmar Thomson Learning, and Michelle Ruelos Cannistraci, Developmental Editor, for their enthusiastic support during the preparation of this book. I also appreciate the outstanding work of the production staff, including Larry Main, Christopher Chien, and David Arsenault at Delmar Thomson Learning.

I would also like to express my thanks for the many useful comments and suggestions provided by colleagues who reviewed this text during the course of its development, especially to David Delker, Kansas State University; Lee Harrison, State University of New York, Albany; Rick Henderson, DeVry Institute of Technology, Kansas City, MO; Daniel Metzger, Monroe College; Michael Miller, DeVry Institute of Technology, Phoenix, AZ; Chris Panayiotou, Indian River College; Carlo Sapijaszko, DeVry Institute of Technology, Alberta, Canada; Gilbert Seah, DeVry Institute of Technology, Ontario, Canada; Roman Stemprok, University of North Texas; and Jamie Zipay, DeVry Institute of Technology, Long Beach, CA.

Finally, I am grateful to my wife, Su-Jane and my sons, Craig and Derek, for their encouragement, tolerance, and support during the entire preparation of this book.

Han-Way Huang
May, 2000

About the Author

Dr. Han-Way Huang received his M.S. and Ph.D. degrees in Computer Engineering from Iowa State University. He has taught microprocessor and microcontroller applications extensively for 14 years. Before teaching at Minnesota State University, Mankato, Dr. Huang worked for four years in the computer industry. In addition to this book, Dr. Huang has also authored a book on the Intel 8051 microcontroller, *Using the MCS-51 Microcontroller* from Oxford University Press.

1

Introduction to
Motorola 68HC11

1.1 Objectives

After completing this chapter you should
be able to

- define or explain the following terms:
  computer, processor, microprocessor,
  microcontroller, hardware, software, cross
  assembler, cross compiler, RAM, SRAM,
  ROM, EPROM, EEPROM, flash memory,
  byte, nibble, bus, KB, MB, mnemonic,
  opcode, and operand

- explain the differences between the
  immediate, direct, extended, indexed,
  relative, and inherent addressing modes

- disassemble machine code into mnemonic
  assembly language instructions

- explain the 68HC11 instruction execution
  cycles

1.2   What is a Computer?

A computer is made up of hardware and software. The computer hardware consists of four main components: (1) a processor, which serves as the computer's "brain," (2) an input unit, through which programs and data can be entered into the computer, (3) an output unit, on which computational results can be displayed, and (4) memory, in which the computer software programs and data are stored. Figure 1.1 shows a simple block diagram of a computer. The processor communicates with memory and input/output (I/O) devices through a set of signal lines referred to as a *bus*. The common bus actually consists of three buses: a *data* bus, an *address* bus, and a *control* bus.

1.2.1   The Processor

The processor, which is also called the *central processing unit* (CPU), can be further divided into three major parts:

Registers. A register is a storage location in the CPU. It is used to hold data and/or a memory address during execution of an instruction. Access to data in registers is faster than access to data in memory. Registers play an essential role in the efficient execution of programs. The number of registers varies greatly from computer to computer.

Arithmetic logic unit. The arithmetic logic unit (ALU) is the computer's numerical calculator and logical operation evaluator. The ALU receives data from main memory and/or registers, performs a computation, and, if necessary, writes the result back to main memory or registers. Today's fastest personal computer can perform hundreds of millions of additions in one second.

Control unit. The control unit contains the hardware instruction logic. The control unit decodes and monitors the execution of instructions. The control unit also acts as an arbiter as various portions of the computer system compete for the resources of the CPU. The activities of the CPU are synchronized by the system clock. All CPU activities are measured by *clock cycles*. The clock rates of modern microprocessors have reached more than 800 MHz by the time of this writing, where

1 MHz = 1 million ticks (or cycles) per second

The period of a 1 MHz clock signal is 1 μs (10^{-6} second). The control unit also maintains a register called a *program counter* (PC), which controls the memory address of the next instruction to be executed. During the execution of an instruction, the presence of overflow, an addition carry, a subtraction borrow, and so forth, are flagged by the system and stored in another register called a *status register*. The resultant flags are then used by the programmer for program control and decision-making.

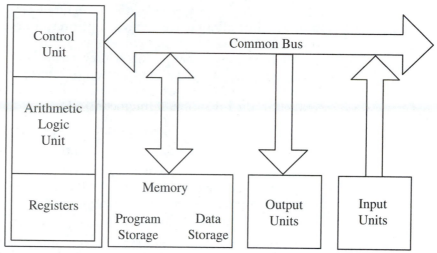

Figure 1.1 ■ Computer Organization

WHAT IS A MICROPROCESSOR?

The processor in a very large computer is built from a number of integrated circuits. A *microprocessor* is a processor packaged in a single integrated circuit. A *microcomputer* is a computer that uses a microprocessor as its CPU. Early microcomputers are quite simple and slow. However, many of today's desktop microcomputers have become very sophisticated and are even faster than many large computers manufactured only a few years ago.

Microprocessors come in 4-bit, 8-bit, 16-bit, 32-bit, and 64-bit models. The number of bits refers to the number of binary digits that the microprocessor can manipulate in one operation. A 4-bit microprocessor, for example, is capable of manipulating 4 bits of information in one operation. Four-bit microprocessors are used for the electronic control of relatively simple machines. Some pocket calculators, for example, contain 4-bit microprocessors.

The access of main memory takes much longer than the period of the CPU control clock signal. To enhance the CPU performance, many 32-bit and 64-bit microprocessors contain on-chip high-speed *cache memory*. The cache memory has a copy of the most recently accessed instructions and data. A cache memory is much smaller than the main memory. However, due to the fact that program execution demonstrates locality behavior in its references to instructions and data (CPU tends to access a small area of the memory in a short period of time), a relatively small cache can achieve very high hit ratio. Since a cache memory can satisfy the memory requests from the CPU most of the time, adding cache memory on the CPU chip improves the processor performance dramatically.

Because processors and input/output have very different characteristics and speeds, peripheral chips are required to interface I/O devices to the microprocessor. For example, the integrated circuit M6821 is designed to interface a parallel device such as a printer or seven-segment display to the Motorola M6800 8-bit microprocessors.

Microprocessors have been widely used since their invention. It is not exaggerating to say that the invention of microprocessors has revolutionized the electronics industry. However, the following limitations of microprocessors led to the invention of microcontrollers:

- A microprocessor requires external memory to execute programs.
- A microprocessor cannot directly interface to I/O devices; peripheral chips are needed.
- Glue logic (such as address decoders and buffers) is needed to interconnect external memory and peripheral interface chips to the microprocessor.

Because of these limitations, a microprocessor-based design cannot be made as small as might be desirable. The invention of microcontrollers not only eliminated most of these problems but also simplified the hardware design of microprocessor-based products.

WHAT IS A MICROCONTROLLER?

A *microcontroller* is a computer implemented on a single very large scale integration (VLSI) chip. A microcontroller contains everything contained in a microprocessor along with one or more of the following components:

- memory
- timer
- analog-to-digital converter
- digital-to-analog converter
- direct memory access (DMA) controller
- parallel I/O interface (often called a parallel port)
- serial I/O interface
- memory component interface circuitry

The Motorola 68HC11 is an 8-bit microcontroller family developed in 1985. The 68HC11 microcontroller family has more than fifty members, and the number is still increasing. The microcontrollers in this family differ mainly in the size of their on-chip memories and in their I/O capabilities. The characteristics of different memory technologies will be discussed shortly. As shown in Figure 1.2, the 68HC11A8 has the following features:

- 256 bytes on-chip static random access memory (SRAM)
- 512 bytes on-chip electrically-erasable, programmable read-only memory (EEPROM)
- 8-KB on-chip read-only memory (ROM)

- three input-capture functions (ICi, i = 1, 2, 3)
- five output-compare functions (OCi, i = 1,..., 5)
- an 8-bit pulse accumulator circuit
- a serial communication interface (SCI)
- an 8-channel, 8-bit analog-to-digital converter
- a serial peripheral interface (SPI)
- a real-time interrupt (RTI) circuit
- a computer operating properly (COP) watchdog system

These functions will be discussed in detail in subsequent chapters.

APPLICATIONS OF MICROCONTROLLERS

Since their introduction microcontrollers have been used in every application that we can imagine. They are used as controllers for displays, printers, keyboards, modems, charge card phones, and home appliances such as refrigerators, washing machines, and microwave ovens. They are also used to control the operation of automobile engines and machines in factories. Today, most homes have one or more microcontroller-controlled appliances.

1.2.2 Memory

Memory is where software programs and data are stored. A computer may contain semiconductor, magnetic, and/or optical memory. Only semiconductor memory will be discussed in this book. Semiconductor memory can be further classified into two major types: *random-access memory* (RAM) and *read-only memory* (ROM).

RANDOM-ACCESS MEMORY

Random-access memory is *volatile* in the sense that it cannot retain data in the absence of power. RAM is also called *read/write memory* because it allows the processor to read from and write into it. Both read and write accesses to a RAM chip take roughly the same amount of time. The microprocessor can temporarily store or write data into RAM, and it can later read that data back. Reading memory is nondestructive to the contents of the memory location. Writing memory is destructive. When the microprocessor writes data to memory, the old data is written over and destroyed.

There are two types of RAM technologies: *static RAM* (SRAM) and *dynamic RAM* (DRAM). Static RAM uses from four to six transistors to store one bit of information. As long as power is stable, the information stored in the SRAM will not be degraded. Dynamic RAM uses one transistor and one capacitor to store one bit of information. The information is stored in the capacitor in the form of an electric charge. The charge stored in the capacitor will leak away over time, so a periodic refresh operation is needed to maintain the contents of DRAM.

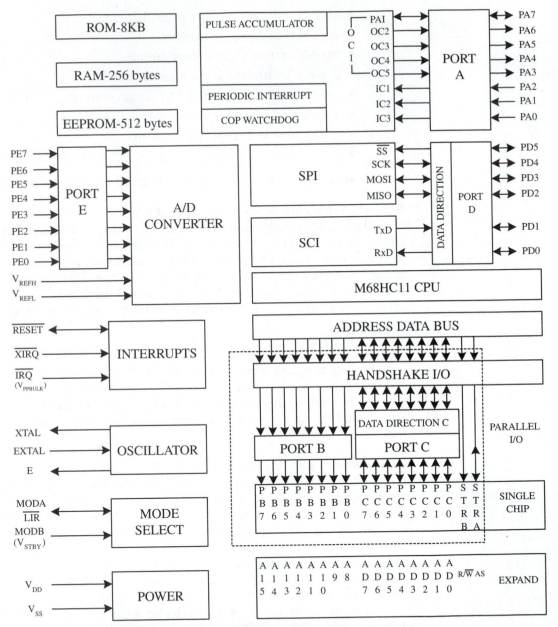

Figure 1.2 ■ 68HC11A8 block diagram (redrawn with permission of Motorola)

RAM is mainly used to store *dynamic* programs and data. A computer user often wants to run different programs on the same computer, and these programs usually operate on different sets of data. The programs and data must therefore be loaded into RAM from hard disk or other secondary storage, and for this reason they are called dynamic.

READ-ONLY MEMORY

ROM is *nonvolatile*. If power is removed from ROM and then reapplied, the original data will still be there. However, as its name implies, ROM data can only be read. If the processor attempts to write data to a ROM location, ROM will not accept the data, and the data in the addressed ROM memory location will not be changed. There are many different kinds of ROM technologies in use today:

Masked-programmed read-only memory (MROM) is a type of ROM that is programmed when it is manufactured. The semiconductor manufacturer places binary data in the memory according to the request of the customer. To be cost-effective, many thousands of MROM memory chips, each containing a copy of the same data (or program), must be sold.

Programmable read-only memory (PROM) is a type of read-only memory that can be programmed in the field (often by the end user) using a device called a PROM programmer or a PROM *burner*. Once a PROM has been programmed, its contents cannot be changed. PROMs are fuse-based; i.e., end users program the fuses to configure the contents of the memory.

Erasable programmable read-only memory (EPROM) is a type of read-only memory that can be erased by subjecting it to strong ultraviolet light. The circuit design of EPROM requires us to erase the contents of a location before we can write a new value into it. A quartz window on top of the EPROM integrated circuit permits ultraviolet light to be shone directly on the silicon chip inside. Once the chip is programmed, the window can be covered with dark tape to prevent gradual erasure of the data. If no window is provided, the EPROM chip becomes one-time programmable (OTP) only. Many microcontrollers incorporate on-chip one-time programmable EPROM. EPROM is often used in prototype computers, where the software may be revised many times until it is perfected. EPROM does not allow erasure of the contents of an individual location. The only way to make change is to erase the entire EPROM chip and reprogram it. The programming of an EPROM chip is done electrically by using a device called an *EPROM programmer*. Today, most programmers are universal in the sense that they can program many different types of devices including EPROM, EEPROM, flash memory, and *programmable logic devices*.

Electrically erasable programmable read-only memory (EEPROM) is a type of nonvolatile memory that can be erased by electrical signals and reprogrammed. Like EPROM, the circuit design of EEPROM also requires us to erase the contents of a memory location before we can write a new value into it. EEPROM allows each individual location to be erased and reprogrammed. Unlike EPROM, EEPROM can be erased and programmed using the same programmer. However, EEPROM pays the price for being so flexible in its erasability. The cost of an EEPROM chip is much higher than that of an EPROM chip of comparable density.

Flash memory was invented to incorporate the advantages and avoid the drawbacks of both EPROM and EEPROM technologies. Flash memory can be erased and reprogrammed in the system without using a dedicated programmer. It achieves the density of EPROM, but it does not require a window for erasure. Like EEPROM, flash memory can be programmed and erased electrically. However, it does not allow the erasure of an individual memory location—the user can only erase the entire chip. Today, more and more microcontrollers are incorporating on-chip flash memory for storing programs and static data.

1.3 The Computer's Software

A computer is useful because it can execute programs. Programs are known as *software*. A program is a set of instructions that the computer hardware can execute. The program is stored in the computer's memory in the form of binary numbers called *machine instructions*. For example, the 68HC11 machine instruction

0001 1011

adds the contents of accumulator A and accumulator B together and leaves the sum in accumulator A.

10000110 00000001

places the value 1 in accumulator A.

00111101

multiplies the values in accumulators A and B and leaves the product in accumulators A and B.

Several tasks are very difficult in developing programs in machine language:

1. *Program writing*. The programmer will need to memorize the binary pattern of each machine instruction, which can be very challenging because a microprocessor may have several hundred different machine instructions, and each machine instruction may have different length. Constant table lookup will be necessary if the programmer cannot memorize every binary pattern. On the other hand, programmers are forced to work on program logic at a very low level because every machine instruction implements only a very primitive operation.

2. *Program debugging*. Whenever there are errors, it is extremely difficult to trace the program because the program consists of only sequences of 0s and 1s. A programmer will need to identify each machine instruction and then think about what operation is performed by that instruction. This is not an easy task.

3. *Program maintenance*. Most programs will need to be maintained in the long run. A programmer who did not write the program will have a hard time reading the program and following the program logic.

Assembly language was invented to simplify the programming job. An *assembly program* consists of assembly instructions. An *assembly instruction* is the mnemonic representation of a machine instruction. For example, in the 68HC11:

> ABA stands for "add the contents of accumulator B to accumulator A." The corresponding machine instruction is 00011011.
>
> DECA stands for "decrement the contents of accumulator A by 1." The corresponding machine instruction is 01001010.

A programmer no longer needs to scan through the 0s and 1s in order to identify what instructions are in the program. This is a significant improvement over machine language programming.

The assembly program that the programmer enters is called *source program* or *source code*. A software program called an *assembler* is then invoked to translate the program written in assembly language into machine instructions. The output of the assembly process is called *object code*. It is a common practice to use a *cross assembler* to assemble assembly programs. A cross assembler is an assembler that runs on one computer but generates machine instructions that will be executed by another computer that has a different instruction set. In contrast, a *native assembler* runs on a computer and generates machine instructions to be executed by machines that have the same instruction set. The Motorola freeware as11 is a cross assembler that runs on an IBM PC or Apple Macintosh and generates machine code that can be downloaded into a 68HC11-based computer for execution.

There are several drawbacks to programming in assembly language:

- The programmer must be very familiar with the hardware organization of the computer on which the program is to be executed.

- A program (especially a long one) written in assembly language is extremely difficult to understand for anyone other than the author.

- Programming productivity is not satisfactory for large programming projects because the programmer needs to work on the program logic at a very low level.

For these reasons, high-level languages such as Fortran, PASCAL, C, C++, and Java were invented to avoid the problems of assembly language programming. High-level languages are very close to plain English and hence a program written in a high-level language becomes easier to understand. A statement in high-level language often needs to be implemented by tens of assembly instructions. The programmer can now work on the program logic at a much higher level, which makes the programming job much easier. A program written in a high-level language is also called a *source program*, and it requires a software program called a *compiler* to translate it into machine instructions. A compiler compiles a program into *object code.* Just as there are cross assemblers, there are *cross compilers* that run on one computer but translate programs into machine instructions to be executed on a computer with a different instruction set.

High-level languages are not perfect, either. One of the major problems with high-level languages is that the machine code compiled from a program written in a high-level language cannot run as fast as its equivalent in the assembly language. For this reason, many time-critical programs are still written in assembly language.

C language has been used extensively in microcontroller programming in the industry, which can be proved by the fact that most microcontroller software tool developers provide C cross compilers. Both C and assembly language will be used throughout this text. The C programs in this text are compiled by the Imagecraft C cross-compiler and tested on Motorola and Axiom evaluation boards.

A *text editor* is used to develop a program using a computer. A text editor allows the user to type, modify, and save the program source code in a text file.

1.4   The 68HC11 CPU Registers

The 68HC11 microcontroller has many registers. These registers can be classified into two categories: CPU registers and I/O registers. CPU registers are used solely to perform general-purpose operations such as arithmetic, logic, and program flow control. I/O registers are mainly used to control the operations of I/O subsystems and record the status of I/O operations, etc. I/O registers are treated as memory locations when they are accessed. CPU registers do not occupy the 68HC11 memory space.

The CPU registers of the 68HC11 are shown in Figure 1.3 and are listed below. Some of the registers are 8-bit and some are 16-bit.

General-purpose accumulators A and B. Both A and B are 8-bit registers. Most arithmetic functions are performed on these two registers. These two accumulators can also be concatenated to form a single 16-bit accumulator that is referred to as the D accumulator.

Index registers IX (or X) and IY (or Y). These two registers are used mainly in addressing memory operands. However, they are also used in several arithmetic operations.

Stack pointer (SP). A stack is a first-in-last-out data structure. The 68HC11 has a 16-bit stack pointer that initially points to the location above the top element of the stack, as shown in Figure 1.4. The stack will be discussed in chapter 3.

Program counter (PC). The address of the next instruction to be executed is specified by the 16-bit program counter. The 68HC11 fetches the instruction one byte at a time and increments the PC by 1 after fetching each instruction byte. After the execution of an instruction, the PC is incremented by the number of bytes of the executed instruction.

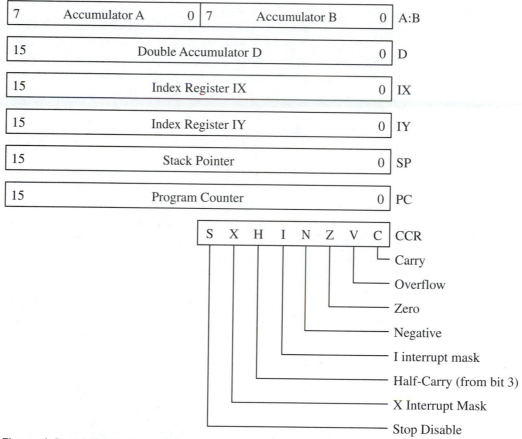

Figure 1.3 ■ MC68HC11 programmer's model

Condition code register (CCR). This 8-bit register is used to keep track of the program execution status, control the execution of conditional branch instructions, and enable/disable the interrupt handling. The contents of the CCR register are shown in Figure 1.3. The function of each condition code bit will be explained in later sections and chapters.

All of these registers are accessible to the programmer.

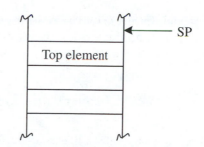

Figure 1.4 ■ 68HC11 stack structure

1.5   Memory Addressing

Memory consists of a sequence of directly addressable "locations." A memory location is referred to as an *information unit*. A memory location can be used to store data, instruction, the status of peripheral devices, etc. An information unit has two components: its *address* and its *contents*, shown in Figure 1.5.

Figure 1.5 ■ The components of a memory location

Each location in memory has an address that must be supplied before its contents can be accessed. The CPU communicates with memory by first identifying the location's address and then passing this address on the address bus. This is similar to the fact that a mailman needs an address in order to deliver a letter. The data are transferred between memory and the CPU along the data bus (see Figure 1.6). The number of bits that can be transferred on the data bus at one time is called the *data bus width* of the processor.

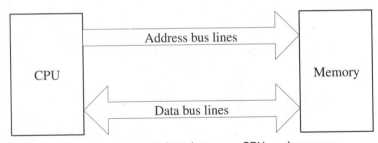

Figure 1.6 ■ Transferring data between CPU and memory

The 68HC11 has an 8-bit data bus and can access only one memory byte at a time. The 68HC11 has an address bus of 16 signal lines and can address up to 2^{16} (65636) different locations. The accessible memory addresses are in the range from 0000_{16} to $FFFF_{16}$. The size of memory is measured in bytes. Each byte has 8 bits. A 4-bit quantity is called a *nibble*. To simplify the quantification of memory, the unit *kilobyte* (KB) is often used. K is given by the following formula:

$K = 1024 = 2^{10}$

Another frequently used unit is *megabyte* (MB), which is given by the following formula:

$M = K^2 = 1024 \times 1024 = 1048576$

1.6 The 68HC11 Addressing Modes

A 68HC11 instruction consists of one or two bytes of opcode and zero to three bytes of operand information. Instructions that use indexed addressing mode with Y as the index register have two bytes of opcode. All other instructions have only one byte of opcode.

Addressing modes are used to specify the operands needed in an instruction. Six addressing modes are provided in the 68HC11: *immediate, direct, extended, indexed* (with either of two 16-bit index registers and an 8-bit offset), *inherent,* and *relative.* The immediate mode specifies an operand to be operated on. Direct, extended, and indexed modes are used to specify the address of a memory operand. The relative mode is used to specify the branch target for conditional branch instructions.

Each of the addressing modes (except for immediate and inherent modes) results in an internally generated, double-byte value referred to as the *effective address*. This value appears on the address bus during the external memory reference portion of the instruction.

The following paragraphs describe each of the addressing modes. In these descriptions, the effective address is used to specify the address of the memory location, from which the argument is fetched, at which the argument is to be stored, or from which execution is to proceed.

Address and data values are represented in binary format inside the computer. However, a large binary number is not easy for a human being to deal with, so decimal and hexadecimal formats are often used instead. Octal numbers are also used in some cross assemblers. In this text, we will use a notation that adds a prefix to a number to indicate the base used in the number representation. The prefixes for binary, octal, decimal, and hexadecimal numbers are given in Table 1.1. This method is also used in Motorola microcontroller and microprocessor manuals. The word "hexadecimal" has a shorthand called "hex." We will use **hex** instead of **hexadecimal** from now on.

Base	Prefix
binary	%
octal	@
decimal	(nothing)*
hexadecimal	$
*Note: Some assemblers use '&'	

Table 1.1 ■ Prefix for number representation

1.6.1   Immediate (IMM)

In the immediate addressing mode, the actual argument is contained in the byte or bytes immediately following the instruction opcode. The number of bytes matches the size of the register. In assembly language syntax, an immediate value is preceded by a # character. The following instructions illustrate the immediate addressing mode:

    LDAA #22

loads the decimal value 22 into the accumulator A.

    ADDA #@32

adds the octal value 32 to accumulator A.

    LDAB #$17

loads the hex value 17 into accumulator B.

    LDX #$1000

loads the hex value 1000 into the index register X, where the upper byte of X receives the value of $10 and the lower byte of X gets the value of $00.

1.6.2   Direct Mode (DIR)

In the direct addressing mode, the least significant byte of the effective address of the instruction operand appears in the byte following the opcode. The high-order byte of the effective address is assumed to be $00 and is not included in the instruction. This limits use of the direct mode to operands in the $0000-$00FF area of the memory. Instructions using the direct mode execute one clock cycle faster than their counterparts using the extended mode. The following instructions illustrate the direct addressing mode:

    ADDA $00

adds the value stored at the memory location with the effective address $0000 to accumulator A.

    SUBA $20

subtracts the value stored at the memory location with the effective address $0020 from accumulator A.

    LDD $10

loads the contents of the memory locations at $0010 and $0011 into double accumulator D, where the contents of the memory location at $0010 are loaded into accumulator A and those of the memory location at $0011 are loaded into accumulator B.

1.6.3 Extended Mode (EXT)

In the extended addressing mode, the effective address of the operand appears explicitly in the two bytes following the opcode:

    LDAA $1003

loads the 8-bit value stored at the memory location with effective address $1003 into accumulator A.

    LDX $1000

loads the 16-bit value stored at the memory locations with the effective addresses $1000 and $1001 into the index register X. The byte at $1000 will be loaded into the upper byte of X and the byte at $1001 will be loaded into the lower byte of X.

    ADDD $1030

adds the 16-bit value stored at the memory locations with the effective addresses $1030 and $1031 to double accumulator D.

1.6.4 Indexed Mode (INDX, INDY)

In the indexed addressing mode, one of the index registers (X or Y) is used in calculating the effective address. Thus the effective address is variable and depends on the current contents of the index register X (or Y) and a fixed, 8-bit unsigned offset contained in the instruction. Because the offset byte is unsigned, only positive offsets in the range from 0 to 255 can be represented. If no offset is specified, the machine code will contain $00 in the offset byte. For example,

    ADDA 10,X

adds the value stored at the memory location pointed to by the sum of 10 and the contents of the index register X to accumulator A.

Each of the following instructions subtracts the value stored at the memory location pointed to by the contents of index register X from accumulator A:

    SUBA 0,X
    SUBA ,X
    SUBA X

Please note that the third format is not acceptable to the Motorola as11 freeware assembler.

You probably wonder why the index addressing mode is useful. Programs often need to change the address part of an instruction as the program runs. The index addressing mode requires part of the address to be placed in the index register in the microprocessor. The contents of the index register can be changed by the program, thus changing the effective address of the instruction operand while the offset byte of the instruction remains unchanged. Using the index addressing mode also shortens the instruction by one byte if the effective address is higher than the hex value $FF.

1.6.5   Inherent Mode (INH)

In the inherent mode, everything needed to execute the instruction is encoded in the opcode. The operands are CPU registers and thus are not fetched from memory. These instructions are usually one or two bytes.

    ABA

adds the contents of accumulator B to accumulator A.

    INCB

increments the value of accumulator B by 1.

    INX

increments the value of the index register X by 1.

1.6.6   Relative Mode (REL)

The relative addressing mode is used only for branch instructions. Branch instructions, other than the branching versions of the bit-manipulation instructions, generate two machine-code bytes, one for the opcode and one for the *branch offset*. The branch offset is the distance relative to the first byte of the instruction immediately following the branch instruction. The branch offset has a range of −128 to +127 bytes. When the branch is taken, the branch offset is added to the program counter to form the effective address. The source program specifies the destination of any branch instruction by its absolute address, given as either a numerical value or a symbol or expression that can be numerically evaluated by the assembler.

In Figure 1.7, the 68HC11 will branch to execute the instruction DECB if the Z bit in the CCR register is 1 when the instruction BEQ $e164 is executed. A better way to specify the branch target is to use a symbolic label. Figure 1.8 is an improvement to the example in Figure 1.7.

Address	Opcode	Operand	
	BEQ	$e164	
$e100	ADDA	#10	
	...		$64 bytes
$e164	DECB		
	...		

Figure 1.7 ■ Example of this relative addressing mode. The opcode byte of the instruction DECB is $64 bytes away from the opcode byte of the instruction ADDA #10

Address	Opcode	Operand
	BEQ	there
	ADDA	#10
	...	
there	DECB	
	...	

Figure 1.8 ■ Using a label to specify the branch target

1.7   A Sample of 68HC11 Instructions

A 68HC11 instruction consists of one or two bytes of opcode and zero to three bytes of operand information. The opcode specifies the operation to be performed. A 68HC11 instruction can have from zero to three operands. One of the operands is used both as a source and as the destination of the operation. The operand information is represented by one of the addressing modes.

1.7.1   The LOAD Instruction

LOAD is the generic name of a group of instructions that place a value or copy the contents of a memory location (or memory locations) into a register. Most 68HC11 arithmetic and logical instructions include a register as one of the operands. Before a meaningful operation can be performed, a value must be placed in the register. The LOAD instruction places or copies a value from a memory location into a register. The 68HC11 has LOAD instructions to load values into accumulator A, accumulator B, double accumulator D, stack pointer SP, index register X, and index register Y.

For example, the following instruction loads the decimal value 10 into accumulator A:

LDAA #10

where, the # character indicates that the value that follows (that is, 10) is to be placed into accumulator A.

The following instruction copies the contents of the memory location at $1000 into accumulator A:

LDAA $1000

A more extensive sample of LOAD instructions is given in Table 1.2.

When two consecutive memory bytes are loaded into a 16-bit register, the contents of the memory location at the lower address are loaded into the upper half of the register, while the contents of the memory location at the higher address are loaded into the lower half of the register. When the contents of a 16-bit register are saved in the memory, the upper byte of the register is saved at the lower address while the lower byte is saved at the higher address.

Instruction	Meaning	Addressing mode
LDAA #10	Place the decimal value 10 (hex A) into accumulator A.	immediate
LDAA $1000	Copy the contents of the memory location at $1000 into accumulator A.	extended
LDAB #10	Place the decimal value 10 into accumulator B.	immediate
LDAB $1000	Copy the contents of the memory location at $1000 into accumulator B.	extended
LDD #10	Place the decimal value 10 into double accumulator D.	immediate
LDD $1000	Copy the contents of the memory locations at $1000 & $1001 into the upper and lower bytes of double accumulator D, respectively.	extended
LDS #255	Place the decimal value 255 (hex FF) into the stack pointer SP.	immediate
LDS $1000	Copy the contents of the memory location at $1000 and $1001 into the upper and lower bytes of register SP, respectively.	extended
LDX #$1000	Place the hex value $1000 into index register X.	immediate
LDX $1000	Copy the contents of the memory locations at $1000 and $1001 into the upper and lower bytes of index register X, respectively.	extended
LDY #1000	Place the decimal value 1000 (hex 3E8) into index register Y.	immediate
LDY $1000	Copy the contents of the memory locations at $1000 and $1001 into the upper and lower bytes of index register Y, respectively.	extended

Table 1.2 ■ A sample of LOAD instructions

Example 1.1

▼

Write an instruction to place the decimal value 1023 (or $3FF) into the stack pointer SP.

Solution: The following instruction will place the decimal value 1023 into the stack pointer SP:

LDS #1023

The binary representation of the decimal value 1023 is 11111111111_2. After the execution of this instruction, the 16-bit stack pointer SP contains the value of 0000001111111111_2.

In order to unify the representation, we will use the following notations throughout this book:

[reg]: refers to the contents of the register **reg**. reg can be any one of the following:

A, B, D, X, Y, SP, or PC.

[addr]: refers to the contents of the memory location at address **addr**.

mem[addr]: refers to the memory location at address **addr**.

▲

Example 1.2

▼

Write an instruction to load the contents of the memory locations at $0000 and $0001 into double accumulator D. Initially, the contents of D and the memory locations at $0000 and $0001 are $1010, $20, and $30, respectively. Show the new values in these registers after the execution of the instruction.

Solution: To load two consecutive memory bytes into the double accumulator D, we need to specify only the address of the most significant byte, that is, $0000, in this example. The instruction is:

    LDD $0000

The contents of D and memory locations $0000 and $0001 before and after execution of the instruction are as follows:

Before execution of LDD $0000	After execution of LDD $0000
[D] = $1010 [$0000] = $20 [$0001] = $30	[D] = $2030 [$0000] = $20 [$0001] = $30

▲

1.7.2   The ADD Instruction

ADD is the generic name of a group of instructions that perform the addition operation. The ADD instruction is one of the most important arithmetic instructions in the 68HC11. The ADD instruction can have either two or three operands. In a three-operand ADD instruction, the C flag of the condition code register is always included as one of the source operands. Three-operand ADD instructions are used mainly in multiprecision arithmetic, which will be discussed in chapter 2. The ADD instruction has the following constraints:

- The ADD instruction can specify at most one memory location as a source operand.
- The memory operand can be used only as a source operand.
- The destination operand must be a register (it can be A, B, X, Y, or D).
- The register specified as the destination operand must also be used as a source operand.

For example,

ADDA #20

adds the decimal value 20 (hex $14) to the contents of accumulator A and places the result in accumulator A.

ADDA $40

adds the contents of the memory location at $40 to the contents of accumulator A and places the result in accumulator A.

ADCA $00

adds the carry bit (in CCR) and the contents of the memory location at $00 to accumulator A and places the result in accumulator A. More examples of ADD instructions are given in Table 1.3.

The ADD instructions that specify one of the index registers as the destination are mainly used in address calculation, not for general-purpose 16-bit addition. The instruction **ABX** adds the contents of accumulator B to the lower byte of the index register X. If there is a carry out, it will be added to the upper byte of the index register X. The instruction **ABY** is similar except that the destination is the index register Y.

Instruction	Meaning	Addressing mode
ABA	Add accumulator B and accumulator A and store the sum in A.	inherent
ABX	Add accumulator B and the index register X and store the sum in X.	inherent
ABY	Add accumulator B and the index register Y and store the sum in Y.	inherent
ADCA #12	Add the decimal value 12 and the C flag in the CCR register to accumulator A and store the sum in A.	immediate
ADCA $20	Add the contents of the memory location at $20 and the carry flag in the CCR register to accumulator A and store the sum in A.	direct
ADCB #12	Add the decimal value 12 and the C flag in the CCR register to accumulator B and store the sum in B.	immediate
ADCB $20	Add the contents of the memory location at $20 and the C flag in the CCR register to accumulator B and store the sum in B.	direct
ADDA #12	Add the decimal value 12 to accumulator A and store the sum in A.	immediate
ADDA $20	Add the contents of the memory location at $20 to accumulator A and store the sum in A.	direct
ADDB #12	Add the decimal value 12 to accumulator B and store the sum in B.	immediate
ADDB $20	Add the contents of the memory location at $20 to accumulator B and store the sum in B.	direct
ADDD #0012	Add the decimal value 12 to double accumulator D and store the sum in double accumulator D.	immediate
ADDD $0020	Add the 16-bit value stored at memory locations $20 and $21 to double accumulator D and store the sum in double accumulator D.	direct

Table 1.3 ■ A sample of ADD instructions.

Example 1.3

▼

Write an instruction sequence to add the contents of the memory locations at $10 and $20 and leave the sum in accumulator A.

Solution: This problem can be solved by loading the contents of one of the memory locations into accumulator A and then adding the contents of the other memory location into accumulator A, as is done by the following instructions:

    LDAA $10
    ADDA $20

The first instruction loads the contents of the memory location at $10 into accumulator A. The second instruction then adds the contents of the memory location at $20 to accumulator A.

▲

1.7.3   The SUB Instruction

SUB is the generic name of a group of instructions that perform the subtraction operation. Like the ADD instruction, the SUB instruction can have either two or three operands. The three-operand SUB instruction includes the C flag of the CCR register as one of the source operands. Three-operand SUB instructions are mainly used in multiprecision arithmetic. The SUB instruction has the following constraints:

- The SUB instruction can specify at most one memory location as a source operand.
- The memory operand can be used only as a source operand.
- The destination operand must be an accumulator (it can be either A, B, or D).
- The register specified as the destination operand is also used as a source operand.

For example,

    SUBA #10

subtracts the decimal value 10 from accumulator A and leaves the difference in accumulator A.

    SUBB $10

subtracts the contents of the memory location at $10 from accumulator B and leaves the difference in accumulator B.

Example 1.4

▼

Write an instruction sequence to subtract the value of the memory location at $00 from that of the memory location at $30 and leave the difference in accumulator B.

Solution: We need to load the contents of the memory location at $30 into accumulator B and then subtract the contents of the memory location at $00 directly from accumulator B. The appropriate instructions are:

LDAB $30
SUBB $00

More examples of SUB instructions are given in Table 1.4.

Instruction	Meaning	Addressing mode
SBA	Subtract the value of accumulator B from accumulator A and store the difference in accumulator A.	inherent
SBCA #10	Subtract the decimal value 10 and the C bit of the CCR register from accumulator A and store the difference in A.	immediate
SBCA $20	Subtract the contents of the memory location at $20 and the C flag from accumulator A and store the difference in A.	direct
SBCB #10	Subtract the decimal value 10 and the C bit of the CCR register from accumulator B and store the difference in B.	immediate
SBCB $20	Subtract the contents of the memory location at $20 and the C flag from accumulator B and store the difference in B.	direct
SUBA #10	Subtract the decimal value 10 from accumulator A and store the difference in A.	immediate
SUBA $20	Subtract the contents of memory location at $20 from accumulator A and store the difference in A.	direct
SUBB #10	Subtract the decimal value 10 from accumulator B and store the difference in B.	immediate
SUBB $20	Subtract the contents of memory location at $20 from accumulator B and store the difference in B.	direct
SUBD #$0010	Subtract the hex value 10 from double accumulator D and store the difference in D.	immediate
SUBD $0020	Subtract the 16-bit value stored at memory locations at $0020 and $0021 from double accumulator D and store the difference in D.	extended

Table 1.4 ■ A sample of SUB instructions

▲

1.7.4   The STORE Instruction

STORE is the generic name of a group of instructions that store the contents of a register into a memory location or memory locations. The 68HC11 has six STORE instructions. The STORE instruction allows the contents of accumulator A, accumulator B, the stack pointer SP, index register X, or index register Y to be stored at one or two memory locations. The destination must be a memory location.

For example,

STAA $10

stores the contents of accumulator A in the memory location at $10.

STAB $10

stores the contents of accumulator B in the memory location at $10.

STD $10

stores the upper and lower eight bits of double accumulator D in the memory locations at $10 and $11, respectively.

STX $2000

stores the upper and lower eight bits of index register X in the memory locations at $2000 and $2001.

Example 1.5

Write an instruction sequence to add the contents of the memory locations at $00 and $01 and then store the sum in the memory location at $10.

Solution: This problem can be solved in three steps:

Step 1
Load the contents of the memory location at $00 into accumulator A.

Step 2
Add the contents of the memory location at $01 to accumulator A.

Step 3
Store the contents of accumulator A in the memory location at $10.
The appropriate instructions are:

```
LDAA    $00    ; load the contents of the memory location at $00 into A
ADDA    $01    ; add the contents of the memory location at $01 to A
STAA    $10    ; store the sum in A at the memory location at $10
```

Example 1.6

Write an instruction sequence to swap the contents of the memory locations at $00 and $10.

Solution: To swap, we need to load the contents of the memory at $00 and $10 into accumulator A and B, respectively, and then store the contents of A and

B in the memory locations at $10 and $00, respectively. The following instructions would be used:

```
LDAA    $00    ; load the contents of $00 into A
LDAB    $10    ; load the contents of $10 into B
STAA    $10    ; store A into $10
STAB    $00    ; store B into $00
```

▲

1.8   The 68HC11 Machine Code

We have learned that each 68HC11 instruction consists of one to two bytes of opcode and zero to three bytes of operand information. In this section, we will look at the machine codes of a sample of 68HC11 instructions, the disassembly of machine instructions, and instruction execution timing.

1.8.1   A Machine Code Sequence

The basic assembly language or machine code instructions can be sequenced to perform calculations, as we have seen in previous examples. Consider the following high-level language statements:

```
I := 29
L := I + M
```

Assume that the variables I, L, and M refer to memory locations $00, $01, $02, respectively. The first statement assigns the value 29 to variable I, and the second statement assigns the sum of variables I and M to variable L. The high-level language statements translate to the following equivalent assembly language and machine instructions:

Assembly instructions	Machine instructions (in hexadecimal format)
LDAA #29	86 1D
STAA $00	97 00
ADDA $02	9B 02
STAA $01	97 01

Note that the decimal number 29 is equivalent to hex value $1D. Assume that these four instructions are stored in consecutive memory locations starting at $C000. Then the contents of the memory locations from $C000 to $C007 are as follows:

Address	Machine code
$C000	86
$C001	1D
$C002	97
$C003	00
$C004	9B
$C005	02
$C006	97
$C007	01

If the memory locations at $02 contains $20, then the memory location at $01 is assigned the value $1D + $20 = $3D. Figure 1.9 shows the changes in the values stored in the memory locations and in accumulator A when the instructions are executed. (These values are represented in hex format.)

1.8.2   Decoding Machine Language Instructions

The process of decoding (disassembling) a machine language instruction is more difficult than assembling it. The opcode is the first one or two bytes of an instruction. By decomposing its bit pattern, the assembly instruction mnemonic and the addressing mode of the operand can be identified.

To illustrate the disassembling process, we will use Table 1.5. In Table 1.5, the left column shows the machine code byte while the right column shows the instruction format corresponding to the machine code byte.

Address	Before program execution	After program execution
$00	???	1D
$01	???	3D
$02	20	20
A	???	3D

Figure 1.9 ■ Changes in the contents of memory locations and accumulator A after program execution

Example 1.7

A segment of program machine code contains the following opcode and addressing information:

96 30 8B 07 97 30 96 31 8B 08 97 31

Machine Code	Assembly Instruction Format	
01	NOP	
86	LDAA	IMM
96	LDAA	DIR
C6	LDAB	IMM
D6	LDAB	DIR
CC	LDD	IMM
DC	LDD	DIR
8B	ADDA	IMM
9B	ADDA	DIR
CB	ADDB	IMM
DB	ADDB	DIR
C3	ADDD	IMM
D3	ADDD	DIR
97	STAA	DIR
D7	STAB	DIR
DD	STD	DIR

Note:
1. IMM is a one-byte immediate value for instructions that involve A or B, and it is a two-byte immediate value for instructions that involve D.
2. DIR stands for a one-byte direct address between $00 and $FF.

Table 1.5 ■ Machine opcodes and their corresponding assembly instructions

Using the machine opcodes and corresponding assembly instructions in Table 1.5, decode the given machine code into assembly instructions.

Solution: The process of decoding the machine language instruction begins with the opcode byte 96.

a. The opcode byte 96 corresponds to the following LOAD instruction format:

LDAA DIR

To complete the decoding of this instruction, the byte that immediately follows 96 (that is, 30) should be included. Therefore, the machine code of the first instruction is 96 30. The corresponding assembly instruction is LDAA $30.

b. The opcode of the second instruction is 8B, which corresponds to the following ADD instruction format:

ADDA IMM

To decode this instruction completely, the byte that immediately follows 8B (that is, 07) should be included. The machine code of the second instruction is thus 8B 07. The corresponding assembly instruction is ADDA #07.

c. The opcode of the third instruction is 97, which corresponds to the following STORE instruction format:

```
STAA    DIR
```

Including the byte that immediately follows 97 (that is, 30), we see that the machine code of the third instruction is 97 30. The corresponding assembly instruction is STAA $30.

Continuing in this manner, we can decode the remaining machine code bytes into the following assembly instructions:

```
LDAA    $31
ADDA    #08
STAA    $31
```

▲

A program that can disassemble machine code into assembly instructions is called *a disassembler.* A disassembler can be used to translate the machine code in ROM into assembly instructions.

1.8.3   The Instruction Execution Cycle

In order to execute a program, the microprocessor or microcontroller must access memory to fetch instructions or operands. The process of accessing a memory location is called a *read cycle,* the process of storing a value in a memory location is called a *write cycle,* and the process of executing an instruction is called an *instruction execution cycle.*

When executing an instruction, the 68HC11 performs a combination of the following operations:

- one or multiple read cycles to fetch instruction opcode byte(s) and addressing information
- one or two read cycles required to fetch the memory operand(s) (optional)
- the operation specified by the opcode
- one or two write cycles to write back the result to either a register or a memory location (optional)

We will illustrate the instruction execution cycle using the LOAD, ADD, and STORE instructions shown in Table 1.6. The details of the data transfer on the buses are included to illustrate the read/write cycles. Assume the program counter PC is set at $C000, the starting address for the machine instructions. The contents of the memory locations at $2000 and $3000 are $19 and $37, respectively.

Assembly language instructions	Memory location address	Machine code
LDAA $2000	$C000	B6 20 00
ADDA $3000	$C003	BB 30 00
STAA $2000	$C006	B7 20 00

Table 1.6 ■ A sample of instructions

Instruction 1 LDAA $2000

Execution of this instruction involves the following steps:

Step 1
The value in PC ($C000) is placed on the address bus with a request to "read" the contents of that location.

Step 2
The eight-bit value at the location $C000 is the instruction opcode byte $B6. This value is placed on the data bus by the memory hardware and returned to the processor, where the control unit begins interpretation of the instruction. On the read cycle, the control unit causes the PC to be incremented by 1, so it now points to location $C001. Figure 1.10 shows the opcode read cycle.

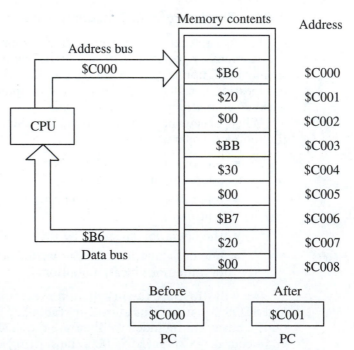

Figure 1.10 ■ Instruction 1—opcode read cycle

Step 3

The control unit recognizes that the LOAD instruction requires a two-byte value for the operand address. This is found in the two bytes immediately following the opcode byte (at locations $C001 and $C002). Therefore, two read cycles are executed, and the value of the PC is incremented by 2. The PC has a final value of $C003, and the address $2000 is stored in an internal register (invisible to the programmer) inside the CPU. Figure 1.11 shows the address read cycles.

Step 4

The actual execution of the LOAD instruction requires an additional read cycle. The address $2000 is put on the address bus with a read request. The contents of memory location $2000 are placed on the data bus and stored in accumulator A, as shown in Figure 1.12.

Instruction 2 ADDA $3000

The PC initially has the value of $C003. Three read bus cycles are required to fetch the second instruction from memory. The execution cycle for this instruction involves the following steps:

Step 1

Fetch the opcode byte at $C003. At the end of this read cycle, the PC is incremented to $C004 and the opcode byte $BB has been fetched. The

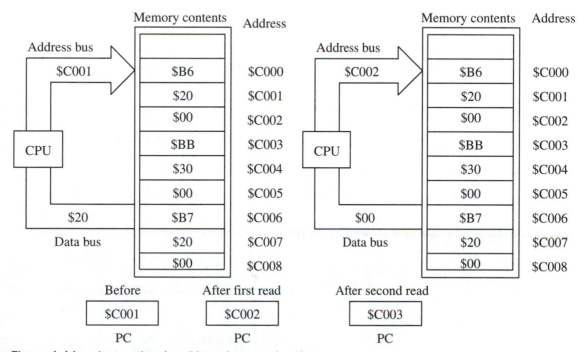

Figure 1.11 ■ Instruction 1—address byte read cycles

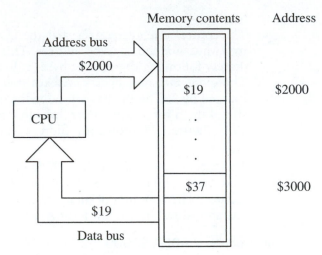

Figure 1.12 ■ Instruction 1—execution read cycles

control unit recognizes that this version of the ADD instruction requires two more read cycles to fetch the extended address. These two read cycles are performed in the following two steps.

Step 2
Fetch the upper extended address byte ($30) from the memory location at $C004. The PC is then incremented to $C005.

Step 3
Fetch the lower extended address byte ($00) from the memory location at $C005. The PC is then incremented to $C006.

Step 4
Execution of this instruction requires an additional read cycle to read in the operand at location $3000. The control unit places the value $3000 on the address bus to fetch the contents of memory location $3000.

Step 5
The returned value $37 is added to accumulator A. The accumulator now has the value $50 ($19 + $35 = $50).

Instruction 3 STAA $2000

As was the case with the previous two instructions, three read cycles are required to fetch this instruction from memory. The PC initially has the value $C006. The execution cycle for this instruction involves the following steps:

Step 1
Fetch the opcode byte at the location $C006. At the end of the read cycle, the PC is incremented to $C007 and the opcode byte $B7 has

been fetched. The control unit recognizes that this version of the STORE instruction requires two more read cycles to fetch the extended address. These two read cycles are performed in the next two steps.

Step 2

Fetch the upper extended address byte ($20) from memory location at $C007. The PC is then incremented to $C008.

Step 3

Fetch the lower extended address byte ($00) from the memory location at $C008. The PC is then incremented to $C009.

Step 4

The purpose of this instruction is to store the contents of accumulator A in memory, so the control unit places the extended address $2000 on the address bus, and the value in accumulator A ($50) is written into the memory location at $2000. Figure 1.13 shows the execution write cycle.

1.8.4   Instruction Timing

The 68HC11 Reference Manual lists the CPU execution time for every 68HC11 instruction (in number of E clock cycles). The 68HC11 internal operations and external read and write cycles are controlled by the E clock signal. The E clock signal is derived by dividing the external crystal oscillator output signal by 4.

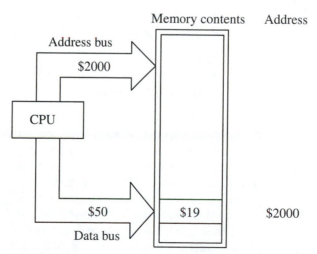

Figure 1.13 ■ Instruction 3—execution write cycle

1.9   Summary

A computer is made up of hardware and software. The computer hardware consists of four main components: (1) a processor (called CPU), (2) an input unit, (3) an output unit, and (4) memory. The processor can be further divided into three major parts: (1) registers, (2) arithmetic logic unit, and (3) the control unit. The system clock synchronizes the activities of the CPU. The clock rates of modern microprocessors can be as high as several hundred MHz. The control unit maintains a register called *program counter* (PC), which controls the memory address of the next instruction to be executed. During the execution of an instruction, the presence of overflow, an addition carry, a subtraction borrow, and so forth, are flagged by the system and stored in another register called the *status register*.

A microprocessor is a processor fabricated on a single integrated circuit. A microcomputer is a computer that uses a microprocessor as its CPU. Although microprocessors have been widely used since their invention, there are several limitations that led to the invention of microcontrollers. First, a microprocessor requires external memory to execute programs. Second, a microprocessor cannot interface directly to I/O devices; peripheral chips are needed. Third, glue logic is needed to interconnect external memory and peripheral interface chips to the microprocessor.

A microcontroller is a computer implemented on a single very large scale integration (VLSI) chip. It contains everything contained in a microprocessor along with one or more of the following components:

- memory
- timer
- analog-to-digital converter
- digital-to-analog converter
- direct memory access (DMA) controller
- parallel I/O interface
- serial I/O interface
- memory component interface circuitry

Memory is where software programs and data are stored. Semiconductor memory chips can be classified into two major categories: random-access memory (RAM) and read-only memory (ROM).

There are many different types of ROM. MROM is a type of ROM that is programmed when it is fabricated. PROM is a type of ROM that can be programmed in the field by the end user. EPROM is a type of ROM that is programmed electrically and erased by ultraviolet light. EEPROM is a type of ROM that can be programmed and erased electrically. EEPROM can be erased in one location, in one row, or in bulk in one operation. Flash memory can be erased and programmed electrically. However, flash memory can only be erased in bulk.

Programs are known as software. A program is a set of instructions that the computer hardware can execute. Programmers write a program in some kind of programming language. Only machine language was available during the early days of computers. A machine language program consists of a sequence of machine instructions. A machine instruction is a combination of 0s and 1s that informs the CPU to perform some specific operation. Using machine language to write programs is very difficult and hence the assembly language was invented to improve the productivity of programmers. Programs written in assembly language consist of a sequence of assembly instructions. An assembly instruction is the mnemonic representation of some machine instruction. Programs written in assembly language are difficult to understand and programming productivity is not high. High-level languages such as FOR-TRAN, COBOL, PASCAL, C, C++, and JAVA were invented to avoid the drawback of the assembly language. Programs written in assembly language or high-level languages are called *source code*. Source code must be translated before it can be executed. The translator of a program written in assembly language is called an *assembler*, whereas the translator of a program in a high-level language is called a *compiler*.

The 68HC11 has a few CPU registers, including accumulators A and B, the condition code register (CCR), index registers X and Y, the stack pointer SP, and the program counter (PC).

A memory location has two components: its contents and its address. When accessing a memory location, the CPU sends out the address on the address bus and the memory component will place the requested value on the data bus.

A 68HC11 instruction consists of one or two bytes of opcode and zero to three bytes of operand information. The operands of an instruction are specified by addressing modes. The 68HC11 provides the following addressing modes:

- immediate
- direct
- extended
- indexed
- relative

A programmer may occasionally need to decode the machine instructions into assembly instructions for debugging or other purposes. The first step is to identify the format of the instruction that an opcode corresponds to. The second step is to identify the remaining operand bytes. The third step is to combine the opcode byte(s) and operand byte(s) into an instruction.

The execution cycle of an instruction includes the following steps:

- Perform one or multiple read cycles to fetch the instruction.
- Perform the required read cycles to fetch the memory operands (optional).
- Perform the operation specified by the opcode.
- Write back the result to a register or a memory location (optional).

1.10　Exercises

E1.1 What is a processor? What sections of a computer make up a processor?

E1.2 What makes a microprocessor different from the processors used in large computers?

E1.3 What makes a microcontroller different from the microprocessor used in a PC?

E1.4 How many bits of data are stored in each memory location of a microcontroller trainer built around the 68HC11 microcontroller?

E1.5 How many different memory locations can the 68HC11 microcontroller address?

E1.6 Why must every computer have some nonvolatile memory?

E1.7 What are the differences between MROM, PROM, EPROM, EEPROM and flash memory? For what type of application is each most suitable?

E1.8 What is the difference between source code and object code?

E1.9 What register is used to keep track of the address of the next instruction to be executed?

E1.10 Convert 5K, 13K, and 24K into decimal representation.

E1.11 Write an instruction sequence to swap the contents of accumulators A and B. Hint: use a memory location as a swap buffer.

E1.12 Write an instruction sequence to subtract 4 from memory locations $00 to $02.

E1.13 Write an instruction sequence to add the contents of memory locations $10,$11,$12 and store the sum at memory location $15.

E1.14 Write an instruction sequence to swap the contents of memory locations at $11 and $12.

E1.15 Write an instruction sequence to place 33 at memory locations $11, $12, and $13.

E1.16 Write an instruction sequence that performs the operations equivalent to those performed by the following high-level language statements:

```
I := 10;
J := 20;
K := J - I;
```

Assume variables I, J, and K are located at $01, $04, and $10, respectively.

E1.17 Translate the following assembly instructions into machine instructions using Table 1.5.

```
LDAA    #$20
ADDA    $10
LDAB    #$90
STD     $00
```

E1.18 Translate the following assembly instructions into machine instructions using Table 1.5.

```
LDAA    #00
LDAB    $01
STD     $60
LDAB    $03
STD     $62
```

E1.19 Disassemble the following machine code into 68HC11 assembly instructions using Table 1.5.

```
96 40 8B F8 97 40 96 41 8B FA 97 41
```

E1.20 Disassemble the following machine code into 68HC11 assembly instructions using Table 1.5.

```
DC 00 8B FE CB FD DD 00
```

E1.21 Determine the number of read and write cycles performed during the execution of the following instructions:

```
LDX #$1000
LDS #$FF
SUBA $2000
ADDD 0,Y
```

E1.22 Determine the number of read and write cycles performed during the execution of the following instructions:

```
LDAA $00
LDAB $01
STAB $00
LDAB $02
STAB $01
STAA $02
```

E1.23 Determine the contents of the memory locations at $00, $01, and $02 after the execution of the following instruction sequence, given that [$00] = $11, [$01] = 22, and [$02] = 33.

```
LDAA $02
ADDA $01
STAA $01
ADDA $00
STAA $00
```

E1.24 Refer to Appendix B to find the execution time (in E clock cycles) of the following instructions.

```
LDAA #$00
STAA $1000
LDAB 10,Y
ADDA 9,X
STD $2000
```

E1.25 Assume that the E clock signal frequency is 2 MHz. You are given the following instructions:

```
LOOP    PSHA
        PULA
        PSHA
        PULA
        DEX
        BNE     LOOP
```

Find

1. The period of the E clock signal in seconds.

2. The total execution time (in number of E clock cycles and in seconds) of these six instructions.

3. The number of times that these six instructions must be executed in order to create a time delay of one second.

2

68HC11 Assembly Programming

2.1 Objectives

After completing this chapter you should be able to

- use assembler directives to allocate memory blocks, define constants, etc.

- write assembly programs to perform simple arithmetic operations

- write program loops to perform repetitive operations

- communicate with evaluation boards such as EVB, EVBU, and CMD-11A8 using a PC

- enter commands to display and modify the registers and memory locations of the evaluation board

- enter and download programs onto the evaluation board for execution

- create delays of any length using program loops

2.2   Introduction

Assembly language programming is a method of creating instructions that are the symbolic equivalent of machine code. The syntax of each instruction is structured to allow direct translation to machine code.

This chapter begins a formal study of Motorola 68HC11 assembly language programming. Although chapter 1 included several basic instructions, this chapter introduces the formal rules of program structure, specification of variables and data types, and the syntax rules for program statements. The development of this chapter will follow the standards set out in the M68HC11 reference manual. The opcode mnemonics, register names, and directives are fairly standard. However, particular assemblers may permit considerable variation in symbol names (such as uppercase and lowercase letters), operand expressions, delimiters, and so forth. This text will note some of the typical variations, and the specific ones permitted by the assembler on your system will be listed in your manual.

The rules for the Motorola portable cross assembler (PASM) will be followed in this chapter. Of all cross assemblers, the Motorola freeware as11 is especially popular in universities. Notes on the use of this cross assembler will be made at the appropriate places. All assembly programs in this text are compatible with this freeware.

Evaluation boards are indispensable for learning the use of the microcontroller and for product development. Three evaluation boards will be introduced in this chapter: the Motorola 68HC11 EVB and EVBU and the Axiom CMD-11A8. The as11 freeware will be discussed in this chapter, and the C compiler from Imagecraft will be discussed in chapter 4.

2.3   Assembly Language Program Structure

Let us begin by examining a C program and its equivalent assembly language code. Many of the details of the assembly language code will not be clear on the first reading. However, it will serve as an introduction to the general structure of an assembly language program and some of the specific rules of syntax. The following C program assigns values to two variables and performs a simple arithmetic computation:

```
main ( )
{
    char i, j, k;        /* i, j, & k are 8-bit integer variables */
    i = 75;
    j = 10;
    k = i + j - 6;
}
```

The equivalent assembly language code is shown below. It illustrates the structure of an assembly language program and includes directives, comments, and symbolic addresses. Line numbers are added for reference.

```
(1) * Data storage declaration
(2)        ORG     $00
(3) i      RMB     1           ; variable i
(4) j      RMB     1           ; variable j
(5) k      RMB     1           ; variable k
(6)* Program instruction section
(7) start  ORG     $C000       ; starting address of programs
(8)        LDAA    #75         ; initialize i to 75
(9)        STAA    i           ;    "
(10)       LDAA    #10         ; initialize j to 10
(11)       STAA    j           ;    "
(12)       ADDA    i           ; compute i + j
(13)       SUBA    #6          ; compute i + j – 6
(14)       STAA    k           ; store i + j – 6 at k
(15)       END
```

Before we examine the structure of assembly language programs, you should note the following key concepts in the program:

- Lines (1) and (6) are full-line comments, while lines (3)-(5) and (7)-(14) contain in-line comments that explain the function (or operation) of the corresponding assembler directive (or instruction). Comments are used extensively in assembly language programs to provide documentation.

- Lines (2) and (7) introduce the directive ORG. A directive is an instruction that tells the assembler to perform a support function such as reserving memory. ORG sets up the location counter, and it must be followed by instructions or some other directives such as RMB or FCB. The functions of RMB and FCB will be explained in section 2.4.

- Lines (3)-(5) allocate data space. The symbolic addresses i, j, and k are given as labels. The directive **RMB 1** (reserve memory byte) tells the assembler to reserve one byte of data storage. The value of this byte is not initialized.

- Lines (3)-(5) and (7) begin with a label. A label is a symbolic name for a memory address.

- Lines (9), (11), (12), and (14) use the symbolic addresses i, j, i, and k.

- Line (15) is the END directive. It signals the assembler to stop assembling instructions and interpreting directives. It must be the last statement of every 68HC11 assembly program.

2.3.1   The Global View of a 68HC11 Assembly Program

An assembly language program is divided into four sections that contain the main program components. In some cases these sections can be mixed to provide better algorithm design.

- Assembler directives: These instructions are supplied to the assembler by the user; they define data and symbols, allocate data storage locations, set assembler and linking conditions, and specify output format. Assembler directives do not produce machine code.

- Assembly language instructions: These instructions are 68HC11 instructions; some are defined with labels.

- Comments: There are two types of comments in an assembly program. The first type is used to explain the function of a single instruction or directive. The second type explains the function of a group of instructions or directives or a whole routine. Adding comments makes a program more readable. The rules for forming comments will be explained shortly.

- The END directive: This is the last statement in a 68HC11 assembly language source code, and it causes termination of the assembly process. The AS11 assembler ignores the END directive.

2.3.2   The Local View of a 68HC11 Assembly Program

Each line of a 68HC11 assembly program, excluding certain special constructs, comprises of four distinct fields. Some of the fields may be empty. The order of these fields is:

1. Label
2. Operation
3. Operand(s)
4. Comment

THE LABEL FIELD

Labels are symbols defined by the user to identify memory locations in the program or data areas of the assembly module. For most instructions and directives, the label is optional. The rules for forming a label are:

- A label must begin with a letter (A-Z, a-z), and the letter can be followed by letters, digits (0-9), or special symbols. Some assemblers permit special symbols (such as a period, a dollar sign, or an underscore), some permit only uppercase or lowercase letters, and others will internally convert all letters to either uppercase or lowercase and thus will not distinguish between them. The as11 is case-sensitive.

- Most assemblers restrict the number of characters in a label name (usually to the most significant eight characters), but others allow many more.
- A label can be defined in one of two ways: either (1) the name starts in column 1 and is separated from the rest of the line by a space character, or (2) the name starts in any column and terminates with a colon (:). The Motorola freeware as11 requires the label to start from column 1, and no colon character is required.

The Motorola freeware allows the following characters to be included in a label:

[a-z][A-Z]_.[0-9]$

where . and _ count as nondigits and $ counts as a digit to avoid confusion with hex constants. All characters of a label are significant, and upper- and lowercase characters are distinct. The maximum number of characters in a label is currently set at fifteen.

Example 2.1

Valid and invalid labels

The following instructions contain valid labels:

```
a. START   ADDA   #1        ; label begins in column 1
b. PRINT:   JSR    HEXOUT    ; label is terminated by a colon (:)
            JMP    START     ; instruction references label "START"
```

The following instructions contain invalid labels:

```
c. GO OUT  ADDA   $00       ; a blank is included in the label
d.  LOOP   ADDB   $10       ; label begins in column 2 and so a colon is required
```

THE OPERATION FIELD

This field contains the mnemonic names for machine instructions and assembler directives. If a label is present, the opcode or directive must be separated from the label field by at least one space. If there is no label, the operation field must be at least one space from the left margin.

Example 2.2

Examples of operation fields

```
         ADDA   #$20      ; ADDA is the instruction mnemonic
   ZERO  EQU    0         ; equate directive EQU occupies the operation field
```

THE OPERAND FIELD

If an operand field is present, it follows the operation field and is separated from the operation field by at least one space. The operand field may contain operands for instructions or arguments for assembler directives. The following examples include operand fields:

```
TRUE    EQU    1      ; 1 is the operand field
        LDAB   0,X    ; 0,X is the operand field
LOOP:   BNE    NEQ    ; NEQ is the operand field
```

THE COMMENT FIELD

The comment field is optional and is added mainly for documentation. The comment field is ignored by the assembler. A comment may be inserted in one of three ways:

1. If an asterisk (*) is the first printable character in the line, a comment may be inserted at the beginning of a line. In this case, the entire line is a comment—an instruction or directive preceded by an asterisk will not be recognized. The asterisk generally does not have to be in the first column; however, the Motorola freeware as11 requires it to be in the first column.

2. A comment may follow the operation and operand fields of an instruction or directive; in this case it is preceded by at least one space.

3. If the first non–white-space character in a line is an exclamation mark (!), the rest of the line is a comment. However, the Motorola freeware does not recognize comments starting with !.

A comment cannot occur on a line containing only a label field because the first word of the comment would be interpreted as an opcode. In this text, we will use ";" as the first character of a comment after the operand field. The first two types of comments are illustrated in the following examples:

a. ADDA   #2   ; add 2 to accumulator A

b. ABC: an invalid comment—no opcode is present

c. * The whole line is a comment

Example 2.3

▼

Identify the four fields in the following source statement:

```
LOOP:   ADDA   #40       ; add 40 to accumulator A
```

Solution: The four fields of the source statement are as follows:

a. "LOOP" is a label
b. "ADDA" is an instruction mnemonic
c. "#40" is the operand
d. "; add 40 to accumulator A" is a comment

▲

2.4   A Sample of Assembler Directives

Assembler directives look just like instructions in an assembly language program, but they tell the assembler to do something other than create the machine code for an instruction. The available directives vary with the assembler. Interested readers should refer to the appropriate assembler user's manual. In this section, the most often used assembler directives will be discussed. Most of them are supported by the 68HC11 freeware. Statements enclosed in brackets [ ] are optional.

The **END** statement is the assembler directive that is used to end a program to be processed by the assembler. The program has the following form:

```
(your program)
END
```

The END directive indicates the logical end of the source program. Any statement following the END directive is ignored. A warning message will be printed if the END directive is not included in the source code; however, the program will still be assembled correctly if there is no other error in the program. The as11 freeware assembler ignores the END directive.

When writing an assembly program, the programmer needs to allocate storage in one way or another. To *allocate* storage means to find room for a variable or a program in memory. The assembler uses a *location counter* to keep track of the place to put data or an instruction during the assembling process. The **ORG** directive is used to set the value of the location counter, thus telling the assembler where to put the next byte it generates after the ORG directive. The syntax of this directive is as follows:

```
ORG <expression>
```

For example, the sequence

```
ORG    $C000
LDAB   #$FF
```

will put the opcode byte for the instruction *LDAB #$FF* at location $C000.

There are several directives that allow us to reserve an area of memory. The *reserve memory byte* directive **RMB** reserves a block of memory whose size is specified by the number that follows the directive. The syntax of this directive is as follows:

```
[<label>]    RMB    <expression> [<comment>]
```

For example, the statement

```
BUFFER   RMB    100
```

allocates 100 (decimal) bytes for some data and lets the programmer refer to it using the label BUFFER. The bytes in this area can be accessed by using the label for the RMB directive with an offset. For example, to load the first byte into accumulator A, use LDAA BUFFER; to load the tenth byte into accumulator A, use LDAA BUFFER + 9; etc. We can also use an ORG directive to tell the assembler where to reserve a block of memory. For example, the sequence

```
         ORG    $00
BUFFER   RMB    100
```

will reserve a block of 100 bytes starting at the address of $00.

Sometimes it is desirable to give initial values to the reserved memory area. There are five directives that can give initial values to the reserved memory block: BSZ, FCB, FDB, FCC, and DCB.

The *block storage of zeros* directive BSZ causes the assembler to allocate a block of bytes and assign each byte the initial value of zero. The number of bytes allocated is given by the *expression* in the operand field. The standard format of this directive is

```
[<label>]   BSZ    <expression> [<comment>]
```

where <expression> specifies the number of bytes to allocate for storage. The expression must not contain undefined or forward references and must be an absolute expression.

For example,

```
LINELEN   EQU    80
BUFFER    BSZ    LINELEN         ; allocate a storage buffer of 80 characters
```

The *form constant byte* directive FCB will put a byte in memory for each argument of the directive. The standard format of this directive is as follows:

```
[<label>]   FCB    <expression> [,<expression>, ... ,<expression>] [<comment>]
```

For example, the statement

```
ABC   FCB    $11,$22,$33
```

will initialize three bytes in memory to

```
$11
$22
$33
```

and will tell the assembler that ABC is the symbolic address of the first byte, whose initial value is $11. The programmer can also force these bytes to a particular address by adding the ORG directive. For example, the sequence

```
        ORG   $2000
ABC     FCB   $11,$22,$33
```

will initialize the contents of memory locations at $2000, $2001, and $2002 to $11, $22, and $33, respectively.

The *form double byte* directive FDB will initialize two consecutive bytes for each argument. The format of this directive is as follows:

[<label>]   FDB   <expression>[,<expression>, ... ,<expression>] [<comment>]

For example, the directive

```
ABC     FDB   $11,$22,$33
```

will initialize six consecutive bytes in memory to

```
$00
$11
$00
$22
$00
$33
```

and will tell the assembler that ABC is the symbolic address of the first byte of this area, whose value is $00. The FDB is especially useful for putting addresses in memory so that they can be picked up into an index register. Suppose that OC1_ISR is the label of the first instruction of a program and that it has the value of $D000. Then the directive

```
FDB OC1_ISR
```

will generate the following two bytes in memory:

```
$D0
$00
```

The *form constant character* directive FCC will generate the code bytes for the letters in the arguments of the directive, using the ASCII code. All letters to be coded and stored are enclosed in double quotes. The format of this directive is as follows:

[<label>]   FCC   "<string>" [<comment>]

For example, the directive

```
ALPHA   FCC   "DEF"
```

will generate the following values in memory:

```
$44
$45
$46
```

and will let the assembler know the label ALPHA refers to the address of the first letter, which is stored as the byte $44. A character string to be output is often stored in memory using this directive.

The *define-constant-block* directive DCB will reserve an area of memory and initialize each byte to the same constant value. The format of this directive is as follows:

```
[<label>]   DCB    <length>,<value>
```

For example, the directive

```
SPACE   DCB    80,$20
```

will generate a line of 80 space characters and tell the assembler that the label "SPACE" refers to the first space character. The value $20 is the ASCII code of the space character.

The DCB directive is not supported by the as11 freeware assembler. Instead, as11 provides the directive FILL, which serves the same purpose as the DCB directive. The syntax of the FILL directive is

```
[<label>]    FILL    <value>,<length>
```

For example, the following directive

```
ones    FILL    1,40
```

will force the freeware assembler to fill each of the 40 memory locations starting from the label *ones* with a 1.

A program will be more readable and maintainable if symbolic names are used instead of numbers in most situations—this can be achieved by using the *equate* directive. The equate directive EQU allows us to use a symbolic name in place of a number. The syntax of the equate directive is as follows:

```
<label>   EQU    <expression> [<comment>]
```

For example, the directive

```
ROM   EQU    $E000
```

tells the assembler that the value $E000 is to be substituted wherever ROM appears in the program.

2.5   Flowcharts

After writing programs for a while, you will learn the importance of program documentation. Adding comments to a program is one method of program documentation. Another form of program documentation is the flowchart, which describes the logical flow of a program.

Figure 2.1 shows the flowchart symbols used in this book. The terminal symbol is used at the beginning and the end of each program. When it is used

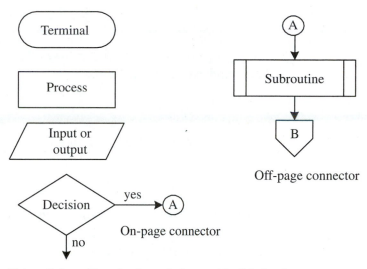

Figure 2.1 ■ Flowchart symbols used in this book

at the beginning of a program, the word **Start** is written inside it. When it is used at the end of a program, it contains the word **Stop.**

The *process box* tells us what must be done at this point in the program execution. The operation specified by the process box could be to shift accumulator A to the right one place, decrement the index register by 1, etc.

The *input/output box* is used to represent data that either enters or exits the computer.

The *decision box* contains a question that can be answered either yes or no. A decision box has two exits, also marked yes and no. The computer will take one action if the answer is yes and will take a different action if the answer is no.

The *on-page connector* indicates that the flowchart continues elsewhere on the same page. The place where it is continued will have the same label as the on-page connector. The *off-page connector* indicates that the flowchart continues on another page. To find where the flowchart continues, you need to look at the following pages of the flowchart to find the matching off-page connector.

Normal flow on a flowchart is from top to bottom and from left to right. Any line that does not follow this normal flow should have an arrowhead on it.

2.6 Writing Programs to Do Arithmetic

In this section, we will use small programs that perform simple calculations to demonstrate how a program is written.

Example 2.4

Write a program to add the values of three memory locations ($00, $01, and $02) and save the sum at $03.

Solution: This problem can be solved by the following steps:

Step 1
Load the contents of the memory location at $00 into accumulator A.

Step 2
Add the contents of the memory location at $01 to accumulator A.

Step 3
Add the contents of the memory location at $02 to accumulator A.

Step 4
Save the contents of accumulator A at $03.

These steps are shown in the flowchart in Figure 2.2. The program is as follows:

```
LDAA    $00        ; load the contents of the memory location at $00 into A
ADDA    $01        ; add the contents of the memory location at $01 to A
ADDA    $02        ; add the contents of the memory location at $02 to A
STAA    $03        ; save the sum at $03
```

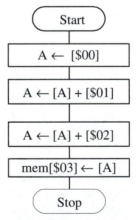

Figure 2.2 ■
Flowchart for adding the contents of three memory locations

Example 2.5

▼

Write a program to subtract six from three 8-bit numbers stored at $00, $01, and $02.

Solution: In the 68HC11, a memory location cannot be the designation of an ADD or a SUB instruction. Therefore, three steps must be followed to add or subtract a number to or from a memory location:

Step 1
Load the memory contents into an accumulator.

Step 2
Add (subtract) the number to (from) the accumulator.

Step 3
Store the contents of the accumulator back to the memory location.

The program is as follows:

```
LDAA    $00        ; load the first number into A
SUBA    #06        ; subtract 6 from the first number
STAA    $00        ; store the decremented value at $00
LDAA    $01        ; load the second number into A
SUBA    #06        ; subtract 6 from the second number
STAA    $01        ; store the decremented value at $01
LDAA    $02        ; load the third number into A
SUBA    #06        ; subtract 6 from the third number
STAA    $02        ; store the decremented value at $02
END
```

▲

2.6.1   The Carry/Borrow Flag

So far we have been working with one-byte hex numbers because these are the largest numbers that can fit into accumulators A and B. However, programs can also be written to add and subtract numbers that contain two or more bytes. Arithmetic performed in an 8-bit microprocessor on numbers that are larger than one byte is called *multiprecision arithmetic*. Multiprecision arithmetic makes use of the condition code register (CCR).

Bit 0 of the CCR register is the carry (C) flag. It can be thought of as a temporary ninth bit that is appended to any 8-bit register. The C flag enables us to write programs to add and subtract hex numbers that are larger than one byte. For example, consider the following two instructions:

```
LDAA #$93
ADDA #$8B
```

These two instructions add the numbers $93 and $8B.

$93
+ $8B
———
$11E

The result is $11E, a 9-bit number, which is too large to fit into the 8-bit accumulator A. When the 68HC11 executes these two instructions, the lower eight bits of the answer, $1E, are placed in accumulator A. This part of the answer is called the *sum*. The ninth bit is called a *carry*. A carry of 1 following an addition instruction sets C flag of the CCR register to 1. A carry of 0 following an addition instruction will clear the C flag to 0. For example, execution of the following two instructions

```
LDAB #$23
ADDB #$44
```

will clear the C flag to 0 because the carry resulting from this addition is 0. In summary,

- If the addition produces a carry of binary 1, the carry flag is set to 1.
- If the addition produces a carry of binary 0, the carry flag is set to 0.

2.6.2 Multiprecision Addition

Multiprecision addition is the addition of numbers that are larger than one byte in an 8-bit computer. The numbers $2524E8 and $B23456 are both three-byte numbers. To find their sum using the 68HC11 microcontroller, we have to write a program that uses multiprecision addition. If we add these two numbers on paper we get:

```
  1   1
  $2424E8
+ $BC3456
  ———————
  $E0593E
```

Starting from the right,

Step 1
Add $8 and $6, which gives the sum $E.

Step 2
Add $E and $5, which gives a result of $13, or a sum of $3 and a carry of $1. The carry is written above the column of numbers on the immediate left. This transfers the carry from the low byte of the number to its middle byte.

Step 3
Add $4 and $4 and the carry $1, which gives a sum of $9.

Step 4
Add $2 and $3, which gives a sum of $5.

Step 5
Add $4 and $C, which gives a sum of $10, or a sum of $0 and a carry of $1. The carry is written above the column of numbers on the immediate

left. This transfers the carry from the second-to-most-significant digit of the number to its most significant digit.

Step 6
Add $2 and $B and the carry $1, which gives a sum of $E.

The important point is that the carry had to be transferred from the low byte of the sum to its high byte. The following program can add these three-byte numbers and store the high byte of the sum at $00, the middle byte at $01, and the low byte at $02:

```
LDAA    #$E8      ; load the low byte of the first number into A
ADDA    #$56      ; add the low byte of the second number to A
STAA    $02       ; save the low byte of the sum
LDAA    #$24      ; load the middle byte of the first number into A
ADCA    #$34      ; add the middle byte of the second number and C flag to A
STAA    $01       ; save the middle byte of the sum
LDAA    #$25      ; load the high byte of the first number into A
ADCA    #$B2      ; add the high byte of the second number and C flag to A
STAA    $00       ; save the high byte of the sum
END
```

Note that the LOAD and STORE instructions do not affect the value of the C flag (otherwise, the program would not work). The 68HC11 has the ADDD instruction that can add two 16-bit numbers and leave the sum in double accumulator D. However, this is not to say that the 68HC11 has a 16-bit ALU because all operations are performed in eight bits at a time. Using the ADDD instruction, the above instruction sequence can be changed to

```
LDD     #$24E8    ; load the lower two bytes of the first number into D
ADDD    #$3456    ; add the lower two bytes of the second number to D
STD     $01       ; save the lower two bytes of the sum
LDAA    #$25      ; load the high byte of the first number into A
ADCA    #$B2      ; add the high byte of the second number and C flag to A
STAA    $00       ; save the high byte of the sum
END
```

Example 2.6

▼
──

Write a program to add two 4-byte numbers stored at $00~$03 and $04~$07 and store the sum at $10~$13.

Solution: The addition should start from the least significant byte and then proceed to the most significant byte. Since the ADDD instruction can add 16-bit numbers, we will use it in this example. The program is as follows:

```
LDD     $02       ; place the lower 16 bits of the first number in D
ADDD    $06       ; add the lower 16 bits of the second number to D
```

```
STD     $12     ; save the lower 16 bits of the sum at $12-$13
LDAA    $01     ; place the second byte of the first number in A
ADCA    $05     ; add the second byte of the second number and C flag to A
STAA    $11     ; save the second byte of the sum at $10-$11
LDAA    $00     ; place the highest byte of the first number in A
ADCA    $04     ; add the highest byte of the second number and C flag to A
STAA    $10     ; save the highest byte of the sum at $10-$11
```

2.6.3   Subtraction and the C flag

The C flag also enables the 68HC11 to borrow from the high byte to the low byte during a multiprecision subtraction. Consider the following subtraction problem:

$2D
- $89

We are attempting to subtract a larger number from a smaller number. Subtracting $9 from $D is not a problem:

$2D
- $89
    4

Now we need to subtract $8 from $2. To do this, we need to borrow from somewhere. The 68HC11 borrows from the C flag, thus setting the C flag. When we borrow from the next higher digit of a hex number, the borrow has a value of decimal 16. After the borrow from the C flag, the problem can be completed:

$2D
- $89
 $A4

When the 68HC11 executes a subtract instruction, it always borrows from the C flag. The borrow is either 1 or 0. The C flag operates as follows during a subtraction:

- If the 68HC11 borrows a 1 from the C flag during a subtraction, the C flag is set to 1.
- If the 68HC11 borrows a 0 from the C flag during a subtraction, the C flag is set to 0.

2.6.4   Multiprecision Subtraction

Multiprecision subtraction is the subtraction of numbers that are larger than one byte for an 8-bit microcontroller. To subtract the hex number $4518D9 from $56342A, the 68HC11 has to perform multiprecision subtraction:

$56342A
– $4518D9

Like multiprecision addition, multiprecision subtraction is performed one byte at a time, beginning with the low byte. The following two instructions can be used to subtract the least significant two bytes:

```
LDAB    #$2A
SUBB    #$D9
```

Since a larger number is subtracted from a smaller one, there is a need to borrow, causing the C flag to be set to 1. The contents of accumulator B should be saved before the high bytes are subtracted. Let's save the low byte of the difference at $02:

```
STAB    $02
```

When the middle bytes are subtracted, the borrow 1 has to be subtracted from the middle byte of the result. In other words, we need a "subtract with borrow" instruction. There is such an instruction, but it is called *subtract with carry*. It comes in two versions, a SBCA instruction for accumulator A and SBCB instruction for accumulator B. The instructions to subtract the middle bytes are:

```
LDAB    #$34
SBCB    #$18
```

We need also to save the middle byte of the result at $01 with the following instruction:

```
STAB    $01
```

The high bytes can be subtracted using the similar instructions, and the complete program with comments is as follows:

```
LDAB    #$2A    ; load the low byte of the minuend into B
SUBB    #$D9    ; subtract the low byte of the subtrahend from B
STAB    $02     ; save the low byte of the difference
LDAB    #$34    ; load the middle byte of the minuend into B
SBCB    #$18    ; subtract the middle byte of the subtrahend and carry from B
STAB    $01     ; save the middle byte of the difference
LDAB    #$56    ; load the high byte of the minuend into B
SBCB    #$45    ; subtract the high byte of subtrahend and carry from B
STAB    $00     ; save the high byte of the difference
END
```

The 68HC11 has a 16-bit subtract instruction called SUBD. Using this instruction, we can replace the first six instructions of the previous program with the following instructions:

```
LDD     #$342A
SUBD    #$18D9
STD     $01
```

The resultant program becomes

```
LDD    #$342A    ; load the lower two bytes of the minuend into D
SUBD   #$18D9    ; subtract the lower two bytes of the subtrahend from D
STD    $01       ; store the lower two bytes of the difference at $01~02
LDAB   #$56      ; load the high byte of the minuend into B
SBCB   #$45      ; subtract the high byte of the subtrahend and carry from B
STAB   $00       ; save the high byte of the difference
END
```

Example 2.7

▼

Write a program to subtract the hex numbers stored at $10~$13 from the hex number stored at $20~$23 and save the difference at $30~$33.

Solution: We will perform the subtraction from the lowest byte toward the highest byte as follows:

```
LDD    $22    ; load the lower two bytes of the minuend into D
SUBD   $12    ; subtract the lower two bytes of the subtrahend from D
STD    $32    ; save the lower two bytes of the difference
LDAA   $21    ; load the second-to-highest byte of the minuend into A
SBCA   $11    ; subtract the second-to-highest byte of the subtrahend and
              ; C flag from A
STAA   $31    ; save the second-to-highest byte of the difference
LDAA   $20    ; load the highest byte of the minuend into A
SBCA   $10    ; subtract the highest byte of the subtrahend and C flag from
              ; A
STAA   $30    ; save the second to lowest byte of the difference
END
```

▲

2.6.5   Binary-Coded Decimal Addition

Although virtually all computers work internally with binary numbers, the input and output equipment generally uses decimal numbers. Since most logic circuits only accept two-valued signals, the decimal numbers must be coded in terms of binary signals. In the simplest form of binary code, each decimal digit is represented by its binary equivalent. For example, 12.34 is represented by

0001 0010 0011 0100

This representation is called *binary-coded decimal*. If the BCD format is used, it must be preserved during arithmetic processing.

The principal advantage of the BCD system is the simplicity of input/output conversion; its principal disadvantage is the complexity of arithmetic processing. The choice between binary and BCD depends on the type of problems the system will be handling.

The 68HC11 microcontroller can add only binary numbers—not decimal numbers. The following program appears to cause the 68HC11 to add the decimal numbers 12 + 34 and store the sum in the memory location at $0000:

```
LDAB #$12
ADDB #$34
STAB $00
```

The program performs the following addition:

$$
\begin{array}{r}
\$12 \\
+ \ \$34 \\
\hline
\$46
\end{array}
$$

When the 68HC11 executes this program, it adds the numbers according to the rules of binary addition and produces the sum $46. This is the correct BCD answer, because the result represents the decimal sum of 12 + 34. In this example, the 68HC11 gives the appearance of performing decimal addition. However, a problem occurs when the 68HC11 adds two BCD digits and generates a sum greater than nine. Then the sum is incorrect in the decimal number system, as the following three examples illustrate:

$$
\begin{array}{ccc}
\$12 & \$25 & \$29 \\
+ \ \$08 & + \ \$37 & + \ \$49 \\
\hline
\$1A & \$5C & \$72
\end{array}
$$

The answers to the first two problems are obviously erroneous in the decimal number system because the hex digits A and C are not between 0 and 9. The answer to the third example appears to contain valid BCD digits, but in the decimal system 29 plus 49 equals 78, not 72; this example involves a carry from the lower nibble to the higher nibble.

In summary, a sum in the BCD format is incorrect if it is greater than $9 or if there is a carry to the next higher nibble. Incorrect BCD sums can be adjusted by adding $6 to them. To correct the examples,

1. Add $6 to every sum digit greater than 9.
2. Add $6 to every sum digit that had a carry of 1 to the next higher digit.

Here are the problems with their sums adjusted:

$$
\begin{array}{ccc}
\$12 & \$25 & \$29 \\
+ \ \$08 & + \ \$37 & + \ \$49 \\
\hline
\$1A & \$5C & \$72 \\
+ \ \$ \ 6 & + \ \$ \ 6 & + \ \$ \ 6 \\
\hline
\$20 & \$62 & \$78
\end{array}
$$

The fifth bit of the condition code register is the *half-carry,* or H flag. A carry from the low nibble to the high nibble during addition is a half-carry. A half-carry of 1 during addition sets the H flag to 1, and a half-carry of 0 during addition clears it to 0. If there is a carry from the high nibble during addition, the C flag is set to 1, which indicates that the high nibble is incorrect. A $6 must be added to the high nibble to adjust it to the correct BCD sum.

Fortunately, we don't need to write instructions to detect illegal BCD sums following a BCD addition. The 68HC11 provides a *decimal adjust accumulator A* instruction, DAA, which takes care of all these detailed detection and correction operations. The DAA instruction monitors the sums of BCD additions and the C and H flags and automatically adds $6 to any nibble that requires it. The rules for using the DAA instruction are:

- The DAA instruction can be used only for BCD addition. It does not work for subtraction or hex arithmetic.
- The DAA instruction must be used immediately after one of the three instructions that leaves their sum in accumulator A. (These three instructions are ADDA, ADCA, and ABA).
- The numbers added must be legal BCD numbers to begin with.

Example 2.8

▼

Write a program to add the BCD numbers stored at memory locations $00 and $01 and save the sum at $02.

Solution:

```
LDAA    $00     ; load the first BCD number into A
ADDA    $01     ; perform the addition
DAA             ; decimal adjust the sum in A
STAA    $02     ; save the sum
END
```

▲

2.6.6 Multiplication and Division

The 68HC11 provides one multiply and two divide instructions. The MUL instruction multiplies the 8-bit unsigned binary value in accumulator A by that in accumulator B to obtain a 16-bit unsigned result in double accumulator D. The upper byte of the product is in accumulator A, and the lower byte of the product is in accumulator B.

Example 2.9

▼

Multiply $45 by $24 and store the product at $00 and $01.

Solution: Before a multiplication is performed, the operands must be loaded into accumulators A and B, as is done in the following program:

```
LDAA   #$45      ; load the multiplicand into A
LDAB   #$24      ; load the multiplier into B
MUL
STD    $00       ; save the product
END
```

The unsigned multiply procedure also allows multiprecision operations. In multiprecision multiplication, the multiplier and the multiplicand must be broken down into 8-bit chunks, and several 8-bit by 8-bit multiplications must be performed. Assume we want to multiply a 16-bit hex number **M** by another 16-bit hex number **N**. To illustrate the procedure, we will break M and N down as follows:

$M = M_H M_L$
$N = N_H N_L$

where M_H and N_H are the upper eight bits of M and N, respectively, and M_L and N_L are the lower eight bits. Four 8-bit by 8-bit multiplications are performed, and then the partial products are added together as shown in Figure 2.3. The procedure to add these four partial products is as follows:

Step 1
Allocate four bytes to hold the product. Assume these four bytes are located at P, P + 1, P + 2, and P + 3.

Step 2
Generate the partial product $M_L N_L$ (in D) and save it at locations P + 2 and P + 3.

Step 3
Generate the partial product $M_H N_H$ (in D) and save it at locations P and P + 1.

Step 4
Generate the partial product $M_H N_L$ (in D) and add it to memory locations P + 1 and P + 2. The C flag may be set to 1 after this addition.

Step 5
Add the C flag to memory location P.

Step 6
Generate the partial product $M_L N_H$ (in D) and add it to memory locations P + 1 and P + 2. The C flag may be set to 1 after this addition.

Step 7
Add the C flag to memory location P.

Example 2.10

▼

Write a program to multiply the 16-bit numbers stored at M~M + 1 and N ~ N + 1, respectively and store the product at memory locations P ~ P + 3.

Solution: The highest, the next-to-highest, the next-to-lowest, and the lowest byte of the product are to be stored at P, P + 1, P + 2, and P + 3, respectively. The following program is a direct translation of the previous multiprecision multiplication algorithm:

```
P      RMB    4        ; four bytes reserved to store the product
M      RMB    2        ; reserved to hold the multiplicand
N      RMB    2        ; reserved to hold the multiplier


       LDAA   M + 1    ; load M_L into A
       LDAB   N + 1    ; load N_L into B
       MUL
       STD    P + 2    ; store the partial product M_L N_L
       LDAA   M        ; load M_H into A
       LDAB   N        ; load N_H into B
       MUL
       STD    P        ; store the partial product M_H N_H
       LDAA   M        ; load M_H into A
       LDAB   N + 1    ; load N_L into B
       MUL             ; generate the partial product M_H N_L
*The following instructions add M_H N_L to memory locations at P + 1 and P + 2
       ADDD   P + 1
       STD    P + 1
*The following three instructions add the C flag to memory location at P
       LDAA   P
       ADCA   #0
       STAA   P


       LDAA   M + 1    ; load M_L into A
       LDAB   N        ; load N_H into B
       MUL             ; generate the partial product M_L N_H
*The following instructions add M_L N_H to memory locations at P + 1 and P + 2
       ADDD   P + 1
```

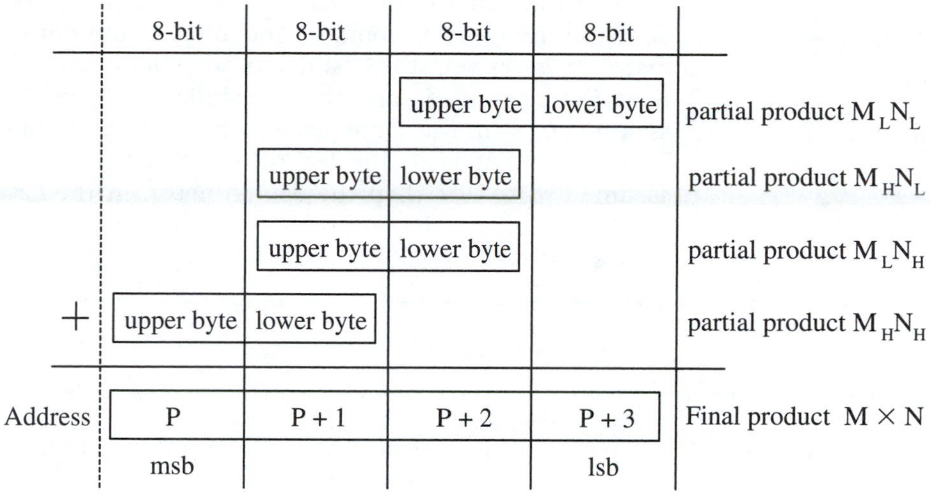

Note: msb stands for most significant byte and lsb for least significant byte

Figure 2.3 ■ 16-bit by 16-bit multiplication

```
        STD     P + 1
*The following three instructions add the C flag to memory location at P
        LDAA    P
        ADCA    #0
        STAA    P
        END
```

The integer divide instruction IDIV performs an unsigned integer divide operation of the 16-bit numerator in double accumulator D by the 16-bit denominator in index register X and sets the condition code accordingly. The quotient is placed in X, and the remainder is placed in D. Both the quotient and the remainder are integer numbers. In the case of division-by-zero, the quotient is set to $FFFF, and the remainder is indeterminate.

Example 2.11

Divide $4567 by $1234 and leave the quotient in X and remainder in D.

Solution:

```
        LDD     #$4567      ; load the dividend in D
        LDX     #$1234      ; load the divisor in D
        IDIV
        END
```

The fractional divide instruction FDIV performs an unsigned fractional divide of the 16-bit numerator in double accumulator D by the 16-bit denominator in index register X and sets the condition code accordingly. The quotient is placed in X, and the remainder is placed in D. The radix point is assumed to be in the same place for both the numerator and the denominator. The radix point is to the left of bit 15 for the quotient. The numerator is assumed to be less than the denominator. In the case of division-by-zero or overflow (i.e., the denominator is less than or equal to the numerator), the quotient is set to $FFFF, and the remainder is indeterminate.

▲

Example 2.12

▼

Write an instruction to divide the fractional number $.2345 by $.5326.

Solution: Before the fractional division can be performed, we need to load these two numbers into D and X, respectively.

```
LDD    #$2345
LDX    #$5326
FDIV
END
```

Since most arithmetic operations can be performed only on accumulators, the contents of accumulator D and index register X must be swapped so that further operations can be performed on the quotient. The 68HC11 provides two swap instructions:

- The XGDX instruction exchanges the contents of accumulator D and index register X.
- The XGDY instruction exchanges the contents of accumulator D and index register Y.

The following example illustrates the application of instructions IDIV and XGDX.

Example 2.13

▼

Write a program to convert the 16-bit binary number stored at $00-$01 to BCD format and store the result at $02-$06. Store each BCD digit in one byte.

Solution: A binary number can be converted to BCD format using repeated division by 10. The largest 16-bit binary number corresponds to decimal 65535, which has 5 decimal digits. The first division by 10 computes the least significant digit and should be stored at memory location $06, the second division by 10 computes the second least significant digit, and so on. The program is as follows:

```
LDD     $00      ; place the 16-bit number in D
LDX     #10
IDIV
STAB    $06      ; save the least significant digit
XGDX
LDX     #10
IDIV             ; compute the second-to-least significant digit
STAB    $05      ; save the second-to-least significant BCD digit
XGDX
LDX     #10
IDIV             ; compute the middle BCD digit
STAB    $04      ; save the middle BCD digit
XGDX
LDX     #10
IDIV             ; compute the second-to-most significant and most
                 ; significant digits
STAB    $03      ; save the second-to-most significant digit
XGDX             ; place the most significant digit in B
STAB    $02      ; save the most significant digit
END
```

2.7   Program Loops

A computer is very good at performing repetitive operations. We can write a program loop to tell the computer to execute the same series of instructions many times. A *finite loop* is a program loop that will be executed only a finite number of times, while an *endless loop* is one in which the computer stays forever.

We will consider four major variants of the looping mechanism:

- **Do** statement **S** forever

    This is an endless loop in which statement S is repeated forever. In some applications, we might add the statement "IF C then EXIT" to leave the endless loop.

- **For** $i = n1$ **to** $n2$ **do** S or **For** $i = n2$ **downto** $n1$ **do** S

    Here, the variable i is the *loop counter*, which keeps track of the number of remaining times statement S is to be executed. The loop counter can be incremented (in the first case) or decremented (in the second case). Statement *S* is repeated *n2 – n1 + 1* times. The value *n2* is assumed to be no smaller than *n1*. If there is concern that the relationship *n1* ≤ *n2* may not hold, then it must be checked at the beginning of the loop. Five steps are required to implement a FOR loop:

Step 1
Initialize the loop counter.

Step 2
Perform the specified operations and increment (or decrement) the loop counter.

Step 3
Compare the counter with the limit.

Step 4
Exit if the value of the loop counter is greater (or less) than the upper (lower) limit.

Step 5
Return to step 2.

- WHILE *C* DO *S*
  Whenever a *While* construct is executed, the logical expression is evaluated first. If it yields a true value, statement S is carried out. If it does not yield a true value, statement S will not be executed. The action of a *While* construct is illustrated in Figure 2.4.

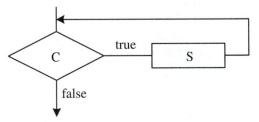

Figure 2.4 ■ The WHILE . . . DO Looping construct

- REPEAT *S* UNTIL *C*
  Statement S is first executed, then the logical expression C is evaluated. If C is true, the next statement will be executed. Otherwise, statement S will be executed again. The action of this construct is illustrated in Figure 2.5. Statement S will be executed at least once.

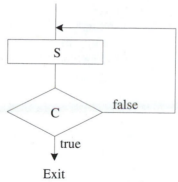

Figure 2.5 ■ The REPEAT . . .
UNTIL looping
construct

To implement a finite loop in 68HC11 assembly language, we must use one of the conditional branch instructions. When executing the conditional branch instruction, the 68HC11 decides whether the branch should be taken by checking the values of condition flags.

2.7.1   The Condition Code Register

The contents of the condition code register are shown in Figure 2.6. The shaded characters are condition flags that reflect the status of an operation. The meanings of these condition flags are as follows:

- C, the carry flag
  Whenever a carry is generated as the result of an operation, this flag will be set to 1. Otherwise, it will be cleared to 0.

- V, the overflow flag
  Whenever the result of a two's complement arithmetic operation is out of range, this flag will be set to 1. Otherwise, it will be set to 0. The V flag is set to 1 when the carry from the most significant bit and the second most significant bit differ as the result of an arithmetic operation.

- Z, the zero flag
  Whenever the result of an operation is zero, this flag will be set to 1. Otherwise, it will be set to 0.

- N, the negative flag
  Whenever the most significant bit of the result of an operation is 1, this flag will be set to 1. Otherwise, it will be set to 0. This flag indicates that the result of an operation is negative.

■ H, the half-carry flag
  Whenever there is a carry from the lower four bits to the upper
  four bits as the result of an operation, this flag will be set to 1.
  Otherwise, it will be set to 0.

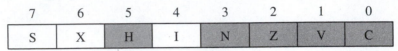

Figure 2.6 ■ Condition code register

2.7.2   Conditional Branch Instructions

A conditional branch instruction is required to implement a finite loop.
The format of a conditional branch instruction is:

[<label>]   Bcc rel   [<comment>]

where

cc   is one of the condition codes listed in Table 2.1.

rel is the distance of the branch from the first byte of the instruction that follows the Bcc
       instruction. We normally use a label instead of a value to specify the branch target.

The 68HC11 also has an unconditional branch instruction that is often
used in a program loop. The format of this instruction is:

[<label>]   BRA rel   [<comment>]

As you can see in Table 2.1, some of the comparisons are unsigned while
others are signed. Unsigned comparisons should be applied to quantities that
are never negative. Memory address is an example of non-negative quantity.

Signed comparisons should be applied to quantities that can be both positive and negative. Voltages and temperatures are examples of signed quantities.

The format for using one of the conditional branch instructions to implement a loop is shown in Figure 2.7. There is no conditional branch instruction
that checks the H flag to determine whether a branch should be taken. Most
of the conditional branch instructions check only one condition flag, as shown
in the following list:

■ C flag   BCC—branch if C = 0
            BCS—branch if C = 1
            BLO—branch if C = 1
            BHS—branch if C = 0
■ Z flag   BEQ—branch if Z = 1
            BNE—branch if Z = 0
■ N flag   BPL—branch if N = 0

Condition code	Meaning
CC	carry clear
CS	carry set
EQ	equal to 0
GE	greater than or equal to 0 (signed comparison)
GT	greater than 0 (signed comparison)
HI	higher (unsigned comparison)
HS	higher or same (unsigned comparison)
LE	less than or equal to 0 (signed comparison)
LO	lower (unsigned comparison)
LS	lower or same (unsigned comparison)
LT	less than 0 (signed comparison)
MI	minus (signed comparison)
NE	not equal to 0
PL	plus (signed comparison)
VC	overflow bit is clear
VS	overflow bit is set

Table 2.1 ■ Branch condition code

$$\text{BMI—branch if } N = 1$$

■ V flag $\quad$ BVS—branch if $V = 1$

$$\text{BVC—branch if } V = 0$$

The following conditional branch instructions check more than one condition flag to determine whether the branch should be taken:

- BGE: branch if $(N \oplus V) = 0$
- BGT: branch if $(Z + (N \oplus V)) = 0$
- BHI: branch if $(C + Z) = 0$
- BLE: branch if $(Z + (N \oplus V)) = 1$
- BLS: branch if $(C + Z) = 1$
- BLT: branch if $(N \oplus V) = 1$

where $\oplus$ stands for the exclusive-or operation.

$$\text{LOOP: instruction X}$$
$$.$$
$$.$$
$$.$$
$$\text{Bcc LOOP}$$

where cc is one of the condition codes

Figure 2.7 ■ Program loop format

Example 2.14

▼

For which of the following conditional branch instructions will the branch be taken after the execution of the instruction CMPA #10? Accumulator A contains 5.

a. BCC target
b. BPL target
c. BNE target
d. BCS target
e. BMI target
f. BHS target

Solution: The 68HC11 subtracts 10 from 5 when executing the instruction CMPA #10. The result is negative. After execution of the instruction, the N and C flags are set to 1, while the V and Z flags are cleared to 0. Therefore, instructions (c), (d), and (e) will cause the branch to be taken.

▲

2.7.3   Decrementing and Incrementing Instructions

A program loop requires a counter, often called the *loop count*, to keep track of the number of times the loop remains to be executed. The loop count can be either decremented or incremented by 1. The following instructions decrement a register or a memory location:

- DECA: subtract one from the contents of accumulator A.
- DECB: subtract one from the contents of accumulator B.
- DEC opr: subtract one from the contents of the memory location *opr*, which must be specified in either the extended or the index addressing mode.
- DES: subtract one from the stack pointer SP.
- DEX: subtract one from index register X.
- DEY: subtract one from index register Y.

These instructions increment a register or memory location:

- INCA: add one to the contents of accumulator A.
- INCB: add one to the contents of accumulator B.
- INC opr: add one to the contents of the memory location *opr*, which must be specified in either the extended or the index addressing mode.
- INS: increment the value of the stack pointer by 1.
- INX: increment the value of index register X by 1.
- INY: increment the value of index register Y by 1.

Example 2.15

Write a program to compute $1 + 2 + ... + N$ (say $N = 20$) and save the sum at $00.

Solution: The procedure for computing the sum of the integers from 1 to 20 is illustrated in Figure 2.8. We will use accumulator B for the loop count *i* and accumulator A for the variable *sum*. The assembly program that implements this algorithm is as follows:

```
N       EQU    20         ; loop count limit
        LDAB   #0         ; initialize the loop count to 0
        LDAA   #00        ; initialize sum to 0
again   INCB              ; increment the loop count
        ABA               ; add i to sum
        CMPB   #20        ; compare loop count with the upper limit
        BNE    again
        STAA   $00        ; save the sum
        END
```

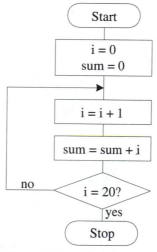

Figure 2.8 ■ Flowchart for computing $1 + 2 + ... + 20$

2.7.4 Compare Instructions

The 68HC11 has a group of compare instructions that are dedicated to setting the condition flags. Compare instructions and conditional branch instructions are often used together to implement a program loop. Compare instructions are executed to change the condition flags of the condition code register, which are then checked by conditional branch instructions to determine whether a branch should be taken. The compare instructions are listed in Table 2.2. The functions of these compare instructions are as follows:

- CBA: subtracts the value in accumulator B from that in accumulator A and sets condition flags accordingly; the contents of both A and B are unchanged.
- CMPA: subtracts the value of a memory location (or immediate value) from accumulator A and sets the condition flags accordingly; both the contents of A and the memory location are not modified.
- CMPB: subtracts the value of a memory location (or immediate value) from accumulator B and sets the condition flags accordingly; both the contents of B and the memory location are not modified.
- CPD: subtracts the 16-bit value stored at the specified memory locations (or just the immediate value) from accumulator D and sets the condition flags accordingly; the contents of D and the memory location are not changed.
- TST: subtracts 0 from the value stored at the specified memory location and sets the condition flags accordingly without changing the contents of the memory location.
- TSTA: subtracts 0 from the accumulator A and sets condition flags accordingly without changing the value in A.
- TSTB: subtracts 0 from the accumulator B and sets condition flags accordingly without changing the value in B.

Instruction format			Function
[<label>]	CBA	[<comment>]	compare A to B
[<label>]	CMPA <opr>	[<comment>]	compare A to a memory location or value
[<label>]	CMPB <opr>	[<comment>]	compare B to a memory location or value
[<label>]	CPD <opr>	[<comment>]	compare D to a memory location or value
[<label>]	TST <opr>	[<comment>]	test a memory location for negative or zero
[<label>]	TSTA	[<comment>]	test A for negative or zero
[<label>]	TSTB	[<comment>]	test B for negative or zero
[<label>]	CPX <opr>	[<comment>]	compare X to a memory location or value
[<label>]	CPY <opr>	[<comment>]	compare Y to a memory location or value

opr is specified in one of the following addressing modes: EXT, INDX, INDY, IMM (not applicable to TST opr), or DIR (not applicable to TST opr)

Table 2.2 ■ 68HC11 compare instructions

- CPX: subtracts the specified value or the value of the specified memory location from X and sets the condition flags accordingly without changing the value in X.
- CPY: subtracts the specified value or the value of the specified memory location from Y and sets the condition flags accordingly without changing the value in Y.

Example 2.16

Write a program to determine the largest of the twenty 8-bit numbers stored at $00-$13. Save the largest number at $20.

Solution: The procedure for finding the largest element of an array is illustrated in Figure 2.9. Accumulator A will be used to hold the temporary largest element. Accumulator B will be used as the loop counter, and index register X will be used as the pointer to the array element. The program should stop when the loop counter B is equal to the array count minus one. The program is as follows:

```
N       EQU     $13
        ORG     $C000       ; starting address of the program
        LDAB    #1
        LDAA    $00         ; set [$00] as the temporary max of the array
        LDX     #$01        ; set X to point to location at $01
```

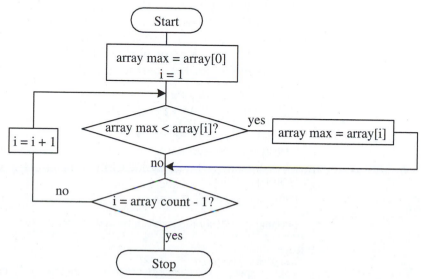

Figure 2.9 ■ Flowchart for finding the largest element in an array

```
again      CMPA    0,X        ; compare A with the next element
           BHS     chkend     ; do we need to update the temporary array max?
           LDAA    0,X        ; update the temporary array max
chkend     CMPB    #N         ; compare the loop count with loop limit
           BEQ     exit       ; are we done?
           INX                ; not yet, move the array pointer
           INCB               ; increment the loop count
           BRA     again
exit       STAA    $20        ; save the array max
           END
```

2.7.5 Special Conditional Branch Instructions

The 68HC11 provides two special conditional branch instructions for implementing program loops. The syntax of the first special conditional branch instruction is

[<label>]   BRCLR (opr)(msk)(rel)   [<comment>]

where

opr   specifies the memory location to be checked and must be specified using either the direct or the index addressing mode.

msk   is an 8-bit mask that specifies the bits of the specified memory location to be checked. The bits to be checked correspond to those bit positions that are ones in the mask.

rel   is the branch offset and is specified in the relative mode

This instruction tells the 68HC11 to perform the logical AND of the memory location specified by *opr* and the mask supplied by the instruction and then to branch if the result is zero (only if all bits corresponding to ones in the mask byte are zeros in the tested byte).

For example, for the instruction sequence

```
           LDX     #$1000
here       BRCLR   $30,X %10000000 here
           LDAA    $31,X
```

the 68HC11 will continue to execute the second instruction if the most significant bit of the memory location at $1030 is 0. Otherwise, the instruction *LDAA $31,X* will be executed.

The syntax of the second special conditional branch instruction is

[<label>]   BRSET (opr)(msk)(rel)   [<comment>]

where

opr   specifies the memory location to be checked and must be specified using either the direct or the index addressing mode.

msk   is an 8-bit mask that specifies the bits of the specified memory location to be checked. The bits to be checked correspond to those bit positions that are ones in the mask.

rel   is the branch offset and is specified in the relative mode

This instruction tells the 68HC11 to perform the logical AND of the specified memory location inverted and the mask supplied in the instruction and then to branch if the result is zero (only if all bits corresponding to ones in the mask byte are ones in the tested byte).

For example, for the instruction sequence

```
loop    ADDA    #01
        ...
        BRSET   $20 %11110000 loop
        LDAB    $01
```

the branch will be taken if the most significant four bits of the memory locations at $20 are all ones.

Example 2.17

Write a program to compute the sum of the odd numbers in an array with twenty 8-bit numbers. The starting address of the array is $00. Save the sum at $20 and $21.

Solution: The address of the last element of the given array is $13. The sum may be larger than eight bits, and it will be stored in memory locations at $20 and $21. The algorithm for solving this problem is as follows:

Step 1
Initialize the sum to 0. Use index register X as the pointer to the array.

Step 2
Check the least significant bit of the memory location pointed to by X. If it is a one, then add it to the sum. Because the sum and the array element are of different lengths, addition proceeds in two steps. In the first step, the array element is added to the lower byte of the sum. In the second step, the carry resulting from the first step is added to the upper byte.

Step 3
Compare [X] to $13. If [X] equals to $13, then exit. Otherwise, increment X by 1 and go to step 2.

The program is as follows:

```
N       EQU     $13
        ORG     $00
array   FCB     ......              ; 20 integer elements of the array
        ORG     $20
SUM     RMB     2                   ; two bytes used to hold sum
```

```
              ORG     $C000                   ; starting address of the program
              LDAA    #$00
              STAA    SUM
              STAA    SUM + 1
              LDX     #array                  ; place the starting address of the array in X
     loop     BRCLR   0,X $01 chkend          ; test to see if the element is odd
              LDD     SUM
              ADDB    0,X
              ADCA    #0                      ; add carry to the most significant byte
              STD     SUM                     ; update the sum
     chkend   CPX     #array + N              ; compare with the address of the last array element
              BHS     exit
              INX                             ; move to the next number
              BRA     loop
     exit     SWI                             ; jump back to BUFFALO monitor in EVB
              END
```

The third instruction of this program uses the expression SUM + 1, which stands for the address of the byte after SUM. Most assemblers support this expression.

▲

2.7.6 Instructions for Variable Initialization

When writing a program, we often need to initialize a variable to zero. The 68HC11 has three instructions for this purpose. They are

[<label>]   CLR   opr   [<comment>]

where **opr** stands for a memory location specified using either the extended or the index addressing mode. The memory location is initialized to zero by this instruction.

[<label>]   CLRA   [<comment>]

Accumulator A is set to zero by this instruction.

[<label>]   CLRB   [<comment>]

Accumulator B is set to zero by this instruction.

2.8 Shift and Rotate Instructions

Shift and *Rotate* instructions are useful for bit field operations, and they can be used to speed up the integer multiply and divide operations if one of the operands is a power of 2. A shift/rotate instruction shifts/rotates the operand by one bit. The 68HC11 has shift instructions that can operate on accumulators

A, B, and D or a memory location. A memory operand (specified as **opr** in the following paragraphs) must be specified using the extended or the index-addressing mode. Both left and right shifts/rotates are available. The format and function of each shift/rotate instruction are described in the following paragraphs.

The following arithmetic shift left instructions operate on 8-bit operands:

```
[<label>]   ASL    opr   [<comment>]      ; memory location opr is the operand
[<label>]   ASLA         [<comment>]      ; accumulator A is the operand
[<label>]   ASLB         [<comment>]      ; accumulator B is the operand
```

The operation of these three instructions is illustrated in the following diagram:

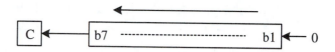

where C is the carry flag. The bit 0 is vacated and is filled with a 0.

The arithmetic shift left instruction can also operate on the D accumulator:

```
[<label>]   ASLD   [<comment>]
```

The operation of this instruction is as follows:

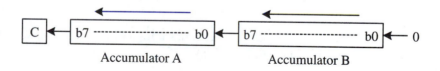

Accumulator A   Accumulator B

The following arithmetic-shift-right instructions operate on 8-bit operands:

```
[<label>]   ASR    opr   [<comment>]      ; memory location opr is the operand
[<label>]   ASRA         [<comment>]      ; accumulator A is the operand
[<label>]   ASRB         [<comment>]      ; accumulator B is the operand
```

The operation of these three instructions is shown in the following diagram:

The sign of the operand is maintained because the sign bit is duplicated after the shift right operation. This feature can be used to divide a signed number by a power of 2.

The following logical-shift-left instructions operate on 8-bit operands:

```
[<label>]   LSL    opr  [<comment>]      ; memory location opr is the operand
[<label>]   LSLA        [<comment>]      ; accumulator A is the operand
[<label>]   LSLB        [<comment>]      ; accumulator B is the operand
```

The operation of these three instructions is as follows:

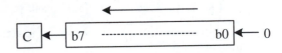

The 68HC11 also has three logical-shift-right instructions that operate on 8-bit operands:

```
[<label>]   LSR    opr  [<comment>]      ; memory location opr is the operand
[<label>]   LSRA        [<comment>]      ; accumulator A is the operand
[<label>]   LSRB        [<comment>]      ; accumulator B is the operand
```

The operation of these three instructions is as follows:

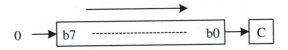

The 68HC11 also has a logical-shift-right instruction that operates on double accumulator D:

```
[<label>]   LSRD        [<comment>]
```

The operation of this instruction is as follows:

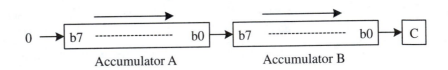

The 68HC11's rotate instructions operate only on 9-bit operands because they include the C flag in the rotate operation. Only one place is rotated at a time.

The following are *rotate left* instructions:

```
[<label>]   ROL    opr  [<comment>]      ; memory location opr is the operand
[<label>]   ROLA        [<comment>]      ; accumulator A is the operand
[<label>]   ROLB        [<comment>]      ; accumulator B is the operand
```

The operation of these three instructions is as follows:

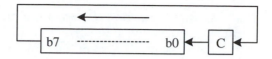

The following are rotate right instructions:

[<label>]	ROR	opr	[<comment>]	; memory location *opr* is the operand
[<label>]	RORA		[<comment>]	; accumulator A is the operand
[<label>]	RORB		[<comment>]	; accumulator B is the operand

The operation of these three instructions is as follows:

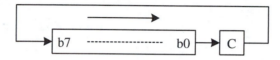

Example 2.18

Compute the new values of accumulator A and the C flag after execution of the instruction *ASLA*. Assume that the original value in A is $74 and the C flag is 1.

Solution: The operation of this instruction is shown in Figure 2.10.

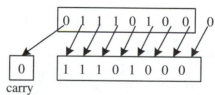

carry

Figure 2.10 ■ Operation of the ASLA instruction

The result is as follows:

Original value	New value
[A] = 01110100 C = 1	[A] = 11101000 C = 0

Example 2.19

▼

Compute the new values of the memory location at $00 and the carry flag after execution of the instruction *ASR $00*. Assume that the original value of the memory location at $00 is $F6 and that the C flag is 1.

Solution: The operation of this instruction is shown in Figure 2.11.

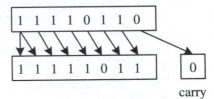

Figure 2.11 ■ Operation of the
instruction ASR $00

The result is:

Before ASR $00	After ASR $00
[$00] = 11110110	[$00] = 11111011
C = 1	C = 0

Example 2.20

▼

If the memory location at $00 contains $F6 and the C flag is 1, what are the new values of this memory location and the C flag after the execution of the instruction LSR $00?

Solution: The operation of this instruction is illustrated in Figure 2.12.

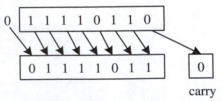

Figure 2.12 ■ Operation of the
instruction LSR $00

The result is:

Before LSR $00	After LSR $00
[$00] = 11110110 C = 1	[$00] = 01111011 C = 0

Example 2.21

Compute the new values of accumulator B and the C flag after execution of the instruction ROLB. Assume the original value of B is $BE and C = 1.

Solution: The operation of this instruction is illustrated in Figure 2.13.

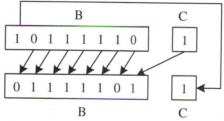

Figure 2.13 ■ Operation of the instruction ROLB

The result is

Before ROLB	After ROLB
[B] = 10111110 C = 1	[B] = 01111101 C = 1

Example 2.22

Compute the new values of accumulator B and the C flag after execution of the instruction RORB. Assume the original value of B is $BE and C = 1.

Solution: The operation of this instruction is illustrated in Figure 2.14.

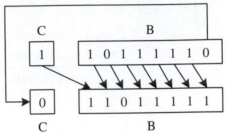

Figure 2.14 ■ Operation of the
instruction RORB

The result is

Before RORB	After RORB
[B] = 1011110	[B] = 1101111
C = 1	C = 0

Example 2.23

Write a program to count the number of ones contained in the 16-bit number stored at $00 and $01 and save the result in $02.

Solution: The logical-shift-right instruction is available for double accumulator D. We can load this 16-bit value into D and shift it to the right sixteen times or until it becomes zero. The algorithm of this program is as follows:

Step 1
Initialize *count* to 0.

Step 2
Place the 16-bit value in D.

Step 3
Shift D to the right.

Step 4
If the C flag is 1, increment *count* by 1.

Step 5
If [D] = 0 then stop; otherwise go to step 3.

The program is as follows:

```
                ORG     $C000
                LDAA    #00
                STAA    $02          ; initialize count to 0
                LDD     $00          ; place the 16-bit value in D
again           LSRD                 ; shift D to the right logically
                BCC     testzero     ; branch if carry is 0
                INC     $02          ; increment count by 1
testzero        CPD     #00          ; compare with 0
                BNE     again        ; if [D] ≠ 0, then continue
                END
```

Sometimes we need to shift a number larger than 16 bits. However, the 68HC11 does not have an instruction that does this. Suppose the number has k bytes and that the most significant byte is located at *loc*. The remaining k − 1 bytes are located at loc + 1, loc + 2, ..., loc + k − 1, as shown in diagram 2.15.

Figure 2.15 ■ k bytes to be shifted

The shift-one-bit-to-the-right logically operation is shown in Figure 2.16.

Figure 2.16 ■ Shift-one-bit-to-the-right operation

As shown in Figure 2.16,

- Bit 7 of each byte will receive bit 0 of the byte on its immediate left with the exception of the most significant byte, which will receive a 0.
- Each byte will be shifted to the right by one bit. Bit 0 of the least significant byte will be shifted out and lost.

This operation can therefore be implemented as follows:

Step 1

Shift the byte at *loc* to the right one place (using the LSR <opr> instruction).

Step 2

Rotate the byte at *loc + 1* to the right one place (using the ROR <opr> instruction).

Step 3

Repeat step 2 for the remaining bytes.

By repeating this procedure, the given k-byte number can be shifted to the right as many bits as desired. Shift multibyte number to the left can be done using a similar method.

Example 2.24

Write a program to shift the 32-bit number stored at $20~$23 to the right four places.

Solution: The most significant, second-to-most-significant, second-to-least-significant, and the least significant bytes of this number are located at $20, $21, $22, and $23, respectively. The following instruction sequence implements the algorithm that we just described:

```
        LDAB    #4          ; set up the loop count
        LDX     #$0020      ; use X as a pointer
again   LSR     0,X
        ROR     1,X
        ROR     2,X
        ROR     3,X
        DECB
        BNE     again
```

2.9 Program Execution Time

Instruction execution in the 68HC11 is controlled by the E clock signal. The E clock signal is derived from an external crystal oscillator. A 68HC11 instruction may take from 2 to 41 E clock cycles to execute. The execution time of each 68HC11 instruction can be found in Appendix D. Knowledge of the execution time of each instruction and the clock period allows a time delay of any length to be created. The execution times of a sample of instructions are shown in Table 2.3.

Instruction	Execution time (unit is E clock cycle)
BNE <rel>	3
DECB	2
DEX	3
LDAB (imme)	2
LDX (imme)	3
NOP	2

Table 2.3 ■ Execution times of a few instructions

Example 2.25

Write a program to create a time delay of 100 ms. Assume the frequency of the E clock signal is 2 MHz.

Solution: A time delay requires the execution of a program loop or loops. Let **x** be the execution time of the program loop, and let **T** be the desired delay. Then the loop count **N** can be calculated from the following equation:

$N = T \div x$

For example, the execution time of the following four instructions is 10 E clock cycles (5 μs):

```
again    NOP              ; 2 E cycles
         NOP              ; 2 E cycles
         DEX              ; 3 E cycles
         BNE      again   ; 3 E cycles
```

To create a 100-ms time delay, this loop must be executed 20,000 times:

100 ms ÷ 5 μs = 0.1 sec ÷ 0.000005 sec = 20000

Therefore the following instruction sequence will create a time delay of 100 ms:

```
         LDX    #20000
again    NOP
         NOP
         DEX
         BNE    again
```

The actual execution time of this program segment is three E clock cycles more than 100 ms because the instruction LDX #20000 takes three E clock cycles to execute. This instruction is the overhead for setting up the delay loop. A longer delay can be created by using nested program loops.

Example 2.26

Write a program to create a delay of 10 seconds.

Solution: We need to write a two-layer loop that includes the loop in the previous example and executes it for the following number of times:

10 sec ÷ 100 ms = 100

The program is as follows:

```
(1)          LDB    #100        ; use B as the outer loop count
(2) outer    LDX    #20000      ; use X as the inner loop count
(3) inner    NOP
(4)          NOP
(5)          DEX                ; decrement the inner loop count
(6)          BNE    inner
(7)          DECB               ; decrement outer loop count
(8)          BNE    outer
(9)          END
```

This program also has overhead that makes the time delay slightly over 10 seconds. As explained earlier, it takes 5 μs to execute instructions (3) to (6). These four instructions are repeated 2,000,000 times to create a time delay of 10 seconds. However, as you have learned, program loops repeat a sequence of instructions, while certain other instructions are needed to set up the program loops. Those instructions are often not included in the calculation of execution time just for simplicity—they are the overhead of creating program loops. The overhead in this example is created by the instructions listed in Table 2.4. Therefore, the total overhead is 401 μs. To create an exact time delay, these instructions must be taken into account when setting up the program loop. More accurate time delays can be created by using timer functions, which will be studied in chapter 8.

Instruction	Execution time (in E clock cycles)	Times of execution	Execution time
LDX #20000	3	100	150 μs
LDAB #100	2	1	1 μs
DECB	2	100	100 μs
BNE OUTER	3	100	150 μs

Table 2.4 ■ Execution time of overhead instructions

2.10   68HC11 Development Tools

There are many development tools for microcontroller-based products. These tools can be classified into two categories: software tools and hardware tools. Software tools include text editors, communication (terminal emulator) programs, cross assemblers, cross compilers, simulators, debuggers, and integrated development software. Hardware tools include evaluation boards, oscilloscopes, logic analyzers, in-circuit emulators, etc. In this book, we will discuss software tools and several evaluation boards.

2.10.1   Software Tools

TEXT EDITORS

The text editor is a software program that allows the user to enter and edit program files. Many readers will probably use the Microsoft EDIT editor that comes with the Microsoft DOS, Windows 95/98, or Windows NT operating system. The Notepad program can also be used. Although it is not a sophisticated text editor, the EDIT editor is adequate for assembly programming. To invoke the EDIT editor, simply type **edit <filename>** at the DOS command prompt.

In the editing session, the user can invoke the commands listed in Table 2.5 by simultaneously pressing the **alt** key and the first character of the category. For example, to open a file, press **alt** and **f** simultaneously to display the commands in the file category and then press **o.** The EDIT editor will then display a *dialog box* that asks the user to enter the name of the file to be opened. To invoke commands in the edit category, press **alt** and **e** simultaneously, followed by the character (**t** for cut, **p** for paste, **c** for copy, or **e** for clear) representing the edit command. The other commands can be invoked in the same way. Before invoking the Cut and Copy commands, the user must specify a block of text to be cut and copied by pressing the shift key and an arrow key simultaneously. There are four arrow keys: right, left, up, and down. The right and down arrow keys are used more often.

The EDIT editor comes with the computer system without extra charge. For a software developer who mainly programs in high-level languages, the EDIT editor is probably not adequate. He or she would probably want to use a sophisticated editor that provides functions such as automatic keyword completion, automatic indentation, parenthesis matching, and syntax checking. These functions can boost the productivity of a high-level language programmer significantly.

CROSS ASSEMBLERS AND COMPILERS

Cross assemblers and cross compilers generate the executable code that is placed in the ROM, EPROM, or EEPROM of a 68HC11-based product. Since cross assemblers and compilers have considerable variations in many areas, we encourage readers to refer to the vendors' reference manuals.

The Motorola freeware cross assembler as11 is used in many universities and 68HC11 developers mainly because it is free and easy to use. The freeware as11 assembles a 68HC11 assembly program into executable code in the S-record format so that it can be downloaded into an evaluation board for execution. To invoke as11, enter the following command at the DOS command prompt:

as11 filename.asm

where *asm* is the file name extension. After this command is entered and if there is no syntax error, the user should see a new file, *filename.s19*, generated in the current working directory.

In the last ten years C language has become very popular in microcontroller applications. There are many commercial C compilers available. We choose the C compiler from ImageCraft to test all the C programs in this text for its ease of use and low price. A C language tutorial and laboratory procedure will be presented in chapter 4.

COMMUNICATION PROGRAMS

There are several options for communicating to the 68HC11 evaluation boards. The HyperTerminal program allows us to communicate to the 68HC11 evaluation boards. The ImageCraft C compiler for Windows 95/98 is itself an integrated development environment. It includes an editor, linker, C compiler, assembler, and a terminal program. The terminal program included in the ImageCraft C compiler also allows us to communicate with the 68HC11 evaluation board.

USING HYPERTERMINAL

HyperTerminal is available in Windows 95/98 and NT. According to this author's experience, it works better under Windows 95/98.

To bring up the HyperTerminal window, click on the following buttons in the specified order:

Start -> Programs -> Accessories -> Communications -> HyperTerminal

Once this is done, a pop-up window for HyperTerminal will appear as shown in Figure 2.17.

Double click the *Hypertrm.exe* icon within the HyperTerminal window. This action will bring up a connection dialogue window as shown in Figure 2.18. You are asked to enter a connection name. Type in a name that you like, for example, *terminal* and then click the *OK* button. A window for selecting a connection will be brought up as shown in Figure 2.19. Set the field *Connecting using* to *direct to com1* (or com2) and then click on OK. The result is shown in Figure 2.20. After clicking the OK button, another dialogue box will appear in which to set the baud rate, data format, and flow control. The correct setting for the EVB, EVBU, and CMD-11A8 evaluation boards is shown in Figure 2.21. Click on the OK button and you will see the HyperTerminal window as shown in Figure 2.22.

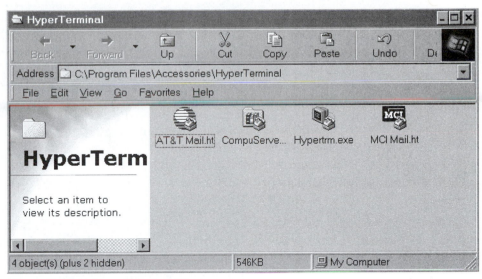

Figure 2.17 ■ HyperTerminal Window

THE 68HC11 SIMULATOR

A simulator allows the user to execute microcontroller programs without having actual hardware. It uses computer memory to represent microcontroller registers and memory locations and to interpret each microcontroller instruction, and it saves execution results in the computer memory. The simulator also allows the user to set the contents of memory locations and

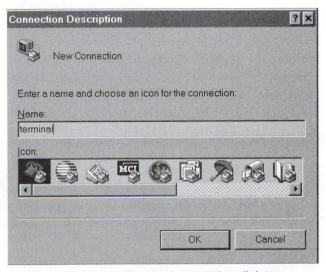

Figure 2.18 ■ HyperTerminal connection dialogue

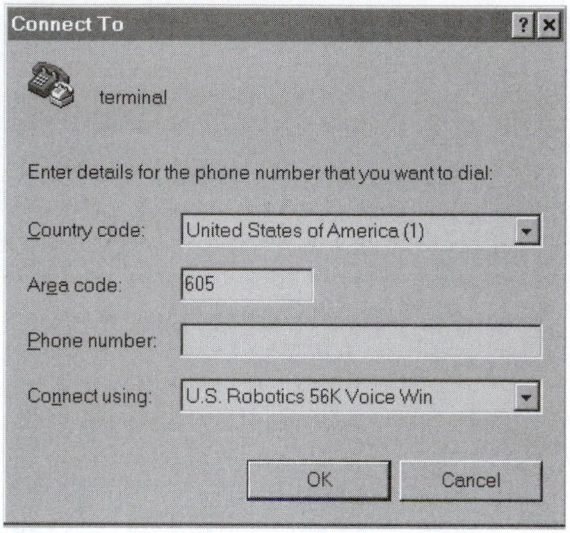

Figure 2.19 ■ Connect To dialogue window

registers before the simulation run starts. However, software simulations have limitations—the most severe one is that they cannot effectively simulate I/O behavior. Motorola has a shareware simulator (sim11) for the 68HC11 that can be downloaded from its Web page (at *www.mot.com).*

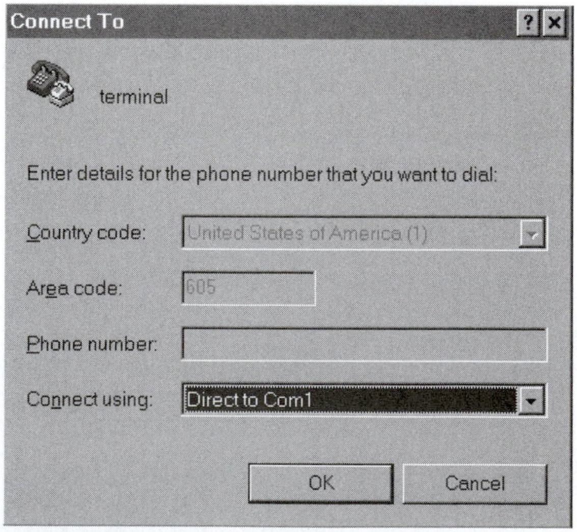

Figure 2.20 ■ Connect To dialogue

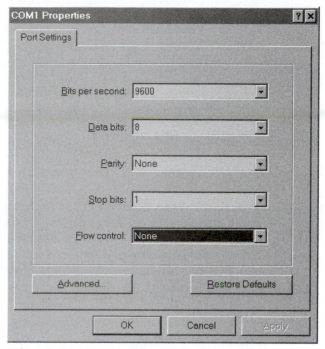

Figure 2.21 ■ Port setting dialogue session

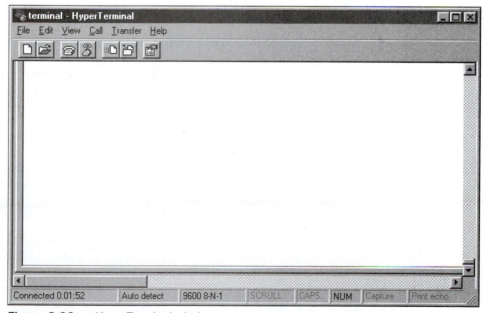

Figure 2.22 ■ HyperTerminal window

SOURCE-LEVEL DEBUGGER

The source-level debugger is a program that allows you to find problems in your code at the high-level language (such as C) or assembly language level. A debugger may have the option to run your program on the evaluation board or simulator. Like a simulator, a debugger can display the contents of registers, internal memory, external memory, and program code in separate windows. With a debugger, all debugging activities are done at the source level. You can see the value of a variable after a statement has been executed. You can set a breakpoint at a statement in a high-level language. A source-level debugger requires a lot of computation. A debugger may run slowly if it needs to simulate the microcontroller instruction set instead of using the actual hardware to run the program.

A source-level debugger needs to communicate with the monitor program on the evaluation board in order to display the contents of CPU registers and memory locations, set or delete breakpoints, trace program execution, etc. Since the monitor programs on different evaluation boards may not be the same, a source-level debugger may be used only with one type of evaluation board.

INTEGRATED DEVELOPMENT TOOLS

Ideally, integrated development software would provide an environment that combines the following software tools so that we can perform all development activities without needing to exit any program:

- a text editor
- a cross assembler and/or compiler
- a simulator
- a source-level debugger
- a communication program
- a front-end interface to an evaluation board

An integrated development tool usually costs dearly.

2.10.2   68HC11 Evaluation Boards

Motorola and several other companies have developed evaluation boards to facilitate the prototyping and evaluation of 68HC11-based designs. Most of these single-board computers have a resident *monitor* program that can, in response to the user's request, provide the following services:

- display the contents of registers and/or memory locations
- modify the contents of registers and/or memory locations
- set program execution breakpoints where program execution will be stopped
- trace program execution

- download the user's program from a PC or other computer
- allow the user to enter short programs
- perform other functions and services

The monitor program takes control of the evaluation board when the power is first turned on. It will initialize the I/O registers and the contents of some memory locations, and it may perform some power-on testing. It may also provide many other functions and services that can be invoked by the user program.

We will look at the Motorola EVB and EVBU and the Axiom CMD-11A8 evaluation boards in this book. These boards are inexpensive and widely used. These boards have the same monitor—*Buffalo monitor*. The Buffalo monitor provides a one-line assembler/disassembler, which allows the user to enter short programs onto the board for direct execution.

THE 68HC11EVB

The block diagram and the photograph of the EVB are shown in Figures 2.23 and 2.24, respectively. The details of the communication ports will be discussed in chapter 9.

The microcontroller unit (MCU) used in the EVB computer is a 68HC11A1. An 8-MHz crystal oscillator is used to generate the 2-MHz E clock signal to control the operation of the 68HC11A1. The MCU is configured to operate in the expanded mode, which makes ports B and C unavailable to the user. Port B and C pins are restored and made available to the user by adding

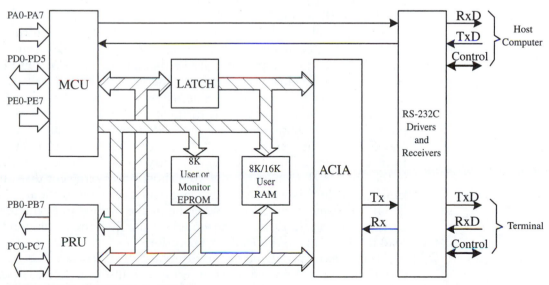

Figure 2.23 ■ EVB block diagram

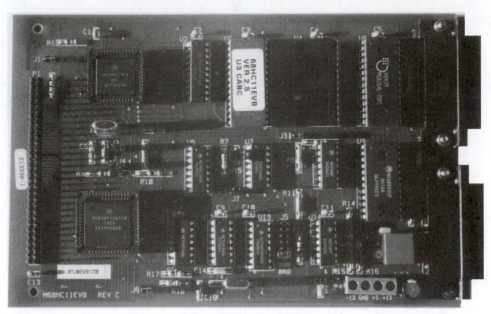

Figure 2.24 ■ 68HC11 EVB board (Reprinted with permission of Motorola)

the 68HC24 *port replacement unit* (PRU). The Buffalo monitor program resides in the 8KB EPROM. By default, the EVB provides 8KB of SRAM, ranging from $C000 to $DFFF, and it can be expanded to 16KB by adding another 8KB SRAM. The EVB has two communication ports: one is the *terminal port*, the other is the *host port.* A personal computer (PC) communicates with the EVB via the terminal port.

The idea of providing a host port can be traced back to the 68000-based ECB computer designed by Motorola in 1979. In those days, the price of a PC was very high. Most companies used either a mainframe or a minicomputer as a host to run development software. The ECB computer had a terminal port and a host port. Design engineers used a terminal to talk to the ECB computer through the terminal port, while the host port was connected to the host computer. When the engineer needed to edit or assemble the program, he or she set the ECB computer to the *transparent mode,* in which characters coming in from the terminal port were sent directly to the host port by the ECB computer and the characters coming from host port were sent directly to the terminal port. In this way the design engineer could log in the host computer to do all the development work. When he or she needed to evaluate the program on the ECB computer, he or she would set the ECB to normal mode and download the program to the ECB for execution.

Today most people do not use the host port because PCs are so inexpensive that there is no need to use a terminal and run the cross assembler/compiler on a mini- or mainframe computer.

EVB CONFIGURATION

The memory space allocation is often described by a map called the *memory map.* By examining the memory map, we can identify the memory blocks occupied by different devices and registers and also figure out quickly what memory blocks are still available for allocation.

The EVB memory map has a single-map design. User RAM resides at different memory locations from the MCU ROM, as shown in Figure 2.25. User RAM is an area where we can write and test programs. The internal RAM is 256 bytes. However, the area from $0040 to $00FF is used by the Buffalo monitor, so the user is restricted to the area from $0000-$003F. Small programs that should be retained when power is turned off can be placed in the 512-byte EEPROM.

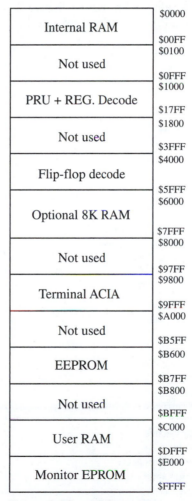

Figure 2.25 ■ EVB memory
map diagram

The Buffalo monitor provides a self-contained operating environment. The user can interact with the monitor through predefined commands entered from a terminal. The commands supported by the Buffalo monitor are listed in Table 2.5.

USING BUFFALO COMMANDS

Before using the EVB, the user must either invoke the HyperTerminal program or the terminal program bundled with the ImageCraft C compiler. We will use only the HyperTerminal provided by Windows 95/98 in this chapter. The terminal program bundled with the ImageCraft C compiler will be discussed in chapter 4.

Suppose you have started the HyperTerminal program and brought out the screen as shown in Figure 2.22. Press the enter key, and the Buffalo monitor will put out the command prompt on the monitor screen as shown below:

```
>_
```

Command	Description
ASM [<address>]	assembler/disassembler
BF <address 1><address 2><data>	block-fill memory with data
BR [-][<address>]	breakpoint set
BULK	bulk-erase EEPROM
BULKALL	bulk-erase EEPROM + CONFIG register
CALL [<address>]	execute subroutine
G [<address>]	execute program
HELP	display monitor commands
LOAD <host download command>	download (S-records) via host port
LOAD <T>	download (S-records) via terminal port
MD [<address1>[<address2>]]	dump memory to terminal
MM [<address>]	memory modify
MOVE <address1><address2>[<destination>]	move memory to a new location
P	proceed/continue
RM [p, y, x, a, b, c, s]	register modify
T [<n>]	trace $1-$FF instructions
TM	enter transparent mode
VERIFY <host download command>	compare memory to download data via host port
VERIFY <T>	compare memory to download data via terminal port

where <> enclose syntactical variable
      [ ] enclose optional fields
      [ ]... enclose repeated optional fields
All numbers are entered in hex format

Table 2.5 ■ EVB monitor commands

where > is the Buffalo monitor prompt and the underscore character is the blinking cursor.

The command

    BF <addr1><addr2><data>

allows the user to fill the block of memory locations from *addr1* to *addr2* with the value given by *data.* The command

    MD [<addr1>[<addr2>]]

displays the contents of memory locations; either one, two, or no addresses can be specified in this command. If no address is specified, the Buffalo monitor will display the memory locations whose addresses were given in the previous MD command. If only one address is specified, the Buffalo monitor will display contents of nine rows of memory locations, starting from the address specified in the command. Each row contains 16 memory locations in hex format.

In the following examples, boldface letters indicate commands or responses entered by the user, while normal letters represent messages displayed by the Buffalo monitor program. The **Enter** key is required by all commands but will not be shown.

Example 2.27

▼

Use Buffalo commands to fill the memory locations from 00 to $3F of the demo boards with the value $FF and display them on the monitor screen.

Solution: At the Buffalo monitor prompt enter the command **BF 00 3F FF.** Wait until the next monitor prompt appears, and then enter the command **MD 00 3F.** You will see the following on your screen:

    >BF 00 3F FF
    >MD 00 3F
    0000 FF FF FF FF FF FF FF FF FF FF FF FF FF FF FF FF
    0010 FF FF FF FF FF FF FF FF FF FF FF FF FF FF FF FF
    0020 FF FF FF FF FF FF FF FF FF FF FF FF FF FF FF FF
    0030 FF FF FF FF FF FF FF FF FF FF FF FF FF FF FF FF

The command **MM [<address>]** allows the user to modify the contents of the specified memory location. The Buffalo monitor will display the current contents of that memory location, and the user should enter the new value to its right. This command allows you to modify the contents of one location at a time. The monitor prompt will reappear when the command is completed.

▲

Example 2.28

▼

Set the value of the memory location at $00 to 0 and display the contents of the memory locations from $00 to $0F on the screen.

Solution: Enter the command **MM 00** at the Buffalo prompt. The monitor will display the message **0000 FF** on the screen, as shown below:

```
>MM 00
0000 FF_
```

The cursor (underscore character) will be blinking to the right of the second line. Type the new value (00) and press the Enter key at the cursor:

```
0000 FF 00
```

You can then verify the new value of the memory location $0000 by entering the command MD 00 0F. The whole session should look this:

```
>MM 00
0000 FF 00
>MD 00 0F
0000 00 FF FF FF FF FF FF FF FF FF FF FF FF FF FF FF
```

▲

Sometimes the user needs to set the values of CPU registers before running the program. The command

```
RM [p,x,y,a,b,c,s]
```

allows the user to display and modify the contents of all CPU registers. The letters inside the brackets stand for program counter PC, index register X, index register Y, accumulator A, accumulator B, condition code register CCR, and stack pointer SP, respectively. If the user does not specify the register in the command, then PC is selected by default. If the user changes his or her mind and decides not to modify the register, he or she can press the Enter key.

Example 2.29

▼

Use a Buffalo command to set the value of the stack pointer to $DFFF.

Solution: You should use the command **RM S.** The Buffalo monitor will display the contents of all the CPU registers on one line and the contents of the stack pointer on the next line, and then it will wait for the new value to be entered by the user. The whole session should look like this:

```
>RM S
P-C000 Y-FFFF X-FFFF A-FF B-FF C-D0 S-004A
S-004A DFFF
>
```

You can verify the contents of the stack pointer by reentering the command **RM.**

▲

Example 2.30

▼

Use the RM command to set the contents of Y to $0001 and the contents of B to $00.

Solution: First enter the command **RM Y.** The Buffalo command will display the contents of all the CPU registers on one line and the contents of Y on the next line, and then it will wait for you to enter the new value. Enter the new value and press the space bar to display the next register. The whole session will look like this:

```
>RM Y
P-C000 Y-FFFF X-FFFF A-FF B-FF C-D0 S-003F
Y-FFFF 0001<space bar>
X-FFFF <space bar>
A-FF <space bar>
B-FF 00 <Enter key>
```

▲

After learning how to set and display the contents of memory locations and CPU registers, the next step is to learn how to enter and run a program on the EVB. Note that the user program should be entered in the area from $C000 to $DFFF ($2000 to $3FFF for CMD-11A8). There are two ways to enter programs into a demo board:

1. *Use the EVB assembler/disassembler.* In this method, the command

    ASM [<addr>]

   is used to enter instructions directly into the SRAM of the demo board. For example, to enter the instruction **ADDA #$20** at $C000, the user first types **ASM C000** to the right of the Buffalo prompt >. The Buffalo monitor will display the current instruction at $C000, then display the assembler prompt (the > character indented four columns from the leftmost column) and wait for the user to enter the new instruction. The instruction entered by the user will be assembled into machine code and displayed in the following line. The procedure can be continued as long as the user has more instructions to be entered. The whole process is something like this:

```
>ASM C000
C000 STOP $FFFF
  >ADDA #20
  8B 20
C002 STOP $FFFF
  >
```

When the complete program has been entered, the user can press the **Control** and **A** keys simultaneously to exit the assembler; the Buffalo monitor prompt will reappear.

2. *Download from a PC.* The following steps will download a program into the EVB computer:

Step 1
Use a text editor (such as EDIT editor) to enter the program as a text file.

Step 2
Invoke the cross assembler to assemble the program (type the command **as11 filename.asm**).

Step 3
Invoke the linker to link the object code files created in Step 2 (this step is not necessary when using the Motorola freeware).

Step 4
Invoke the software that converts the executable code into the S-record format (not needed if the Motorola freeware is used). S-record is a common data format defined by Motorola so that program codes generated by assemblers and compilers from different vendors can be loaded into the same hardware for execution.

Step 5
Use the command **LOAD T** to download the executable code onto the demo board. This step requires a specific procedure dictated by the communication program. If you are using the HyperTerminal program, select the command **Send Text File...** under the Transfer menu as shown in Figure 2.26. A dialog box as shown in Figure 2.27 will appear and will ask you to specify the file name (or select from the listed files) to be downloaded. Enter the file name with full path (or click the file to be downloaded) and use the mouse to click the OK button. After the file is downloaded, the Buffalo prompt will reappear.

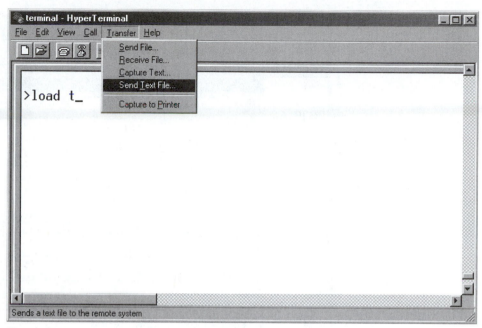

Figure 2.26 ■ Download file to the EVB board

After the program has been downloaded (or entered) into the RAM, you can run it. To start executing the resident program, type the following command (startaddress is the starting address of your program):

>G startaddress

It is desirable to return the CPU control back to the Buffalo monitor when your program has completed execution. This can be done by adding the SWI instruction as the last instruction of your program or by setting a breakpoint to the memory location next to the last byte of your program.

If a program has bugs and cannot execute correctly, the user needs to locate the bugs and fix them. There are two ways to identify the problems:

- Use the *breakpoint set* (BR) command to examine the program execution at any suspected locations. Up to four breakpoints can be set at one time. The contents of the CPU registers will be displayed at the breakpoint. Program execution can be continued from a breakpoint by using the **P** (proceed) command. Breakpoints can be deleted using the command **BR –** when they are not needed. A breakpoint cannot be set at a non-existent memory location.

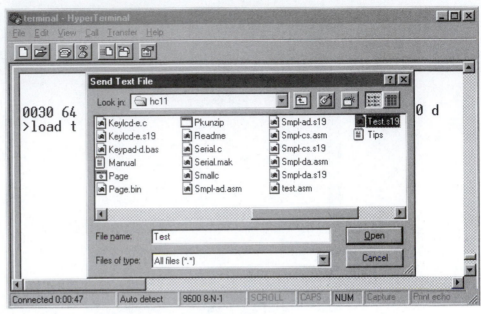

Figure 2.27 ■ Select a file to be downloaded

- Use the *trace* (T) command to monitor program execution on an instruction-by-instruction basis. The contents of the CPU registers will be displayed at the completion of each instruction. You must set the address (PC) of the instruction to be traced before issuing this command. One or several instructions can be traced at the same time.

Example 2.31

▼

Enter Buffalo commands to view the current breakpoints, set two new breakpoints at $DF00 and $DF20, and then delete the breakpoint at $DF20.

Solution: Breakpoints can only be set in RAM. To display the current breakpoints, type the command **BR** followed by **Enter** key. Up to four breakpoints can be set at one time. If no breakpoints have been set, the screen should look like this:

```
>BR
0000 0000 0000 0000
>
```

To set breakpoints at $DF00 and $DF20, type the command **BR DF00 DF20** then press **Enter** key. After this command, you should see the following display on the screen:

```
>BR DF00 DF20
DF00 DF20 0000 0000
>
```

To delete the breakpoint at $DF20, type the command **BR –DF20** followed by **Enter.** You should then see the following message on the screen:

```
>BR –DF20
DF00 0000 0000 0000
>
```

To delete all breakpoints, type **BR -.**

Example 2.32

Use Buffalo command to enter the following instructions, starting from $C000, and trace the execution result of these instructions.

```
LDX #$1000
LDS #$3F
CLRA
STAA $10
```

Solution: You can use the direct method to enter these instructions. First use the command **ASM C000** to specify where to enter the instructions. After entering the four instructions, press the **Control** and **A** keys simultaneously to exit the assembler mode. The whole process will look this:

```
>ASM C000
C000 STX $F000        ; this line may be different on your demo board
    >LDX #1000
    CE 10 00
C003 STX $FFFF        ; this line may be different on your demo board
    >LDS #3F
    8E 00 3F
C006 STX $FFFF        ; this line may be different on your demo board
    >CLRA
    4F
C007 STX $FFFF        ; this line may be different on your demo board
    >STAA 10
    97 10
C009 STX $FFFF
>
>
```

In the trace mode, the Buffalo monitor executes one instruction at a time and displays the contents of CPU registers. Before tracing the program, we need to set the program counter to the address of the first instruction to be traced. To trace these four instructions, simply type the command **T 4.** You should see the following message on the screen:

```
>RM P
P-FFFF Y-FFFF X-FFFF A-FF B-FF C-D0 S-004A
P-FFFF C000
>T 4
Op-CE
P-C003 Y-FFFF X-1000 A-FF B-FF C-C0 S-004A
Op-8E
P-C006 Y-FFFF X-1000 A-FF B-FF C-C0 S-003F
Op-4F
P-C007 Y-FFFF X-1000 A-00 B-FF C-C4 S-003F
Op-97
P-C009 Y-FFFF X-1000 A-00 B-FF C-C4 S-003F
>
```

THE 68HC11 EVBU

The 68HC11 EVBU is designed as a low-cost tool for debugging user-assembled code and evaluating 68HC11A8, 711E9, 811A8, and 811E2 micro-controller unit (MCU) devices. The EVBU includes the following features:

- One-line assembler/disassembler
- Host computer downloading capability
- MC68HC11 MCU-based debugging/evaluating circuitry
- MC68HC68T1 real-time clock + RAM with serial interface peripheral circuitry
- RS-232C compatible terminal I/O port
- Wire-wrap area for custom interfacing
- Single (+5 V) input power source requirements

The 68HC11E9 MCU on the EVBU is configured to operate in the single-chip mode by the factory. However, it can also be reconfigured to operate in one of the other three operation modes of the 68HC11: expanded mode, boot-strap mode, and special test mode. The operation modes of the 68HC11 will be discussed in detail in chapter 5. A photograph and block diagram of the EVBU are shown in Figures 2.28 and 2.29, respectively.

The EVBU memory map has a single-map design reflecting the resident 68HC11E9. The 68HC11E9 has 12KB of ROM, 512 bytes of SRAM, and 512 bytes of EEPROM on the MCU chip. The memory map of the EVBU is shown

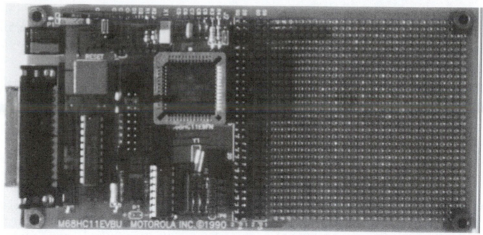

Figure 2.28 ■ 68HC11EVBU board (Reprinted with permission of Motorola)

in Figure 2.30. The on-chip SRAM locations $0048-$00FF are used by the Buffalo monitor, so approximately 328 bytes are available to the user. Only the 512 bytes of EEPROM and 328 bytes of SRAM are available for program development. To use the on-chip 512 bytes EEPROM, we need to reconfigure the EVBU's MCU to operate in bootstrap mode.

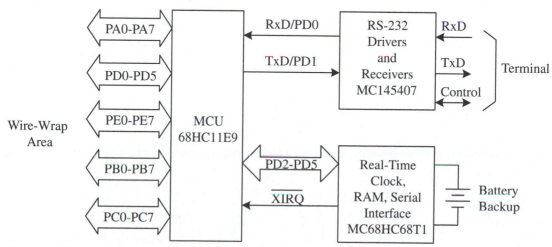

Figure 2.29 ■ EVBU block diagram (Redrawn with permission of Motorola)

The operation of the EVBU is also controlled by the Buffalo monitor program (a later version than that in the EVB). Therefore, in addition to the commands supported by the EVB, the EVBU supports two more commands:

STOPAT <address>

and

XBOOT <address1> [<address2>]

The STOPAT command causes a user program to be executed one instruction at a time until the specified address is encountered. This command should be used only when the current PC value is known. The XBOOT command loads/transfers a block of data from address1 through address2 via the serial communication interface (SCI) to another MC68HC11 MCU device that has been reset in the bootstrap mode. A leading control character $FF is sent prior to sending the data block. The procedure is complicated. Interested readers should read the EVBU user's manual to find out the detail. Another option for programming the on-chip EEPROM is to use the software **PCbug11** that Motorola bundled with the EVBU board. Running on a PC, PCbug11 requires the EVBU to operate in the bootstrap mode and communicates to the EVBU

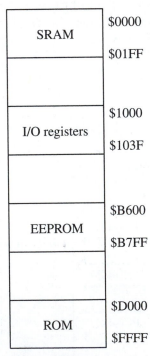

Figure 2.30 ■
EVBU memory map

via the SCI subsystem of the 68HC11. The EVBU can be used to program the 68HC711E9 and 68HC811E2 along with the PCbug11. The procedure for programming the 68HC711E9 will be detailed in chapter 5.

All Buffalo commands that we discussed earlier also work on the EVBU board. Suppose you type and assemble the following program that adds the sum of even numbers between 2 and 20 and saves the sum at location $14:

```
           ORG    $00
           FCB    1,2,3,4,5,6,7,8,9,10,11,12,13,14,15,16,17,18,19,20
    sum    RMB    1
           ORG    $100          ; starting address of the program on the EVBU
           LDY    #20
           LDX    #0000         ; set X to point to the first element of the array
           CLR    sum           ; initialize the sum to 0
    again  LDAA   0,X
           LSRA                 ; shift the bit 0 of A into carry flag
           BCS    chend         ; branch if the number is odd
           LDAA   sum
           ADDA   0,X
           STAA   sum
    chend  INX
           DEY
           BNE    again
           SWI
           END
```

Next, download this program onto the EVBU board. Use the **md** command to display the contents of the memory locations from 0 to $1F, the result will be similar to that in Figure 2.31.

Now run the program by entering the command **g 100.** Display the contents of the memory location at $14 by entering the command **md 10 1F.** The result should be similar to that in Figure 2.32. Readers should see the contents of the memory location $14 to be equal to $6E (110 in decimal).

Since the on-chip SRAM available to the programmer consists of two noncontiguous areas ($00-$47 and $100-$1FF), we will need to use jump instruction to connect the program segments in these two regions. Care should be taken to avoid any errors that can happen when using this approach.

The EVBU is a product being phased out due to its limited size of the SRAM available for program development. The reader may want to look for other alternative low-cost evaluation boards such as the CME-11E9-EVBU from Axiom Manufacturing. This evaluation board (costs less than $90 at the time of this writing) also uses the 68HC11E9 as the MCU and has 32KB of external SRAM and 8KB of external EEPROM that allows the user to perform program development and prototyping. It also comes with a power adaptor and serial connector cable.

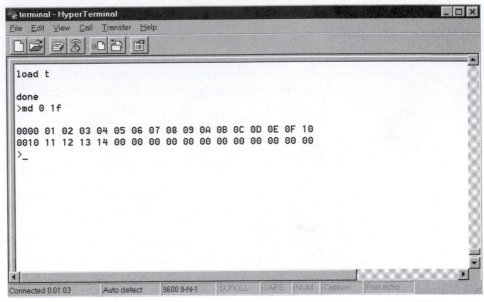

Figure 2.31 ■ Download the program that adds even numbers onto the EVBU board

THE AXIOM CMD-11A8

The CMD-11A8 is manufactured by Axiom Manufacturing, a company based at Richardson, Texas. As shown in Figure 2.33, the CMD-11A8 incorporates the following components:

- a 68HC11A1 microcontroller
- an 8-KB SRAM
- two 8-KB EEPROM
- an i8255 parallel interface chip
- a R65C51 serial communication interface chip

In addition, the CMD-11A8 implements

- two communication ports: COM1 and COM2
- one LCD connector
- one keyboard connector
- one keypad connector
- one i8255 I/O port connector: the signals of the three i8255 ports are available from this connector
- one CPU bus port: 34 68HC11 signals are available from this port

The COM1 is the communication port that we will use to communicate to the PC for program development.

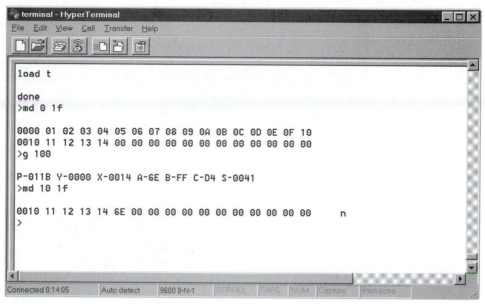

```
terminal - HyperTerminal

File   Edit   View   Call   Transfer   Help

load t

done
>md 0 1f

0000 01 02 03 04 05 06 07 08 09 0A 0B 0C 0D 0E 0F 10
0010 11 12 13 14 00 00 00 00 00 00 00 00 00 00 00 00
>g 100

P-011B Y-0000 X-0014 A-6E B-FF C-D4 S-0041
>md 10 1f

0010 11 12 13 14 6E 00 00 00 00 00 00 00 00 00 00 00      n
>

Connected 0:14:05       Auto detect      9600 8-N-1      SCROLL   CAPS   NUM   Capture   Print echo
```

Figure 2.32 ■ Memory contents of $14 ($6E) after the execution of the program

Figure 2.33 ■ CMD-11A8 demo board (Reprinted with permission
of Axiom Manufacturing)

Since the CMD-11A8 also incorporates the Buffalo monitor, all of the commands that we discussed earlier also apply to the CMD-11A8 evaluation board. The SRAM covers the address range of $2000-$3FFF. Our program should be downloaded into this area for execution. The procedure for downloading programs into the CMD-11A8 is identical to that for EVB.

Axiom Manufacturing also carries several other evaluation boards for the 68HC11. Some of them are fairly inexpensive (less than $80). Interested readers may want to contact them to find out more details (see www.axman.com).

2.11 Summary

An assembly program consists of four major parts: assembler directives, assembly language instructions, the END directive, and comments. A statement of an assembly program consists of four fields: label, operation code, operands, and comment. Assembly directives supported by the Motorola freeware assembler as11 are discussed in detail.

The 68HC11 instructions are explained category by category. Simple program examples are used to demonstrate the applications of different instructions. The 68HC11 is an 8-bit microcontroller. Therefore it can perform only 8-bit arithmetic. Numbers longer than eight bits must be manipulated using multiprecision arithmetic.

Microcontrollers are good at performing repetitive operations. Repetitive operations are implemented by using program loops. There are two types of program loops: *endless loop* and *finite loop*. There are four major variants of the looping mechanism:

- DO statement S forever
- For i = n1 to n2 DO S or For i = n2 down to n1 DO S
- While C DO S
- Repeat S until C

In general, the implementation of program loops requires:

1. the initialization of loop counter (or condition)
2. performing the specified operations
3. comparing the loop counter with the limit (or evaluating the condition)
4. making a decision regarding whether the program loop should be continued

The condition code register and many 68HC11 instructions are provided to implement the program loops.

The shifting and rotating instructions are useful for bit field manipulation. Some integer multiply and divide operations can also be sped up by using these instructions if one of the operands is a power of 2.

The 68HC11 uses E clock signal to control the execution of instructions. By repeating the execution of a sequence of instructions, any amount of time delay can be generated.

Microcontroller-based design requires software and hardware tools. Software tools include text editors, communication programs, cross assemblers/compilers, simulators and debuggers. Many vendors provide integrated development tools that consists of a text editor, a cross assembler, a cross compiler, a simulator, and even a source-level debugger.

Hardware development tools include oscilloscopes, logic analyzers, in-circuit emulators, and evaluation boards. We examined three evaluation boards in detail: the EVB and EVBU from Motorola and the CMD-11A8 from Axiom Manufacturing.

2.12   Exercises

E2.1 Identify the four fields of the following instructions:

```
a.         LDAA    #$20        ; initialize A to $20
b. loop    BNE     next        ; branch if not equal
```

E2.2 Find the valid and invalid labels in the following assembly statements, and explain why the invalid labels are not valid.

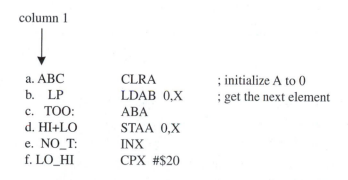

```
column 1

a. ABC        CLRA              ; initialize A to 0
b.  LP        LDAB  0,X         ; get the next element
c.  TOO:      ABA
d. HI+LO      STAA  0,X
e. NO_T:      INX
f. LO_HI      CPX  #$20
```

E2.3 Write assembler directives to reserve 100 bytes from $D000 to $D063 and initialize them to 0.

E2.4 Write assembler directives to build a table to hold the ASCII codes of the capital letters A-Z. The table should be stored in memory locations $D000 to $D019.

E2.5 Write assembler directives to store the following message in memory locations starting from $00: Welcome to Electrical Engineering Department!

E2.6 Write a program to compute the sum of $DA03 and $934A and save the result at $00 and $01.

E2.7 Write a program to compute the sum of two 4-byte numbers stored at $D000 - $D003 and $D004-$D007 and store the sum at $D010-$D013.

E2.8 Write a program to subtract the 3-byte number stored at $00-$02 from the 3-byte number stored at $03-$05 and save the difference at $10-$12.

E2.9 Write a program to subtract the 6-byte number stored at $D000-$D005 from the 6-byte number stored at $D020-$D025 and save the difference at $D030-$D035.

E2.10 Write a program to compute the average of an array of ten 8-bit numbers. The array is stored at $D000-$D009. Save the result at $00.

E2.11 Find the values of accumulator A and B after the execution of the MUL instruction if they originally contain the values

a. $33 and $80, respectively
b. $7C and $55, respectively

E2.12. Which of the conditional branch instructions in the following list will cause the branch to be taken if the condition flags $N = C = 1$ and $Z = V = 0$.

a. BCC target   b. BNE target
c. BGE target   d. BLS target
e. BMI target   f. BCS target
g. BLT target

E2.13. Write a program to swap the last element of an array with the first element, the next-to-last element with the second element, etc. Assume that the array has thirty 8-bit elements and that the array starts at $D000.

E2.14 Write a program to compute the sum of the positive numbers of an array with $30 8-bit numbers. Store the sum at $00-$01. The array starts at $D000.

E2.15 Generate the machine code for the following conditional branch instructions, if each of the instructions occurs at $C100. Let ALPHA be the address $C090, and let BETA be the address $C150.

a. BNE ALPHA
b. BGT BETA

E2.16. Find the values of condition flags N, Z, V, and C in the CCR register after the execution of each of the following instructions, given that [A] = $50 and the condition flags are $N = 0$, $Z = 1$, $V = 0$, and $C = 1$.

a. TSTA   b. CMPA #$60
c. ADDA #$67   d. SUBA #$70
e. LSLA   f. RORA

E2.17 Find the values of condition flags N, Z, V, and C in the CCR register after the execution of each of the following instructions, given that [B] = $00, and the condition flags $N = 0$, $C = 0$, $Z = 1$, and $V = 0$.

a. TSTB   b. ADDB #$30
c. SUBB #$7F   d. LSLB
e. ROLB   f. ADDB #$CF

E2.18 Determine the branch instruction and the offset relative to the PC from the following machine code.

a. 2B 80   b. 2D 20
c. 25 E0   d. 2F F0

E2.19 Write a program to compute the average of the square of all elements of an array with twenty 8-bit numbers. The array is stored at $00-$13. Save the result at $21-$22.

E2.20 Write a program to compute the product of two 3-byte numbers stored at $00-$02 and $03-$05 and save the product at $10-$15.

E2.21 Write a program to compute the sum of the array elements that are a multiple of 4. The array has forty 8-bit elements and starts at $D000. Store the sum at $00.

E2.22 Determine the number of times that the following loop will be executed.

```
        LDAA    #%10000000
loop    LSRA
        STAA    $00
        ADDA    $00
        BMI     loop
        .
        .
        .
```

E2.23 Write a program to shift the 24-bit number stored at $10-$12 three places to the left.

E2.24 Write a program to create a time delay of twenty seconds.

E2.25 Write a program to find the number of odd elements in an array of N elements. The starting address of the array is $D000.

E2.26 Write a program to count the number of elements in an array that are smaller than 15. The array starts at $D000 and has N 8-bit elements. (Hint: use the BRCLR instruction.)

E2.27 Write a program to count the number of elements in an array whose bits 3, 4, and 6 are ones (Hint: use the BRSET instruction.)

2.13   Lab Exercises and Assignments

L2.1 Turn on the PC and start the HyperTerminal program to connect to the EVB, EVBU, or CMD-11A8 evaluation board. Then perform the following operations:

a. Enter a command to set the contents of the memory locations from $00 to $3F to 00

b. Enter a command to display the contents of the memory locations from $00 to $3F to verify the previous operation.

L2.2. Enter monitor commands to display the breakpoints, set new break-points at

    a. $D000 and $D100 for EVB.
    b. $3000 and $3100 for CMD-11A8
    c. $160 and $$170 for EVBU

and then delete all breakpoints.

L2.3. Enter monitor commands to set the contents of accumulator A and B to $00 and $01, respectively.

L2.4 Invoke the one-line assembler/disassembler to enter the following instructions to the evaluation board, starting from address $C000 (start from $2000 for CMD-11A8 and $100 for EVBU), and trace through the program:

```
LDAA    #12
LDAB    #08
MUL
STD     $10
SWI
```

L2.5. Use a text editor to enter the following assembly program as a file with the file name *learn1.asm*:

```
         ORG    $00
sum      RMB    1
arcnt    RMB    1
         ORG    $C000      ; set to $2000 ($100) for CMD-11A8 (EVBU)
         LDAA   #20
         STAA   arcnt
         LDX    #array
         LDAA   #0
         STAA   sum
again    LDAA   0,X
         LSRA
         BCC    chend
         LDAA   sum
         ADDA   0,X
         STAA   sum
chend    INX
         DEC    arcnt
         BNE    again
         SWI
         ORG    $D000      ; set to $3000 ($20) for CMD-11A8 (EVBU)
array    FCB    1,2,3,4,5,6,7,8,9,10,11,12,13,14,15,16,17,18,19,20
         END
```

After entering the program, do the following:

a. Assemble the program.
b. Download the S-record file (file name *learn1.s19*) to the evaluation board.
c. Display the contents of the memory locations from $00 to $0F and $D010 to $D030
   (check $00-$0F and $3000-$3030 ($20-$40) for CMD-11A8 (EVBU)).
d. Execute the program.
e. Display the contents of the memory location at $00.

Note: This program adds all odd numbers in the given array and stores the sum at $00.

L2.6 Write a program to compute the sum of the integers from 1 to 100 and store the sum at $10 and $11.

L2.7 Write a program to divide each element of an array by 4. The array has twenty 8-bit elements and is stored at $D000-$D013 ($3000-$3013 for CMD-11A8, $00-$13 for EVBU). To test the program, define an array using the directive FCB.

L2.8 Write a program to determine how many elements in an array are divisible by 8. The array has twenty 8-bit elements and is stored at $00-$13. Store the result at $20.

3

Data Structures and Subroutine Calls

3.1 Objectives

After completing this chapter you should be able to

- access stack elements and manipulate the stack data structure

- manipulate array, matrix, and string data structures

- write subroutines

- make subroutine calls

- invoke EVB I/O routines to input data from the keyboard and output messages to the PC monitor screen

3.2   Introduction

The main function of a computer is to manipulate information. Programs are written for this purpose. From an information processing standpoint, a program can be considered to have two parts: *data structures* and *algorithms*. In order to manipulate information efficiently, we need to study data structures to learn how information is organized and how elements of particular structures can be manipulated. Common operations applied to data structures include adding and deleting elements, traversing and searching a data structure, etc. A complete study of data structures and their associated operations is beyond the scope of this book. This chapter will study only *stacks, arrays,* and *strings:*

> *Stacks.* A stack is a data structure with a top and a bottom. Elements can be added or removed only at the top of a stack. A stack is a last-in-first-out (LIFO) data structure.
>
> *Arrays.* An array is an ordered set of elements of the same type. The elements of the array are arranged so that there is a zeroth, first, second, and so forth. An array may be one-, two-, or multidimensional. A vector is a one-dimensional array. A matrix is a two-dimensional array. Only one- and two-dimensional arrays will be discussed in this chapter.
>
> *Strings.* A string is a sequence of characters terminated by a special character such as a NULL (ASCII code 0) or EOT (ASCII code 4). A computer system often needs to output a string to inform users of something.

The algorithms for processing the data structures are also an indispensable part of a program. Algorithms are often represented in procedure steps or flowcharts. We will use both in this book.

3.3   The Stack

Conceptually, a stack is a list of data items whose elements can be accessed from only one end. A stack data structure has a top and a bottom. The operation that adds a new item to the top is called *push.* The top element can be removed by performing an operation called *pull* or *pop.* Physically, a stack is a reserved area in main memory where programs agree to perform only push and pull operations. The structure of a stack is shown in Figure 3.1. Depending on the microprocessor, a stack has a stack pointer that points to the top element or to the memory location above the top element.

The 68HC11 has a 16-bit stack pointer (SP) to facilitate the implementation of the stack data structure. The stack pointer points to the memory location immediately above the top byte of the stack. By convention, the stack

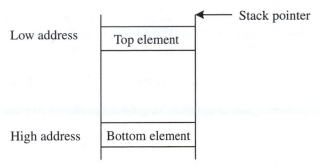

Figure 3.1 ■ Diagram of stack

grows from high addresses toward lower addresses. An area in main memory is allocated for use as the stack area. If the stack grows into memory locations with addresses lower than the stack buffer, a stack overflow error occurs; if the stack is pulled too many times, a stack underflow occurs. The on-chip SRAM of the 68HC11 is often used to implement the stack.

The various push and pull instructions available for the 68HC11 are listed in Table 3.1. A push instruction writes data from the source to the stack and then decrements the SP. There are four push instructions: PSHA, PSHB, PSHX, and PSHY.

Mnemonic	Function
PSHA	Push A into the stack
PSHB	Push B into the stack
PSHX	Push X into the stack
PSHY	Push Y into the stack
PULA	Pull A from the stack
PULB	Pull B from the stack
PULX	Pull X from the stack
PULY	Pull Y from the stack

Table 3.1 ■ The 68HC11 push and pull instructions

Example 3.1

▼

Suppose that

[A]   = $33
[SP] = $00FF
[B]   = $20

What will be the contents of the top byte of the stack before and after the execution of PSHA? What will be the contents of the top two bytes of the stack if PSHB is also executed?

Solution: The contents of the stack before and after the execution of PSHA and PSHB are shown in Figure 3.2.

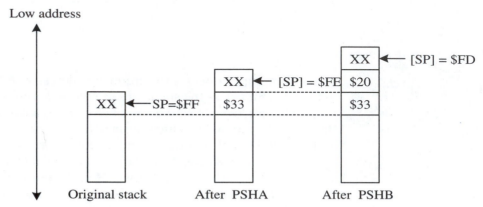

Figure 3.2 ■ Stack frame after two push operations

Pull instructions are the inverse of pushes. First, the stack pointer is incremented. Then the designated register is loaded from the stack at the address contained in the stack pointer. There are also four pull instructions: PULA, PULB, PULX, and PULY. Some microprocessors have pop instructions, which are equivalent to pull instructions.

Before the stack can be used, the stack pointer must be initialized using the LDS instruction. All modes except the inherent addressing mode can be used with the LDS instruction. When necessary, the stack pointer can be saved by using the STS instruction. The memory location to hold the stack pointer can be specified in the direct, extended, or index addressing mode. Either the on-chip SRAM area or the external RAM can be used as the stack area. External memory is preferred if a large stack is needed. When running programs with subroutine calls on EVB, EVBU, or CMD-11A8 demo board, make sure that you have initialized the stack pointer properly.

The 68HC11 has instructions that facilitate access to variables in the stack, since the stack may contain variables that the user program needs to

Instruction	Operation
DES	Decrement the stack pointer by 1
INS	Increment the stack pointer by 1
LDS	Load the contents of a memory location or immediate value into SP
STS	Store the contents of the stack pointer in a memory location
TSX	Load the contents of the stack pointer plus 1 into index register X
TSY	Load the contents of the stack pointer plus 1 into index register Y
TXS	Load the contents of index register X minus 1 onto SP
TYS	Load the contents of index register Y minus 1 onto SP

Table 3.2 ■ Instructions for the stack pointer

access during its execution. By making index register X or Y point to the top element of the stack, the program can access the in-stack variables using the index-addressing mode. The following instructions will load the value in SP plus one into the specified index register:

- TSX: Load the contents of the stack pointer plus one into X.
- TSY: Load the contents of the stack pointer plus one into Y.

Additional instructions related to the stack pointer are listed in Table 3.2. Suppose the top three bytes (from low address to high address) of the stack are $11, $22, $33. Then after the execution of the TSX instruction, the index register points to $11, as illustrated in Figure 3.3. It is now very straightforward to access the top three bytes in the stack. For example, the middle byte (value $22) can be loaded into A by using the instruction LDAA 1,X. This technique will be used extensively throughout this chapter.

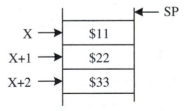

Figure 3.3 ■ The top three bytes of the stack

Example 3.2

Write an instruction sequence to load the top element of the stack into accumulator A and the ninth byte from the top of the stack into accumulator B.

Solution   We need to load the stack pointer into an index register so that the index-addressing mode can be used to access the stack elements. The following instruction sequence does this and accesses the top (0th) and the ninth stack elements:

```
TSX
LDAA    0,X        ; place the zeroth element of the stack in A
LDAB    9,X        ; place the ninth element of the stack in B
```

3.4   Indexable Data Structures

Vectors and matrices are indexable data structures. A *vector* is a sequence of elements in which each element is associated with an index *i* that can be used to access it. Conceptually, a vector is a one-dimensional data structure. To make address calculation easy, the first element is usually associated with the index 0 and each successive element with the next integer, but you can change the index origin of the array to 1 if you are willing to modify the routines. The elements of a vector have the same precision.

The assembler directive FCB (or FDB, depending on the length of the element) can be used to define a vector. Suppose that the vector VEC has elements 1, 2, 3, 4, 5, and 6. It can be defined as follows:

```
VEC    FCB    1,2,3,4,5,6
```

The zeroth element, which is 1, is referred to as VEC(0). To access the *ith* element, use the following instructions:

```
LDAB    #i         ; load the index i into B
LDX     #VEC       ; load the vector base address into X
ABX                ; compute the address of VEC(i)
LDAA    0,X        ; load the contents of VEC(i) into A
```

The array, which was used in chapter 2, is also a vector data structure.

Example 3.3

Sequential search. There are N 16-bit numbers stored in memory. A search key is stored at $00 and $01. Write a sequential search program to find the first occurrence of the key in the array. Store the address of the first element that matches the key at locations $02-$03. Store −1 if the key is not found.

Solution: The logic flow of the sequential search algorithm is illustrated in Figure 3.4. To implement the sequential search algorithm, we will

- use double accumulator D to hold the search key
- use index register Y as the loop count
- use index register X to step through the array elements

The program is as follows:

```
N           equ     20                  ; array count
notfound    EQU     -1


            ORG     $00
loop_cnt    RMB     1
key         FDB     xxxx                ;
result      RMB     2
            ORG     $C000               ; starting address of the program
            LDAA    #N
            STAA    loop_cnt            ; initialize loop count to N
            LDD     #notfound           ; store –1 in result
            STD     result              ;       "
            LDD     key                 ; place key in D
            LDX     #array              ; use X to point to the array element
loop        CPD     0,X
            BEQ     found               ; found the key?
            INX                         ; move array pointer
            INX                         ;       "
            DEC     loop_cnt            ; decrement the loop count
            BNE     loop                ; are we done?
```

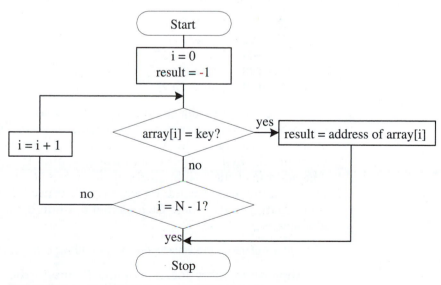

Figure 3.4 ■ Flowchart for sequential search

```
         JMP    stop
found    STX    result
stop     SWI                  ; return to BUFFALO monitor
array    FDB    $20,$30,$202,$1,$200,$10,$39,$47,$59,$23,$45,$61,$130,$440
         FDB    $190,$240,$89,$45,$410,$390

END
```

A *matrix* is a vector whose elements are vectors of the same length. We normally think of a matrix as a two-dimensional pattern, as in

$$\text{MAT} = \begin{matrix} 1 & 2 & 3 & 4 & 5 \\ 6 & 7 & 8 & 9 & 10 \\ 11 & 12 & 13 & 14 & 15 \\ 16 & 17 & 18 & 19 & 20 \\ 21 & 22 & 23 & 24 & 25 \end{matrix}$$

A matrix can be stored in row-major order, as shown in the following examples:

```
mat1    FCB    1,2,3,4,5
        FCB    6,7,8,9,10
        FCB    11,12,13,14,15
        FCB    16,17,18,19,20
        FCB    21,22,23,24,25
```

or it can be stored in column-major order, which is created as follows:

```
mat2    FCB    1,6,11,16,21
        FCB    2,7,12,17,22
        FCB    3,8,13,18,23
        FCB    4,9,14,19,24
        FCB    5,10,15,20,25
```

In the following discussion, matrices are assumed to have N rows and M columns, and the notation $(i, j)th$ is used to refer to the matrix element located at the intersection of the ith row and the jth column. To facilitate address calculation, the first element in a row or a column is associated with the index 0. The address of the (i, j)th matrix element can be computed by using a polynomial equation that depends on which order is used. For example, in a row-major order matrix *mat1* where each element is one byte, the address of the (i, j)th element is

address of the (i,j)th element = (i x M) + j + address of mat1(0,0)

Suppose memory locations i and j contain the row and column indices, respectively. The following instruction sequence computes the address of the element mat1(i, j) and leaves the address in index register X:

```
LDAA    i           ; place row index in A
LDAB    #M          ; put the column dimension in B
MUL                 ; compute i x M
ADDB    j           ; compute i x M + j
ADCA    #0          ;       "
ADDD    #mat1       ; compute i x M + j + address of mat1(0,0)
XGDX                ; place the address in X
```

For a similar matrix *mat2* defined in column major order, the address of (i, j)th element is

address of the (i, j)th element = (j x N) + i + address of mat2(0,0)

The address of mat2(i, j) can be calculated in a similar method:

```
LDAA    j
LDAB    #N
MUL                 ; compute j x N
ADDB    i
ADCA    #0          ; compute j x N + i
ADDD    #mat2       ; compute j x N + i + address of mat2(0,0)
XGDX                ; place the address in X
```

The following program computes the sum of two matrices MA and MB and creates a new matrix MC. Each element of the new matrix is the sum of the corresponding elements of MA and MB.

Example 3.4

Matrix sum. Write a program to compute the sum of two 8×8 matrices. The names and starting addresses of these two matrices are MA and MB and that of the resultant matrix is MC. These three matrices are stored in memory in row major order. Each element of the matrix is one byte.

Solution: To solve this problem, we must add together all the corresponding elements of two matrices. The following steps through every pair of (i, j)th elements, adds them together, and stores the result in the corresponding element of matrix MC.

```
N       EQU     8       ; column and row sizes of the matrix
        ORG     $00
i       RMB     1       ; row index
j       RMB     1       ; column index
buf     RMB     1       ; used to hold one element
disp    RMB     2       ; used to hold the value of N x i + j

        ORG     $C000   ; starting address of the program
        LDAA    #N      ; start from the last row
```

```
                    STAA    i               ; initialize i
        next_row    DEC     i               ; compute the next row index
                    LDAB    #N              ; start from the last column back to zeroth column
                    STAB    j               ;           "
        next_col    DEC     j
                    LDAA    i               ; place row index in A
                    LDAB    #N
                    MUL                     ; compute N x i
                    ADDB    j               ; compute N x i + j
                    ADCA    #0              ;           "
                    STD     disp            ; save the value of N x i + j
                    ADDD    #MA             ; compute N x i + j + address of MA(0, 0)
                    XGDX                    ; put the address in X
                    LDAA    0,X             ; load MA(i, j) in A
                    STAA    buf
                    LDD     disp
                    ADDD    #MB             ; compute the address of MB(i, j)
                    XGDX                    ; put the address in X
                    LDAA    0,X
                    ADDA    buf             ; add MA(i, j) and MB(i, j)
                    STAA    buf             ; save the sum temporarily
     * the following three instructions compute the address of MC(i, j) and leave it in X
                    LDD     disp
                    ADDD    #MC
                    XGDX

                    LDAA    buf
                    STAA    0,X
                    TST     j               ; have we processed a complete row?
                    BNE     next_col        ; if not, branch back to the next column
                    TST     i               ; have we processed all of the rows?
                    BNE     next_row        ; not yet, and continue
                    END
```

In mathematics, the *transpose* M^T of a matrix M is the matrix obtained by writing the rows of M as the columns of M^T. To obtain the transpose of a matrix, the (i, j)th element must be swapped with the (j, i)th element for each i and j from 0 to N − 1.

Example 3.5

Matrix transpose program. Write a program to transpose an N × N matrix.

Solution: An N × N matrix can be divided into three parts, as shown in Figure 3.5. The elements on the diagonal do not need to be swapped. When swap-

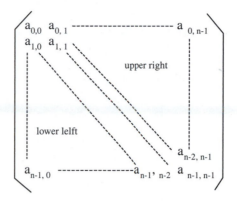

Figure 3.5 ■ Matrix breakdown

ping the matrix elements, the row index i runs from 0 to $N - 2$. For each i, the column index j runs from $i + 1$ to $N - 1$. The procedure for transposing a matrix is illustrated in the flowchart in Figure 3.6.

The following assembly program computes the transpose of a given matrix:

```
N        EQU    8
ILIMIT   EQU    N-2              ; upper limit of i
JLIMIT   EQU    N-1              ; upper limit of j
         ORG    $00
i        RMB    1                ; used for index i
j        RMB    1                ; used for index j
         ORG    $D000            ; starting address of the matrix
MAT      FCB    01,02,03,04,05,06,07,08
         FCB    09,10,11,12,13,14,15,16
         FCB    17,18,19,20,21,22,23,24
         FCB    25,26,27,28,29,30,31,32
         FCB    33,34,35,36,37,38,39,40
         FCB    41,42,43,44,45,46,47,48
         FCB    49,50,51,52,53,54,55,56
         FCB    57,58,59,60,61,62,63,64
         ORG    $C000
         CLR    i                ; initialize i to 0
row_loop LDAA   i
         INCA
         STAA   j                ; j starts from i + 1
* the following seven instructions compute the address of element MAT(i, j) and leave
* it in X
col_loop LDAA   i
         LDAB   #N
         MUL                     ; compute N x i
         ADDB   j                ; compute N x i + j
         ADCA   #0               ;            "
```

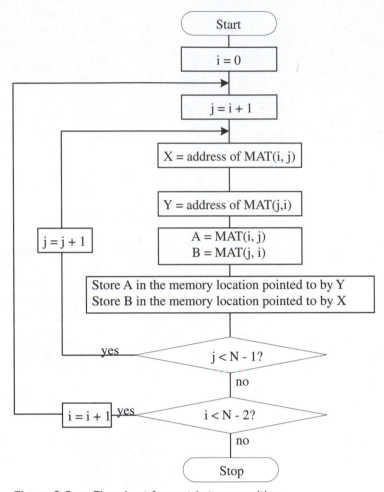

Figure 3.6 ■ Flowchart for matrix transposition

```
            ADDD    #MAT
            XGDX                    ; place address in X
* the following seven instructions compute the address of element MAT (j, i) and leave
* it in Y
            LDAA    j
            LDAB    #N
            MUL                     ; compute N x j
            ADDB    i               ; compute N x j + i
            ADCA    #0              ;         "
            ADDD    #MAT
            XGDY                    ; place address in X
* the following four instructions swap MAT(i, j) and MAT(j, i)
            LDAA    0,X
```

```
                    LDAB     0,Y
                    STAA     0,Y
                    STAB     0,X

                    LDAB     j
                    INC      j
                    CMPB     #JLIMIT        ; is j = N – 1?
                    BNE      col_loop
                    LDAA     i
                    INC      i
                    CMPA     #ILIMIT        ; is i = N – 2?
                    BNE      row_loop
                    END
```

3.5   Strings

A string is a sequence of characters terminated by a NULL (ASCII code 0) or other special character such as EOT (ASCII code $04). Common operations applied to strings include string concatenation, character and word counting, string matching, substring insertion, substring deletion, etc.

Example 3.6

Append one string to the end of another string
Write a program that will append *string1* to the end of *string2*.

Solution: To append a string to the end of another string, we need to make sure that there is enough space at the end of the target string (string2 in this example). There are two steps in this operation:

Step 1
Find the end of the target string (string2).

Step 2
Copy the source string (string1).

The program is as follows:

```
                    ORG      $C000
                    LDX      #string2
again               LDAA     0,X
                    BEQ      copy           ; reach the end of string2 yet?
                    INX                     ; move the string pointer
                    BRA      again
```

```
copy        LDY     #string1
copy_loop   LDAA    0,Y              ; get one character from string1
            STAA    0,X              ; store it
            BEQ     done             ; reach the NULL character?
            INX                      ; move string2 pointer
            INY                      ; move string1 pointer
            BRA     copy_loop
done        NOP
            ...
            ORG     $D000
string1     FCC     "......."
            FCB     0
string2     FCC     "......"
            FCB     0
            END
```

Example 3.7

String character and word count

Write a program to count the number of characters and words contained in a given string. Words are separated by one or more white spaces—space, tab, carriage return, and line-feed characters are white spaces. A word may consist of only one character. White spaces are included in the character count, whereas the NULL character is not.

Solution: Let the character count, word count, and current character be represented by *char_cnt, wd_cnt,* and *curr_char*, respectively. The starting address of the string to be processed is at string_X. The pointer to the current character (*char_ptr*) is X. An empty string consists of a NULL character.

The logic flow of this problem is illustrated in Figure 3.7. In Figure 3.7, every non-null character causes character count to be incremented by 1. A new word is identified by the occurrence of a nonwhite character. Whenever a new word is encountered, we need to step through the remaining nonwhite characters until the next white character is encountered.

```
tab         EQU     $09              ; ASCII code of horizontal tab
sp          EQU     $20              ; ASCII code of space character
CR          EQU     $0D              ; ASCII code of carriage return
LF          EQU     $0A              ; ASCII code of line feed

            ORG     $00
char_cnt    RMB     1                ; character count
wd_cnt      RMB     1                ; word count
            ORG     $D000
```

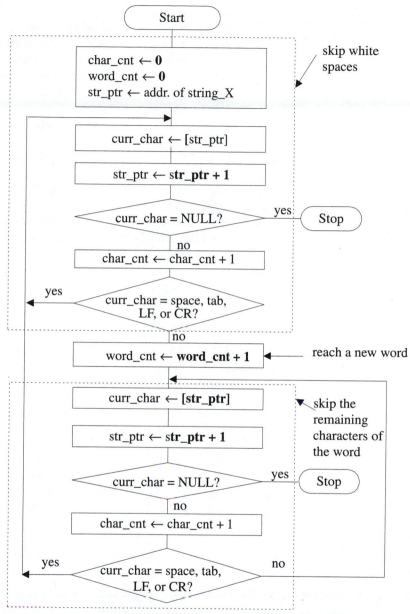

Figure 3.7 ■ Flowchart for character count and word count program

```
string_X    FCC    "........."        ; the string to be processed
            FCB    0                  ; null character to terminate the string


            ORG    $C000              ; starting address of the program
            LDX    #string_X
            CLR    char_cnt
```

```
                CLR     wd_cnt
string_lp       LDAB    0,X             ; read a character
                BEQ     done            ; is this the end of the string?
                INC     char_cnt        ; increment character count
                INX                     ; move the string pointer
* the following eight instructions skip the spaces between words
                CMPB    #sp
                BEQ     string_lp       ; skip the space character
                CMPB    #tab
                BEQ     string_lp       ; skip the tab character
                CMPB    #CR
                BEQ     string_lp       ; skip the carriage return
                CMPB    #LF
                BEQ     string_lp       ; skip the line feed character
* a nonwhite character is the start of a new word
                INC     wd_cnt
wd_loop         LDAB    0,X
                BEQ     done
                INC     char_cnt
                INX                     ; move the string pointer
* the following eight instructions check the end of a word
                CMPB    #sp
                BEQ     string_lp
                CMPB    #tab
                BEQ     string_lp
                CMPB    #CR
                BEQ     string_lp
                CMPB    #LF
                BEQ     string_lp
                BRA     wd_loop         ; a nonwhite character is part of a word
done            NOP
                SWI                     ; return to the BUFFALO monitor
                END
```

Example 3.8

Word Searching
Write a program to search a specified word from a string.

Solution: Assume the string and the word are stored at *string_X* and *word_X*, respectively. The basic operation of the program is to identify the start of a new word and compare it with the word to be searched. If the specified word is found, then stop. Otherwise, continue until the end of the string. The memory location *search* will be set to 1 if the word is found; otherwise, it will be cleared to 0. The given word must be compared character by character, and the comparison must continue one character beyond the last character of the given word if the word occurs in the string. There are three possible outcomes:

Case 1

The word is not the last word in the string. Comparison of the last characters will yield "not equal."

Case 2

The word is the last word in the given string, but there are one or more white spaces between the last word and the null character. Comparison of the last characters will again given the result "not equal."

Case 3

The word is the last word in the given string, and it is followed by the null character. In this case, the comparison result for the last character is "equal."

The logic flow of the program is shown in Figure 3.8.

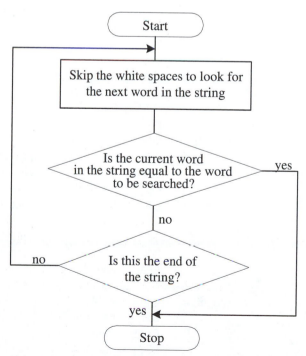

Figure 3.8 ■ Flowchart of the word search program

The program is as follows:

```
tab           EQU      $09
sp            EQU      $20
CR            EQU      $0D
LF            EQU      $0A
NULL          EQU      $00
              ORG      $00
search        RMB      1                       ; indicate if the word is found in the string
              ORG      $C000                   ; starting address of the program
              CLR      search
              LDX      #string_X
loop          LDAB     0,X                     ; get a character from the string
              INX                              ; move the string pointer
* the following ten instructions skip white spaces to search for the next word in the
* string
              TSTB
              BEQ      done                    ; is this the end of the string
              CMPB     #sp                     ; skip the space character
              BEQ      loop                    ;         "
              CMPB     #CR                     ; skip the CR character
              BEQ      loop                    ;         "
              CMPB     #LF                     ; skip the LF character
              BEQ      loop                    ;         "
              CMPB     #tab                    ; skip the tab character
              BEQ      loop                    ;         "
* the first nonwhite character is the beginning of a new word to be compared
              LDY      #word_X
              LDAA     0,Y                     ; get a character from the word
              INY
next_ch       CBA
              BNE      end_of_wd               ; check the next word
              CMPA     #NULL                   ; is this the end of the word
              BEQ      matched
              LDAA     0,Y                     ; get the next character from word_X
              LDAB     0,X                     ; get the next character from string_X
              INX                              ; move the string pointer
              INY                              ; move the word pointer
              BRA      next_ch                 ; compare the next character
* the following ten instructions check to see if the end of the given word is reached
end_of_wd     CMPA     #NULL
              BNE      next_wd                 ; if not the end of the given word, then not matched
              CMPB     #CR
              BEQ      matched
              CMPB     #LF
```

```
                    BEQ     matched
                    CMPB    #tab
                    BEQ     matched
                    CMPB    #sp
                    BEQ     matched
* the following instructions skip the unmatched word in the string
next_wd             LDAB    0,X
                    BEQ     done          ; stop if this is the end of the string
                    INX
                    CMPB    #CR
                    BEQ     jmp_loop      ; the label loop is too far to use conditional branch
                    CMPB    #LF
                    BEQ     jmp_loop
                    CMPB    #tab
                    BEQ     jmp_loop
                    CMPB    #sp
                    BEQ     jmp_loop
                    BRA     next_wd
jmp_loop            JMP     loop
matched             LDAB    #1
                    STAB    search
done                SWI
                    END
```

3.6 Subroutines

Good program design is based on the concept of modularity—the partitioning of program code into subroutines. A main module contains the logical structure of the algorithm, while smaller program units execute many of the details.

The principles of program design in high-level language apply even more to the design of assembly language programs. Begin with a simple main program whose steps clearly outline the logical flow of the algorithm, and then assign the execution details to subroutines. Of course, subroutines may themselves call other subroutines. Figure 3.9 shows an example of the partitioning of a structured program code.

3.7 Subroutine Call and Return

A subroutine is a sequence of instructions stored in memory at a specified address. The subroutine can be called from various places in the program. When a subroutine is called, the address of the next instruction is saved and program control passes to the called subroutine, which then executes its

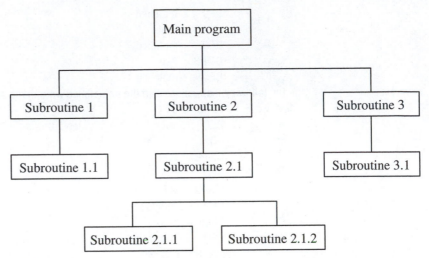

Figure 3.9 ■ A structured program

instructions. The subroutine terminates with a return instruction directing program control back to the instruction following the one that called the subroutine. Figure 3.10 illustrates this process.

The situations in which we will use a subroutine include the following:

- There are several places in a program that require the same computation to be performed with or without parameters. Using subroutine calls can shorten our program.

- The program is complex. If we divide a big program into smaller ones and solve each smaller problem in a subroutine, the problem will be easier to manage and understand, and it will also be easier to maintain.

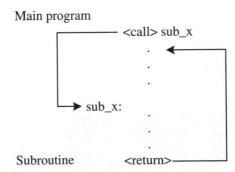

Figure 3.10 ■ Subroutine processing

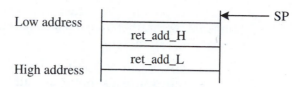

ret_add_H: upper byte of the return address
ret_add_H: lower byte of the return address

Figure 3.11 ■ The top two bytes of the stack after the execution of BSR or JSR

The 68HC11 provides two instructions for calling subroutines:

```
[<label>]   BSR   <rel>    [<comment>]
[<label>]   JSR   <opr>    [<comment>]
```

where

<rel>   is the offset to the subroutine.
<opr>   is the address of the subroutine and is specified in the DIR, EXT, or indexed addressing mode.

In terms of program control, these instructions are similar to the Bcc (branch on condition) and JMP (jump) instructions and have the same syntax. The BSR and JSR instructions, however, provide for automatically saving the program counter on the stack.

The address of the first byte of the instruction that follows the BSR or JSR instruction is saved in the stack, as shown Figure in 3.11. Upon return from the subroutine, the saved return address is pulled from the stack and placed in the program counter. Program execution then continues.

The BSR instruction can use only the relative addressing mode to specify the subroutine address. The range of the branch is limited to −128 to 127 bytes. The JSR instruction can use the direct, extended, or index addressing modes to specify the subroutine address; thus the subroutine being called can be as far as 64KB away from the JSR instruction.

The last instruction of a subroutine is the RTS instruction, which causes the top two bytes of the stack to be pulled and loaded into PC. Program execution will then continue from the new PC value.

3.8   Issues in Subroutine Calls

In the following discussion we will refer to the program or routine that makes the subroutine call as the *caller* and the subroutine called by other routines as the *callee*. We say that a subroutine is *entered* when it is being executed. There are three important issues to consider during a subroutine call:

- *Parameter passing.* We often want the subroutine to perform computations using parameters passed to it. The following methods are often used in passing parameters:

  1. *Use registers.* In this method, the parameters are placed in registers before a subroutine call is made. This method is most convenient when there are only a few parameters to be passed to the callee; it is also the easiest method and has the lowest overhead in accessing the passed parameters.

  2. *Use the stack.* The parameters are pushed onto the stack before a subroutine call is made. This method allows us to make recursive subroutine calls (i.e., a subroutine calls itself). Most compilers use this method when some parameters cannot be passed in registers. To avoid running out of stack space, either the caller or the callee must clean up the stack space used for passing parameters after the computation is completed.

  3. *Use global memory.* Global memory is accessible to both the caller and callee. The caller simply places the parameters in global memory, and the callee will be able to access them.

- *Result returning.* The result of a computation can be returned by three methods:

  1. *Use registers.* This method is most convenient when there are only a few bytes of values to be returned. When this method is used, the caller may need to save registers before making subroutine call. In the 68HC11, registers A, B, X, and Y can be used to hold arguments or results. The carry bit in the condition code register can be used to pass a 1-bit result that can be used in instructions like BCC or BCS. If there are only a few bytes to be returned, using registers is the best method.

  2. *Use the stack.* The caller creates a hole in the stack by decrementing the stack pointer before making a subroutine call. The callee saves the computational result in the hole before returning to the caller.

  3. *Use global memory.* Global memory is accessible to both the caller and the callee. The callee simply saves the result in the designated locations.

- *Allocation of local variables.* In addition to the parameters passed to it, a subroutine may need memory locations to hold loop counters and working buffers that hold temporary variables and results. These variables are useful only during execution of the subroutine and are called *local variables* because they are local in the subroutine. Local variables are often allocated in the stack so that they are not accessible to the caller.

There are two methods for allocating local variables:

1. Use as many DES instructions as needed if fewer than six bytes are needed. This approach takes less time if no more than five bytes are needed for local variables.

2. Use the following instruction sequence if more than five bytes are needed:

```
TSX
XGDX
SUBD #N          ; allocate N bytes
XGDX
TXS              ; move the stack pointer up by N bytes
```

It takes 18 E clock cycles to allocate six bytes in the stack using the DES instruction, while the instruction sequence for the second method takes only 16 E clock cycles for any number of bytes. The TXS instruction decrements the value in X by 1 and places it in the stack pointer. A similar operation can be performed on index register Y using the TYS instruction.

The space allocated to local variables must be deallocated before the subroutine returns to the caller. There are two corresponding methods for deallocating local variables:

1. Use as many INS instructions as needed if fewer than six bytes are to be deallocated. This approach takes less time if no more than five bytes are to be deallocated.

2. Use the following instruction sequence if there are six or more bytes to be deallocated.

```
TSX
XGDX
ADDD #N
XGDX
TXS              ; move down the stack pointer by N bytes
```

It takes 18 E clock cycles to deallocate six bytes from the stack using the INS instruction, while it takes only 16 E clock cycles to deallocate any number of bytes using this instruction sequence.

3.9 The Stack Frame

The stack is used heavily during a subroutine call: the caller may pass parameters using the stack, and the callee may need to save registers and allocate local variables in the stack. The region in the stack that holds incoming parameters, the subroutine return address, local variables, and saved registers is referred to as the *stack frame*. Some microprocessors dedicate a register to the management of the stack frame—the register is called the *frame pointer*. (The 68HC11, however, does not have a register dedicated to the function of

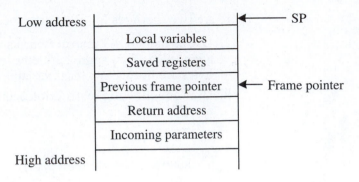

Figure 3.12 ■ A stack frame

the frame pointer.) Since the stack frame is created during a subroutine call, it is also called the *activation record* of the subroutine. The stack frame exists as long as the subroutine is not exited. An example of a stack frame is shown in Figure 3.12.

The frame pointer primarily facilitates access to local variables and incoming parameters because the stack pointer may change during the life of a subroutine. The frame pointer points to a fixed location in the stack and can avoid this problem. A negative offset relative to the frame pointer is used to access a local variable, and a positive offset relative to the frame pointer is used to access an incoming parameter. Since the 68HC11 does not have a dedicated frame pointer, we will not use the term "frame pointer" in the following discussion. We can use either TSX or TSY so that index register X or Y points to the top byte of the stack. The index-addressing mode can then be used to access all the values in the stack frame. The stack frame of a 68HC11 subroutine is shown in Figure 3.13.

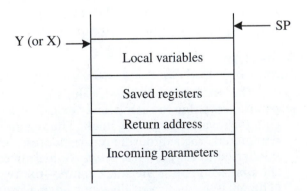

Figure 3.13 ■ Example of 68HC11 stack frame

Example 3.9

Draw the stack frame for the following program segment after the ninth instruction of the subroutine MAT_REV has been executed.

```
        LDAA #N
        PSHA
        LDX #MAT
        PSHX
        BSR MAT_REV
        .
        .
        .
MAT_REV
   (1)  PSHB
   (2)  PSHA
   (3)  PSHX
   (4)  PSHY
   (5)  TSX
   (6)  XGDX
   (7)  SUBD #6
   (8)  XGDX
   (9)  TXS
        ...
```

Solution: Two parameters are passed to MAT_REV. These two parameters occupied three bytes. Six bytes are allocated for local variables. Two 8-bit and two 16-bit registers are saved on the stack. The resultant stack frame is shown in Figure 3.14.

When a subroutine pushes registers onto the stack, it must restore them by pulling their old values from the stack. Since the stack is a first-in-last-out data structure, the register that was last pushed onto the stack must be the first one to be pulled. For example, if a subroutine has the following push instructions at its entry point:

```
PSHA
PSHB
PSHX
PSHY
```

then it must have the following pull instructions before it returns:

```
PULY
PULX
PULB
PULA
```

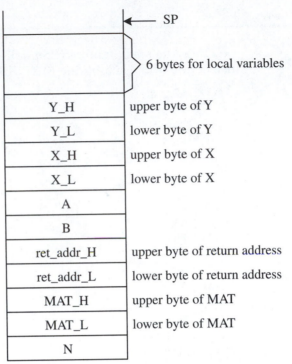

Figure 3.14 ■ Stack frame for Example 3.9

3.10   Examples of Subroutine Calls

The examples in this section illustrate parameter passing, result returning, and local variable allocation.

3.10.1   Using Registers to Pass Parameters

In the following example, registers are used to pass parameters and to return the result. The subroutine does not allocate local variables in the stack.

Example 3.10

Write a subroutine to compute the average of an array with N 8-bit elements, and write an instruction sequence to call the subroutine. The array starts at ARRAY. Use registers to pass parameters and return the average in accumulator B.

Solution: The caller passes the array count N and array base address in registers A and X, respectively, and would use the following instructions to make the subroutine call:

```
LDX    #ARRAY
LDAA   #N
BSR    AVERAGE
```

The array count **N** is needed for computing array average and hence must be saved in the stack. We push a 0 into the stack so that we can use an LDX instruction to load the array count into X and compute the array average. The resulted stack frame is shown in Figure 3.15. The logic flow of the program is shown in Figure 3.16.

We will use index register as the array pointer and use Y as the loop count. The subroutine that computes the array average is as follows:

```
average   PSHA                ; save the array count
          CLRA
          PSHA
          TSY                 ; Y points to the top of the stack
          LDAB   1,Y          ; place array count in B
          CMPB   #1
          BLO    exit         ; is array count equal to 0?
          BHI    do           ; does the array have more than one element?
          LDAB   0,X          ; return array[0]
          BRA    exit
do        CLRA
          XGDY                ; place array count in Y
          CLRA                ; initialize sum to 0
          CLRB                ;        "
again     ADDB   0,X          ; add an element to the sum
          ADCA   #0           ; add carry to the upper byte of the sum
```

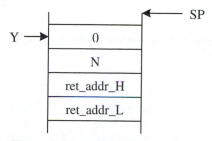

Figure 3.15 ■ Stack frame of example 3.10

```
            INX                    ; move the array pointer
            DEY                    ; decrement the loop count
            BNE     again          ; is this the end of the loop?
            TSY
            LDX     0,Y            ; load N into X
            IDIV                   ; compute the array average
            XGDX                   ; place array average in B
    exit    PULA                   ; clean up the stack
            PULA                   ;      "
            RTS
```

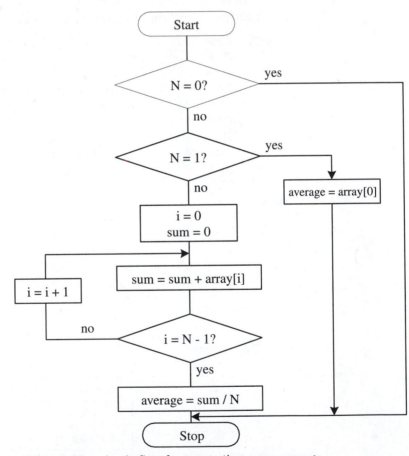

Figure 3.16 ■ Logic flow for computing array average

3.10.2 Using the Stack to Pass Parameters

The user would use the following instruction sequence to push parameters onto the stack and call the subroutine SUB1:

```
LDAA    #N          ; N is the argument to SUB1
PSHA                ; push the argument N onto the stack
BSR     SUB1
```

If the caller does not save any registers, it could use the following instruction sequence to retrieve the argument from the stack:

```
TSX                 ; X points to the top byte of the stack
LDAA    2,X         ; get the argument
```

When using the stack to pass parameters, either the caller or the callee must be responsible for cleaning up the stack after the subroutine has been executed. Otherwise, the stack space might be used up and cause our program to crash.

Example 3.11

▼

Write a subroutine to find the largest element of an array, and write an instruction sequence to call this subroutine. The following parameters are passed in the stack:

- *array*: the starting address of the given array
- *arcnt*: the array count
- *amax*: address of the memory location to hold the largest element of the array

Solution: The caller would include the following instruction sequence to pass parameters, make the subroutine call, and restore the stack:

```
    .
    .
    .
    LDX     #array
    PSHX
    LDAA    #arcnt
    PSHA
    LDX     #amax
    PSHX
    BSR     MAX
    TSX                 ; clean up the stack
    LDAB    #5          ;      "
    ABX                 ;      "
```

TXS   ;   "
.
.
.

The program logic of this subroutine is illustrated in Figure 3.17. The registers that will be used in the subroutine must be saved. The stack frame and the offset of each parameter from the top byte of the stack are shown in Figure 3.18. The subroutine is as follows:

```
array    EQU    11        ; offset of array base address from the top of the stack
arcnt    EQU    10        ; offset of array count from the top of the stack
armax    EQU    8         ; offset of array max from the top of the stack
         ORG    $C000     ; starting address of the program
```

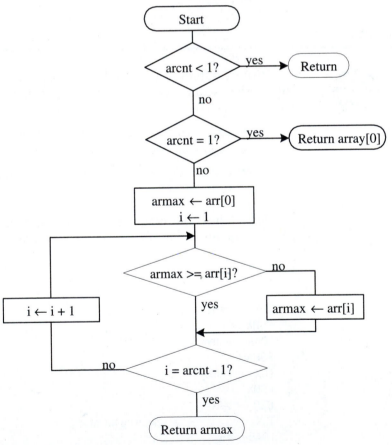

Figure 3.17 ■ Logic flow of the array max subroutine

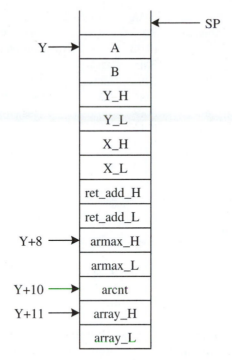

Each slot of the stack is one byte.
The suffix _H specifies the upper byte of a parameter.
The suffix _L specifies the lower byte of a parameter.

Figure 3.18 ■ Stack frame for the array max
subroutine

```
MAX     PSHX                    ; save all CPU registers
        PSHY                    ;       "
        PSHB                    ;       "
        PSHA                    ;       "
        TSY                     ; Y points to the top byte of the stack
        LDX     array,Y         ; place the array base address in X
        LDAB    arcnt,Y         ; load the array count in B
        LDAA    0,X             ; assign the first element as the temporary max
        CMPB    #1              ; check array count
        BLO     exit            ; return if array count is 0
        BHI     start           ; look for array max if the array count is > 1
        BRA     done
start   INX                     ; set X to point to the second element
        DECB                    ; set loop limit to arcnt - 1
again   CMPA    0,X             ; compare array element with the temporary MAX
        BGE     noswap          ; should MAX be updated?
```

```
                    LDAA    0,X             ; update MAX
        noswap      INX                     ; move X to point to the next array element
                    DECB                    ; decrement loop count
                    BNE     again
        done        LDX     armax,Y
                    STAA    0,X             ; save the array MAX
        exit        PULA                    ; restore registers
                    PULB                    ;       "
                    PULY                    ;       "
                    PULX                    ;       "
                    RTS
```

Example 3.12

Write a program to compute the greatest common divisor (gcd) of two 16-bit integers. The caller of this subroutine pushes these two integers onto the stack, and this subroutine returns the gcd in double accumulator D.

Solution: Let these two integers be *n1* and *n2*. Assume n1 is smaller than n2. Swap them if this is not the case. We need to use an integer *i* as the testing number and set gcd to be 1 at the beginning. The testing number *i* is in the range from 2 to n1. The algorithm for finding the gcd is as follows:

Step 1
If n2 < n1 then swap n1 and n2.

Step 2
i = 2, gcd = 1.

Step 3
If (n1 = 1) or (n2 = 1) return.

Step 4
If both n1 and n2 can be divided by i then gcd ← i.

Step 5
If i = n1 then stop. Otherwise, i ← i + 1. Go to Step 4.

To implement this algorithm, we will need two local variables. They are *i* and *gcd*. Both are 16 bits. The caller of this subroutine will use the following instruction sequence to make the call:

```
        ...
        LDX     #n2
        PSHX
        LDX     #n1
        PSHX
        JSR     find_gcd
```

```
        INS                        ; deallocate stack space
        INS                        ;     "
        INS                        ;     "
        INS                        ;     "
        ...
```

We will allocate local variables in the stack. The stack frame of this program is shown in Figure 3.19.

The *find_gcd* subroutine is as follows:

```
n1_dis      EQU     8
n2_dis      EQU     10
gcd_dis     EQU     0
i_dis       EQU     2

find_gcd    PSHY                        ; save Y
            TSY                         ; Y points to the top of the stack
            DES                         ; allocate space for local variables
            DES                         ;     "
            DES                         ;     "
            DES                         ;     "
            LDX     #1
            STX     gcd_dis,Y           ; initialize gcd to 1
            LDD     n2_dis,Y            ; if n2 = 1 then gcd = 1
            CPD     #1                  ;     "
            BEQ     done                ;     "
            LDD     n1_dis,Y            ; if n1 = 1 then gcd = 1
            CPD     #1                  ;     "
            BEQ     done                ;     "

            CPD     n2_dis,Y            ; compare n1 with n2
            BLS     start               ; if n1 is smaller then start to compute gcd
```

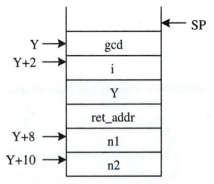

Figure 3.19 ■ Stack frame for
gcd program

```
                       LDX     n2_dis,Y      ; n1 is larger than n2 then swap them
                       STD     n2_dis,Y      ;         "
                       STX     n1_dis,Y      ;         "
          start        LDX     #2
                       STX     i_dis,Y       ; start i from 2
          again        LDD     n1_dis,Y      ; test if n1 is divisible by i
                       IDIV                  ;         "
                       CPD     #0            ;         "
                       BNE     next_i        ; if i can't divide n1 then check next i
                       LDD     n2_dis,Y      ; test if n2 is divisible by i
                       LDX     i_dis,Y       ;         "
                       IDIV                  ;         "
                       CPD     #0            ;         "
                       BNE     next_i        ; if i can't divide n2 then check next i
                       LDD     i_dis,Y       ; i can divide both n1 and n2
                       STD     gcd_dis,Y     ; update the gcd
          next_i       LDX     i_dis,Y
                       CPX     n1_dis,Y
                       BEQ     done          ; we are done if i equals n1
                       INX                   ; increment i
                       STX     i_dis,Y       ;         "
                       JMP     again
          done         INS                   ; deallocate local variables
                       INS                   ;         "
                       INS                   ;         "
                       INS                   ;         "
                       PULY
                       RTS
```

▲

Example 3.13

▼

Write a subroutine that can divide a 32-bit number into another 32-bit number.

Solution: An easy way to perform a 32-bit by 32-bit division is to use repeated subtraction. The hardware required to perform the repeated subtraction is illustrated in Figure 3.20. The algorithm is as follows:

Step 1
Shift the register pair (R, Q) one bit left.

Step 2
Subtract register P from register R, put the result back to R if the result is non-negative.

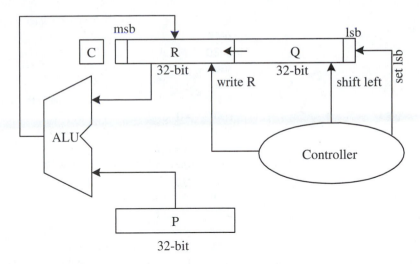

Figure 3.20 ■ Hardware for 32-bit by 32-bit division

Step 3

If the result of step 2 is negative, then set the least significant bit of Q to 0. Otherwise, set the least significant bit of Q to 1.

Perform this division procedure n times, and the quotient and the remainder will be left in register Q and R, respectively. Before the division steps are performed, the dividend and divisor must be loaded into register Q and register P, respectively. Register R must be cleared to 0.

We will implement this algorithm as a subroutine. The caller will allocate eight bytes in the stack to hold the quotient and remainder and then push the dividend and the divisor onto the stack before calling this subroutine. This routine will allocate thirteen bytes in the stack to be used as local variables. The stack frame of this subroutine is shown in Figure 3.21. The subroutine is as follows:

```
buf       EQU    0          ; used as buffer space
R         EQU    5          ; offset of R from the top byte of the stack
Q         EQU    9          ; offset of Q from the top byte of the stack
divisor   EQU    21         ; offset of divisor from the top byte of the stack
dividend  EQU    25         ; offset of dividend from the top byte of the stack
i         EQU    4          ; offset of loop index i from the top byte of the stack
local     EQU    13

div32     PSHA
          PSHB
          PSHX
          PSHY
```

```
                TSX                        ; allocate local variables
                XGDX                       ;         "
                SUBD    #local             ;         "
                XGDX                       ;         "
                TXS                        ;         "
                TSY                        ; Y points to the top byte of the stack
                LDD     #0
                STD     R,Y                ; initialize register R to 0
                STD     R + 2,Y            ;         "
                LDD     dividend,Y         ; transfer dividend to Q
                STD     Q,Y                ;         "
                LDD     dividend + 2,Y     ;         "
                STD     Q + 2,Y            ;         "
                LDAA    #32                ; initialize i to 32
                STAA    i,Y                ;         "
loop            LSL     Q + 3,Y            ; shift register pair (R, Q) to the left one place
                ROL     Q + 2,Y            ;         "
                ROL     Q + 1,Y            ;         "
                ROL     Q,Y                ;         "
                ROL     R + 3,Y            ;         "
                ROL     R + 2,Y            ;         "
                ROL     R + 1,Y            ;         "
                ROL     R,Y                ;         "
                LDD     R + 2,Y
                SUBD    divisor + 2,Y      ; subtract the divisor P from R
                STD     buf + 2,Y          ;         "
                LDAA    R + 1,Y            ;         "
                SBCA    divisor + 1,Y      ;         "
                STAA    buf + 1,Y          ;         "
                LDAA    R,Y                ;         "
                SBCA    divisor,Y          ;         "
                BCS     smaller            ; is [R] – divisor < 0?
                STAA    R,Y                ; store the difference back in R register
                LDD     buf + 2,Y          ;         "
                STD     R + 2,Y            ;         "
                LDAA    buf + 1            ;         "
                STAA    R + 1,Y            ;         "
                LDAA    Q + 3,Y
                ORAA    #01                ; set the lsb of Q register to 1
                STAA    Q + 3,Y            ;         "
                BRA     looptest
smaller         LDAA    Q + 3,Y            ; set the lsb of Q register to 0
                ANDA    #$FE               ;         "
                STAA    Q + 3,Y            ;         "
looptest        DEC     i,Y                ; decrement loop count
                BNE     loop
```

```
LDD     R,Y
STD     rem,Y                  ; put the remainder in the hole to be returned
LDD     R + 2,Y                ;      "
STD     rem + 2,Y              ;      "
LDD     Q,Y                    ; put the quotient in the hole to be returned
STD     quo,Y                  ;      "
LDD     Q + 2,Y               ;      "
STD     quo + 2,Y             ;      "
TSX                            ; deallocate local variables
XGDX                           ;      "
ADDD    #local                 ;      "
XGDX                           ;      "
TXS                            ;      "
PULY                           ; restore registers
PULX                           ;      "
PULB                           ;      "
PULA                           ;      "
RTS
```

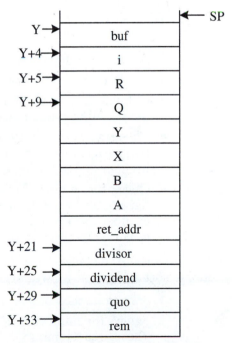

Figure 3.21 ■ Stack frame of 32-bit
by 32-bit division
subroutine

Example 3.14

Write a subroutine to sort an array of 8-bit numbers using the bubble sort method, and also write an instruction sequence to call this routine. The starting address of the array is *arr*, and the array count is *N*. Use the stack to pass parameters to this routine.

Solution: The caller would use the following instruction sequence to call the *bubble* subroutine:

```
.
.
.
LDX     #arr_base
PSHX
LDAA    #N
PSHA
JSR     bubble
...
```

The algorithm of bubble sort consists of *N − 1* iterations. Let iterations be labeled from *0 to N − 2*. The following operations will be performed in iteration i:

- N − 1 − i comparisons are performed in iteration ***i***.
- Set an in-order flag to 1 at the beginning of the iteration.
- Compare every pair of adjacent elements starting from array elements with indices ranging from 0 to N − 1 − i.
- The array can be sorted in ascending or descending order. Swap two adjacent elements if they are not in the right order. Clear the in-order flag to 0 whenever two elements are swapped.
- Check the in-order flag at the end of an iteration. If the in-order flag remains to be 1, then stop the comparison. The fact that the in-order flag remains to be 1 implies that no swapping is done because the array is already in order.

The logic of the bubble sort is illustrated in Figure 3.22. To implement the bubble sort algorithm, we will need the following local variables:

1. *iteration*: the number of iterations remained to be performed.
2. *inner*: the number of comparisons remained to be performed in the current iteration.
3. *buf*: buffer memory for swapping array elements.
4. *in_order*: a flag that indicates that if the array is in order.

We need to allocate four bytes to be used for local variables. The resultant stack is shown in Figure 3.23.

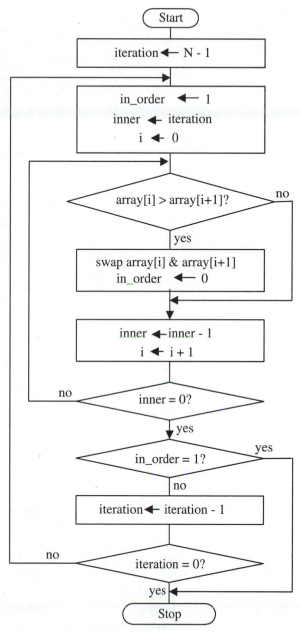

Figure 3.22 ■ Logic flow of bubble sort

arr	EQU	13	; offset of array base from the top of the stack
arcnt	EQU	12	; offset of array count from the top of the stack
buf	EQU	3	; offset of buffer from the top of the stack
in_order	EQU	2	; offset of the in-order flag from the stack top

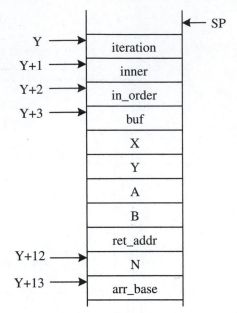

Figure 3.23 ■ Stack frame for bubble sort

inner	EQU	1	; offset of the inner loop count from the stack top
iteration	EQU	0	; offset of the outer loop count from the stack top
true	EQU	1	
false	EQU	0	

```
bubble    PSHB
          PSHA
          PSHY
          PSHX
          DES                      ; allocate space for local variables
          DES                      ;    "
          DES                      ;    "
          DES                      ;    "
          TSY                      ; set Y to point to the top of the stack
          LDAA    arcnt,Y
          DECA                     ; compute the number of iterations to be performed
          STAA    iteration,Y      ;    "
ploop     LDAA    #true            ; set the in-order flag to true
          STAA    in_order,Y
          LDX     arr,Y            ; place the starting address of the array in X
          LDAA    iteration,Y
          STAA    inner,Y          ; initialize inner loop count
cloop     LDAA    0,X              ; get one element
```

```
                CMPA    1,X             ; compare to the next element
                BLE     looptest
* the following five instructions swap the two adjacent elements
                STAA    buf,Y           ; swap two adjacent elements
                LDAA    1,X             ;      "
                STAA    0,X             ;      "
                LDAA    buf,X           ;      "
                STAA    1,X             ;      "
                LDAA    #false          ; reset the in-order flag
                STAA    in_order,Y      ;      "
looptest        INX                     ; move the array pointer
                DEC     inner,Y
                BNE     cloop
                TST     in_order,Y
                BNE     done
                DEC     iteration,Y
                BNE     ploop
* the following four instructions deallocate space allocated to local variables
done            INS
                INS
                INS
                INS
                PULX                    ; restore registers
                PULY                    ;      "
                PULA                    ;      "
                PULB                    ;      "
                RTS
```

3.11 Input and Output Routines

The Buffalo monitor has commands that allow the user to modify and examine the contents of memory locations. However, it would be more convenient if the executing program could read data directly from the terminal (or PC) keyboard and output results directly to the monitor screen. The Buffalo monitor also provides a set of I/O routines that can be called by the user program. These I/O routines are listed in Table 3.3. Of course, we can also write our own I/O routines. However, we will delay the discussion of writing I/O routines until chapter 9. In this section, we will learn how to call the I/O routines in Table 3.3 to read data from the keyboard and output data to the monitor screen.

The I/O routines in Table 3.3 are in ROM. A jump table allows the user program to call these I/O routines by jumping to the desired entry of the table. Each entry is a 3-byte jump instruction. The first byte is the opcode of the JMP

Routine Name	Function
UPCASE	converts the character in accumulator A to uppercase
WCHEK	tests the character in A and returns with the Z bit set if the character is a white space (space, comma, tab)
DCHEK	tests the character in A and returns with the Z bit set if the character is a delimiter (carriage return or whitespace)
INIT	initializes I/O device
INPUT	reads I/O device
OUTPUT	writes I/O device
OUTLHLF	converts left nibble of A to ASCII and outputs to terminal port
OUTRHLF	converts right nibble of A to ASCII and outputs to terminal port
OUTA	outputs the ASCII character in A
OUT1BYT	converts the binary byte at the address in index register X to two ASCII characters and outputs them; returns address in index register X pointing to next byte
OUT1BSP	converts the binary byte at the address in index register X to two ASCII characters and outputs them followed by a space; returns address in index register X pointing to next byte
OUT2BSP	converts two consecutive binary bytes starting at address in index register X to four ASCII characters and outputs the characters followed by a space; returns address in index register X pointing to next byte
OUTCRLF	outputs ASCII carriage return followed by a line feed
OUTSTRG	output string of ASCII bytes pointed to by address in index register X until character is an end-of-transmission ($04)
OUTSTRG0	same as OUTSTRG, except that leading carriage returns and line feeds are skipped
INCHAR	inputs ASCII character to A and echoes back; this routine loops until character is actually received

Table 3.3 ■ Buffalo monitor I/O routines

Address	Instruction
$FFA0	JMP UPCASE
$FFA3	JMP WCHEK
$FFA6	JMP DCHEK
$FFA9	JMP INIT
$FFAC	JMP INPUT
$FFAF	JMP OUTPUT
$FFB2	JMP OUTLHLF
$FFB5	JMP OUTRHLF
$FFB8	JMP OUTA
$FFBB	JMP OUT1BYT
$FFBE	JMP OUT1BSP
$FFC1	JMP OUT2BSP
$FFC4	JMP OUTCRLF
$FFC7	JMP OUTSTRG
$FFCA	JMP OUTSTRG0
$FFCD	JMP INCHAR

Table 3.4 ■ 68HC11 Buffalo monitor I/O routine jump table

instruction. The second and third bytes are the address of the actual subroutine that performs the I/O operation. To invoke one of the I/O routines, execute a JSR instruction to the corresponding address given in Table 3.4. For example, to output the ASCII character contained in accumulator A, use the following instruction:

```
JSR    $FFB8
```

To output a string, use the following instruction sequence:

```
LDX    #string      ; string is the starting address of the string
JSR    $FFC7
```

Example 3.15

Write an instruction sequence to output the following string on the screen using the I/O function provided by the Buffalo monitor:

Please make a choice (1, 2, 3, 4):

Solution: We need to define this string using the assembler directive FCC. A string must be terminated by an end-of-transmission character (ASCII code is 04).

```
msg         FCC      "Please make a choice (1, 2, 3, 4):"
            FCB      04
put_string  EQU      $FFCA          ; use mnemonic to represent the hex address
```

To output the string, use the following instruction sequence:

```
LDX    #msg
JSR    put_string
```

Example 3.16

Write a subroutine to input a string from the keyboard and save the string in a buffer pointed by index register X. The input string is terminated by the carriage-return character.

Solution: This subroutine will keep reading characters from the keyboard until the carriage-return character is read.

```
getchar     EQU    $FFCD          ; use mnemonic to represent hex address
CR          EQU    $0D            ; ASCII code of carriage return
EOT         EQU    $04
get_string  JSR    getchar
            CMPA   #CR            ; have we read a carriage-return character?
```

```
              BEQ     done
              STAA    0,X              ; save the character in the buffer
              INX                      ; move the buffer pointer
              BRA     get_string       ; continue
    done      LDAA    #EOT             ; use end-of-transmission to terminate the string
              STAA    0,X              ;      "
              RTS
```

▲

3.12   Summary

This chapter clearly defines the most commonly used data structures, including strings, arrays, stacks, and queues. Conceptually, a stack is a list of data items whose elements can be accessed from only one end. A stack has a top and a bottom. The most common operations that can be applied to a stack include push and pop (or pull). The push operation adds a new item to the top of the stack. The pop operation removes the top element of the stack. Physically, a stack is a reserved area in main memory where programs perform only push and pop operations. The 68HC11 has a 16-bit register SP that points to the byte immediately above the top byte of the stack. The 68HC11 provides instructions for pushing and popping accumulators A and B and index registers X and Y.

An array is an indexable data structure. It consists of a sequence of elements in which each element is associated with an index *i* that can be used to access it. The index of an array often starts from 0 to simplify the address calculation. An array can be one-dimensional or multidimensional. A two-dimensional array is also called a matrix. A matrix consists of rows and columns and can be stored in row-major order or column-major order. The notation (i, j) is used to specify the matrix element that is stored at the intersection of ith row and jth column. The row and column numbers often start from 0 to simplify the address calculation.

A string is a sequence of characters terminated by a null (ASCII code 0) or other special character such as EOT (ASCII code $04). Common operations applied to strings include concatenation, word or substring insertion, substring deletion, character and word counting, word or string searching, etc.

Good program design is based on the concept of modularity—the partitioning of programming code into subroutines. A main module contains the logical structure of the algorithm, while smaller program units execute many of the details.

The principles of program design in high-level languages apply even more to the design of assembly programs. Begin with a simple main program whose steps clearly outline the logical flow of the algorithm, and then assign the execution details to subroutines. Subroutines may themselves call other subroutines.

The 68HC11 provides the instructions BSR and JSR for making subroutine calls and the instruction RTS for returning from the subroutine. The BSR and

JSR instructions will save the return address in the stack before jumping to the subroutine. The RTS instruction pops the top two bytes from the stack into the program counter and continues the instruction execution from there.

Issues related to subroutine calls include parameter passing, result returning, and local variables allocation.

Parameters can be passed in registers, program memory, the stack, or the global memory. The result can be returned in registers, the stack, or the global memory. Local variables must be allocated in the stack so that they are not accessible to the caller and other program units. Local variables come into being only when the subroutine is being executed. The 68HC11 provides instructions to facilitate the access of variables in the stack.

Several examples are provided to illustrate the writing of subroutines, parameter passing, result returning, and local variable allocation.

The Buffalo monitor provides I/O functions (available in all three demo boards that we discussed in chapter 2) so that we can invoke them to read data from the keyboard or output data to the monitor screen. These functions can greatly facilitate the program development process.

3.13   Exercises

E3.1 Write an instruction sequence to set up the top six bytes of the stack as follows:

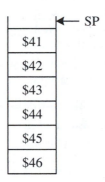

E3.2 Write an instruction sequence to load the tenth element (associated with index 9) of an array into accumulator A. The base address of the array is ARR. The first element is associated with index 0.

E3.3 Write a program to find the median of the sorted array VECT with N 8-bit elements. The median of an array is the middle element of the array after the array is sorted. If n is odd, the median is **arr**[n/2]. If n is even, then the median is the average of elements **arr**[n/2 − 1] and **arr**[n/2]. Place the median in A. The array index starts from 0.

E3.4 Write a program to perform a binary search on a sorted array. The sorted array is located at ARR and has N 8-bit numbers. The key for searching is stored at KEY. The program leaves the address of the element that matches the key in X or leaves $FFFF in X if no element matches the key. Assume that the array is sorted in ascending order. The binary search algorithm is as follows:

Step 1
Initialize variables *max* and *min* to *N – 1* and 0, respectively.

Step 2
If *max* < *min*, then exit. No element matches the key.

Step 3
Let mean = (min + max)/2.

Step 4
If *key* equals ARR[*mean*], key is found, exit.

Step 5
If *key* < ARR[*mean*], then set *max* to *mean – 1* and go to Step 2.

Step 6
If *key* > ARR[*mean*], then set *min* to *mean + 1* and go to Step 2.

E3.5 Compute the addresses of the (2, 3)th, (4, 9)th, and (10, 8)th elements of a 12×12 matrix of 8-bit elements. The given matrix is stored in column major order at $D000-$D08F.

E3.6 Compute the addresses of the (4, 1)th, (5, 2)th, and (7, 5)th elements of a 9×9 matrix of 16-bit elements. The matrix is stored in row major order starting from $D100.

E3.7 Write a program to multiply two 8×8 matrices MA and MB, made up of 8-bit elements, and create an 8×8 matrix MC with 16-bit elements to hold the result. Assume that MA, MB, and MC are the starting addresses of the matrices and that all matrices are stored in row major order.

E3.8 Write a program to manipulate an $N \times N$ matrix so that

1. all elements on the diagonal are reset to 0
2. all elements in the upper right are divided by 2
3. all elements in the lower left are multiplied by 2

The base address of the matrix is MA. MA is stored in memory in row major order.

E3.9 The label ARR is the starting address of an array of twenty-five 8-bit elements. Trace the following code sequence and describe what the subroutine TESTSUB does.

```
LDX     #ARR
LDAA    #25
BSR     testsub
  .
  .
  .
```

```
testsub    PSHA
           DECA
           PSHX
           LDAB    0,X
           INX
again      CMPB    0,X
           BLE     next
           LDAB    0,X
next       INX
           DECA
           BNE     again
           PULX
           PULA
           RTS
```

E3.10 Write a subroutine that can generate a time delay from one to sixty seconds and a program to call the subroutine. The program writes a 1 to memory location at $1004 and calls the subroutine to create a delay of one second; it then writes a 2 to the same location and calls the subroutine to delay for two seconds. The same operation is repeated until the delay is sixty seconds, and then the program starts all over again. Repeat this operation forever. The delay in seconds should be passed to the subroutine in accumulator A.

E3.11 Draw the stack frame after the execution of the instruction with the label *kkk* in the following instruction sequence:

```
address    instruction
           LDAB    #$10
           LDAA    #$20
           PSHA
           LDX     #000
           PSHX
$C040      BSR     XYZ
                   .
                   .
                   .
XYZ        PSHB
           PSHA
           PSHX
           TSX
           XGDX
           SUBD    #10
           XGDX
kkk        TXS
                   .
                   .
                   .
```

E3.12 Draw the stack frame and enter the value of each stack slot (if it is known) at the end of the following instruction sequence:

```
            DES
            DES
            CLRB
            LDAA        #10
            PSHA
            LDAA        #$C0
            PSHA
            LDX         #$D000
            PSHX
            BSR         SUB1
            .
            .
            .

SUB1        PSHB
            PSHY
            TSX
            XGDX
            SUBD        #12
            XGDX
            TXS
            .
            .
            .
```

E3.13 Convert the 16×16 multiplication program into a subroutine so that it can be called by other programs.

E3.14 Write a subroutine to concatenate two strings P and Q to form a new string R, and also write the instruction sequence to call this subroutine. P, Q, and R are terminated by the end-of-transmission (EOT) character. P, Q, and R are the starting addresses of these three strings and are passed to the subroutine in the stack. The subroutine will output the strings P, Q, and R and the following three messages on the screen:

"Source string 1 is:"	\<string P\>
"Source string 2 is:"	\<string Q\>
"Result string is:"	\<string R\>

E3.15 Write a subroutine to compare two strings. The starting addresses of these two strings are passed in the stack. If the strings are equal, the subroutine returns a 1 to the caller in accumulator A. Otherwise, it returns a 0.

E3.16 Convert the character and word count program in Example 3.8 into a subroutine. Parameters are passed to the subroutine in the stack. The starting address of the string is passed in the stack. The caller of this subroutine will allocate two bytes in the stack for returning the character and word counts. The character count will be below the word count in the stack.

E3.17 Convert the word-matching program in Example 3.8 into a subroutine. Parameters are passed to the subroutine in the stack. The starting address of the string is pushed into the stack first, followed by the starting address of the word to be searched. If the word is found, return a 1 in A. Otherwise, return a 0 in A.

E3.18 Write a subroutine that determines if an integer is a prime number. The number to be tested is passed in double accumulator D. The original number is returned in double accumulator D if the given number is a prime. Otherwise, a 0 is returned in D.

E3.19 Write a subroutine that will insert a string to another string. The starting addresses of both the target string and the string to be inserted are passed in the stack.

E3.20 Write a subroutine that will compute the least common multiple of two 16-bit integers. The caller of this subroutine pushes these two 16-bit integers into the stack. The result will be returned in the stack. The caller will make a 4-byte hole in the stack before pushing those two integers.

3.14   Lab Exercises and Assignments

L3.1 Write a subroutine that finds and returns the maximum odd element of an array of N 8-bit integers. The base address of the array and the array count are passed to this subroutine in index register X and accumulator B, respectively. The result should be returned in accumulator B. Return a 0 when the array has no odd numbers. Also write a main program to test this subroutine. Invoke Buffalo I/O routines to output the result. Output the following message if there is the maximum odd element:

The maximum odd element of the array is xxxx.

Otherwise, output the following message:

The array has no odd numbers!

L3.2 Write a subroutine to convert the hex number in accumulator A to two BCD digits and return the result in A. The number to be converted is assumed to be less than 100. Also write a main program to test your subroutine. Include the following steps:

1. Call the Buffalo I/O routines OUTLHLF and OUTRHLF to output the hex number in accumulator A.
2. Call the Buffalo I/O routine OUTCRLF to output the carriage return and line feed pair.

3. Call the hex-to-BCD subroutine to perform the conversion.

4. Call the OUTLHLF and OUTRHLF routines to output the result.

5. Call the Buffalo I/O routine OUTCRLF to output the carriage return and line feed pair.

6. Test the subroutine using an array of ten integer numbers.

L3.3 Write a subroutine to swap the first row of a matrix with the last row, the second row with the second-to-last row, etc. The matrix is N by M and is stored in row-major order. The starting address of the matrix and the matrix dimensions (N and M) are passed to this routine in the stack in this order: matrix base address, N, M (that is, the matrix base address is pushed onto the stack first, followed by N and M). The matrix elements are 8-bit. Write a main program to test this routine, including the following features:

1. Call the Buffalo I/O routine OUTSTRG to output the message "The original matrix is as follows:" and call the routine OUTCRLF to move the screen cursor to the next line.

2. Call the Buffalo monitor I/O routines OUTLHLF and OUTRHLF to output the matrix elements, one row per line. Elements should be separated by one space.

3. Call the matrix swap subroutine to swap the rows of the matrix.

4. Call the I/O routine OUTSTRG to output the message "The swapped matrix is as follows:" and call the routine OUTCRLF to move the screen cursor to the next line.

5. Call the Buffalo I/O routines OUTLHLF and OUTRHLF to output the matrix elements, one row per line.

L3.4 Write a subroutine that determines if a number is a prime number and a main program to test it. The main program calls the subroutine to identify those primes that are in the range from 500 to 1000 and invokes Buffalo I/O routines to display them on the monitor screen. Displays ten numbers in each row and each number is displayed in four digits.

4

C Language Programming

4.1 Objectives

After completing this chapter you should be able to

- explain the overall structure of a C language program

- use appropriate operators to perform desired operations in C language

- understand the basic data types and expressions of C language

- write program loops

- write functions and make subroutine calls in C language

- use arrays and pointers for data manipulation

- perform basic I/O operations in C language

- use ImageCraft C compiler to compile your C programs

- use the integrated development environment of ICC11 compiler

4.2   Introduction to C

This chapter is not intended to provide a complete coverage of the C language. Instead, it provides only a summary of those C language constructs that will be used in this book. It should be adequate to deal with the 68HC11 interfacing programming if you fully understand the contents of this chapter.

C language is gradually replacing assembly language in many embedded applications because it has several advantages over assembly language. The most important one is that it allows us to program at a level higher than assembly language; thus programming productivity is greatly improved.

A C program, whatever its size, consists of functions and variables. A function contains statements that specify the operations to be performed. The types of statements in a function could be *declaration, assignment, function call, control,* and *null.* A variable stores a value to be used during the computation. The *main( )* function is required in every C program and is the one to which control is passed when the program is executed. A simple C program is as follows:

(1)	#include <stdio.h>	*—include information about standard library*
(2)	/* this is where program execution begins */	
(3)	main ( )	*—defines a function named* **main** *that receives* *—no argument values*
(4)	{	*—statements of main are enclosed in braces*
(5)	int a, b, c;	*—defines three variables of type* **int**
(6)	a = 3;	*—assigns 3 to variable* **a**
(7)	b = 5;	*—assigns 5 to variable* **b**
(8)	c = a + b;	*—adds* **a** *and* **b** *together and assigns it to* **c**
(9)	printf(" a + b = %d \n", c);	*—calls library function printf to print the result*
(10)	return 0;	*—returns 0 to the caller of main*
(11)	}	*—the end of* **main** *function*

The first line of the program,

#include <stdio.h>

causes the file **stdio.h** to be included in your program. This line appears at the beginning of many C programs.

The second line is a comment. A comment explains what will be performed and will be ignored by the compiler. A comment in C language starts with /* and ends with */. Comments make a program more readable.

The third line **main( )** is where program execution begins. The opening brace on the fourth line marks the start of the **main( )** function's code. The fifth line declares 3 integer variables **a, b,** and **c.** In C, all variables must be declared before they can be used.

The sixth line assigns 3 to the variable **a.** The seventh line assigns 5 to the variable **b.** The eighth line computes the sum of variables **a** and **b** and assigns it to the variable **c.** You will see that assignment statements are major components in our C programs.

The ninth line calls the library function **printf** to print the string **a + b =** followed by the value of **c** and move the cursor to the beginning of the next line. The tenth line returns 0 to the caller of **main( ).** The closing brace in the eleventh line ends the main ( ) function.

4.3 Types, Operators, and Expressions

Variables and constants are the basic objects manipulated in a program. Variables must be declared before they can be used. A variable declaration must include the name and type of the variable and may optionally provide its initial value. The name of a variable consists of letters and digits. You can also include the underscore character "_" to improve the readability of long variables. Don't begin variable names with underscore, however, since library routines often use such names. Upper and lower case letters are distinct. Keywords such as *do, if, while, int, etc.,* are reserved: you cannot use them as variable names.

4.3.1 Data Types

There are only a few basic data types in C: *char, int, float,* and *double.* A variable of type *char* can hold a single byte of data. A variable of type *int* is an integer, which is normally the natural size for a particular machine. The type *float* refers to a 32-bit, single-precision, floating-point number. The type *double* represents a 64-bit, double-precision, floating-point number. In addition, there are a number of qualifiers that can be applied to these basic types. *Short* and *long* apply to integers. These two qualifiers will modify the lengths of integers. In the ImageCraft C compiler, *short* is 16 bits and *long* is 32 bits. The qualifiers *signed* and *unsigned* may be applied to data types *char* and *integer.* The types *float* and *double* will not be used in this text.

4.3.2 Declarations

All variables must be declared before use. A declaration specifies a type and contains a list of one or more variables of that type, as in

```
int i, j, k;
char cx, cy;
```

A variable may also be initialized when it is declared, as in

```
int i = 0;
char echo = 'y';    /* the ASCII code of letter y is assigned to variable echo. */
```

4.3.3   Constants

There are four kinds of constants: integers, characters, floating-point numbers, and strings. A character constant is an integer, written as one character within single quotes, such as 'x'. A character constant is represented by the ASCII code of the character. A string constant is a sequence of zero or more characters surrounded by double quotes, as in

"68HC11 is made by Motorola"

or

""   /* an empty string */

Each individual character in the string is represented by its ASCII code.

An integer constant such as 3241 is an *int*. A long constant is written with a terminal l or L, as in 44332211L. The following constant characters are predefined in C language:

\a	alert (bell) character	\\	backslash
\b	backspace	\?	question mark
\f	formfeed	\'	single quote
\n	newline	\"	double quote
\r	carriage return	\ooo	octal number
\t	horizontal tab	\xhh	hexadecimal number
\v	vertical tab		

As in assembly language, a number in C can be specified in different bases. The method to specify the base of a number is to add a prefix to the number. The prefixes for different bases are:

base	prefix	example
decimal	none	1234
octal	0	04321
hexdecimal	0x	0x45

4.3.4   Arithmetic Operators

There are seven arithmetic operators:

- \+   add and unary plus
- \–   subtract and unary minus
- *   multiply
- /   divide
- %   modulus (or remainder)

++   increment

--   decrement

The expression

a % b

produces the remainder when a is divided by b. The % operator cannot be applied to float or double. The **++** operator adds 1 to the operand, and the **− −** operator subtracts 1 from the operand. The / operator truncates the quotient to integer when both operands are integers.

4.3.5   Bitwise Operators

C provides six operators for bit manipulations; these may only be applied to integral operands, that is, *char, short, int,* and *long,* whether *signed* or *unsigned.*

&   AND

| OR

^   XOR

~   NOT

>>   right shift

<<   left shift

The **&** operator is often used to clear one or more bits to zero. For example,

PORTD   = PORTD & 0xBD;   /* PORTD is 8 bits */

clears bits 6 and 2 of PORTD to 0.

The | operator is often used to set one or more bits to 1. For example,

PORTB   = PORTB | 0x40;   /* PORTB is 8 bits */

sets the bit 6 of PORTB to 1.

The XOR operator can be used to toggle a bit. For example,

abc   = abc ^ 0xF0;   /* abc is of type char */

toggles the upper four bits of the variable *abc.*

The >> operator shifts the involved operand to the right for the specified number of places. For example,

xyz = abc >> 3;

shifts the variable *abc* to the right three places and assigns it to the variable *xyz.*

The << operator shifts the involved operand to the left for the specified number of places. For example,

xyz = xyz << 4;

shifts the variable *xyz* to the left four places.

The assignment operator = is often combined with the operator. For example,

```
PORTD = PORTD & 0xBD;
```

can be rewritten as

```
PORTD &= 0xBD;
```

The statement

```
PORTB = PORTB | 0x40;
```

can be rewritten as

```
PORTB |= 0x40;
```

4.3.6 Relational and Logical Operators

Relational operators are used in expressions to compare the values of two operands. If the result of the comparison is true, then the value of the expression is 1. Otherwise, the value of the expression is 0. Here are the relational and logical operators:

==	equal to
!=	not equal to
>	greater than
>=	greater than or equal to
<	less than
<=	less than or equal to
&&	and
||	or
!	not (one's complement)

Here are some examples of relational and logical operators:

```
if (!(ADCTL & 0x80))
    statement₁;          /* if bit 7 is 0, then execute statement₁ */

if (i < 10 && i > 0)
    statement₂;          /* if 0 < i < 10 then execute statement₂ */

if (a1 == a2)
    statement₃;          /* if a1 = a2 then execute statement₃ */
```

4.4 Control Flow

The control-flow statements specify the order in which computations are performed. In C language, the semicolon is a statement terminator. Braces { and } are used to group declarations and statements together into a *compound statement,* or *block,* so that they are syntactically equivalent to a single statement.

4.4.1   If-Else Statement

The syntax of an *if-else statement* is

```
if (expression)
    statement₁
else
    statement₂
```

The **else** part is optional. The *expression* is evaluated; if it is true (non-zero), *statement₁* is executed. If it is false, *statement₂* is executed. Here is an example of the *if-else* statement:

```
if (a != 0)
    r = b;
else
    r = c;
```

A more concise way to accomplish the same thing is to write

```
r = (a != 0)? b : c;
```

4.4.2   Multiway Conditional Statement

A multiway decision can be expressed as a cascaded series of *if-else* statements. Such series looks like this:

```
if (expression₁)
    statement₁
else if (expression₂)
    statement₂
else if (expression₃)
    statement₃
    ...
else
    statementₙ
```

Here is an example of a three-way decision:

```
if (abc > 0) return 5;
else if (abc = 0) return 0;
else return –5;
```

4.4.3   Switch Statement

The *switch* statement is a multiway decision based on the value of a control expression. The syntax of the switch statement is

```
switch (expression) {
    case const_expr₁:
        statement₁;
        break;
    case const_expr₂:
        statement₂;
        break;
    ...
    default:
        statementₙ;
}
```

As an example, consider the following program fragment:

```
switch (i) {
    case 1: printf("%");
        break;
    case 2: printf("%%");
        break;
    case 3: printf("%%%");
        break;
    case 4: printf("%%%%");
        break;
    case 5: printf("%%%%%");
}
```

The number of % characters printed is equal to the value of **i.** The break keyword forces the program flow to drop out of the switch statement so that only the statements under the corresponding *case-label* are executed. If any break statement is missing, then all the statements from that case-label until the next break statement within the same switch statement will be executed.

4.4.4 ■ For-Loop Statement

The syntax of a for-loop statement is

```
for (expr1; expr2; expr3)
    statement
```

where, *expr1* and *expr3* are assignments or function calls, and *expr2* is a relational expression. For example, the following *for loop* computes the sum of the squares of integers from 1 to 9:

```
sum = 0;
for (i = 1; i < 10; i++)
    sum = sum + i * i;
```

The following *for loop* prints out the first ten odd integers:

```
for (i = 1; i < 20; i++)
    if (i % 2) printf("%d ", i);
```

4.4.5   While Statement

The syntax of a *while statement* is

```
while (expression)
    statement
```

The *expression* is evaluated. If it is non-zero, *statement* is executed and *expression* is re-evaluated. This cycle continues until *expression* becomes zero, at which point execution resumes after *statement*. The *statement* may be a *null* statement. A null statement does nothing and is represented by a semicolon.

Consider the following program fragment:

```
int_cnt = 5;
while (int_cnt);
```

The CPU will do nothing before the variable *int_cnt* is decremented to zero. In microprocessor applications, the decrement of *int_cnt* is often done by external events such as interrupts.

4.4.6   Do-While Statement

The *while* and *for* loops test the termination condition at the beginning. By contrast, the *do-while* statement tests the termination condition at the end of the statement; the body of the statement is executed at least once. The syntax of the statement is

```
do
    statement
while (expression);
```

The following *do-while* statement displays the integers 9 down to 1:

```
int digit = 9;
do
    printf("%d ", digit--);
while (digit >= 1);
```

4.5   Input and Output

Input and output facilities are not part of C language itself. However, input and output are fairly important in application. The ANSI standard defines a set of library functions that must be included so that they can exist in compatible form on any system where C exists. Some of them deal with file input and output. Others deal with text input and output. In this section we will look at the following four input and output functions:

1. int *getchar* ( ). This function returns a character when it is called. The following program fragment returns a character and assigns it to the variable *xch*:

   ```
   char xch;
   xch = getchar ();
   ```

2. int *putchar* (int). This function outputs a character on the standard output device. The following statement outputs the letter **a** from the standard output device:

   ```
   putchar ('a');
   ```

3. int *puts* (const char *s). This function outputs the string pointed to by **s** on the standard output device. The following statement outputs the string ***Welcome to USA!*** from the standard output device:

   ```
   puts ("Welcome to USA! \n");
   ```

4. int *printf* (*formatting string*, arg$_1$, arg$_2$, ..., arg$_n$). This function converts, formats, and prints its arguments on the standard output under control of *formatting string*. *arg$_1$, arg$_2$, ..., arg$_n$* are arguments that represent the individual output data items. The arguments can be written as constants, single variable or array names, or more complex expressions. The formatting string is composed of individual groups of characters, with one character group associated with each output data item. The character group corresponding to a data item must start with **%**. In its simplest form, an individual character group will consist of the percent sign followed by a *conversion character* indicating the type of the corresponding data item.

Multiple character groups can be contiguous, or they can be separated by other characters, including white space characters. These "other" characters are simply transferred directly to the output device where they are displayed. Several of the more frequently used conversion characters are listed in Table 4.1. Between the % and the conversion character there may be, in order:

- A minus sign, which specifies left adjustment of the converted argument.
- A number that specifies the minimum field width. The converted argument will be printed in a field at least this wide. If necessary it will be padded on the left (or right, if left adjustment is called for) to make up the field width.
- A period, which separates the field width from precision.

Conversion character	Meaning
c	data item is displayed as a single character
d	data item is displayed as a signed decimal number
e	data item is displayed as a floating-point value with an exponent
f	data item is displayed as a floating-point value without an exponent
g	data item is displayed as a floating-point value using either e-type or f-type conversion, depending on value; trailing zeros, trailing decimal point will not be displayed
i	data item is displayed as a signed decimal integer
o	data item is displayed as an octal integer, without a leading zero
s	data item is displayed as a string
u	data item is displayed as an unsigned decimal integer
x	data item is displayed as a hexadecimal integer, without the leading 0x

Table 4.1 ■ Commonly used conversion characters for data output

- A number, the precision, that specifies the maximum number of characters to be printed from a string, or the number of digits after the decimal point of a floating-point value, or the minimum number of digits for an integer.
- An h if the integer is to be printed as a *short,* or *l* (letter ell) if as a *long.*

Several valid printf calls are as follows:

```
printf ("this is a challenging course! \n");        /* outputs only a string */
printf ("%d %d %d", x1, x2, x3);                    /* outputs variables x1, x2, x3 using mini-
                                                       mal number of digits with one space sepa-
                                                       rating each value */

printf("Today's temperature is %4.1d \n", temp);    /* display the string Today's temperature
                                                       is followed by the value of temp. Display
                                                       one fractional digit and use at least four
                                                       digits for the value. */
```

4.6   Functions and Program Structure

Every C program consists of one or more functions. If a program consists of multiple functions, their definitions cannot be embedded within another. The same function can be called from several different places within a program. Generally, a function will process information passed to it from the calling portion of the program and return a single value. Information is passed to the function via special identifiers called *arguments* (also called *parameters*) and returned via the *return* statement. Some functions, however, accept information but do not return anything (for example, the library function *printf*).

The syntax of a function definition is as follows:

return_type function_name (declarations of arguments)
{
    declarations and statements
}

The declaration of an argument in the function definition consists of two parts: the *type* and the *name* of the variable. The return type of a function is *void* if it does not return any value to the caller. An example of a function that converts a lowercase letter to uppercase letter is as follows:

```
char lower2upper (char cx)
{
    if (cx >= 'a' && cx <= 'z') return (cx - ('a' - 'A'));
    else return cx;
}
```

A character is represented by its ASCII code. A letter is in lowercase if its ASCII code is between 97 (0x61) and 122 (0x7A). To convert a letter from low-ercase to uppercase, subtract its ASCII code by the difference of the ASCII codes of letters **a** and **A.** To call a function, simply put down the name of the function and replace the argument declarations by actual arguments or values and terminate it with a semicolon.

Example 4.1

Write a function to test whether an integer is a prime number.

Solution: The integer 1 is not a prime number. A number is prime if it is indivisible by any integer between 2 and its half.

```
/* this function returns a 1 if a is prime. Otherwise, it returns a 0. */
char test_prime (int a)
{
    int i;
    if (a == 1) return 0;
    for (i = 2; i < a/2; i++)
        if ((a % i) == 0) return 0;
    return 1;
}
```

Example 4.2

Write a program to find out the number of prime numbers between 100 and 1000.

Solution: We can find the number of prime numbers between 100 and 1000 by calling the function in Example 4.1 to find out if a number is prime.

```
#include <stdio.h>
char test_prime (int a);     /* prototype declaration for the function test_prime */

main ( )
{
    int i, prime_count;
    prime_count = 0;
    for (i = 100; i <= 1000; i++) {
        if (test_prime(i))
            prime_count ++;
    }
    printf("\n The total prime numbers between 100 and 1000 is %d\n", prime_count);
}
char test_prime (int a)
{
    int i;
    if (a == 1) return 0;
    for (i = 2; i < a/2; i++)
        if ((a % i) == 0) return 0;
    return 1;
}
```

A function cannot be called before it has been defined. This dilemma is solved by using the function prototype statement. The syntax for a function prototype statement is as follows:

return_type function_name (declarations of arguments);

The statement

char test_prime (int a);

before *main( )* is a function prototype statement.

▲

4.7   Pointers and Arrays

4.7.1   Pointers and Addresses

A *pointer* is a variable that holds the address of a variable. Pointers are used frequently in C, as they have a number of useful applications. For example, pointers can be used to pass information back and forth between a function and its reference (calling) point. In particular, pointers provide a way to return multiple data items from a function via function arguments. Pointers also permit references to other functions to be specified as arguments to a given function. This has the effect of passing functions as arguments to the given function.

Pointers are also closely associated with arrays and therefore provide an alternate way to access individual array elements.

The syntax for declaring a pointer type is

```
type_name *pointer_name;
```

For example,

```
int *ax;
```

declares that the variable *ax* is a pointer to integer.

```
char *cp;
```

declares that the variable *cp* is a pointer to character.

To access the value pointed to by a pointer, use the *dereferencing* operator *. For example,

```
int        a, *b;    /* b is a pointer to int */
...
a = *b;
```

assigns the value pointed to by **b** to variable **a.**

We can assign the address of a variable to a pointer by using the unary operator **&.** The following example shows how to declare a pointer and how to use & and *:

```
int x, y;
int *ip;

ip = &x;    /* assigns the address of the variable x to ip */
y = *ip;    /* y gets the value of x */
```

Example 4.3

Write the bubble sort function to sort an array of integers.

Solution: The algorithm for bubble sort is already described in chapter 3. Here is the C language version:

```
void   swap (int *, int *);
void bubble (int a[], int n) /* n is the array count */
{
    int i, j;
    for (i = 0; i < n - 1; i++)
        for (j = n - 1; j > i; j--)
```

```
                    if (a[j - 1] > a[j])
                        swap (&a[j - 1], &a[j]);
    }
    void swap (int *px, int *py)
    {
        int temp;
        temp = *px;
        *px = *py;
        *py = temp;
    }
```

4.7.2   Arrays

Many applications require the processing of multiple data items that have common characteristics (e.g., a set of numerical data, represented by x_1, x_2, ..., x_n). In such situations it is more convenient to place data items into an *array*, where they will all share the same name. The individual data items can be characters, integers, floating-point numbers, and so on. They must all be of the same type and the same storage class.

Each array element is referred to by specifying the array name followed by one or more *subscripts*, with each subscript enclosed in brackets. Each subscript must be expressed as a nonnegative integer. Thus the elements of an n-element array x are x[0], x[1], ..., x[n − 1]. The number of subscripts determines the dimensionality of the array. For example, x[i] refers to an element of an one-dimensional array. Similarly, y[i][j] refers to an element of a two-dimensional array. Higher dimensional arrays can be formed by adding additional subscripts in the same manner. However, higher dimensional arrays are not used very often.

In general, a one-dimensional array can be expressed as

data-type array_name[expression];

A two-dimensional array is defined as

data-type array_name[expr1][expr2];

An array can be initialized when it is defined. This is a technique used in table lookup, which can speed up the computation process. For example, a data acquisition system that utilizes an analog-to-digital converter can use table lookup technique to speed up the conversion process. An example will be given in chapter 11.

4.7.3   Pointers and Arrays

In C, there is a strong relationship between pointers and arrays. Any operation that can be achieved by array subscripting can also be done with pointers. The pointer version will in general be faster but somewhat harder to understand.

For example,

```
int ax[20];
```

defines an array **ax** of 20 elements.

The notation ax[i] refers to the i-th element of the array. If ip is a pointer to an integer, declared as

```
int *ip;
```

then the assignment

```
ip = &ax[0];
```

makes ip to contain the address of ax[0]. Now the statement

```
x = *ip;
```

will copy the contents of ax[0] into x.

If *ip* points to ax[0], then ip+1 points to ax[1], and ip+i points to ax[i], etc.

4.7.4   Passing Arrays to a Function

An array name can be used as an argument to a function, thus permitting the entire array to be passed to the function. To pass an array to a function, the array name must appear by itself, without brackets or subscripts, as an actual argument within the function call. When declaring a one-dimensional array as a formal argument, the array name is written with a pair of empty square brackets. The size of the array is not specified within the formal argument declaration. If the array is two-dimensional, then there should be two pairs of empty brackets following the array name.

The following program outline illustrates the passing of an array from the main portion of the program to a function.

```
int average (int n, int arr[]);
main ( )
{
    int n, avg;              /* variable declaration */
    int arr[50];            /* array definition */

    ...

    avg = average(n, arr);   /* function call */
    ...
}
int average (int k, int brr[])   /* function definition */
{
    ...
}
```

Within *main* we see a call to the function *average.* This function call contains two actual arguments—the integer variable *n* and the one-dimensional integer array *arr.* Note that *arr* appears as an ordinary variable within the function call.

In the first line of the function definition, we see two formal arguments, call *k* and *brr.* The formal argument declarations establish *k* as an integer variable and *brr* as a one-dimensional integer array. Note that the size of *brr* is not defined in the function definition.

As formal parameters in a function definition,

```
int brr[];
```

and

```
int *brr;
```

are equivalent.

4.8   Miscellaneous Items

4.8.1   External Variables

A variable defined inside a function is an *internal variable* of that function. *External variables* are defined outside of any function and are thus potentially available to many functions. Because external variables are globally accessible, they provide an alternative to function arguments and return values for communicating data between functions. Any function may access an external variable by referring to it by name, if the name has been declared somehow.

External variables are also useful when two functions must share some data, yet neither calls the other.

4.8.2   Scope Rules

The functions and external variables that make up a C program need not all be compiled at the same time; the source text of the program may be kept in several files, and previously compiled routines may be loaded from libraries.

The scope of a name is the part of the program within which the name can be used. For a variable declared at the beginning of a function, the scope is the function in which the name is declared. Local (internal) variables of the same name in different functions are unrelated.

The scope of an external variable or a function lasts from the point at which it is declared to the end of the file being compiled. Consider the following program segment:

```
...
void f1 (...)
{
    ...
}
int a, b, c;
void f2 (...)
{
    ...
}
```

Variables a, b, and c are accessible to function f2 but not to f1.

When a C program is split into several files, it is convenient to put all global variables into one file so that they can be accessed by functions in different files. Functions residing in different files that need to access global variables must declare them as external variables. In addition, we can place the prototypes of those functions in one file so that they can be called by functions in other files.

The following example is a skeletal outline of a two-file C program that makes use of external variables:

In file1:

```
extern int  xy;
extern long arr[];
main ( )
{
    ...
}
void foo (int abc) { ... }
long soo (void) { ... }
```

In file2:

```
int xy;
long arr[100];
```

4.9   Using the ImageCraft C Compiler (ICC11)

ICC11 is a C cross compiler and development environment for the Motorola family of 68HC11 microcontrollers running on Windows platforms. The compiler accepts the ANSI C language with the following exception:

The supplied library is only a subset of what is defined by the ANSI standard.

ICC11 includes an IDE (Integrated Development Environment) with a project builder, a syntax aware text editor, a terminal program, plus integration with the command line compiler tools. The tight integration of tools allows fast edit-compile-download development cycle of the 68HC11.

The procedure for using the ICC11 to enter, compile, and download a C program into a 68HC11 demo board for execution is as follows:

Step 1

Invoke the ICC11 by clicking the icon of the ICC11 program. The ICC11 window will appear as shown in Figure 4.1.

Step 2

Click the **file** menu and select **new** to create a new file for entering a new program. The screen should look like Figure 4.2.

Step 3

Type in the new program and save it in an appropriate directory. An example is shown in Figure 4.3.

Step 4

Create a new project by clicking the **Project** menu and select **new.** The screen is shown in Figure 4.4. After this, the screen should change to Figure 4.5. Type in an appropriate project name that you like and click the **OK** button. The result is shown in Figure 4.6.

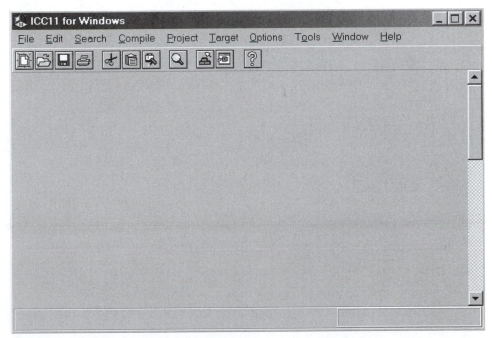

Figure 4.1 ■ ICC11 command window

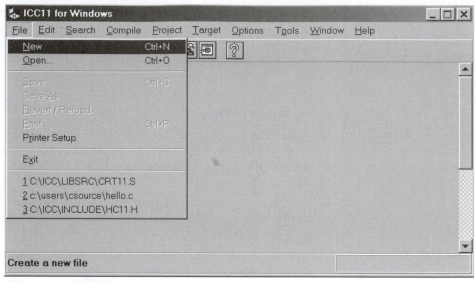

Figure 4.2 ■ Create a new file for entering a program

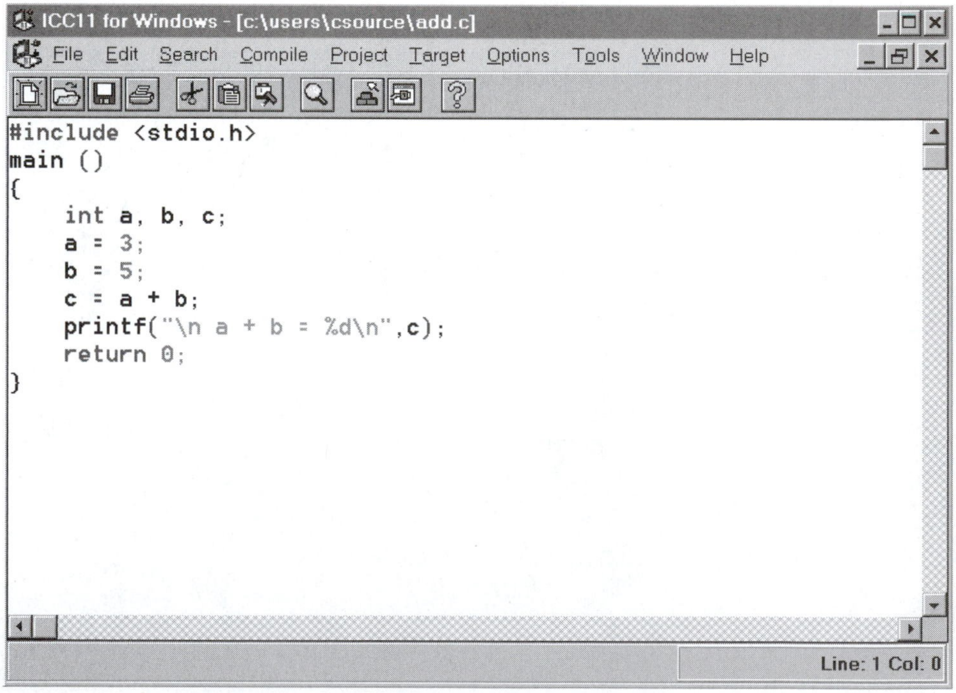

```
#include <stdio.h>
main ()
{
    int a, b, c;
    a = 3;
    b = 5;
    c = a + b;
    printf("\n a + b = %d\n",c);
    return 0;
}
```

Figure 4.3 ■ Entering a new program in ICC11

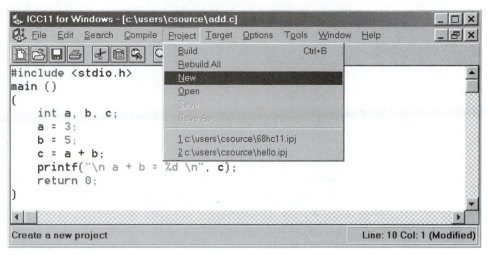

Figure 4.4 ■ Screen for creating a new project

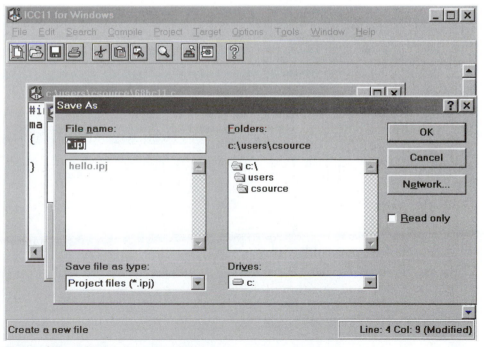

Figure 4.5 ■ Create a new project

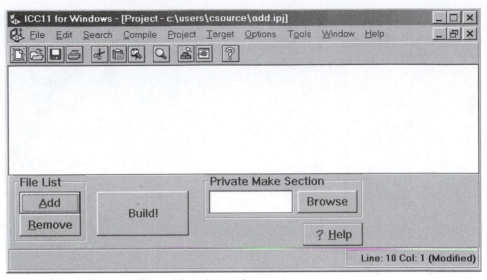

Figure 4.6 ■ Screen after setting the project name

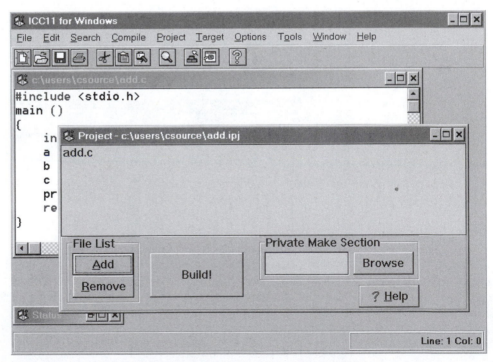

Figure 4.7 ■ Screen after adding add.c into the project add.ipj

Step 5

Add the programs to the newly created project. To add a program to the project, press the **Add** button. A list of files will be displayed, and you can select those files that you want to include in this project. In this example, we will include the file *add.c.* Click on this file and press the **OK** button. The screen is changed to Figure 4.7.

Step 6

Setting appropriate options for the compiler, editor, and terminal. The options available can be seen by pressing the **Options** menu as shown in Figure 4.8.

Step 7

Set compiler preprocessor options. There are three sets of options that need to be set in the compiler: *preprocessor, compiler,* and *linker.* Select **Compiler** in Figure 4.8 and then click on Preprocessor and the screen shown in Figure 4.9a will be brought up. The only compiler preprocessor option that we need to set is the *include* path. Enter the appropriate *include* path. Do not click the **OK** button yet.

Step 8

Set compiler options. The default options for compiler shown in Figure 4.9b are OK.

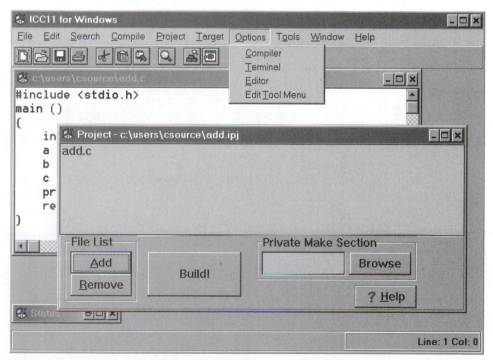

Figure 4.8 ■ Screen for setting ICC11 options

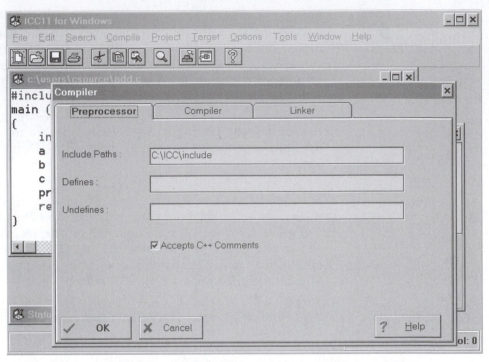

Figure 4.9a ■ Screen for setting compiler preprocessor options

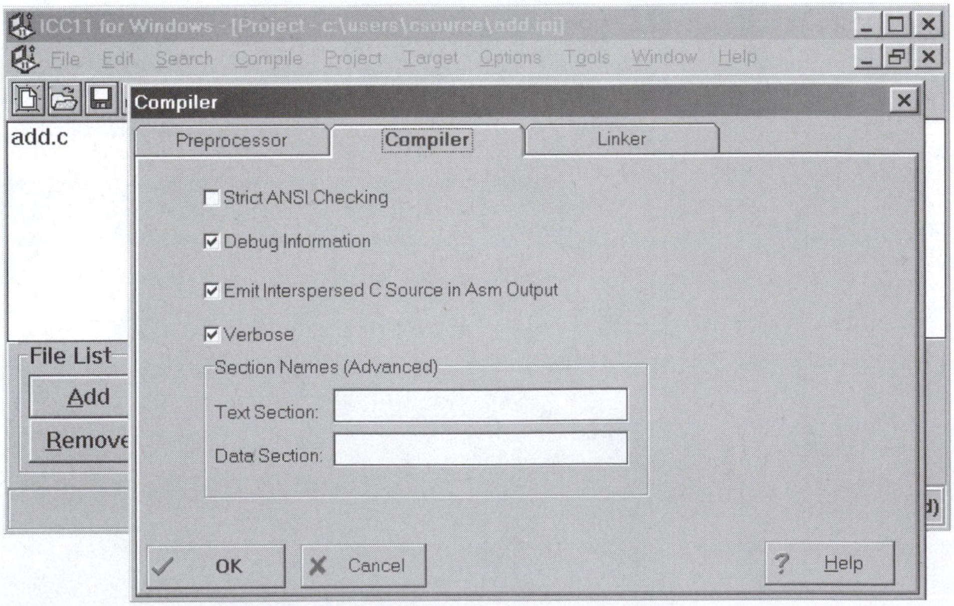

Figure 4.9b ■ Screen for setting compiler options

Step 9

Set linker options. Click the **Linker** button to bring up the linker option screen. The appropriate option values for EVB demo board is shown in Figure 4.9c. The linker options for the CMD11A8 demo board are different from those for the EVB. A good setting is shown in Figure 4.9d. The text section sets the starting address of program code. The program code will be stored from this address toward higher addresses. The stack section defines the starting address of the stack. The stack will grow from higher addresses toward lower addresses. The data section defines the starting address for data section. Data section grows from this address to higher addresses. After compiler and linker options are set, click the **OK** button.

Step 10

Set terminal options. Go back to **Option** menu and select **Terminal.** This will bring up the terminal option screen as shown in Figure 4.10. The **com port** can be **com1** or **com2,** depending on which com port is used to connect the demo board to the PC. *Baud rate* should be set to 9600 for both the EVB and the CMD11A8. *Flow control* should be set to none. **Font** also needs to be set properly. A popular font is Courier or Courier New (shown in Figure 4.11). When you are satisfied with all the options, click the **OK** button and the screen will change back to Figure 4.8.

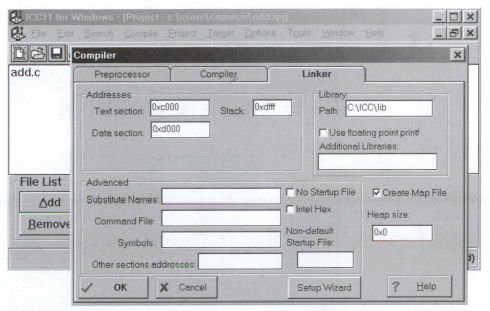

Figure 4.9c ■ Linker options for EVB demo board

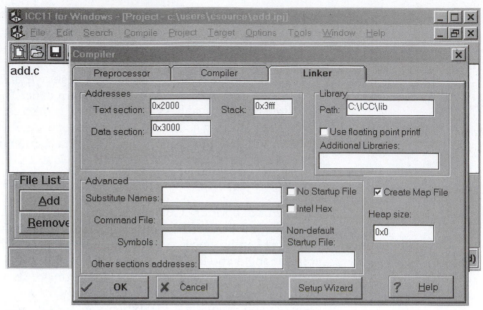

Figure 4.9d ■ Linker options for CMD11A8 demo board

Step 11

Build the project. After setting all the options properly, we want to compile the source code into executable code and download it to demo board for execution. Click the **Build** button, and ICC11 will start to process your source file. The **Build** button will be dimmed before it is done.

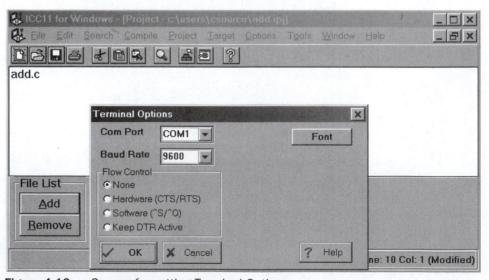

Figure 4.10 ■ Screen for setting Terminal Options

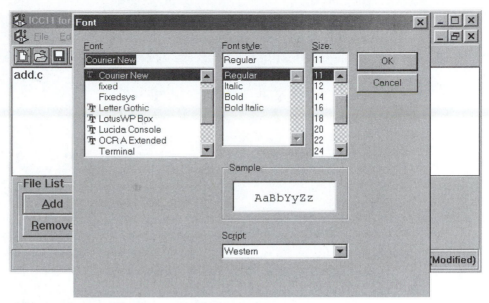

Figure 4.11 ■ Screen for setting font options

When you see the **Build** button change back to bold, ICC11 has completed processing your source code. Minimize both the project and source-code windows and maximize the status window. If there is no error, the status window should look like Figure 4.12.

Step 12
Download the program onto the demo board for execution. Minimize the status window and click the **Target** menu. Select **Terminal** and this should bring up the terminal window as shown in Figure 4.13. Type **load t**

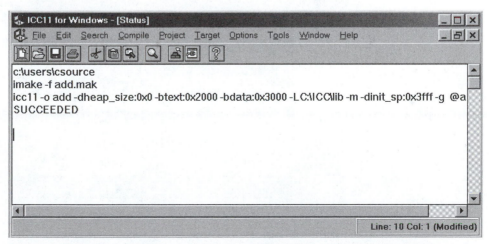

Figure 4.12 ■ Status window after successfully building the project add.ipj

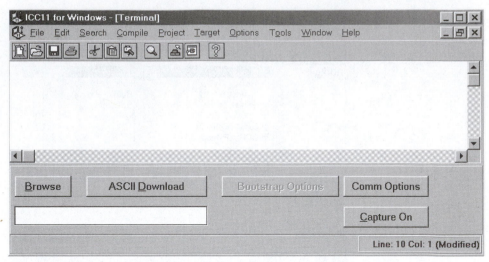

Figure 4.13 ■ ICC11 terminal window

followed by the **enter** key, and then click on the **Browse** button. This will bring up a popup window (shown in Figure 4.14) for you to select a file to download. Select **add.s19** and click the **OK** button. To start download, click the **ASCII Download** button. During the download process you will see a screen like Figure 4.15. The screen looks like Figure 4.16 when the download process is done.

Step 13

Program execution and debugging. ICC11 terminal program allows you to use all the Buffalo commands to debug programs when using an EVB

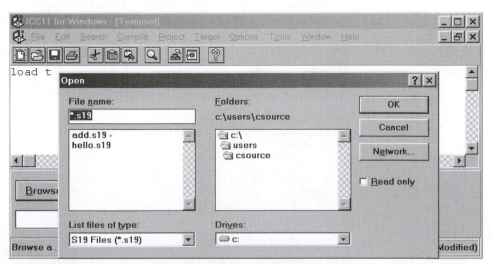

Figure 4.14 ■ Screen for selecting a file to download

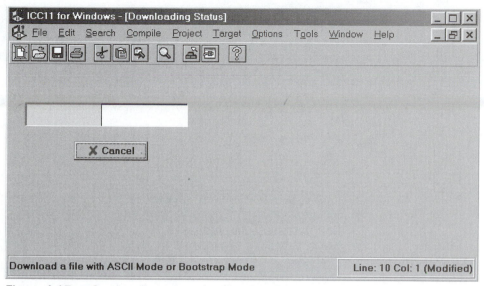

Figure 4.15 ■ Screen of program download process

or CMD11A8 demo board. Type **G 2000** to run this program on the CMD11A8 demo board. The result should look like Figure 4.17.

ICC11 saves the options setting for future uses so you don't need to set the options every time.

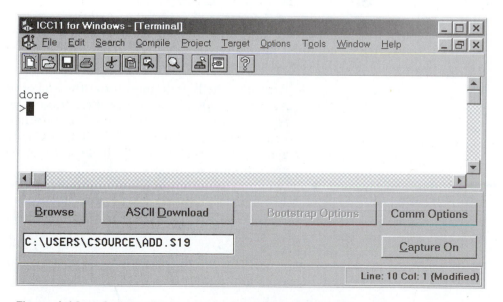

Figure 4.16 ■ Screen after a successful program download

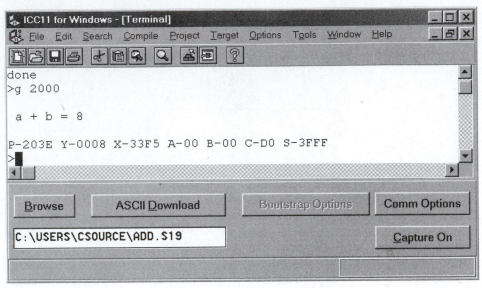

Figure 4.17 ■ Screen after executing the program add.c

4.10   Adjust the ImageCraft C Compiler

There are few things that need to be adjusted before you can use the ICC11 C compiler to generate executable code for your target board (EVB or CMD11A8). They are:

1. Return to the Buffalo monitor at the end of program execution. After your program completes execution, you want the CPU control to return to the Buffalo monitor (or other monitor) so that you can do something else. The ICC11 C compiler links your program to a startup program called crt11.s. A segment of the crt11.s program is as follows:

```
        ...
        ; call user main routine
        jsr _main

_exit::
        bra    _exit
        ...
```

The startup routine calls the main function after finishing its initialization work and then stays in an infinite loop. Because of that, it will not return to the Buffalo monitor. Therefore, we don't know for sure if the program executes correctly. There are several methods to solve this problem:

(1) Add an SWI instruction before the end of your main function. ICC11 allows you to add in-line assembly instructions. An in-line instruction can be added by using the following statement:

asm ("assembly instruction");

For example, entering the following statement will add the SWI instruction to your C program

asm ("SWI");

(2) Replace the *bra _exit* instruction of the crt11.s program with the SWI instruction. If this method is used, then you need to reassemble crt11.s and copy the generated object code to the c:\icc\lib directory. (You can use the ICC11 compiler to compile the crt11.s program.) This is the preferred method because it needs to be done only once.

2. I/O functions. The ICC11 library uses the 68HC11 SCI function to implement *putchar ( )* and *getchar ( )* functions and the *puts* and *printf* functions call putchar ( ) to perform output. This works with the CMD11A8 demo board. However, this won't work with the Motorola EVB because it uses the MC6850 chip to control the terminal port. The getchar ( ) and putchar ( ) functions are implemented in the iochar.c program (located under ..\icc\libsrc). In order to use the putchar and getchar functions, you need to provide your own versions of these two functions and place them into the library:

ilib -a libc11.a iochar.o

An example for *getchar* and *putchar* functions (replace the versions in iochar.c) is provided in the following so that you can use them to perform I/O in the EVB demo board. Add the following definitions to the hc11.h file (located under ..\icc\include) of the ICC11 compiler so that we can use symbols to access ACIA registers:

```
#define ACIA_CTRL    *(unsigned char volatile *)( 0x9800)
#define ACIA_STAT    *(unsigned char volatile *)( 0x9800)
#define ACIA_XMIT    *(unsigned char volatile *)( 0x9801)
#define ACIA_RCV     *(unsigned char volatile *)( 0x9801)
```

```
char getchar ( )
{
    char xch;

    while (!(ACIA_STAT & 0x01));    /* wait until receive data register is full */
    xch = ACIA_RCV & 0x7F;         /* mask out parity bit */
    return xch;
}
void putchar (char xch)
{
    while (!(ACIA_STAT & 0x02));    /* wait until transmit data register is empty */
    ACIA_XMIT = xch;
}
```

There is a keyword *volatile* added between *char* and * in those four *define* statements. The purpose of *volatile* is to force an implementation to suppress optimization that could otherwise occur. For example, for a microcontroller with memory-mapped input/output, a pointer to device register might be declared as a pointer to *volatile*, in order to prevent the compiler from removing apparently redundant references through the pointer.

4.11 Summary

A C program consists of one or more functions and variables. The *main ( )* function is required in every C program. It is the entry point of your program. A function contains statements that specify the operations to be performed. The types of statements in a function could be *declaration, assignment, function call, control,* and *null.*

A variable stores a value to be used during the computation. A variable must be declared before it can be used. The declaration of a variable consists of the name and the type of the variable. There are four basic data types in C: int, char, float, and double. Several qualifiers can be added to the variable declarations. They are short, long, signed, and unsigned.

Constants are often needed in forming a statement. There are four types of constants: integers, characters, floating-point numbers, and strings.

There are seven arithmetic operators: $+$, $-$, $*$, $/$, $\%$, $++$, $--$. There are six bitwise operators: $\&$, $|$, $\wedge$, $\sim$, $>>$, and $<<$. Bitwise operators can only be applied to integers. Relational operators are used in control statements. They are $==$, $!=$, $<$, $>=$, $<$, $<=$, $\&\&$, $\|$, and $!$.

The control-flow statements specify the order in which computations are performed. Control-flow statements include if-else statement, multiway conditional statement, switch statement, for-loop statement, while-statement, and do-while statement.

Every C program consists of one or more functions. If a program consists of multiple functions, their definitions cannot be embedded within another. The same function can be called from several different places within a program. Generally, a function will process information passed to it from the calling portion of the program and return a single value. Information is passed to the function via special identifiers called *arguments* (also called *parameters*) and returned via the *return* statement. Some functions, however, accept information but do not return anything (for example, the library function *printf*).

A *pointer* holds the address of a variable. Pointers can be used to pass information back and forth between a function and its reference (calling) point. In particular, pointers provide a way to return multiple data items from a function via function arguments. Pointers also permit references to other functions to be specified as arguments to a given function. Two operators are related with pointer: * and &. The * operator returns the value of the variable pointed to by the pointer. The & operator returns the address of a variable.

Data items that have common characteristics are placed in an *array*. An array may be one-dimensional or multidimensional. The dimension of an array is specified by the number of square bracket pairs ([]) following the array name. An array name can be used as an argument to a function, thus permitting the entire array to be passed to the function. To pass an array to a function, the array name must appear by itself, without brackets or subscripts. An alternate way to pass arrays to a function is to use pointers.

A variable defined inside a function is an *internal variable* of that function. *External variables* are defined outside of any function and are thus potentially available to many functions. The *scope* of a name is the part of the program within which the name can be used. The scope of an external variable or a function lasts from the point at which it is declared to the end of the file being compiled.

The procedure of using the ImageCraft C compiler (ICC11) is described in detail. The ICC11 is window-driven and is straightforward to use.

4.12   Exercises

E4.1. Write the function *power(int x, char n)* that computes the value of x^n.

E4.2. Write a function that counts the type of characters contained in a character array. The argument to this function is a pointer to a string terminated by a NULL character.

E4.3. Write a function that tests if a given number is a multiple of 8. A 1 is returned if the given number is a multiple of 8. Otherwise, a 0 is returned. The number to be tested is an integer and is an argument to this function.

E4.4. Write a function that computes the greatest common divisor (gcd) of two integers *m* and *n*.

E4.5. What is a function prototype? What is the difference between a function prototype and function declaration?

E4.6. Write a *switch* statement that will examine the value of an integer variable xx and print the following messages, depending on the value assigned to xx.

    (a) Cold, if xx == 1
    (b) Chilly, if xx == 2
    (c) Warm, if xx == 3
    (d) Hot, if xx == 4

E4.7. Write a program to clear the screen and then move the cursor to the middle line and output the message "Microcontroller is fun to use!". Outputting the character "\f" can clear the screen.

E4.8. Write a function that will convert uppercase letters to lowercase.

E4.9. Write a function that calls getchar ( ) to input an integer from keyboard. The program will output the message "Please enter a 16-bit integer: " and wait for the user to enter a signed 16-bit integer. If the user enters an integer larger than can be accommodated by a 16-bit number, the function will reject the number and ask the user to re-enter the number. If any illegal characters have been entered, this program will ask the user to re-enter. The carriage-return key terminates a number.

E4.10. Write a loop that sums the squares of the first 100 odd integers.

E4.11. An *Armstrong number* is a number of n digits that is equal to the sum of each digit raised to the nth power. For example, 153 (which has three digits) equals $1^3 + 5^3 + 3^3$. Write a function to print out all 3-digit Armstrong numbers.

E4.12. Write a C function to perform binary search on a sorted array. The binary search algorithm is given in exercise problem E3.4. The starting address, the key, and the array count are parameters to this function. Both the key and array count are integers.

4.13 Lab Exercises and Assignments

L4.1. Enter, compile, and download the following C program into EVB or CMD-11A8 for execution using the procedure described in section 4.9:
#include <stdio.h>

```
main ( )
{
    int    digit = 0;
    while (digit <= 9)
            printf("%d \n", digit++);
}
```

L4.2. Write a C program that will generate every third integer, beginning with i = 2 and continuing for all integers that are less than 100. Calculate the sum of those integers that are evenly divisible by 5. Print those integers and their sum on the screen.

L4.3. Write a C function that calculates least common multiple (lcm) of two integers m and n. Integers m and n are parameters to this function. Also write a main program to test this function with several pairs of integers. Print out the result as follows:

> The lcm of m and n is xxx.

L4.4. *String input, output, and reverse.* Here you need to write three C functions:

(1) A C function that calls the getchar( ) function to input a string from the keyboard. The caller of this function passes a pointer to a buffer to hold the string. When saving the string in the buffer, this function adds a null character to terminate the string. The user terminates a string by entering a carriage return.

(2) A C function that reverses the string that is passed to it. The only parameter to this function is a pointer to the string. This function will output the original string and the reversed string.

(3) A main function to test the above two functions. The main function outputs a prompt that asks the user to enter a string, calls the first function to read the string from the keyboard, and then calls the second to reverse it.

5

Operation Modes and Memory Expansion

5.1 Objectives

After completing this chapter you should be able to

- set up the 68HC11 operation mode
- make memory space assignment and design an address decoder
- add external memory to the 68HC11
- perform timing analysis for a memory system

5.2   Introduction

The 68HC11 is a family of microcontrollers developed in 1985. Members in the 68HC11 family have the same instruction set and addressing modes but differ in the size of their on-chip memories and in their I/O functions. Each member is named by appending the part number to the 68HC-prefix, for example, 68HC11A1, 68HC11E9, etc. The A series parts (part numbers A8, A1, and A0) are the foundation of the 68HC11 family. Table 5.1 summarizes the members of the 68HC11 family at the time of this writing.

Operation modes, on-chip memories, and external memory expansion methods will be covered in this chapter. The discussion will be based principally on the 68HC11A8.

5.3   A Summary of the 68HC11 Signal Pins

The block diagram and signal pins of the 68HC11A8 are shown in Figure 1.2. The 68HC11A8 is available in the 48-pin dual in-line package (DIP), 52-pin plastic leaded chip carrier (PLCC), and 64-pin quad flat pack (QFP) versions. The 68HC11 has five I/O ports that can interface to I/O devices directly. An I/O port consists of registers and I/O pins. Registers allow the user to set operation parameters, perform data transfer, and check the status of I/O operations. I/O pins allow direct data transfer to and from an I/O device. Most signal pins serve multiple functions. The function of each I/O port will be discussed in chapter 7.

5.4   68HC11 Operation Modes

The 68HC11 has four different operation modes: expanded, single-chip, special bootstrap, and special test. The expanded mode allows the user to add off-chip memory devices and I/O peripheral chips to the microcontroller, whereas these additions are not permitted in the single-chip mode. The special test mode is mainly used during Motorola's internal production testing. The bootstrap mode allows the user to load a program into the on-chip SRAM from the serial communication interface (SCI). The loaded program can perform any function the user wants. One such application is to program the on-chip EEPROM or EPROM.

5.4.1   Single Chip Mode

In the single-chip mode, the 68HC11 functions without external address and data buses. Port B, port C, strobe A (STRA), and strobe B (STRB) pins are available for general-purpose parallel I/O. In this mode, all software needed to control the microcontroller is contained in internal memories.

0.25 ptPart number		EPROM	ROM	EEPROM	RAM	Description
68HC11A0	–	–	–	256		8-bit A/D, 38 I/O, 3MHz mux bus
68HC11A1	–	–	512	256		8-bit A/D, 38 I/O, 3MHz mux bus
68HC11A8	–	8K	512	256		8-bit A/D, 38 I/O, 3MHz mux bus
68HC11D0	–	–	–	192		14 I/O
68HC11D3	–	4K	–	192		32 I/O, 3MHz bus
68HC711D3	4K	–	–	192		32 I/O, 3MHz bus
68HC11ED0	–	–	–	512		SPI, SCI
68HC11E0	–	–	–	512		8-bit A/D, 38 I/O, 4 input capture, 3 MHz mux bus
68HC11E1	–	–	512	512		8-bit A/D, 38 I/O, 4 input capture, 3 MHz mux bus
68HC11E9	–	12K	512	512		8-bit A/D, 38 I/O, 4 input capture, 3 MHz mux bus
68HC811E2	–		2K	512		8-bit A/D, 38 I/O, 4 input capture, 3 MHz mux bus
68HC711E9	12K	–	512	512		8-bit A/D, 38 I/O, 4 input capture, 3 MHz mux bus
68HC711E20	20K	–	–	768		8-bit A/D, 38 I/O, 4 input capture, 3 MHz mux bus
68HC11E20	–	20K	512	768		8-bit A/D, 38 I/O, 4 input capture, 3 MHz mux bus
68HC11F1	–	–	512	1K		8-bit A/D, 30 I/O, 4 MHz nonmux bus
68HC11G0	–	–	–	512		4 PWM, 8-Ch 10-bit A/D, SPI, SCI
68HC11G5	–	16K	–	512		10-bit A/D, 4 PWMs, 62 I/O, 2MHz non-mux bus
68HC11G7	–	24K	–	512		10-bit A/D, 4 PWMs, 62 I/O, 2MHz non-mux bus
68HC711G7	16K	–	–	512		10-bit A/D, 4 PWMs, 62 I/O, 2MHz non-mux bus
68HC11J6	–	16IK	–	512		54 I/O, 2 MHz nonmux bus
68HC711J6	16K	–	–	512		54 I/O, 2 MHz nonmux bus
68HC11K0	–	–	–	768		8-bit A/D, 4 PWMs, 62 I/O, 4 MHz nonmux bus
68HC11K1	–	–	640	768		8-bit A/D, 4 PWMs, 62 I/O, 4MHz nonmux bus
68HC11K3	–	24K	–	768		8-bit A/D, 4 PWMs, 62 I/O, 4MHz nonmux bus
68HC11K4	–	24K	640	768		8-bit A/D, 4 PWMs, 62 I/O, 4 MHz nonmux bus
68HC711K4	24K	–	640	768		8-bit A/D, 4 PWMs, 62 I/O, 4MHz nonmux bus
68HC11KA0	–	–	–	768		4 PWMs, 8-Ch 8-bit A/D, SPI, SCI
68HC11KA1	–	–	640	768		4 PWMs, 8-Ch 8-bit A/D, SPI, SCI
68HC11KA2	–	32K	640	1K		4 PWMs, 8-Ch 8-bit A/D, SPI, SCI
68HC711KA2	32K	–	640	1K		4 PWMs, 8-Ch 8-bit A/D, SPI, SCI
68HC11KA3	–	24K	–	768		4 PWMs, 8-Ch 8-bit A/D, SPI, SCI
68HC11KA4	–	24K	640	768		4 PWMs, 8-Ch 8-bit A/D, SPI, SCI, 51 I/O
68HC11L0	–	–	–	512		8-bit A/D, 46 I/O, 3 MHz mux bus
68HC11L1	–	–	512	512		8-bit A/D, 46 I/O, 3 MHz mux bus
68HC11L5	–	16K	–	512		8-bit A/D, 46 I/O, 3 MHz mux bus
68HC11L6	–	16K	512	512		8-bit A/D, 46 I/O, 3 MHz mux bus
68HC711L6	16K	–	612	512		8-bit A/D, 46 I/O, 3 MHz mux bus
68HC11M2	–	32K	–	1.25M		8-bit A/D, 16-bit math coprocessor, 62 I/O, nonmux bus
68HC711M2	32K	–	–	1.25M		8-bit A/D, 16-bit math coprocessor, 62 I/O, nonmux bus
68HC711N4	–	24K	640	768		8-bit A/D, 16-bit math coprocessor, 62 I/O, nonmux bus
68HC11N4	24K	–	640	768		8-bit A/D, 8-bit D/A, 16-bit math coprocessor, 62 I/O, 4MHz nonmux bus
	–	32K				
68HC11P2	32K	–	640	1K		8-bit A/D, 62 I/O, 4MHz nonmux bus, 2 SCIs
68HC711P2			640	1K		8-bit A/D, 62 I/O, 4MHz nonmux bus, 2 SCIs

Table 5.1 ■ MC68HC11 family members

5.4.2 Expanded Mode

In the expanded mode, the 68HC11 has the capability to access a 64KB address space. This total address space includes the same on-chip memory addresses used in the single-chip mode plus external peripheral and memory devices. The expansion bus consists of ports B and C and control signals AS and R/$\overline{\text{W}}$. Port B functions as the upper eight address pins (A15~A8), and port C functions as the multiplexed data and low address pins (A7/D7,...,A0/D0). The AS signal is used by external memory system to latch the low address signals. All bus cycles, whether internal or external, execute at the E-clock frequency.

5.4.3 Special Bootstrap Mode

The 68HC11A8 has a 192-byte bootstrap ROM where a bootloader program is located. This ROM is enabled only if the 68HC11 is reset in the special bootstrap mode; it appears as internal memory space at locations $BF40~$BFFF. The bootloader program uses the serial communication interface (SCI) to read a 256-byte program into the on-chip RAM at locations $0000~$00FF. After the character for address $00FF is received, control is automatically passed to that program at location $0000. There are almost no limitations on the functions of the programs that can be loaded and executed through the bootstrap process.

The bootloaded program is often used to perform on-chip EEPROM or EPROM programming. The executable code or data to be programmed into the EEPROM or EPROM is often stored in a PC as a file. During the bootloading and programming process, the PC executes a terminal program to communicate with the bootloader and the bootloaded program to send the executable code. The user should create or get a copy of the PC-resident program in order to perform the bootload process or EEPROM and EPROM programming. Readers who are interested in using this technique to program the 68HC11 on-chip EEPROM or EPROM should refer to the Motorola application notes AN1010 and AN1060.

5.4.4 Special Test Mode

The special test mode, a variation of the expanded multiplexed mode, is primarily used during Motorola's internal production testing.

5.4.5 Setting the Operation Mode

The 68HC11 is in the *reset state* when the voltage level on the RESET pin is *low*. The operation mode is set when the 68HC11 exits the reset state. On the rising edge of the $\overline{\text{RESET}}$ signal, the voltage levels of pins MODA and MODB are latched into the HPRIO register, which sets the operation mode of the 68HC11. The voltage levels of pins MODA and MODB and the corresponding operation modes are shown in Table 5.2.

Inputs		
MODB	**MODA**	**Mode description**
1	0	normal single-chip
1	1	normal expanded
0	0	special bootstrap
0	1	special test

1 = high voltage, 0 = low voltage

Table 5.2 ■ Hardware mode-select summary

5.5 Memory Technology

Memory chips are a major component of many digital systems. Thus it is very important for a digital system designer to understand the characteristics of memory chips and know how to interface them to a microprocessor and microcontroller.

5.5.1 Memory Component Types

Semiconductor memories retain the binary information stored in their memory cells for varying amounts of time, depending on the type of memory they have. The different types of memory devices are therefore categorized according to their storage characteristics as volatile, nonvolatile, dynamic, or static.

VOLATILE AND NON-VOLATILE MEMORIES

Volatile memories lose their stored information when power is removed. The internal transistor circuits that store the information in memories of this type require a constant supply voltage to operate. Nonvolatile memory retains stored information even when power to the memory is removed.

ROMs AND RAMs

There are two major types of memories: read only memory (ROM) and random access memory (RAM). Random access memories are also called read-write memories because read and write accesses take roughly an equal amount of time. Within these two major types, additional classifications reflect the operation of their memory devices. ROMs are nonvolatile, and RAMs are normally volatile. Readers can refer to chapter 1 for a discussion of ROMs.

DYNAMIC AND STATIC MEMORIES

Semiconductor random access memories are further classified as dynamic or static.

Dynamic memories are memory devices that require periodic refreshing of the stored information. *Refresh* is the process of restoring binary data stored in a particular memory location. The internal circuitry of dynamic memories uses a very simple capacitive charge storage circuit in which only one transistor and one capacitor are required to store one bit of information. The simplicity of the circuit allows a very large number of memory bits to be fabricated onto the semiconductor chip, but the penalty for the simple circuitry is that the capacitive charge "leaks." The time interval over which each memory location of a DRAM chip must be refreshed at least once in order to maintain its contents is called its *refresh period*. Refresh period typically ranges from a few milliseconds to tens of milliseconds for present-day high-density DRAMs.

Static memories are designed to store binary information without needing periodic refreshes and require the use of more complicated storage circuitry for each bit. Four to six transistors are needed to store one bit of information. Thus static memories store less information per unit area of semiconductor material than do dynamic memories.

5.5.2 Memory Capacity and Organization

Memory devices are often labeled with a shorthand notation, such as 1M × 1 or 128K × 8, which indicates their storage capacity and organization. It is necessary to understand this notation in order to select the appropriate size and type of memory chip.

MEMORY CAPACITY

The capacity of a memory device is the total amount of information that the device can store. The capacity of semiconductor memories is often referred to as their *memory density*. Because they are binary in nature, semiconductor memories are always manufactured in powers of 2. For example, a 64-Kbit memory chip has 2^{16} memory cells of storage, for a total of 65,536 bits. A 1-Mbit memory chip has 2^{20}, or 1,048,576, total memory bits of storage. Table 5.3 lists several common memory chip capacities.

Memory chip	Power of 2	Capacity (bits)
1 Kbit	2^{10}	1,024
4 Kbit	2^{12}	4,096
16 Kbit	2^{14}	16,384
64 Kbit	2^{16}	65,536
256 Kbit	2^{18}	262,144
1 Mbit	2^{20}	1,048,576
4 Mbit	2^{22}	4,194,304
16 Mbit	2^{24}	16,777,216
64 Mbit	2^{26}	67,108,864
256 Mbit	2^{28}	268,435,456

Table 5.3 ■ Semiconductor memory chip capacities

MEMORY ORGANIZATION

Information on a single memory chip can be retrieved (read) or stored (written) one bit at a time or several bits at a time. The term *memory organization* describes the number of bits that can be written or read onto or from a memory chip during one input or output operation. For example, if only one bit can be stored or retrieved at a time, the organization is "by one," written × 1. If eight bits can be read or written at a time, the organization is "by eight," written × 8. Another common organization is by 4, which is referred to as a *nibble-organized memory*. A memory chip that is labeled $m \times n$ has m locations, and each location has n bits. For example, a 128K × 8 DRAM chip has 128K different locations, and each location has eight bits.

Example 5.1

▼

Using the following memory chips, how many SRAM chips will be needed to build a 512KB, 16-bit memory system for a 16-bit microprocessor? The memory is to be designed so that 16-bit data can be accessed in one read/write operation.

(a) 256K x 1 SRAM
(b) 256K x 4 SRAM
(c) 256K x 8 SRAM
(d) 64K x 8 SRAM

Solution:

a. If we use × 1 organization memory chips, sixteen chips will be needed to build a 16-bit-wide memory system. Sixteen 256K × 1 SRAM chips have a capacity of 512 KB.

b. With × 4 organization memory chips, four chips will be needed to build a 16-bit-wide memory system. Four 256K × 4 SRAM chips have a capacity of 512 KB.

c. With × 8 organization memory chips, two chips will be required to build a 16-bit memory system. Two 256K × 8 SRAM chips have a capacity of 512 KB.

d. Two 64K × 8 SRAM chips have a capacity of 128KB, so eight chips will be needed to build a 512KB, 16-bit memory system.

▲

5.5.3 Memory Addressing

Every type of memory chip has external pins or connections that are called *addresses. Addressing* involves the application of a unique combination of high and low logic levels to select a correspondingly unique memory location.

Small-capacity memories can have a unique external connection or pin for each address line. However, as memory capacities grow, it becomes impractical to provide an external pin for each address line. For example, a 1M × 1 bit memory device requires twenty address lines to allow for unique selection of each memory bit. Because external connections must be kept to a minimum so that the package used for the memory chip can be as small as possible, a technique known as *multiplexing* was invented. Multiplexing keeps the external pin count to a minimum. It is a circuit technique that allows the same external connection to be used for two or more different signals that are differentiated according to the time at which the external connection is examined. The most common way to multiplex memory chip pins is to use half the number of address lines that are necessary for unique address selection. Address signals are then divided into *row* and *column* address signals. The memory chip looks for and latches the row address during the first part of the address cycle, then it looks for and latches the column address during the second part of the cycle. Two signal pins are associated with the multiplexed address operation. The *row address strobe*, RAS, is the signal that indicates that row address logic levels are being applied, while the *column address strobe*, CAS, is the signal that indicates that column address logic levels are being applied. Address multiplexing techniques have been used to reduce the pin count and package size in dynamic memory chips for a long time because DRAMs are used in such large quantities in so many digital systems. Other types of memory chips do not use address multiplexing.

5.6   External Memory Expansion for the 68HC11

In the expanded mode, the user can add external memory to the 68HC11. The port B pins are used as the upper address signals (A15-A8), while the port C pins become the *time-multiplexed* lower address and data signals (A7/D7-A0/D0).

Several issues need to be dealt with when adding external memory chips to the 68HC11:

- address space assignment
- address decoding
- timing considerations

5.6.1 Memory Space Assignments

The 68HC11A8 supports 64KB of memory space. Part of this 64KB space is occupied by the on-chip memories and I/O registers listed in Table 5.4.

Address range	Allocation
$0000-$00FF	on-chip SRAM
$1000-$103F	I/O registers
$B600-$B7FF	on-chip EEPROM
$E000-EFFFF	on-chip ROM

Table 5.4 ■ 68HC11A8 on-chip
memory map

The on-chip SRAM and I/O registers can be repositioned to any 4KB page boundary in the 64KB memory map by programming the INIT register. The INIT register is located at $103D, and its contents are:

7	6	5	4	3	2	1	0	
								INIT
RAM3	RAM2	RAM1	RAM0	REG3	REG2	REG1	REG0	located at
0	0	0	0	0	0	0	1	$103D

Value after reset

- RAM3-RAM0: RAM map position
  These four bits, which specify the upper hex digit of the SRAM address, control the position of the SRAM in the memory map. By changing these bits, the SRAM can be repositioned to the beginning of any 4KB page in the memory map. After reset, these four bits are zeros ($0); thus the SRAM is initially positioned at $0000-$00FF. If these four bits are written to ones ($F), the SRAM moves to $F000-$F0FF.

- REG3-REG0: 64-byte register block position
  These four bits, which specify the upper hex digit of the address for the 64-byte block of internal registers, control the position of these registers in the memory map. These bits can be changed to reposition the register block to the beginning of any 4KB page in the memory map. After reset, these four bits are 0001 ($1); therefore, the registers are initially positioned at $1000-$103F. If these four bits are written to ones ($F), the registers move to $F000-$F03F.

The register INIT can be programmed only within the 64 E clock cycles after reset, so such programming is often done in the *reset handling routine.* When relocating the I/O registers, we must make sure that two different devices are not mapped to the same location. However, no physical harm will be caused if two devices are mapped to the same location. If the on-chip ROM and RAM are mapped to the same area, RAM will take priority and ROM will become inaccessible. If the I/O registers are relocated so that they conflict with the RAM and/or ROM, then the I/O registers will take priority and the RAM and/or ROM at those locations become inaccessible. Similarly, if an internal resource conflicts with an external device, no harmful conflict results—data from the external device will not be applied to the internal data bus and cannot interfere with internal read operations.

Example 5.2

▼

Remap the on-chip SRAM to $2000-$20FF and remap the I/O registers to $3000-$303F.

Solution: To remap the SRAM and I/O registers to the specified memory spaces, a value of $23 must be programmed into the INIT register. The following instruction sequence must be placed in the reset handling routine and be executed within 64 E clock cycles after the RESET signal goes to high:

```
SRAM     EQU    $20                 ; value to remap SRAM to $2000-$20FF
IOREG    EQU    $3                  ; value to remap I/O registers to $3000-$303F
REMAP    EQU    SRAM+IOREG
INIT     EQU    $1030
         LDAB   #REMAP
         STAB   INIT
```

The unoccupied memory space can be assigned to external memories. It is most convenient to allocate memory space in blocks of equal size, with each block comprising contiguous memory locations. The major consideration in the memory space allocation is the number of memory locations in the memory chips. The number of memory locations in the memory chip is a power of 2, for example, 2K, 4K, 8K, 16K, and so on. The address decoder can be simplified if all the memory chips have the same capacity.

▲

Example 5.3

▼

Assign the 68HC11 memory space using a block size of 4KB.

Solution: The 64KB memory space can be divided into sixteen 4KB blocks. The address ranges of the blocks are as follows:

Block number	Address range
0	$0000-$0FFF
1	$1000-$1FFF
2	$2000-$2FFF
3	$3000-$3FFF
4	$4000-$4FFF
5	$5000-$5FFF
6	$6000-$6FFF
7	$7000-$7FFF
8	$8000-$8FFF
9	$9000-$9FFF
10	$A000-$AFFF
11	$B000-$BFFF
12	$C000-$CFFF
13	$D000-$DFFF
14	$E000-$EFFF
15	$F000-$FFFF

Example 5.4

Assign the 68HC11 memory space using a block size of 8KB.

Solution:　The 64KB memory space can be divided into eight 8KB blocks. The address ranges of the blocks are:

Block number	Address range
0	$0000-$1FFF
1	$2000-$3FFF
2	$4000-$5FFF
3	$6000-$7FFF
4	$8000-$9FFF
5	$A000-$BFFF
6	$C000-$DFFF
7	$E000-$FFFF

5.6.2   Address Decoding Methods

To transfer data correctly and protect the computer from damage, only one device at a time can be allowed to drive the data bus. The address decoder selects and enables one and only one data transfer device at a time. Eight-bit microprocessors and microcontrollers can use two address-decoding schemes: *full* and *partial address decoding.* A memory component is said to be fully decoded when each of its addressable locations responds to only a single address on the system bus. A memory component is said to be partially decoded when each of its addressable locations responds to more than one address on the system bus.

The 74138 3-to-8 decoder chip and the 74139 2-to-4 decoder chip have been popular in address decoder design. The 74139 is a dual 2-to-4 decoder chip with two identical 2-to-4 decoders. The pin layouts of these two chips are shown in Figure 5.1. In these chips,

- O0-O7, 1Y0-1Y3, and 2Y0-2Y3 are active low decoder outputs.
- E1, E2, and E3 are decoder enable inputs. E1 and E2 are active low, and E3 is active high.
- A2-A0, A1-A0, and B1-B0 are address or select inputs. They are all active high.

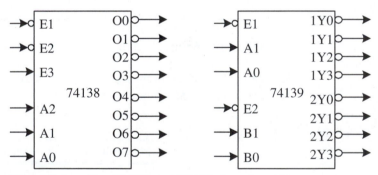

Figure 5.1 ■ The 74138 and 74139 decoder chips

Example 5.5

Use a full decoding scheme to design an address decoder for a computer that has the following address space assignments:

SRAM1: $2000-$3FFF
ROM1: $4000-$5FFF
EEPROM: $6000-$7FFF

SRAM2: $A000-$BFFF
ROM2: $C000-$DFFF

Solution: Since each block of this memory module is 8KB, it will be very straightforward to use the full address decoding scheme to implement the address decoder. A 3-to-8 decoder such as the 74LS138 is perfect for this use (LS stands for low power Schottky technology). The three highest address bits of each module (listed below) are used to select the memory modules:

SRAM1: 001
ROM1: 010
EEPROM: 011
SRAM2: 101
ROM2: 110

The address decoder for this computer system is shown in Figure 5.2.

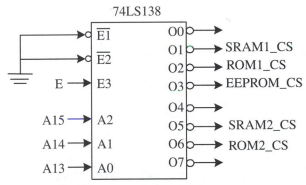

Figure 5.2 ■ Address decoder design for Example 5.5

The reader should also note that the E clock signal is used to control the E3 input. The lower eight address signals are invalid most of the time when the E clock is low and are valid when the E clock is high. Using the E clock as an enable input to the decoder makes sure that no decoder output is asserted when the address signals are not valid. We can also tie the E3 signal to high. There are potential problems in this approach, however, because even if the microcontroller is not driving the address bus, the address inputs to the decoder may have valid values that will assert one of the memory or peripheral chips. If this approach is taken, then memory and peripheral chips must decode the E clock signal directly to make sure they don't respond when the microcontroller is not performing an external bus cycle.

Example 5.6

You have been assigned to design a 68HC11-based product. This product will have 2KB of external EEPROM and 2KB of external SRAM. Use the partial decoding scheme to design a decoder for this product.

Solution: As the designer of this system, you have a lot of freedom to make the design decision. The inexpensive, off-the-shelf dual 2-to-4 74LS139 decoder can be used. The 2KB external EEPROM and SRAM chips have eleven address inputs. The lowest eleven address bits of the 68HC11 are connected directly to the corresponding pins of these two memory components. The highest five address signals are available for controlling the address decoder. One approach is to use the highest two address signals, A15 and A14, as the address inputs to the decoder. The resulting decoding circuitry is shown in Figure 5.3.

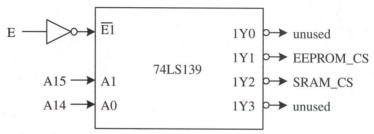

Figure 5.3 ■ Address decoder for Example 5.6

The inverter makes sure that the decoder does not enable any component when the E clock signal is low. The address space allocation is as follows:

EEPROM: $4000-$7FFF
SRAM: $8000-$BFFF

Each component is allocated 16KB of address space. Because address signals A13-A11 are not used in address decoding, each location in the EEPROM or SRAM will respond to eight different addresses. This is an example of *partial decoding*. For example, the following addresses all select the same location in the EEPROM:

$4000, $4800, $5000, $5800, $6000, $6800, $7000, $7800

Partial decoding has the advantage of smaller and simpler decoder circuitry. If partial decoding is not used, this problem would require a 5-to-32 decoder, which is a large package that is not commercially available. However, partial decoding has a significant disadvantage: it prevents full use of the microprocessor's available memory space and produces difficulties if the memory system is expanded at a later time.

When dealing with I/O devices, we need to address the registers inside the I/O device to perform I/O configuration and data transfer. Most I/O devices have very few registers. Allocating memory space in blocks of equal size is not a good idea when there are many I/O devices to be interfaced to the computer system. In this situation, we can use *multilevel decoding* method to perform address decoding for I/O devices. The following example illustrates this method.

Example 5.7

▼

Assume you are designing an 8-bit microcontroller-based system (with 64KB memory space) with two external 8KB SRAM chips, two 8KB EPROM chips, one 8KB flash memory chip, and eight I/O devices. Make the space assignment and design the address decoder. The address 0000H must reside in one of the EPROM chips.

Solution: Since there are thirteen devices to be decoded, a single 3-to-8 decoder is not adequate. We will use a two-level space assignment and use two 74138 decoders to do the address decoding. The address space assignment is shown in Table 5.5, and the decoder circuit is shown in Figure 5.4.

Block number	Address range	Assigned to
0	0000H ~ 1FFFH	not assigned
1	2000H ~ 3FFFH	EPROM2
2	4000H ~ 5FFFH	I/Os
3	6000H ~ 7FFFH	flash memory
4	8000H ~ 9FFFH	SRAM1
5	A000H ~ BFFFH	SRAM2
6	C000H ~ DFFFH	not assigned
7	E000H ~ FFFFH	EPROM1

(a) Level one memory space assignment

Block number	Address range	Assigned to
0	4000H ~ 43FFH	I/O 1
1	4400H ~ 47FFH	I/O 2
2	4800H ~ 4BFFH	I/O 3
3	4C00H ~ 4FFFH	I/O 4
4	5000H ~ 53FFH	I/O 5
5	5400H ~ 57FFH	I/O 6
6	5800H ~ 5BFFH	I/O 7
7	5C00H ~ 5FFFH	I/O 8

(b) Level two memory space assignment

Table 5.5 ■ Two-level memory space assignment for Example 5.7

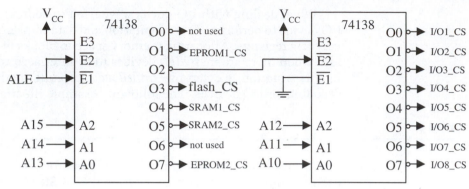

Figure 5.4 ■ An example of two-level decoding

Apply the address signals A15-A13 to the address inputs of the first-level decoder and apply address signals A12-A10 to the address inputs of the second-level decoder. Connect the O2 output to the $\overline{E2}$ input of the second 74138 decoder. The second decoder can operate only when the O2 output of the first 74138 is asserted (low). Most memory devices have chip-enable ($\overline{CE}$) or chip-select ($\overline{CS}$) inputs to control their operation. The decoder outputs will be connected to these inputs to enable/disable the memory chips.

Address decoders can also be implemented using programmable logic devices such as programmable array logic (PAL) devices. The CMD11A8 uses a PAL20V10 to implement its address decoder. Using the PAL device has the following advantages:

1. Flexible address space assignment. A PAL device uses two-level AND-OR logic to implement logic functions, which enables address spaces of different sizes to be assigned to I/O devices and memory components using a single PAL chip.

2. Smaller decoder circuitry can be achieved. When there are many I/O devices and memory components in a system, we will need several off-the-shelf address decoders to achieve the desired address decoding function.

5.7   Bus Cycles

In the normal expanded mode, the 68HC11 can access external memories or I/O devices. It performs a read bus cycle to retrieve data from an external memory location or I/O device, and it performs a write bus cycle to output data. The timing requirements of both the 68HC11 bus cycles and the external devices must be met in order to make the data transfer successful.

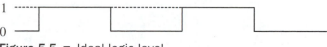

Figure 5.5 ■ Ideal logic level

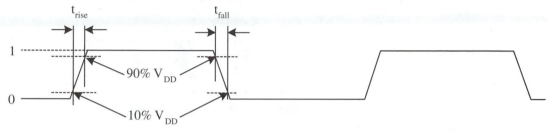

Figure 5.6 ■ Real signal

5.7.1 Conventions of Timing Diagrams

A digital signal has two states, high and low, as shown in Figure 5.5. In Figure 5.5, the transitions from 1 to 0 and from 0 to 1 are instantaneous. In reality, however, a signal transition takes time. The time needed for a signal to go from 10 percent of the power supply voltage (V_{DD}) to 90 percent of the power supply voltage is called the *rise time*, and the time needed for a signal to drop from 90 percent of the power supply voltage to 10 percent of the power supply voltage is called *fall time* (see Figure 5.6).

A single signal is often represented as a set of line segments (for example, see Figure 5.7). The horizontal axis is time and the vertical axis is the magnitude (in volts) of the signal. Multiple signals of the same nature, such as address and data signals, are often grouped together and represented as two parallel lines with crossovers, as illustrated in Figure 5.8. The crossover represents the point when one or multiple signals change values.

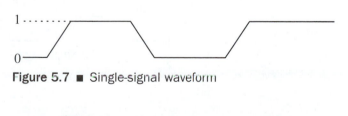

Figure 5.7 ■ Single-signal waveform

Figure 5.8 ■ Multiple-signal waveform

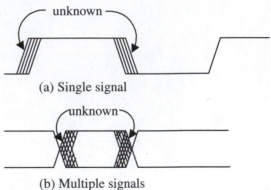

(a) Single signal

(b) Multiple signals

Figure 5.9 ■ Unknown signals

Sometimes a signal value is unknown because the signal is changing. Single and multiple unknown signals are represented by hatched areas in the timing diagram, as shown in Figure 5.9.

Sometimes one or multiple signal lines are not driven (because their drivers are in a high impedance state) and hence cannot be received. An undriven signal is said to be *floating*. Single or multiple floating signals are represented by a value between high and low, as shown in Figure 5.10.

Sometimes there are causal relationships between two or more signals. The causal relationship is often represented by an arrow, as shown in Figure 5.11.

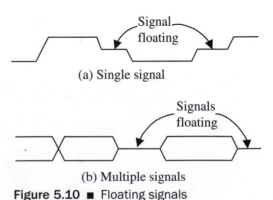

(a) Single signal

(b) Multiple signals

Figure 5.10 ■ Floating signals

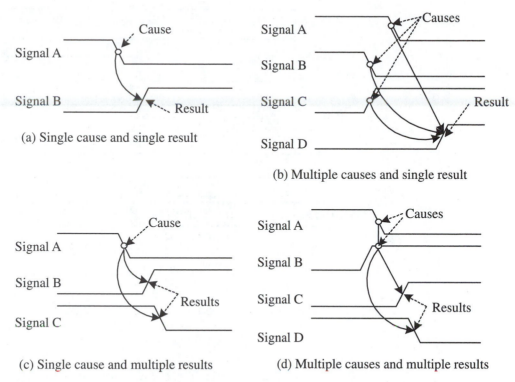

(a) Single cause and single result

(b) Multiple causes and single result

(c) Single cause and multiple results

(d) Multiple causes and multiple results

Figure 5.11 ■ Causal relationships between signals

5.7.2   The 68HC11 Bus Cycle Timing Diagram

The 68HC11 external bus cycle is controlled by the E clock signal and can be performed only in the expanded mode. There are sixteen address and eight data signals. Each read or write bus cycle takes one E clock cycle. A read bus cycle timing diagram is shown in Figure 5.12, and the corresponding timing parameters are shown in Table 5.6. The lower eight address signals and the data signals are time-multiplexed. During the first half of the E clock cycle (E clock is low), the lower eight address signals appear on port C pins. During the second half of the E clock cycle, data is driven on the same port C pins by external devices.

Most memory chips require address signals to be stable during the complete read/write bus cycle. However, the 68HC11 drives the lower eight address signals only during the first half of the E clock cycle. The external memory must latch the lower eight address signals so that they stay stable

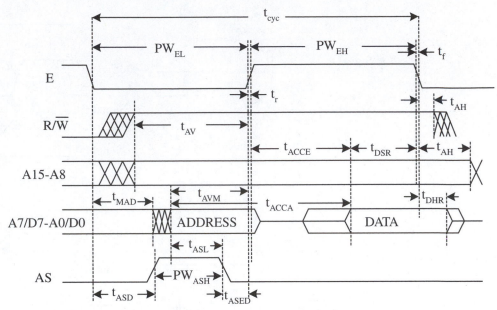

Figure 5.12 ■ MC68HC11 read bus cycle timing diagram

during the whole bus cycle. The 68HC11 provides the address strobe (AS) signal for external devices to latch address signals. The AS pulse is activated during the first half of the E clock cycle. Address signals should be latched by the falling (not rising) edge of the AS signal because they are not valid on the rising edge, as shown in the timing diagram. The low address signals become valid t_{ASL} ns before the falling edge of the AS signal and remain valid for t_{AHL} ns after the falling edge of the AS signal.

For convenience, Motorola measures all timing parameters using 20 percent and 70 percent of the power supply voltage (V_{DD}) as reference points. Many timing parameters are measured relative to the rising and falling edges of the E clock. The phrase "before the falling edge of the E clock" uses the time when the magnitude of the E clock is 0.7 V_{DD} as a reference point. The phrase "after the falling edge of the E clock" uses the time when the E clock is 0.2 V_{DD} as a reference point. Similarly, "before the rising edge of the E clock" refers to the time when the magnitude of the E clock is 0.2 V_{DD}, and "after the rising edge of the E clock" refers to the time when the magnitude of the E clock is 0.7 V_{DD}. For example, the E clock rise time t_r is the amount of time it takes the E clock to rise from 0.2 V_{DD} to 0.7 V_{DD}. The E clock fall time t_f is the amount of time that it takes the E clock to drop from 0.7 V_{DD} to 0.2 V_{DD}. The timing parameter t_{AV}, *the non-multiplexed address to E rise time*, is the delay from the time when the nonmultiplexed address signals A15-A8 become valid until E clock rises to

0.2 V_{DD}. In order to latch data correctly, the 68HC11 requires the external devices to drive valid data t_{DSR} ns before the E clock drops to 0.7 V_{DD} and keeps the data valid for t_{DHR} ns after E clock has dropped to 0.2 V_{DD}. The timing parameter t_{DSR} is called the *read data setup time* and the timing parameter t_{DHR} is called the *read data hold time*. Other timing parameters are defined and measured in a similar way.

The timing diagram of a write bus cycle is shown in Figure 5.13. This timing diagram is very similar to the read bus cycle timing diagram except that the R/$\overline{W}$ signal goes low and data is driven by the 68HC11. The write data is available t_{DDW} ns after the rising edge of the E clock and will stay valid for at least t_{DHW} ns after the falling edge of the E clock.

5.7.3 Adding External Memory to the 68HC11

When adding external memory to the 68HC11, the user has at least six choices of memory technology: DRAM, SRAM, ROM, EPROM, flash memory,

Parameter	Symbol	1.0 MHz		2.0 MHz		3.0 MHz	
		Min	Max	Min	Max	Min	Max
Cycle time	t_{cyc}	1000	—	500	—	333	—
Pulse width, E low	PW_{EL}	477	—	227	—	146	—
Pulse width, E high	PW_{EH}	472	—	222	—	141	—
E and AS rise and fall time	t_r	—	20	—	20	—	20
	t_f	—	20	—	20	—	15
Address hold time	t_{AH}	95.5	—	33	—	26	—
Non-muxed address valid time to E rise	t_{AV}	281.5	—	94	—	54	—
Read data setup time	t_{DSR}	30	—	30	—	30	—
Read data hold time	t_{DHR}	0	145.5	0	83	0	51
Write data delay time	t_{DDW}	—	190.5	—	128	—	71
Write data hold time	t_{DHW}	95.5	—	33	—	26	—
Muxed address valid time to E rise	t_{AVM}	271.5	—	84	—	54	—
Muxed address valid to AS fall	t_{ASL}	151	—	26	—	13	—
Muxed address hold time	t_{AHL}	95.5	—	33	—	31	—
Delay time, E to AS rise	t_{ASD}	115.5	—	53	—	31	—
Pulse width, AS high	PW_{ASH}	221	—	96	—	63	—
Delay time, AS to E rise	t_{ASED}	115.5	—	53	—	31	—
MPU address access time	t_{ACCA}	744.5	—	307	—	196	—
MPU access time	t_{ACCE}	—	442	—	192	—	111
Muxed address delay	t_{MAD}	145.5	—	83	—	51	—

All times are in nanoseconds.
MPU stands for microprocessor unit.

Table 5.6 ■ MC68HC11 A series read/write timing parameters

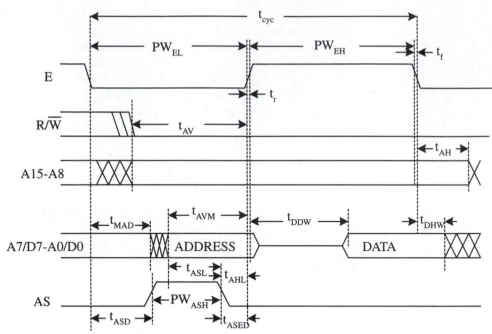

Figure 5.13 ■ MC68HC11 write bus cycle timing diagram

and EEPROM. Because the 68HC11 has a relatively small address space, most external needs can be satisfied by one or two memory chips. DRAMs have the lowest per bit cost. However, the following two factors make it the most expensive memory technology for a digital system that needs only a very small amount of external memory:

1. *Periodic refresh requirement.* DRAM chips use capacitors to store information, but without periodic refresh the information stored in the capacitors will leak away. A timer is needed to generate a periodic refresh request. However, the MPU may request access to the DRAM when the DRAM is requesting the refresh operation—then the memory system must delay the refresh operation because the 68HC11 bus cycle takes one clock cycle and cannot be delayed.

2. *Address multiplexing.* DRAM chips use address multiplexing techniques to reduce the number of address pins. This requires the memory system to include an address multiplex or, further increasing the cost of the memory system.

The control circuit designs for interfacing SRAM, ROM, EPROM, flash memory, and EEPROM to a microprocessor or microcontroller are similar. In the following discussions we will use an 8KB SRAM HM6264A to illustrate the control logic design of an SRAM memory system.

THE HM6264A SRAM

The pin layout of the Hitachi 8KB HM6264A is shown in Figure 5.14. The HM6264A has thirteen address pins to address each of the 8192 locations on the chip, and it uses × 8 configuration, i.e., each location has eight bits. There are two chip enable signals: $\overline{CS1}$ is active low, and CS2 is active high. There are also active low write enable ($\overline{WE}$) and active low output enable ($\overline{OE}$) signals that control the data in and out of the chip. The output enable signal $\overline{OE}$ can be tied to low permanently.

The read cycle timing diagram of the HM6264A is shown in Figure 5.15. The HM6264A comes in three versions, which can be classified according to access time: the first version (HM6264A-10) has a 100-ns access time, the second (HM6264A-12) has a 120-ns access time, and the third (HM6264A-15) has a 150-ns access time. The read cycle timing parameters for the HM6264A-10 and HM6264A-12 are shown in Table 5.7.

Read cycle time is the shortest separation that is allowed between two consecutive memory read accesses. A read access takes at least as long as the

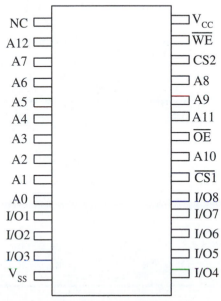

Figure 5.14 ■ Hitachi HM6264A pin assignment

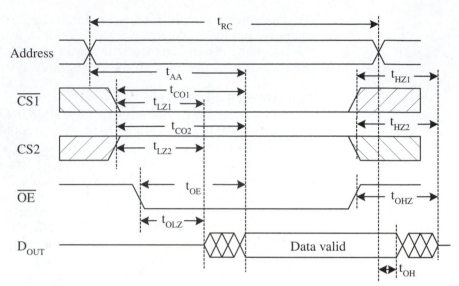

Figure 5.15 ■ HM6264A read cycle timing diagram

read cycle time (t_{RC}). For SRAMs, the read cycle time is as long as the *read access time.* There are four access times in the HM6264A:

> *Address access time* (t_{AA}): the delay time from the moment that the address becomes valid until valid data become available at the data pins (I/O8-I/O1), if all other control signals ($\overline{CS1}$, CS2, and $\overline{OE}$) are asserted.

		HM6264A-10		HM6264A-12	
Parameter	Symbol	Min	Max	Min	Max
Read cycle time	t_{RC}	100	—	120	—
Address access time	t_{AA}	—	100	—	120
CS1 to output valid	t_{CO1}	—	100	—	120
CS2 to output valid	t_{CO2}	—	100	—	120
Output enable to output valid	t_{OE}	—	50	—	60
Output hold from address change	t_{OH}	10	—	10	—
Chip selection to output in low-Z	t_{LZ1}	10	—	10	—
(CS1 and CS2)	t_{LZ2}	10	—	10	—
Output enable to output in low Z	t_{OLZ}	5	—	5	—
Chip selection to output in high Z	t_{HZ1}	0	35	0	40
(CS1 and CS2)	t_{HZ2}	0	35	0	40
Output disable to output high-Z	t_{OHZ}	0	35	0	40

All times are in nanoseconds.

Table 5.7 ■ HM6264A read cycle timing parameters

$\overline{CS1}$ *access time* (t_{co1}): the delay time from the moment that $\overline{CS1}$ becomes valid until valid data become available at the data pins, if all other control signals are asserted and a valid address is applied at the address pins.

CS2 access time (t_{co2}): the delay time from the moment that CS2 becomes valid until valid data become available at the data pins, if all other control signals are asserted and a valid address is applied at the address pins.

$\overline{OE}$ *access time* (t_{oE}): the delay time from the moment that $\overline{OE}$ becomes valid until valid data appear at the data pins, if all other control signals are asserted and a valid address is applied at the address pins.

The address data hold time (t_{oH}) is the length of time that the data from the I/O8-I/O1 pins stay valid after the address changes. There are two *chip deselection to output in high Z delay times* $(t_{HZ1}$ and t_{HZ2}, corresponding to $\overline{CS1}$ and CS2, respectively). Another timing parameter of interest to the memory designer is *output disable to output in high Z* (t_{oHZ}).

There are two write cycle timing diagrams for the HM6264A: in one the $\overline{OE}$ signal is permanently tied to low, while in the other the $\overline{OE}$ signal is pulled to high during the write process. Since the next section contains an example in which the output enable signal is grounded permanently, only the write cycle timing diagram with $\overline{OE}$ grounded will be discussed here. This diagram is shown in Figure 5.16, and the corresponding timing parameters are given in

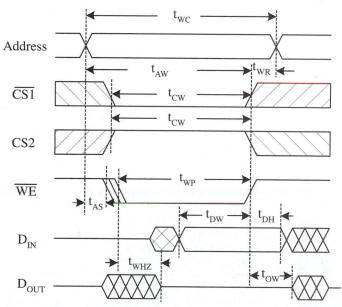

Figure 5.16 ■ HM6264A write cycle timing diagram ($\overline{OE}$ low)

Table 5.8. The HM6264A requires the write data to be valid at least t_{DW} ns before the write signal goes to high and to remain valid at least t_{DH} ns after the rising edge of the write signal. The write pulse width (t_{WP}) must be long enough for data to be written into the memory chip correctly.

INTERFACING THE HM6264A-12 TO THE 68HC11

We will use a 2-MHz (in terms of E clock frequency) 68HC11 to illustrate the design process involved in interfacing an HM6264A-12 to a 68HC11A1. In the 68HC11A1, addresses $1100-$B5FF and $B800-$FFBF are not occupied and can be assigned to the SRAM. The HM6264A-12 has a 120 ns access time, and address spaces $4000-$5FFF are assigned to the HM6264A in this example. The 68HC11 must be configured to operate in the expanded mode. Because the lower address signals (A7-A0) and data signals (D7-D0) are time-multiplexed, a latch is needed to latch the lower address signals so that all address signals can remain stable during the whole access cycle. The 8-bit latch 74F373 is selected for this purpose. An address decoder is needed to enable one and only one external device (memory or interface chip) to respond to the access request from the 68HC11—we will select the 74F138 3-to-8 decoder. Because the address signals are not valid most of the time when the E clock signal is low, the E clock is used to qualify the write access. The circuit connection is shown in Figure 5.17. The output enable signal $\overline{OE}$ to the 74F373 is tied to the ground. The latch enable (LE) is connected to the AS signal from the 68HC11. A7-A0 are latched by the falling edge of the AS signal. The 74F373 has a latch delay time of 11.5 ns. The valid address will appear at pins O7-O0 11.5 ns after the falling edge of the LE (or AS) signal.

Parameter	Symbol	HM6264A-10		HM6264A-12	
		Min	Max	Min	Max
Write cycle time	t_{WC}	100	–	120	–
Chip selection to end of write	t_{CW}	80	–	85	–
Address setup time	t_{AS}	0	–	0	–
Address valid to end of write	t_{AW}	80	–	85	–
Write pulse width	t_{WP}	60	–	70	–
Data to write time overlap	t_{DW}	40	–	50	–
Data hold from write time	t_{DH}	0	–	0	–
Write to output in high-Z	t_{WHZ}	0	35	0	40
Output active from end of write	t_{OW}	5	–	5	–
Write recovery time	t_{WR}	0	–	0	–

All times are in nanoseconds.

Table 5.8 ■ HM6264A write cycle timing parameters

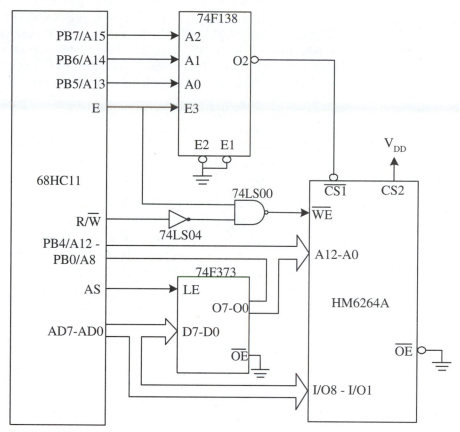

Figure 5.17 ■ Interfacing an 8KB HM6264A to the 68HC11

The upper three address signals A15-A13 are decoded by the 74F138 3-to-8 decoder, which has a propagation delay of 8 ns at room temperature. The lower address signals (A12-A0) are connected to the corresponding address inputs of the HM6264A chip.

This computer system may need to access other external devices, which can be selected using a decoder. However, it is very common for a user program to access a nonexistent memory location. The access of a nonexistent memory location is called a *bus error*. When a bus error occurs, a program cannot execute correctly. The design of the 68HC11 does not allow it to detect a bus error because each bus cycle takes exactly one E clock cycle, even if the bus cycle is accessing a nonexistent memory location. If we OR those decoder outputs that map to nonexistent memory locations and connect the result to either the $\overline{\text{IRQ}}$ or $\overline{\text{XIRQ}}$ input of the 68HC11, the MCU will be interrupted when a nonexistent location is accessed. The user can then identify the problem from the external interrupt signal. Interrupts will be discussed in chapter 6.

The R/$\overline{\text{W}}$ signal is inverted before being NANDed with the E clock signal. The output of the NAND gate becomes the $\overline{\text{WE}}$ input signal to the HM6264A. NANDing the E and W together allows write access to the SRAM only when the address is valid. The output enable input $\overline{\text{OE}}$ is grounded. The HM6264A will output data only when the $\overline{\text{WE}}$ signal is high. The maximum low-to-high and high-to-low propagation delays of both the 74LS00 and the 74LS04 are 15 ns. In addition, the O2 output may become valid only when the E clock is high because the E3 input to the 74F138 is connected to the E clock signal from the 68HC11.

Another way to connect the enable inputs of the 74F138 and the HM6264A is as follows:

1. Tie the $\overline{\text{E1}}$ and $\overline{\text{E2}}$ inputs to the 74F138 to low and pull E3 to high using a pull-up resistor.
2. Connect the O2 output of the 74F138 to the $\overline{\text{CS1}}$ input of the HM6264A, and connect the E clock of the 68HC11 to the CS2 input of the HM6264A.

This circuit connection can achieve the same goal, and the timing verification is similar to that of the previous method.

READ ACCESS TIMING ANALYSIS

To facilitate the read timing analysis, the read cycle timing diagram of the SRAM is overlapped with the 68HC11 read bus cycle timing diagram, as shown in Figure 5.18. During the read access to the HM6264A, control signals CS2 and $\overline{\text{OE}}$ are asserted permanently. The higher address inputs (A12-A8) to the HM6264A become valid 94 ns before the rising edge of the E clock signal. The lower eight address signals are latched by the falling edge of the AS signal, which is valid 53 ns before the rising edge of E. Since the 74F373 has a latch delay of 11.5 ns, A7-A0 will become valid to the HM6264A 41.5 ns before the rising edge of E. Because the address decoder is qualified by the E clock signal, the chip select $\overline{\text{CS1}}$ is asserted (goes to low) 8 ns after the rising edge of the E clock signal. Because it is the last asserted control signal in a read access, the $\overline{\text{CS1}}$ determines the data setup time to the 68HC11. The data from the HM6264A will become valid 128 ns (8 ns + 120 ns) after the rising edge of the E clock signal or 94 ns (PW_{EH} − 128 ns = 222 ns − 128 ns) before the falling edge of the E clock signal. The data setup time requirement is met because the 68HC11 requires only 30 ns.

The 68HC11 drives the high address signals (A15-A8) for 33 ns (t_{AH}) after the falling edge of the E clock signal. The lower address latch 74F373 does not latch a new value until 174 ns (227 − 53) after the falling edge of the E clock. Therefore, the higher address signals determine the actual value of t_{OH}. The $\overline{\text{CS1}}$ is the only control signal that becomes invalid, and it goes high 8 ns after the falling edge of the E clock signal. The data hold time is the smaller of the following two values:

1. The output hold from address change time (t_{OH}). The HM6264A holds data for 10 ns after the address input becomes invalid, which occurs 33 ns after the E clock goes high. Therefore, the data hold for 43 ns (10 ns + 33 ns) after the E clock goes high.

2. Chip selection ($\overline{CS1}$ in this example) to output in high impedance time (t_{HZ1}): 8 to 48 ns.

The HM6264A therefore provides only an 8 to 43 ns data hold time during a read access. The 68HC11 requires a data hold time from 0 to 83 ns (this range

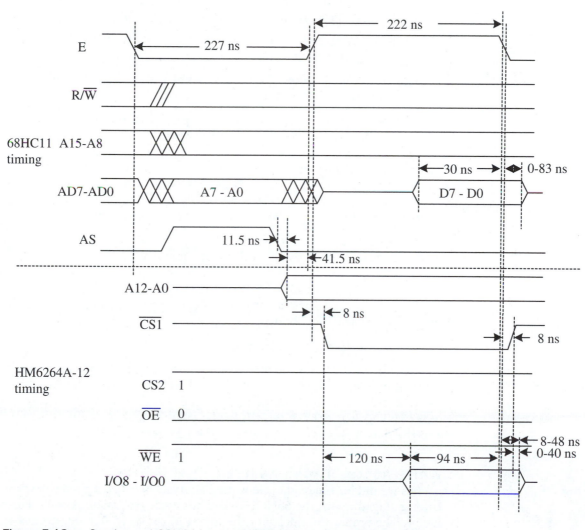

Figure 5.18 ■ Overlapped 68HC11 and HM6264A-12 read timing diagrams

is due to IC fabrication variations). The data hold time requirement is thus violated. However, the 68HC11 does not drive the multiplexed address/data bus until 138 ns after the falling edge of the E clock cycle, and the capacitance of the data bus will hold the data from memory for a short time and thus satisfy the data hold time requirement. The following analysis should give you an idea of the duration over which the data bus capacitance can hold the data at a valid level.

On a printed circuit board, each pin of the multiplexed address/data bus and the ground plane form a capacitor. After the memory chip stops driving the data bus, the charge across the capacitor leaks away via

1. input current into the 68HC11 data pin (on the order of 10μA)
2. input leakage into the memory chip (on the order of 2 μA)
3. other leakage paths on the printed circuit board, depending on the system configuration

The time that it takes the capacitor voltage to degrade from high to low can be estimated as follows. Let

ΔV = the voltage change required for data bus signal to change its value from 1 to 0. For example, for a 5 V power supply, ΔV is equal to 2.5 V.
Δt = the time that it takes the voltage across the capacitor to drop by ΔV.
I = the total leakage current
C = the capacitance of the data bus on the printed circuit board. C varies with the printed circuit board but is in the range of 20 pF per foot. Usually the length of the data bus of a 68HC11 circuit is shorter than 1 foot. However, we will use this value in our calculation.

Then the elapsed time before the data bus signal degrades to an invalid level can be estimated by the following equation:

$\Delta t \approx C\Delta V \div I$

By substituting values into the equation above (and ignoring the third source of leakage), we obtain

$\Delta t \approx C\Delta V \div I \approx 20 \text{ pF} \times 2.5\text{V} \div 12 \text{ μA} = 4 \text{ μs}$

Although the equation above is oversimplified, it does give us some idea about the order of the time over which the charge across the data bus capacitor will hold after the memory chips stop driving the data bus. Suppose the other leakage paths have ten times the current leakage (which is very likely) and the data bus length is 6 inches—then Δt becomes 200 ns.

WRITE ACCESS TIMING ANALYSIS

As we did for the read cycle timing analysis, we overlapped the write access timing diagram with the 68HC11 write bus cycle timing diagram in Figure 5.19. Due to the propagation delay of the inverter, the $R/\overline{W}$ signal becomes stable to the NAND gate input 79 (94 − 15) ns before the E clock rising edge. Because of

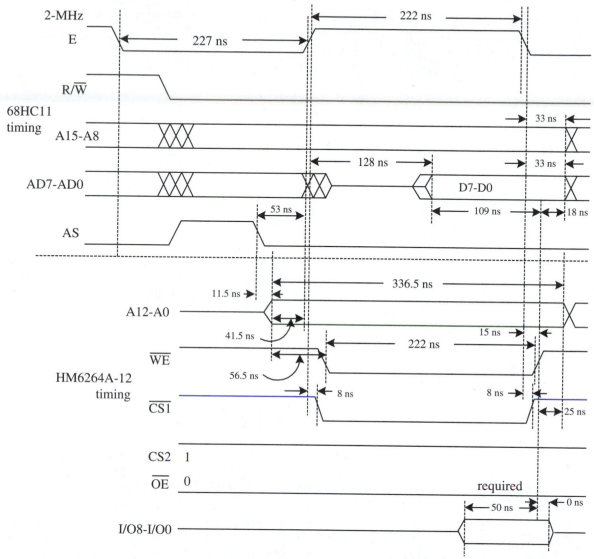

Figure 5.19 ■ Overlapped 68HC11 and HM6264A-12 write cycle timing diagram

the NAND gate propagation delay, the write enable signal $\overline{\text{WE}}$ is valid 15 ns after the rising edge of the E clock and becomes invalid 15 ns after the falling edge of the E clock. The chip enable signal $\overline{\text{CS1}}$ is valid 8 ns after the rising edge of the E clock signal. The write pulse is a delayed version of the E pulse and has a width equal to 222 ns. The SRAM requires the write pulse width to be at least 70 ns. The 68HC11 drives the output data 128 ns after the rising edge of the

Parameter	Required time (ns)	Actual time (ns)
Write cycle time	>= 120	336.5
Write pulse width	>= 70	222
Address setup time	>= 0	76.5
Address valid to end of write	>= 85	318.5
Data valid to end of write	>= 50	109
Data hold time	>= 0	18
Write recovery time	>= 0	24

Table 5.9 ■ Summary of write cycle timing analysis

E clock and hence provides 109 ns ($PW_{EH} - t_{DDW}$ + delay of 74LS00) of data setup time, satisfying the 50 ns requirement. The 68HC11 holds the data valid for 33 ns after the falling edge of the E clock. The HM6264A-12 requires 0 ns of data hold time and hence is satisfied. A summary of the write access timing analysis is shown in Table 5.9.

Example 5.7

▼

Calculate the data setup and hold times provided by the 68HC11 to the HM6264A-12 SRAM during the write access.

Solution: The following analysis is based on the 2 MHz E clock signal. The 68HC11 drives the data pins 128 ns (t_{DDW}) after the rising edge of the E clock signal. The high pulse width of the E clock (PW_{EH}) is 222 ns. The $\overline{WE}$ signal will become invalid 15 ns after the falling edge of the E clock because the propagation delay of the 74LS00 NAND gate is 15 ns. The write data setup time for SRAM is relative to the rising edge (20 percent of V_{DD}) of the $\overline{WE}$ signal and has a value given by the following equation:

$$t_{DW} = PW_{EH} - t_{DDW} + \text{delay of 74LS00}$$
$$= 222 \text{ ns} - 128 \text{ ns} + 15 \text{ ns}$$
$$= 109 \text{ ns}$$

The data hold time is relative to the rising edge of the $\overline{WE}$ signal (70 percent of V_{DD}). The $\overline{WE}$ signal becomes high 15 ns after the falling edge of the E clock. Therefore, the data hold time is the difference of these two values: 33 ns − 15 ns = 18 ns.

▲

Example 5.8

Compute the remaining timing parameters for the write access for the memory system in Figure 5.17.

Solution: The values of the remaining timing parameters are calculated as follows (reader should refer to the timing diagram in Figure 5.19):

■ Write cycle time (t_{wc}). Write cycle time is the period during which the address inputs to the HM6264A are valid. Lower address inputs (A7-A0) to the HM6264A become valid later than the upper address inputs (A12-A8). Lower address signals become valid 41.5 ns before the rising edge of the E clock. The upper address signals remain valid for 33 ns after the falling edge of the E clock signal. The lower address signals will remain valid for at least 174 ns ($PW_{EL} - t_{ASED} = 227$ ns $- 53$ ns) after the E clock's falling edge until the AS signal latches a new value into the address latch. The rise time and fall time of the E clock are 20 ns. Therefore, the actual maximal write cycle time is calculated as follows and is shown in Figure 5.20:

$$
\begin{aligned}
t_{wc} &= 41.5 \text{ ns} + t_r + PW_{EH} + t_f + t_{AH} \\
&= 41.5 \text{ ns} + 20 \text{ ns} + 222 \text{ ns} + 20 \text{ ns} + 33 \text{ ns} \\
&= 336.5 \text{ ns}
\end{aligned}
$$

■ Write pulse width (t_{wp}). Because the write pulse is generated by NANDing the complement of the $R/\overline{W}$ signal and the E clock signals, the duration of the write enable signal (asserted low) will

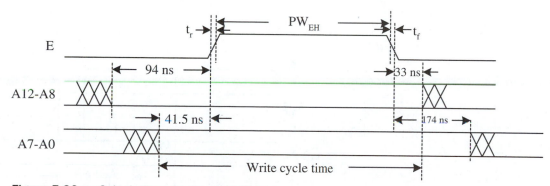

Figure 5.20 ■ Calculation of write cycle time

be determined by the shorter of these two signals. Since the E clock pulse (high period) is shorter than the R/$\overline{\text{W}}$ signal, the write pulse will be equal to the E clock high period, which is 222 ns.

■ Address setup time (t_{AS}). The address setup time is calculated from the moment when all address inputs to the HM6264A become valid until the moment the write enable signal goes low. As we have analyzed before, the address inputs become valid 41.5 ns before the rising edge of the E clock. The write enable signal becomes valid 15 ns after the rising edge of the E clock because the NAND gate has a delay of 15 ns. Therefore, the maximal address setup time can be calculated from the following expression:

$$
\begin{aligned}
t_{AS} \quad &= 41.5 \text{ ns} + t_r + 15 \text{ ns} \\
&= 41.5 \text{ ns} + 20 \text{ ns} + 15 \text{ ns} \\
&= 76.5 \text{ ns}
\end{aligned}
$$

This calculation is illustrated in Figure 5.21.

■ Address valid to end of write (t_{AW}). This parameter is calculated from the moment when all address inputs are valid until the write enable signal becomes invalid ($\geq 0.7 \text{ V}_{DD}$) to the HM6264A. It is the sum of the following terms:

1. latest address signals (A7-A0) valid to rising edge of the E clock delay (41.5 ns)
2. rise time of the E clock (a maximum of 20 ns)
3. E clock pulse high period (222 ns)
4. fall time of E clock (20 ns)
5. propagation delay of NAND gate (15 ns)

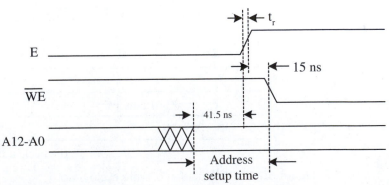

Figure 5.21 ■ Calculation of the address setup time

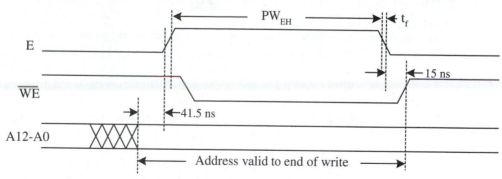

Figure 5.22 ■ Calculation of address valid to end of write

Therefore, the actual value of t_{aw} is 318.5 ns. The calculation process is shown in Figure 5.22.

■ Write recovery time (t_{wR}). Write recovery time is the time delay from the moment that the earliest chip select signal ($\overline{CS1}$ in this example) becomes invalid until the moment that the address signals become invalid. Since the $\overline{CS1}$ signal starts to go high when the E clock begins to fall, the parameter t_{wR} can be calculated from the following expression:

$$t_{wR} = t_{AH} - \text{propagation delay of the decoder}$$
$$= 33 \text{ ns} - 8 \text{ ns}$$
$$= 25 \text{ ns}$$

This calculation is shown in Figure 5.23.

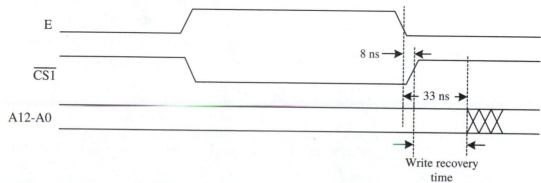

Figure 5.23 ■ Calculation of write recovery time

5.8 Programming 68HC711E9 Devices with PCbug11 and 68HC11EVBU

In the 68HC11 E series, the 68HC711E9, 68HC711E20, and 68HC811E2 are used most often for prototyping because they have on-chip EPROM or EEPROM. Both the EVB and EVBU can be used to program the 711E9 and 811E2. The programmer M68HC711E9PGMR can be used to program all three. The procedures for programming these parts are described in several Motorola engineering bulletin reports. You can find other engineering bulletin reports at this address: *http://mcu.motsps.com/faq/index.html.*

The PCbug11 software is included with the 68HC11 EVBU demo board. The EVBU board is often used to implement 68HC11 design projects. The user can replace the 68HC11E9 with a new MCU. The most common choice is the 68HC711E9 because it contains 12KB of on-chip EPROM and hence is ideal for this purpose. The user can use the EVBU demo board along with the PCbug11 software to program the 68HC711E9 chip. The procedure is as follows:

Step 1

- Before applying the power to the EVBU, remove the jumper from J7 and place it across J3 to ground the MODB pin.
- Place a jumper across J4 to ground the MODA pin. This will force the EVBU into special bootstrap mode on power up.
- Remove the resident MC68HC11E9 MCU from the EVBU.
- Place your MC68HC711E9 in the open socket with the notched corner of the part aligned with the notch on the PLCC socket.
- Connect the EVBU to one of your PC COM ports. Apply +5V volts to V_{DD} and ground to GND on the power connector of your EVBU.

Step 2

From a DOS command line prompt, start PCbug11 with

- C:\PCBUG11\>**pcbug11 -E port = 1** (if the EVBU is connected to COM1)
- C:\PCBUG11\>**pcbug11 -E port = 2** (if the EVBU is connected to COM2)

PCbug11 only supports COM ports 1 and 2. If you have made the proper connections and have a high quality cable, you should quickly get a PCbug11 command prompt. If you do receive a communication fault error, check your cable and board connections. Most PCbug11 communication problems can be traced to poorly made cables or bad board connections.

Step 3

PCbug11 defaults to base 10 for its input parameters; change this to hexadecimal by typing

    control base hex

Step 4

Inform PCbug11 with the address range of the EPROM chip with the following command:

    EPROM D000 FFFF

Step 5

Connect +12 volts (at most 12.5 volts) through a 100-Ω current limiting resistor to pin 18 ($\overline{XIRQ}$ pin) of the P4 connector and enter the following command:

    loads ...\myprog.s19

where, myprog.s19 is the name of the program to be programmed into the EPROM. You need to specify the full path of your file.

Step 6

After the programming operation is complete, PCbug11 will display this message

    Total bytes loaded: $xxxx
    Total bytes programmed: $yyyy

Each ORG directive in your assembly language source will cause a pair of these lines to be generated. For this operation, $yyyy will be incremented by the size of each block of code programmed into the EPROM of the MC68HC711E9.

After seeing this message,

- Remove the programming voltage from P4 connector pin 18 (the $\overline{XIRQ}$ pin).

- At the PCbug11 command prompt type

    verf ...\myprog.s19

If the verify operation fails, a list of addresses that did not program correctly is displayed. Should this occur, you probably need to erase your part more completely. To do so, allow the MC68HC711E9 to sit for at least 45 minutes under an ultraviolet light source. Attempt the programming operation again. If you have purchased devices in plastic packages (one-time programmable parts), you will need to try again with a new, unprogrammed device.

5.9   Summary

The 68HC11 is a family of microcontrollers developed in 1985. Members of the 68HC11 have the same instruction set and addressing modes but differ in the size of their on-chip memories and in their I/O functions.

The 68HC11 has four operation modes: expanded, single-chip, special test, and bootstrap. Applications that do not need external memory will use single-chip mode. In the expanded mode, the 68HC11 can access external memories and memory-mapped I/O devices. The special test is primarily used during Motorola's internal production testing. The 68HC11A8 has a 192-byte bootstrap ROM where a bootloader program is located. This ROM is enabled only if the 68HC11 is reset in the special bootstrap mode. This bootloader program uses the serial communication interface (SCI) to read a 256-byte program into the on-chip RAM at locations $00-$FF. After the program for address $FF is received, control is automatically passed to that program at location $00. This mode is often used to perform on-chip EPROM and EEPROM programming.

Memory chips are a major component of many digital systems. Memory components can be classified according to at least the following characteristics:

1. Volatility. Volatile memories lose their stored information when power is removed. Nonvolatile memories retain stored information even when power to the memory is removed.

2. Accessibility. Some of the memories are read-only while others are read-writable. Memories that are read-writable are also called random-access memories (RAM).

3. Need to perform refreshing operation. Semiconductor memories are further classified as dynamic and static. Dynamic memories are memory devices that require periodic refreshing of the stored information. Refresh is the process of restoring binary data stored in a particular memory location.

The capacity of a memory device is the total amount of information that the device can store. The capacity of semiconductor memories is often referred to as their memory density. Because they are binary in nature, semiconductor memories are always manufactured in powers of 2. Information on a single memory chip can be retrieved (read) or stored (written) one bit at a time or several bits at a time. The term *memory organization* describes the number of bits that can be written or read onto or from a memory chip during one output or input operation.

There are three issues in adding external memory chips to the 68HC11:

- address space assignment
- address decoding
- timing consideration

Several examples have been used to illustrate these three issues.

5.10 Exercises

E5.1. Can the user change the 68HC11 operation mode after the $\overline{\text{RESET}}$ signal goes high?

E5.2. Write a byte into the INIT register to remap the SRAM and register blocks to $4000-$40FF and $C000-$C03F, respectively.

E5.3. How many memory chips are required to build a 1MB, 32-bit wide memory system using the following SRAM chips?

a. 256K $\times$ 1 SRAM chips

b. 256K $\times$ 4 SRAM chips

c. 64K $\times$ 8 SRAM chips

d. 128K $\times$ 16 SRAM chips

E5.4. Refer to Example 5.6. Find the addresses that will select the same location as address $5080.

E5.5. Design an address decoder for a 68HC11-based product containing the following memory modules:

ROM1: 4KB

SRAM1: 4KB

ROM2: 4KB

The reset vector is stored in ROM1. Assign the address space so that addresses $FFFE and $FFFF are covered by ROM1.

E5.6. Design the control circuitry to interface the Am27C64 EPROM to the 2 MHz 68HC11A1. The Am27C64 is a 8192 $\times$ 8 CMOS EPROM. The pin assignment of this chip is shown in Figure 5E.1, and the normal read-access

V_{PP}	1	28	V_{CC}
A12	2	27	$\overline{PGM}$
A7	3	26	NC
A6	4	25	A8
A5	5	24	A9
A4	6	23	A11
A3	7	22	$\overline{OE}$
A2	8	21	A10
A1	9	20	$\overline{CE}$
A0	10	19	DQ7
DQ0	11	18	DQ6
DQ1	12	17	DQ5
DQ2	13	16	DQ4
GND	14	15	DQ3

Figure 5E.1 ■ Am27C64 EPROM pin assignment

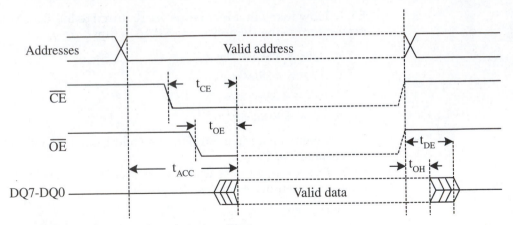

Figure 5E.2 ■ Am27C64 read timing diagram

timing diagram and timing parameters are given in Figure 5E.2 and Table 5E.1. Choose the Am27C64-120, which has a 120-ns access time. Perform read access timing analysis to verify that all timing requirements are satisfied.

E5.7. Suppose that the address inputs A2-A0 of the 74138 decoder are connected in order to the address outputs A15-A13 of the 68HC11 and that the E3 input is connected to A12. E1 and E2 are tied to ground permanently. Determine the address ranges controlled by O_0-O_7.

E5.8. Referring to Figure 5.11, identify the intervals during which signals A15-A8 and A7/D7-A0/D0 are unknown and in high-impedance state.

| Parameter | Symbol | Am27C64 | | | | | | | | | |
| | | -55 | | -70 | | -90 | | -120 | | -150 | |
		Min	Max	Min	Max	Min	Max	Min	Max	Min	Max
Address to output delay	t_{ACC}	–	55	–	70	–	90	–	120	–	150
Chip enable to output delay	t_{CF}	–	55	–	70	–	90	–	120	–	150
Output enable to output delay	t_{OE}	–	35	–	40	–	40	–	50	–	65
Output enable high to output float	t_{DF}	–	25	–	25	–	25	–	30	–	30
Output hold from address, CE, or OE, whichever occurred first	t_{OH}	0	–	0	–	0	–	0	–	0	–

All times are in nanoseconds.

Table 5E.1 ■ EPROM Am27C64 timing parameters

5.11   Lab Exercises and Assignments

L5.1 *Add external SRAM memory to EVBU board.* Follow the following steps:

Step 1
Reconfigure EVBU to operate in expanded mode. This can be done by removing jumpers J3 and J4 (need also cut the trace around jumper J4 if it is not cut yet) and pressing the reset button. The connection of jumpers J3 and J4 are shown in Figure 5L.1. To restore the EVBU back to single chip mode, put back the jumper J4.

Step 2
Add an 8KB SRAM to the EVBU board by following the method described in this chapter. Assign the memory space $8000-$9FFF to this SRAM chip. Use one 8KB SRAM HM6264A (or MCM6264C), one 3-to-8 decoder 74F138 (or 74ALS138), one 2-input NAND gate 74F00 (or 74ALS00), one address latch chip 74F373, one 0.1 μF capacitor (placed between the V_{DD} and ground pins of memory chip), and appropriate IC sockets.

Step 3
Write a memory test program that writes a value into a memory location, compare it with the read back value and displays that fault (when the write data is not the same as the read back value). Test at least the following patterns:

> All 1s (test stuck-at-0 faults)
> All 0s (test stuck-at-1 faults)
> 01010101 (check proper data connections)
> 10101010 (check proper data connections)

You need to test every memory location in this memory chip.

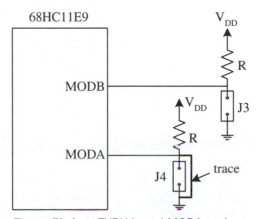

Figure 5L.1 ■ EVBU board MODA and MODB pin connections

Step 4

Verify that the memory system decoder is working properly by running the following program:

```
forever      LDAA    $8000
             BRA     forever
```

The decoder will assert the chip select signal to the SRAM once every seven E clock cycles. Use the oscilloscope to watch the signal. Also test the address $9000 (change $8000 to $9000).

Step 5

Use Buffalo memory-modify (**mm**) command to verify that the memory system is functioning properly (try at least 20 memory locations).

L5.2 *Memory test for the 68HC11 EVB.* Write a program that tests the memory locations immediately after your program until the end of the 8KB SRAM, i.e., at $DFFF. (The EVB external SRAM occupies the space of $C000-$DFFF.) Suppose your program occupies the memory space from $C000 to $CXXX, then test the memory locations from location $CXXX+1 to $DFFF. Use the four patterns mentioned in L5.1. An alternative to this experiment is to place an 8KB SRAM chip in the open IC socket on the EVB board. This SRAM occupies the memory space of $6000-$7FFF. Do the same test to this chip.

L5.3 *Add external SRAM to the CMD-11A8.* The 68HC11A1 MCU on the CMD-11A8 demo board has been configured to operate in the expanded mode. Follow step 2 to step 5 of L5.1. Allocate the memory space $4000-$5FFF to this memory chip. Use the expansion-bus port (a 34-pin connector) of the demo board. Please refer to the CMD-11A8 user manual for the pin assignments.

6

Interrupts and Resets

6.1 Objectives

After completing this chapter you should
be able to

- explain the difference between interrupts
  and resets

- describe the handling procedures for interrupts
  and resets

- raise one of the 68HC11 maskable interrupts
  to highest priority

- enable and disable maskable interrupts

- use the low-power mode to reduce the power
  consumption

- use COP watchdog timer reset to prevent
  software failure

- set up the EVB interrupt vector jump table

6.2   Basics of Interrupts

Interrupts and resets are among the most useful mechanisms that a computer system provides. With interrupts and resets, I/O operations are performed more efficiently, errors are handled more smoothly, and CPU utilization is improved. This chapter will begin with a general discussion of interrupts and resets and then focus on the specific features of the 68HC11 interrupts and resets.

6.2.1   What is an Interrupt?

An *interrupt* is a special event that requires the CPU to stop normal program execution and perform some service to the special event. An interrupt can be generated internally or externally. An external interrupt is generated when external circuitry asserts an interrupt signal to the CPU. An internal interrupt can be generated by the hardware circuitry in the CPU or caused by software errors. In some microcontrollers, timers, I/O interface functions, and the CPU are incorporated on the same chip. These subsystems can generate hardware interrupts to the CPU. Abnormal situations that occur during program execution, such as illegal opcodes, overflow, division by zero, and underflow, are called *software interrupts*. The terms *traps* and *exceptions* are both used to refer to software interrupts.

6.2.2   Why are Interrupts Used?

Interrupts are useful in many applications, such as

- Coordinating I/O activities and preventing the CPU from being tied up during the data transfer process. As we will explain in more detail in later chapters that are dealing with I/O functions, the interrupt mechanism enables the CPU to perform other functions during an I/O activity (when the I/O device is busy). CPU time can thus be utilized more efficiently because of the interrupt mechanism.

- Providing a graceful way to exit from an application when a software error occurs. The service routine for a software interrupt may also output useful information about the error so that it can be corrected.

- Reminding the CPU to perform routine tasks. For example, keeping the time of day is a common function of a computer. Without interrupts, the CPU would need to use program loops to create the delay or check the current time from a dedicated time-of-day chip. However, a timer circuit or a dedicated timer chip can generate interrupts to the CPU so that it can update the current time and perform routine tasks such as updating the system resource status.

In modern computer-operating systems, multiple-user programs are resident in the main memory, and the CPU time is divided into slots of about 10 to 20 ms. The operating system assigns a program to be executed for one time slot. At the end of a time slot or when a program is waiting for the completion of I/O, the operating system takes over and assigns another program to be executed. This technique is called *multitasking.* Because input/output operations are quite slow, CPU utilization is improved dramatically by multitasking because the CPU can execute another program while one program is waiting for the completion of input/output. Multitasking is made possible by the timer interrupt. Multitasking computer systems incorporate timers that periodically interrupt the CPU. On a timer interrupt, the operating system updates the system resource utilization status and switches the CPU from one program to another.

6.2.3   Interrupt Maskability

Some interrupts can be ignored by the CPU while others cannot. Interrupts that can be ignored by the CPU are called *maskable interrupts.* Interrupts that the CPU cannot ignore are called *nonmaskable interrupts.* A program can request the CPU to service or ignore a maskable interrupt by setting or clearing an *enable bit.* When an interrupt is *enabled,* it will be serviced by the CPU. When an interrupt is *disabled,* it will be ignored by the CPU. An interrupt request is said to be *pending* when it is active but not yet serviced by the CPU. A pending interrupt may or may not be serviced by the CPU, depending on whether or not it is enabled.

6.2.4   Interrupt Priority

If more than one interrupt is pending at the same time, the CPU needs to decide which interrupt should receive service first. The solution to this problem is to prioritize all interrupt sources. An interrupt with higher priority always receives service before interrupts at lower priorities. In most microprocessors, interrupt priorities are not programmable.

6.2.5   Interrupt Service

The CPU provides service to an interrupt by executing a program called the *interrupt service routine.* An interrupt is a special event for the CPU. After providing service to an interrupt, the CPU must resume normal program execution. How does the CPU stop execution of a program and resume it later? It does this by saving the program counter value and the CPU status information before executing the interrupt service routine and then restoring the program counter and the CPU status before exiting the interrupt service routine. The complete interrupt service cycle is:

1. When an interrupt occurs, save the program counter value in the stack.
2. Save the CPU status (including the CPU status register and some other registers) in the stack.
3. Identify the cause of interrupt.
4. Resolve the starting address of the corresponding interrupt service routine.
5. Execute the interrupt service routine.
6. Restore the CPU status and the program counter from the stack before exiting the interrupt service routine.
7. Restart the interrupted program.

Another issue is related to the time when the CPU begins to service an interrupt. For all hardware-maskable interrupts, the microprocessor starts to provide service when it completes execution of the instruction being executed when interrupt occurs. For some nonmaskable interrupts, the CPU may start the service without completing the current instruction. Many software interrupts are caused by an error in instruction execution that prevents the instruction from being completed. The service to this type of interrupt is simply to output an error message and abort the program.

For some other types of interrupts, the CPU resolves the problem in the service routine and then re-executes the instruction that caused the interrupt. One example of this type of interrupt is the *page-fault* interrupt in a *virtual-memory operating system.* In a virtual memory operating system, all programs are divided into pages of equal size. The size of a page can range from 512 bytes to 32 KB. When a program is first started, the operating system loads only a few pages (possibly only one page) from secondary storage (generally hard disk) into main memory and then starts to execute the program. Before long, the CPU may need to execute an instruction that hasn't been loaded into the main memory yet. This causes an interrupt called a *page-fault interrupt.* The CPU then executes the service routine for the page-fault interrupt, which brings in the page that caused the page fault into main memory. The program that caused the page is then restarted. The virtual memory operating system gets its name from the fact that it allows execution of a program without loading the entire program into main memory. With a virtual memory operating system, a program that is much larger than the physical main memory can be executed.

6.2.6   Interrupt Vector

The term *interrupt vector* refers to the starting address of the interrupt service routine. In general, interrupt vectors are stored in a table called an *interrupt-vector table.* The interrupt-vector table is fixed for some microprocessors and may be relocated for other microprocessors (for example, the AM29000 family of microprocessors).

The CPU needs to determine the interrupt vector before it can provide service. One of the following methods can be used by a microprocessor to determine the interrupt vector:

1. *Predefined.* In this method, the starting address of the service routine is predefined when the microcontroller is designed. The processor uses a table to store all the interrupt service routines. The Intel MCS-51 microcontroller family uses this approach. Each interrupt is allocated the same number of bytes to hold its service routine. The MCS-51 allocates eight bytes to each interrupt service routine. When the service routine requires more than eight bytes, the solution is to place a jump instruction in the predefined location to jump to the actual service routine.

2. *Fetch the vector from a predefined memory location.* In this approach, the interrupt vector of each interrupt source is stored at a predefined location in the interrupt vector table, where the microprocessor can get it directly. The Motorola 68HC11 uses this approach.

3. *Execute an interrupt acknowledge cycle to fetch a vector number in order to locate the interrupt vector.* During the interrupt acknowledge cycle, the microprocessor performs a read bus cycle, and the external I/O device that requested the interrupt places a number on the data bus to identify itself. This number is called *interrupt vector number.* The CPU can figure out the starting address of the interrupt service routine by using the interrupt vector number. The CPU needs to perform a read cycle in order to obtain it. The Motorola 68000 and Intel ×86 family microprocessors support this method. The Motorola 68000 family of microprocessors also uses the second method.

6.2.7 Interrupt Programming

There are three steps in interrupt programming:

Step 1
Initialize the interrupt vector table. (This step is not needed for microprocessors that have predefined interrupt vectors). This can be done by using the assembler directive ORG (or its equivalent):

```
ORG    $xxxx          ; xxxx is the vector table address
FDB    service_1      ; store the starting address of interrupt source 1
FDB    service_2      ;
   .
   .
   .
FDB    service_n
```

where *service_i* is the starting address of the service routine for interrupt source i. The assembler syntax and the number of bytes needed to store an interrupt vector on your particular microprocessor may be different.

Step 2

Write the service routine. An interrupt service routine should be as short as possible. For some interrupts, the service routine may only output a message to indicate that something unusual has occurred. A service routine is similar to a subroutine—the only difference is the last instruction. An interrupt service routine uses the *return from interrupt* (or *return from exception*) instruction instead of *return from subroutine* instruction to return to the interrupted program. The following is an example of an interrupt service routine (in 68HC11 instructions):

```
IRQ_ISR   LDX   #MSG
          JSR   PUTST                ; PUTST outputs a string pointed by X
          RTI                        ; return from interrupt
MSG:      FCC   "This is an error"
```

The service routine may or may not return to the interrupted program, depending on the cause of the interrupt. It makes no sense to return to the interrupted program if the interrupt is caused by a software error such as division by zero or overflow, because the program is unlikely to generate correct results under these circumstances. In such situations the service routine would return to the monitor program or the operating system instead. Returning to a program other than the interrupted program can be achieved by changing the saved program counter (in the stack) to the desired value. Execution of the *return from exception* instruction will then return CPU control to the new address.

Step 3

Enable the interrupts to be serviced. An interrupt can be enabled by clearing the CPU-level interrupt mask and setting the local interrupt enable bit in the I/O control register. It is a common mistake to forget to enable interrupts when writing interrupt-driven application programs.

6.2.8   Overhead of Interrupts

Although the interrupt mechanism provides many advantages, it also involves some overhead. The overhead of the 68HC11 interrupt includes:

1. Saving the CPU registers and fetching the interrupt vector. As shown in Figure 6.1, this takes twelve E clock cycles for the 68HC11.

2. The execution time of RTI instruction. This instruction restores all the CPU registers that have been stored in the stack by the CPU during the interrupt and takes twelve E clock cycles to complete for the 68HC11.

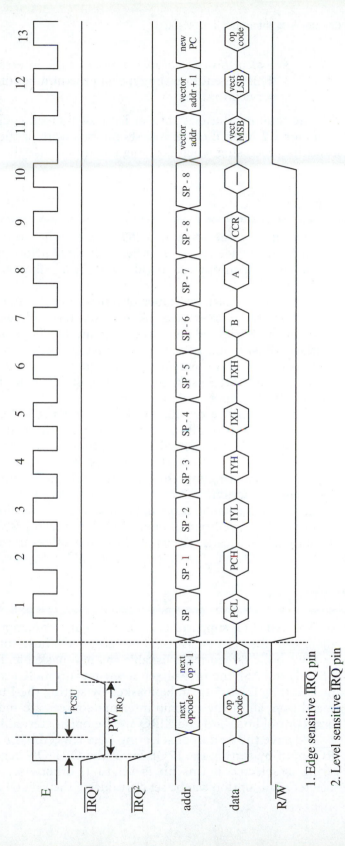

1. Edge sensitive $\overline{\text{IRQ}}$ pin

2. Level sensitive $\overline{\text{IRQ}}$ pin

Figure 6.1 ■ 68HC11 Interrupt timing diagram

3. Execution time of instructions of the interrupt service routine. This depends on the type and the number of instructions in the service routine.

The total overhead is thus at least 24 E clock cycles, which amounts to 12 μs for a 2 MHz E clock. We should be aware of the overhead involved in interrupt processing when deciding whether to use the interrupt mechanism.

6.3   Resets

The initial values of some CPU registers, flip-flops, and the control registers in I/O interface chips must be established before the computer can operate properly. Computers provide a reset mechanism to establish initial conditions.

There are at least two types of resets in each microprocessor: the *power-on reset* and the *manual reset*. A power-on reset allows the microprocessor to establish the initial values of registers and flip-flops and to initialize all I/O interface chips when power to the microprocessor is turned on. A manual reset without power-down allows the computer to get out of most error conditions (if hardware hasn't failed) and reestablish the initial conditions. The computer will *reboot* itself after a reset.

The reset service routine has a fixed starting address and is stored in the read-only memory of all microprocessors. At the end of the service routine, control should be transferred to either the monitor program or the operating system. An easy way to do that is to push the starting address of the monitor program or operating system into the stack and then execute a return from interrupt instruction.

Like nonmaskable interrupts, resets are also nonmaskable. However, resets are different from the nonmaskable interrupts in that no registers are saved by resets because resets establish the values of registers.

6.4   68HC11 Interrupts

The 68HC11 supports sixteen hardware interrupts, two software interrupts, and three resets, as listed in Table 6.1. Those interrupts can be divided into two basic categories, maskable and nonmaskable. In the 68HC11 A series (A0, A1, and A8), the first fifteen interrupts (from SCI serial system to $\overline{IRQ}$ pin interrupt) in Table 6.1 can be masked by setting the I bit of the CCR register. In addition, all of the on-chip interrupt sources are individually maskable by local control bits. $\overline{IRQ}$ and $\overline{XIRQ}$ are the only external interrupt sources.

We have the option of selecting edge triggering or level triggering for the $\overline{IRQ}$ input by setting the IRQE bit of the OPTION register. Falling-edge triggering is selected if this bit is set to 1; otherwise, low-level triggering is selected. This option can be set only during the first 64 E clock cycles after the rising edge of the $\overline{\text{RESET}}$ input and hence must be done in the reset handling routine. By default, level triggering is selected. By choosing edge triggering, we don't need to guarantee the $\overline{\text{IRQ}}$ signal to go high before the $\overline{\text{IRQ}}$ service routine is exited. However, edge triggering does not allow multiple interrupt sources to be wired on this pin. Furthermore, a noise spike on the $\overline{\text{IRQ}}$ pin could be mistaken as a valid $\overline{\text{IRQ}}$ request. Level triggering allows multiple interrupt sources to be wired on the same $\overline{\text{IRQ}}$ pin. However, we need to guarantee that the $\overline{\text{IRQ}}$ signal becomes inactive (goes high) before the $\overline{\text{IRQ}}$ service routine is exited. Otherwise, a single $\overline{\text{IRQ}}$ low pulse might cause $\overline{\text{IRQ}}$ service routine to be executed multiple times.

The software interrupt instruction (SWI instruction) is a nonmaskable interrupt source rather than a maskable interrupt source. The illegal opcode interrupt is also nonmaskable. The interrupt source $\overline{\text{XIRQ}}$ is considered a nonmaskable interrupt because once enabled, it cannot be masked by software; however, it is masked during a reset.

Vector address	Interrupt source	Priority
FFC0, C1	reserved	
.	.	
.	.	
.	.	
FFD4, D5	reserved	
FFD6, D7	SCI serial system	lowest
FFD8, D9	SPI serial transfer complete	
FFDA, DB	pulse accumulator input edge	
FFDC, DD	pulse accumulator overflow	
FFDE, DF	timer overflow	
FFE0, E1	timer output compare 5	
FFE2, E3	timer output compare 4	
FFE4, E5	timer output compare 3	
FFE6, E7	timer output compare 2	
FFE8, E9	timer output compare 1	
FFEA, EB	timer input capture 3	
FFEC, ED	timer input capture 2	
FFEE, EF	timer input capture 1	
FFF0, F1	real timer interrupt	
FFF2, F3	IRQ pin interrupt	
FFF4, F5	XIRQ pin interrupt	
FFF6, F7	SWI	
FFF8, F9	illegal opcode trap	
FFFA, FB	COP failure	
FFFC, FD	COP clock monitor fail	
FFFE, FF	RESET	highest

Table 6.1 ■ 68HC11 interrupt vector address and priority

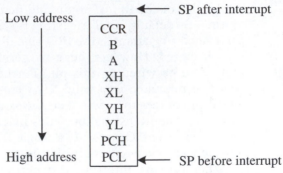

Figure 6.2 ■ 68HC11 interrupt stacking order

6.4.1 The 68HC11 Interrupt Handling Procedure

When an interrupt occurs, the CPU registers are saved in the stack in the order shown in Figure 6.2. After these registers have been saved, the I bit in the CCR register is set to 1 to disable further interrupts. The starting address of the corresponding service routine is fetched from a predefined table, and program execution is resumed from that address. All interrupts and resets are prioritized. Table 6.1 lists the priority of each interrupt and the locations where the interrupt vectors are stored. The starting address of the reset handling routine is also shown in this table.

Example 6.1

▼

The 68HC11 is executing the TAP instruction in line 6 of the following program segment when an IRQ interrupt occurs. List the stack contents immediately before the interrupt service routine is entered.

Line	Location	Opcode	Instruction mnemonics
			ORG $E000
1	E000	C620	LDAB #$20
2	E002	8E00FF	LDS #$00FF
3	E005	CE1000	LDX #$1000
4	E008	4F	CLRA
5	E009	1830	TSY
6	E00B	06	TAP
7	E00C	AB00	ADDA 0, X

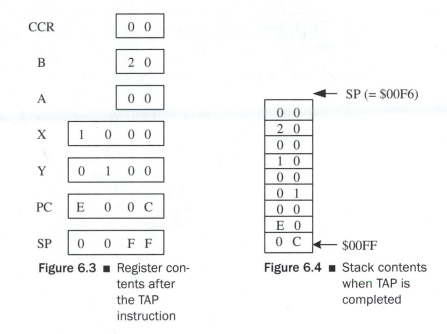

Figure 6.3 ■ Register contents after the TAP instruction

Figure 6.4 ■ Stack contents when TAP is completed

Solution: The interrupt will be serviced after the TAP instruction is completed. The contents of all registers when the TAP instruction is completed are shown in Figure 6.3.

The contents of B are $20 when the first instruction is completed.
The contents of SP are $00FF when the second instruction is completed.
The contents of X are $1000 when the third instruction is completed.
The contents of A are $00 when the fourth instruction is completed.
The contents of Y are $0100 when the fifth instruction is completed.
The contents of CCR are $00 when the sixth instruction is completed.

The contents of the stack will therefore be as shown in Figure 6.4 immediately before the service routine is entered (executed).

6.4.2 Priority Structure of the 68HC11 Maskable Interrupts

The priority of each maskable interrupt source is fixed. However, the user can program the HPRIO register to promote one of the maskable interrupts to highest priority among all maskable interrupts. The contents of the HPRIO register are shown in Figure 6.5.

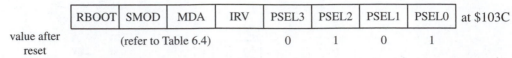

| RBOOT | SMOD | MDA | IRV | PSEL3 | PSEL2 | PSEL1 | PSEL0 | at $103C |

value after reset

| (refer to Table 6.4) | | | | 0 | 1 | 0 | 1 |

Figure 6.5 ■ Contents of the HPRIO register

The last four bits of the HPRIO register select the interrupt source to be promoted, as shown in Table 6.2, while the values of the upper four bits are established during the reset and serve the following purposes:

- RBOOT—Read bootstrap ROM
  This bit is writable only when the SMOD bit equals 1. The bootstrap ROM will be enabled and located from $BF40 to $BFFF when the RBOOT bit is 1, but it will be disabled and not present in the memory map when this bit is 0. This 192-byte mask-programmed ROM contains the firmware required to load a user's program through the serial communication interface into the internal SRAM and jump to the loaded program. In all modes other than the special bootstrap mode, this ROM is disabled and does not occupy space in the 64KB memory map.

- SMOD—Special mode
  This bit may be written to 0 but not back to 1. Its value is established on the rising edge of the $\overline{\text{RESET}}$ signal. The 68HC11 operation mode is determined by the combination of this bit and the MDA bit, as shown in Table 6.3.

PSEL3	PSEL2	PSEL1	PSEL0	Interrupt source promoted
0	0	0	0	timer overflow
0	0	0	1	pulse accumulator overflow
0	0	1	0	pulse accumulator input edge
0	0	1	1	SPI serial transfer complete
0	1	0	0	SCI serial system
0	1	0	1	reserved (defaults to $\overline{\text{IRQ}}$)
0	1	1	0	$\overline{\text{IRQ}}$ pin (external pin)
0	1	1	1	real time interrupt
1	0	0	0	timer input capture 1
1	0	0	1	timer input capture 2
1	0	1	0	timer input capture 3
1	0	1	1	timer output compare 1
1	1	0	0	timer output compare 2
1	1	0	1	timer output compare 3
1	1	1	0	timer output compare 4
1	1	1	1	timer output compare 5

Table 6.2 ■ PSEL3-PSEL0 bits for selecting highest priority interrupts

Operation mode	SMOD	MDA
Normal single-chip	0	0
Normal expanded	0	1
Special bootstrap	1	0
Special test	1	1

Table 6.3 ■ Mode select of 68HC11

- MDA—Mode A select
  This bit is written only when SMOD equals 1. The value of the MDA bit is established on the rising edge of the $\overline{\text{RESET}}$ signal. The 68HC11 operation mode is determined by the combination of this bit and the SMOD bit (see Table 6.3).

- IRV—Internal read visibility
  The IRV bit is written only when the SMOD bit equals 1. This bit is forced to 0 if the SMOD bit equals 0. When the IRV bit is set to 1, data can be driven onto an external bus during internal reads (for example, a read from an on-chip I/O control register or status register). Otherwise, data from internal reads will not be visible during internal reads. The IRV control bit is used during factory testing and sometimes during emulation to allow internal read accesses to be visible on the external data bus.

The establishment of these four bits is shown in Table 6.4. In assigning highest priority interrupts, note that the HPRIO register may be written only when the I bit in the CCR register is a 1. Also note that the interrupt vector address is not affected by reassigning an interrupt source to the highest priority position. The upper four bits may be written only when the 68HC11 is operating in one of the special modes (when SMOD is 1).

6.4.3 Nonmaskable Interrupts

There are three nonmaskable interrupts sources in the 68HC11: an illegal opcode fetch, an SWI instruction, and the $\overline{\text{XIRQ}}$ pin interrupt.

Inputs			Control bits in HPRIO (latched at reset)			
MODB	MODA	Mode description	RBOOT	SMOD	MDA	IRV
1	0	normal single-chip	0	0	0	0
1	1	normal expanded	0	0	1	0
0	0	special bootstrap	1	1	0	1
0	1	special test	0	1	1	1

Table 6.4 ■ Hardware mode selection summary

ILLEGAL OPCODE TRAPS

Some combinations of the opcode byte are not defined, and if one of those combinations is executed, an illegal opcode trap occurs. For this reason, the illegal opcode vector should never be left uninitialized. It is also a good idea to reinitialize the stack pointer after an illegal opcode interrupt so that repeated execution of illegal opcodes does not cause stack overruns.

THE SOFTWARE INTERRUPT INSTRUCTION (SWI)

The SWI instruction is executed in the same manner as any other instruction and takes precedence over all interrupts maskable by the I bit of the CCR register. Registers are saved in the stack as in any other maskable interrupt. The SWI instruction is not masked by the global interrupt mask bits (I and X bits) in the CCR register.

The SWI instruction is commonly used in the debug monitor to implement *breakpoints* and to transfer control from a user program to the debug monitor. A breakpoint in a user program is a memory location where we want program execution to be stopped and information about instruction execution (in the form of register contents) to be displayed. To implement breakpoints, the debug monitor sets up a breakpoint table. Each entry of the table holds the address of the breakpoint and the opcode byte at the breakpoint. The monitor also replaces the opcode byte at the breakpoint with the opcode of the SWI instruction. When the instruction at the breakpoint is executed, it thus causes an SWI interrupt. The service routine of the SWI interrupt will look up the breakpoint table and take different actions depending on whether the saved PC value is in the breakpoint table:

Case 1

The saved PC value is not in the breakpoint table. In this case, the service routine will simply replace the saved PC value (in the stack) with the address of the monitor program and return from interrupt.

Case 2

The saved PC is in the breakpoint table. In this case, the service routine will

1. replace the SWI opcode with the opcode in the breakpoint table
2. replace the saved PC value (in the stack) with the address of the monitor program
3. display the contents of the CPU registers
4. return from the interrupt

An EVB (also true for EVBU and CMD-11A8) user can use SWI as the last instruction in a program. When the SWI instruction is executed, CPU control is returned to the BUFFALO monitor, and we can then enter commands to examine the contents of memory locations or registers or do something else.

THE $\overline{XIRQ}$ PIN INTERRUPT

The $\overline{XIRQ}$ interrupt can be masked by the X bit in the CCR register. Upon reset, the X bit is 1 and the $\overline{XIRQ}$ interrupt is thus disabled. The X bit can be cleared within 64 E clock cycles after reset, but it cannot be cleared and set after that and hence is nonmaskable. When an $\overline{XIRQ}$ interrupt occurs, all registers are saved in the stack, and the X bit in the CCR register is set to 1 to prevent further $\overline{XIRQ}$ interrupts. Because of these characteristics, the $\overline{XIRQ}$ interrupt can be used to detect emergency situations.

Several external interrupts can be tied to the $\overline{IRQ}$ and $\overline{XIRQ}$ pins. In this application, we must choose level-sensitive triggering for the $\overline{IRQ}$ interrupt. The $\overline{XIRQ}$ interrupt is always level triggered.

The option of an edge-triggered $\overline{IRQ}$ interrupt can be set only within the first 64 E clock cycles after reset. If the user does need edge-triggered interrupt capability, four other interrupt sources can be utilized: input-capture functions IC1, IC2, and IC3 and pulse accumulator pin PAI. They will be discussed in chapter 8.

6.5   Low-Power Modes

A good embedded microcontroller should consume as little power as possible. When the microcontroller is performing normal operations, power consumption is unavoidable. However, the microcontroller in an embedded application may not always be performing a useful operation, so it would be ideal if the power consumption could be drastically reduced when the microcontroller is not performing useful operations. This issue is especially important for battery-powered products.

Microcontroller manufacturers are already investigating methods of reducing power consumption when the microcontroller is idle, and the so-called power-down mode has been incorporated in several microcontrollers. The complementary metal oxide semiconductor (CMOS) technology is the dominant technology for implementing microcontrollers and its power consumption is proportional to the operating frequency of the circuit. By slowing down the circuit operation and/or even turning off part of the circuit, the power consumption of these microcontrollers can be dramatically reduced.

The 68HC11 has two programmable low-power modes: WAIT and STOP.

6.5.1   The WAIT Instruction

The WAIT instruction puts the microcontroller in a low-power consumption mode while keeping the oscillator running. Upon execution of a WAIT instruction, the machine state is saved in the stack and the CPU stops instruction execution. The wait state can be exited only through an unmasked interrupt or reset. The amount of power saved depends on the application and also on the circuitry connected to the microcontroller pins.

The WAIT instruction can be used in situations where CPU has nothing to do but wait for the arrival of an interrupt. Because all CPU registers are already saved in the stack, the interrupt response time can be much faster.

6.5.2   The STOP Instruction

The STOP instruction places the microcontroller in its lowest power-consumption mode, provided the S bit in the CCR register is 0. If the S bit is 1, the STOP mode is disabled and STOP instructions have no effect. In the STOP mode, all clocks including the internal oscillator are stopped, causing all internal processing to be halted. Exit from the STOP mode can be accomplished by $\overline{RESET}$ and $\overline{XIRQ}$ interrupt or an unmasked $\overline{IRQ}$.

When the $\overline{XIRQ}$ interrupt is used, the 68HC11 exits from the STOP mode regardless of the state of the X bit in the CCR; however, the actual recovery sequence differs depending on the state of the X bit. If the X bit is 1, then processing will continue with the instruction immediately following the STOP instruction and no $\overline{XIRQ}$ interrupt service routine will be executed. If the X bit is 0, the 68HC11 will perform the sequence that services the $\overline{XIRQ}$ interrupt. A $\overline{RESET}$ will always result in an exit from the STOP mode, and the restart of microcontroller operations is then determined by the reset vector.

6.6   The 68HC11 Resets

The 68HC11 has four sources of resets: an active low external reset pin ($\overline{RESET}$), a power-on reset, a computer operating properly (COP) watchdog timer reset, and a clock monitor failure reset.

6.6.1   External Reset Pin

The $\overline{RESET}$ pin is used to reset the microcontroller and allow an orderly system start-up. When a reset condition is detected, this pin is driven low by an internal device for four E clock cycles and then released; two E clock cycles later, it is sampled. If the pin is still low, it means that an external reset has occurred. If the pin is high, it indicates that the reset was initiated internally by either the computer operating properly (COP) watchdog timer or the clock monitor. An externally generated reset should stay active for at least eight E clock cycles in order to be detected.

It is very important to control reset during power transitions. If the reset line is not held low while V_{DD} is below its minimum operating level, the contents of the on-chip EEPROM could be corrupted; both the EEPROM memories and the EEPROM-based CONFIG register can be affected.

A low-voltage-inhibit (LVI) circuit that holds $\overline{RESET}$ low whenever V_{DD} is below its minimum operating level is required to protect against EEPROM corruption. Figure 6.6 shows an example of reset circuit with manual reset and LVI capability.

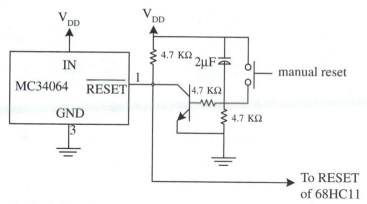

Figure 6.6 ■ A typical external reset circuit

6.6.2   Power-On Reset

The power-on reset occurs when a positive transition is detected on V_{DD}. The power-on reset is used strictly for power turn-on conditions and should not be used to detect drops in power-supply voltage. The power-on circuitry provides 4064-cycle time delay from the time of the first oscillator operation.

In a system where $E = 2$ MHz, the power-on reset lasts about two milliseconds. If the external $\overline{RESET}$ pin is low at the end of the power-on delay time, the 68HC11 remains in the reset condition until the $\overline{RESET}$ pin goes high.

6.6.3   CPU after Reset

After reset, the CPU fetches the reset vector from locations $FFFE and $FFFF ($BFFE and $BFFF if in special bootstrap and special test mode) during the first three cycles and begins instruction execution. The stack pointer and other CPU registers are indeterminate immediately after reset; however, the X and I interrupt mask bits in the CCR register are set to mask any interrupt requests, and the S bit is set to 1 to disable the STOP mode.

All I/O control registers are initialized by reset. The initial value of each control register will be discussed in the appropriate chapters.

6.6.4   Establishing the Mode of Operation

During a reset, the basic mode of operation is established, determining whether the 68HC11 will operate as a self-contained single-chip system or as an expanded system that includes external memory resource. There are also special variations of these two basic modes of operation: the bootstrap mode is a variation of the normal single-chip mode, and the special test mode is a variation of the normal expanded mode. The levels of the two mode select pins during reset determine which of these four modes of operation will be

CR1	CR0	E / 2^{15} divided by	Crystal frequency	
			8 MHz	4 MHz
			Nominal time-out	
0	0	1	16.384 ms	32.768 ms
0	1	4	65.536 ms	131.07 ms
1	0	16	262.14 ms	524.49 ms
1	1	64	1.049 s	2.1 s
			E clock = 2 MHz	1 MHz

Table 6.5 ■ Watchdog rates vs. crystal frequency

selected. The levels of these two mode-select pins also set the upper four bits in the HPRIO register, as shown in Table 6.4.

6.6.5   Computer Operating Properly (COP) Watchdog Timer Reset

The COP watchdog timer system is intended to detect software-processing errors. When the COP system is in use, software is responsible for keeping a free-running watchdog timer from timing out. If the watchdog timer times out, it is an indication that software is no longer being executed in the intended sequence and thus a system reset is initiated.

The COP watchdog timer system is enabled by clearing the NOCOP bit in the CONFIG register, and it is disabled by setting the NOCOP bit. The CONFIG register is implemented in EEPROM and is not as easy to change as a normal control register. Even after the NOCOP bit is changed, the 68HC11 must be reset before the new status becomes effective. In the special test and bootstrap modes, the COP system is initially disabled by the disable resets (DISR) control bit in the TEST1 register. The DISR bit can be written to zero to enable COP resets while the 68HC11 is in the special test or bootstrap mode.

The COP time-out period is set by the COP timer control bits (CR1 and CR0) in the OPTION register. After reset, these bits are both zeros; this combination selects the fastest time-out rate. The 68HC11 internal E clock is first divided by 2^{15} before it enters the COP watchdog system, and then the CR1 and CR0 bits determine a further scaling factor for the watchdog timer (see Table 6.5).

The COP timer must be reset by a software sequence prior to timeout to avoid a COP reset. The software COP reset is a two-step sequence:

Step 1
Write a $55 to the COPRST register (located at $103A). This step arms the COP timer-clearing mechanism.

Step 2
Write a $AA into the same register. This step clears the COP timer.

Any number of instructions can be executed between these two steps as long as both steps are performed in the correct sequence before the COP timer times out.

6.6.6 Clock Monitor Reset

The clock monitor circuit is based on an internal resistor-capacitor (RC) time delay. If no clock edges are detected within this RC time delay, the clock monitor can optionally generate a system reset. The clock monitor function is enabled/disabled by the CME bit (bit 3) in the OPTION register.

The RC time-out may vary from lot to lot and from part to part because of processing variations. An E clock frequency below 10 KHz will definitely be detected as a clock monitor error. An E clock frequency of 200 KHz or higher will prevent clock monitor errors. Any system operating below a 200 KHz E clock frequency should not use the clock monitor function.

The clock monitor is often used as a backup for the COP watchdog system. Because the COP relies on a clock, it is unable to function if the clock stops. In such a system, the clock monitor system could detect clock failure not detected by the COP system.

Another use of the clock monitor reset is to protect against unintentional execution of the STOP instruction. Because applications view the STOP instruction as a serious problem, it causes the microcontroller clock to stop, thus disabling all software execution and on-chip peripheral functions. The stop disable bit (S) in the CCR register is the first line of defense against unwanted STOP instructions. When the S bit is 1, the STOP instruction acts as a no operation instruction, which does not interfere with clock operation. The clock monitor can provide an additional level of protection by generating a system reset if the 68HC11 clocks are accidentally stopped.

If the 68HC11 clocks slow down or stop when the clock monitor is enabled, a system reset is generated. The bidirectional $\overline{RESET}$ pin is driven low to reset the external system and the microcontroller. If the 68HC11 is in the special test or bootstrap mode, resets from COP and clock monitor systems are initially disabled by a 1 in the DISR bit in the TEST1 register. The COP and clock monitor resets can be re-enabled while the 68HC11 is still in the special test or special bootstrap mode by clearing the DISR bit to 0. In normal operation modes, the DISR bit is forced to 0 and cannot be set to 1.

6.7 68HC11 Interrupt Programming

As in any other microprocessor, there are three steps in 68HC11 interrupt programming:

Step 1
Write the interrupt service routines.

Step 2
Set up interrupt-vector table.

Step 3
Set the global and local interrupt-enable bits.

6.7.1   Writing the Interrupt Service Routine

In assembly language, an interrupt service routine should have the following format:

```
xxx_ISR   .
            .
            .
          RTI
```

where xxx_ISR is the label of the first instruction in the interrupt service routine.

Using the ImageCraft C compiler, an interrupt service routine should be written as follows:

```
#pragma interrupt_handler    xxx_ISR( )
void xxx_ISR( )
{
    ...
}
```

The statement "#pragma ..." tells the C compiler that "xxx_ISR" is an interrupt handler so that the C compiler will generate *RTI* instead of *RTS* as the last instruction of the function.

6.7.2   Set Up Interrupt Vector Table

In assembly language, we use assembler directives to set up the interrupt vector table. For example, to add the interrupt vector of $\overline{\text{IRQ}}$ into the table, use the following directives:

```
ORG    $FFF2
FDB    IRQ_ISR      ; IRQ_ISR is the label of the first instruction in IRQ
                    ; service routine
```

We may not want to use all of the possible interrupts in our applications. However, it is a good idea to initialize all interrupt vectors. A common service routine should be written to handle all unwanted interrupts—in case an unwanted interrupt is accidentally triggered, the common service routine can provide an exit. A common form of a common service routine just outputs a message and quits.

The design of the Motorola EVB and EVBU and the Axiom CMD-11A8 evaluation boards do not allow us to write into the default vector-table memory space because the space from $E000 to $FFFF is in EPROM/EEPROM. The

Interrupt vector	Field
serial communication interface (SCI)	$00C4-$00C6
serial peripheral interface (SPI)	$00C7-$00C9
pulse accumulator input edge	$00CA-$00CC
pulse accumulator overflow	$00CD-$00CF
timer overflow	$00D0-$00D2
timer output compare 5	$00D3-$00D5
timer output compare 4	$00D6-$00D8
timer output compare 3	$00D9-$00DB
timer output compare 2	$00DC-$00DE
timer output compare 1	$00DF-$00E1
timer input capture 3	$00E2-$00E4
timer input capture 2	$00E5-$00E7
timer input capture 1	$00E8-$00EA
real time interrupt	$00EB-$00ED
IRQ	$00EE-$00F0
XIRQ	$00F1-$00F3
software interrupt (SWI)	$00F4-$00F6
illegal opcode	$00F7-$00F9
computer operate properly	$00FA-$00FC
clock monitor	$00FD-$00FF

Table 6.6 ■ 68HC11 EVB, EVBU, and CMD-11A8
interrupt vector jump table

EVB, EVBU, and CMD-11A8 reserves 60 bytes ($00C4~$00FF) of the on-chip SRAM as an *interrupt jump table*. Each entry of the jump table consists of three bytes. The first byte must be set to the opcode ($7E) of the JMP instruction, and the next two bytes must be set to the starting address of the corresponding service routine. Each entry (two bytes) of the default vector table of the 68HC11 contains the address of the first byte of the corresponding entry in the interrupt vector jump table. The EVB, EVBU and CMD-11A8 use the same interrupt vector jump table as shown in Table 6.6.

Like the case in assembly language, the setup of interrupt vector table in C language depends on if we are using a monitor (the monitor occupies the default vector table space) or not.

Case 1

No monitor occupies the default vector table space.

Use the following statements to inform C compiler to generate machine code bytes to be placed in the appropriate locations. These statements can be added after your main function or put in a separate file.

```
#pragma abs_address:0xffd6

void (*interrupt_vectors[])(void) =
    {
    SCI_ISR,          /* SCI interrupt service routine */
    SPI_ISR,          /* SPI */
    PAI_ISR,          /* PAIE */
    PAO_ISR,          /* PAO */
    TOF_ISR,          /* TOF */
    TOC5_ISR,         /* TOC5 */
    TOC4_ISR,         /* TOC4 */
    TOC3_ISR,         /* TOC3 */
    TOC2_ISR,         /* TOC2 */
    TOC1_ISR,         /* TOC1 */
    TIC3_ISR,         /* TIC3 */
    TIC2_ISR,         /* TIC2 */
    TIC1_ISR,         /* TIC1 */
    RTI_ISR,          /* RTI */
    IRQ_ISR,          /* IRQ */
    XIRQ_ISR,         /* XIRQ */
    SWI_ISR,          /* SWI */
    ILLOP_ISR,        /* ILLOP */
    COP_ISR,          /* COP */
    CLM_ISR,          /* CLM */
    _start            /* RESET */
    }
#pragma end_abs_address
```

The first statement "#pragma abs_address:0xffd6" tells the C compiler to use absolute address for the following instructions and data while the last statement tells the C compiler to terminate the use of the absolute address.

Each entry in the above table represents the starting address of the service routine (ISR) for each interrupt source. For example, SCI_ISR is the name of the SCI interrupt service routine.

Case 2

A monitor such as BUFFALO is used.

In this case, we need to set up the vector jump table. For example, the following code segment sets up the vector jump table entry for the IRQ interrupt:

```
void IRQ_ISR( );
main ( )
{
    ...
    *(unsigned char *)0xee = 0x7E;      /* 7E is "jmp" */
    *(void (**)())0xef = IRQ_ISR;

    ...
}
```

6.7.3   Enable Interrupts

For most microprocessors, there are two levels of interrupt enabling: global level and local level. The global level interrupt enabling allows the interrupt circuitry to accept interrupt requests. The local level interrupt enabling allows the interrupt requests from a specific source to be generated and processed by the CPU. For example, to enable the timer OC1 interrupt, we need to

1. Clear the I bit of the CCR register—enable interrupt globally.
2. Set the bit 7 of the TMSK1 register—enable OC1 interrupt to be generated

The only exception is that the $\overline{\text{IRQ}}$ interrupt can be enabled by clearing the I bit of the CCR register.

In assembly language,

- Enabling interrupt globally can be done by the instruction **CLI**.
- Disabling interrupt globally can be done by the instruction **SEI**.
- Enabling an interrupt locally can be done by using the BSET instruction. For example, the following instruction sequence will enable the OC1 interrupt:

```
tmsk1   equ     $22
        ldx     #$1000          ; put the starting address of the I/O register
*                               ; block in X
        bset    tmsk1,X $80     ; set the bit 7 of TMSK1 to enable OC1
*                               ; interrupt
```

For ImageCraft C compiler,

- Enabling interrupt globally can be done by adding the following in-line assembly code:

    asm (" cli ");

    or using the macro "INTR_ON( )" defined in hc11.h file.
- Disabling interrupt globally can be done by adding the following in-line assembly code:

    asm (" sei ");

    or using the macro "INTR_OFF ( )" defined in hc11.h file.
- Enabling an individual interrupt can be done by using the "or" operand. For example, the following statements will enable the OC1 interrupt:

    TMSK1 |= 0x80   /* set bit 7 of TMSK1 register to enable OC1 interrupt */

Example 6.2

▼

Write a main program and an interrupt service routine in assembly language. The main program will initialize a variable *count* to 5, enable the $\overline{IRQ}$ interrupt, and stay in a loop to check the value of *count*. When the value of *count* is 0, the main program jumps back to the BUFFALO monitor. The interrupt service routine will decrement the variable *count* and return. The $\overline{IRQ}$ is connected to some external device that generates interrupts from time to time. Assume the voltage level of the $\overline{IRQ}$ pin returns to high before the $\overline{IRQ}$ service is exited.

Solution: The first step is to set up the interrupt vector entry for the $\overline{IRQ}$ interrupt using the following assembler directives:

```
              ORG     $EE
              JMP     IRQ_ISR         ; IRQ_ISR is the interrupt vector for IRQ
              ...
              ORG     $00
count         RMB     1
* set up starting address of program to be run on EVB
              ORG     $C000           ; set to $2000 (or $100) for CMD-11A8 (for EVBU)
              SEI
* set up stack pointer
              LDS     #$DFFF          ; set to $3FFF for CMD-11A8 (set to $47 for EVBU)
              LDAA    #5
              STAA    count           ; set up count
              CLI                     ; enable IRQ interrupt
loop          LDAA    count
              BNE     loop            ; is count equal to 0?
              SWI                     ; jump to BUFFALO monitor

IRQ_ISR
              DEC     count
              RTI
              END
```

▲

Example 6.3

▼

Write a C program for the problem in Example 6.2.

Solution:

```
#include <hc11.h>
int     count;
void IRQ_ISR();
```

```
main ( )
{
    count = 5;                          /* initialize count to 5 */
    *(unsigned char *)0xee = 0x7E;      / * 7E is "jmp" */
    *(void (**)())0xef = IRQ_ISR;
    INTR_ON( );                         /* enable IRQ interrupt */
    while (count)
        ;                               /* wait until count is decremented to 0 */
    INTR_OFF( );                        /* disable IRQ interrupt */
    asm (" swi ");                      /* generate SWI interrupt */
}

#pragma interrupt_handler IRQ_ISR( )
void IRQ_ISR( )
{
    count -= 1;
}
```

6.8   Summary

Interrupt is a special event that requires the CPU to stop normal program execution and provide some service to the event. The interrupt mechanism has many applications, including coordinating I/O activities, exiting from software errors, and reminding CPU to perform routine works, etc.

Some interrupts are maskable and can be ignored by the CPU. Some interrupts are nonmaskable and cannot be ignored by the CPU. Nonmaskable interrupts are often used to handle emergent events such as power failure.

Multiple interrupts may be pending at the same time. The CPU needs to decide which one to service first. The solution to this issue is to prioritize all of the interrupt sources. The pending interrupt with the highest priority will receive the service before other pending interrupts.

The CPU provides service to an interrupt request by executing an interrupt service routine. The current program counter value will be saved in the stack before the CPU executing the service routine so that CPU control can be returned to the interrupted program when the interrupt service routine is completed.

In order to provide service to the interrupt, the CPU must have some way to find out the starting address of the interrupt service routine. There are three methods to know the starting address (called interrupt vector) of the interrupt service routine:

1. Each interrupt vector is predefined when the microprocessor was designed. In this method, the CPU simply jumps to the predefined location to execute the service routine.

2. Each interrupt vector is stored in a predefined memory location. When interrupt occurs, the CPU fetches the interrupt vector from that predefined memory location.

3. The interrupt source provides an interrupt vector number to the CPU so that the CPU can figure out the memory location where the interrupt vector is stored. The CPU needs to perform a read bus cycle to get the interrupt vector number.

There are three steps in the interrupt programming:

Step 1
Initialize the interrupt vector table that holds all the interrupt vectors. This step is not needed for those microprocessors that use the first method to resolve the interrupt vector.

Step 2
Write the interrupt service routine.

Step 3
Enable the interrupt to be serviced.
Reset is a mechanism for

1. Setting up operation mode for the microprocessor.
2. Setting up initial values for control registers.
3. Exiting from software errors and some hardware errors.

The 68HC11 has sixteen hardware interrupt sources and two nonmaskable software interrupts. The 68HC11 has two low-power modes that are triggered by the execution of WAIT and STOP instructions. Power consumption will be reduced dramatically in either low power mode. The 68HC11 has a COP timer reset mechanism to detect the software error. A software program that behaves properly will reset the COP timer before it times out and prevent it from resetting the CPU.

The 68HC11 has a clock monitor reset mechanism that can detect the slowing down or loss of clock signals. Whenever the clock frequency gets too low (lower than 10 KHz), the clock monitor will detect it and reset the CPU.

Examples on interrupt programming are given in section 6.7.

6.9   Exercises

E6.1 What is the name given to a routine that is executed in response to an interrupt?

E6.2 What are the advantages of using interrupts to handle data inputs and output?

E6.3 What are the requirements of interrupt processing?

E6.4 How do you enable other interrupts when the 68HC11 is executing an interrupt service routine?

E6.5 Why would there be a need to promote one of the maskable interrupts to highest priority among all maskable interrupts?

E6.6 Write the assembler directives to initialize the $\overline{IRQ}$ interrupt vector located at $E200.

E6.7 What is the last instruction in most interrupt service routines? What does this instruction do?

E6.8 Suppose the 68HC11 is executing the following instruction segment and the $\overline{IRQ}$ interrupt occurs when the TSY instruction is being executed. What will be the contents of the top ten bytes in the stack?

```
ORG     $E000
LDS     #$FF
CLRA
LDX     #$1000
BSET    10,X $48
LDAB    #$40
INCA
TAP
PSHB
TSY
ADDA    #10
```

E6.9 Suppose the E clock frequency is 4 MHz. Compute the COP watchdog timer timeout period for all the possible combinations of the CR1 and CR0 bits in the OPTION register.

E6.10 Suppose the starting address of the service routine of the timer overflow interrupt is at $D000. Write the assembler directives to initialize its vector table entry on the EVB computer.

E6.11 Write an instruction sequence to clear the X and I bits in the CCR. Write an instruction sequence to set the S, X, and I bits in the CCR.

E6.12 Why does the 68HC11 need to be reset when the power supply is too low?

E6.13 Write the instruction sequence to prevent the COP timer from timing out and resetting the microcomputer.

6.10   Lab Exercises and Assignments

L6.1 *Simple interrupt.* Connect the $\overline{IRQ}$ pin of the demo board to a debounced switch that can generate a negative-going pulse. Write a main program and an $\overline{IRQ}$ service routine. The main program initializes the variable *irq_cnt* to 10, stays in a loop, and keeps checking the value of *irq_cnt*. When *irq_cnt* is decremented to 0, the main program jumps back to monitor. The $\overline{IRQ}$ service routine simply decrements irq_cnt by 1 and returns.

The lab procedure is as follows:

Step 1
Connect the $\overline{\text{IRQ}}$ pin of the demo board to a debounced switch that can generate a clean negative-going pulse.

Step 2
Enter the main program and $\overline{\text{IRQ}}$ service routine, assemble them, and then download them to the single-board computer. Remember to enable the $\overline{\text{IRQ}}$ interrupt in your program.

Step 3
Pulse the switch ten times.

If everything works properly, you should see the BUFFALO monitor prompt after ten pulses applied to the $\overline{\text{IRQ}}$ pin.

7

Parallel I/O Ports

7.1 Objectives

After completing this chapter you should be able to

- define I/O addressing methods
- explain the data transfer synchronization methods between the CPU and I/O interface chips
- explain the data transfer synchronization methods between the I/O interface chip and the I/O device
- explain input and output handshake protocols
- input data from simple switches
- input data from keyboards and keypads
- output data to LED and LCD displays
- explain the operation of the Centronics printer interface
- do I/O programming
- add parallel interface chips to the 68HC11
- do parallel interface chip programming

7.2   Basic I/O Concepts

I/O devices (also called peripheral devices) are pieces of equipment that exchange data with a computer. Examples of I/O devices include switches, light-emitting diodes, cathode-ray tube (CRT) screens, printers, modems, keyboards, and disk drives. The speeds and characteristics of these I/O devices differ significantly from those of the CPU, so they are not connected directly to the microprocessor. Instead, interface chips are used to resolve the differences between the microprocessor and these peripheral devices.

The major function of an interface chip is to synchronize data transfer between the CPU and an I/O device. An interface chip consists of control registers, status registers, data registers, latches, data direction registers, and control circuitry. Control registers allow us to set up parameters for the desired I/O operations. The status register records the progress and status of I/O operations. The data direction register allows us to set the direction of data transfer for bidirectional I/O pins. A data register may hold the new data placed by the input device, or data to be sent to the output device.

In input operations the input device places data in the data register, which holds data until they are read by the microprocessor. In output operations the processor writes data into the data register (on the interface chip), and the data register holds the data until it is fetched by the output device.

An interface chip has data pins that are connected to the microprocessor data bus and I/O port pins that are connected to the I/O device. Since a computer may have multiple I/O devices, the microprocessor data bus can be connected to the data buses of multiple interface chips. Data bus contention may occur if there are multiple devices driving the data bus at the same time. Severe damage may occur if one device is driving the data bus line too high while the other is driving the same bus line too low. This situation is called *bus contention*. To avoid data bus contention, the address decoder circuitry allows only one interface chip at a time to react to the microprocessor data transfer. Each interface chip is designed to have a chip enable signal input or inputs. When the chip enable signal is asserted, the interface chip is allowed to react to the data transfer request from the CPU. Otherwise, the data pins of the interface chip are isolated (or disabled) from the microprocessor data bus. The data pin drivers of the interface chip are normally designed to be tristate devices, so that they remain in a high-impedance state except when they are enabled. When the output pins of a peripheral device are in a high-impedance state, the device is electrically isolated from the data bus.

The simplified block diagram in Figure 7.1 shows a microcomputer system with one RAM module, one ROM module, one input device (not shown) connected to an interface chip, and one output device (not shown) connected to an interface chip. To transfer data to or from one of these modules, the microprocessor sends out an address, and the address decoder asserts one and only one of its outputs to enable one module to respond. Here, the word *module* refers to the set of devices (or a single chip) that is enabled by an address decoder output.

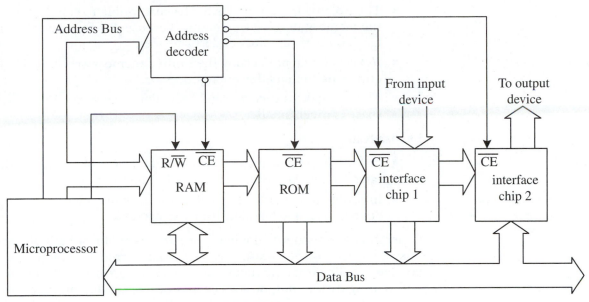

Figure 7.1 ■ A simple computer with RAM, ROM, and input and output ports

Data transfer between the I/O device and the interface chip can proceed bit-by-bit (serial) or in multiple bits (parallel). Data are transferred serially in low-speed devices such as modems and low-speed printers. Parallel data transfer is mainly used by high-speed I/O devices. We will discuss only parallel I/O in this chapter. Serial I/O will be discussed in chapters 9 and 10. Other issues related to I/O operations will be discussed in the following sections.

7.3   I/O Addressing

An interface chip normally has several registers. Each of these registers must be assigned a separate address in order to be accessed. It is possible to have two or more registers share the same address, but in this case some other control signal, such as the R/W̄ signal, must then be used to select one of the registers. This approach is very common in the interface chips designed for 8-bit microprocessors.

Different microprocessors deal with I/O devices differently. Some microprocessors have dedicated instructions for performing input and output operations—this approach is called *isolated I/O*. In this approach, input instructions, such as IN 1, read from the input devices. This input instruction inputs a word from input device 1 into the accumulator. Similarly, the output instruction OUT 4 outputs a word from the accumulator to output device 4. In the isolated I/O method, I/O devices have their own address space, which is separate from the main memory address space. The advantages of this method include:

- The isolated I/O method is not as susceptible to software errors as the memory-mapped I/O method because different instructions are used to access memory and I/O devices.
- I/O devices do not occupy the limited memory space, which is quite small in an 8-bit microprocessor.
- The I/O address decoder can be smaller because the I/O address space is much smaller.

Its disadvantages are:

- Inflexible I/O instructions—most microprocessors that use the isolated I/O method have only a few I/O instructions.
- Inflexible I/O addressing modes—the memory addressing modes are not available for addressing I/O devices.

Other microprocessors do not have separate instructions for input and output. Instead, these microprocessors use the same instructions for reading from memory and reading from input devices, as well as for writing data into memory and writing data to output devices. I/O devices and main memory share the same memory space. This approach is called *memory-mapped I/O*. In this method, an I/O address such as $1000 is considered a byte in memory space. A load instruction such as LDAA $1000 is then an input instruction, and a store instruction such as STAA $1000 is an output instruction. There is no need for a separate input or output instruction. The major advantage of memory-mapped I/O is that all instructions and addressing modes are available for I/O operations, so I/O programming is very flexible. The disadvantages of memory-mapped I/O are:

- Memory-mapped I/O is more susceptible to software errors. We may accidentally write into the main memory during an I/O operation or into an I/O device during a main memory access, thus corrupting the computer system.
- A smaller memory space is available for main memory.

Traditionally, Intel microprocessors use the isolated I/O method, whereas Motorola microprocessors use the memory-mapped I/O method. The 68HC11 microcontroller has no separate I/O instructions and uses memory-mapped I/O exclusively.

7.4 I/O Transfer Synchronization

The role of an interface chip is illustrated in Figure 7.2. The electronics in the I/O device converts electrical signals into mechanical actions or vice versa, but the microprocessor interacts with the interface chip instead of dealing with the I/O device directly.

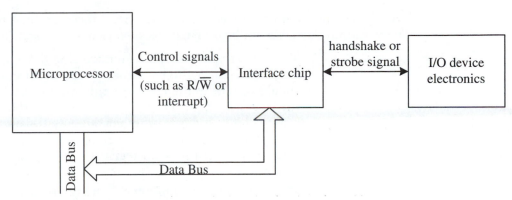

Figure 7.2 ■ The role of an interface chip

When inputting data, the microprocessor reads data from the interface chip, so there must be some mechanism to make sure data are valid when the microprocessor reads it. When outputting data, the microprocessor writes data into the interface chip. Again, there must be some mechanism to make sure that the output device is ready to accept data when the microprocessor outputs it. These two aspects of the I/O synchronization process—synchronization between the interface chip and the microprocessor and synchronization between the interface chip and the I/O device electronics—will be discussed in the following subsections.

7.4.1 Synchronizing the Microprocessor and Interface Chip

To input valid data from an input device, the microprocessor must make sure that the interface chip has correctly latched the data from the input device. There are two ways to do that:

1. *The polling method.* The interface chip uses a status flag to indicate whether it has valid data (normally stored in a data register) for the microprocessor. The microprocessor knows that the interface chip has valid data when the status flag is set to 1. The microprocessor keeps checking (reading) the status flag until the status flag is set to 1, and then it reads the data from the data register. When this method is used, the microprocessor is tied up and cannot do anything else when it is reading the data. However, this method is very simple to implement and is often used when the microprocessor has nothing else to do when waiting for completion of the input.

2. *Interrupt-driven method.* In this method, the interface chip asserts an interrupt signal to the microprocessor when it has valid data in the data register. The microprocessor then executes the service routine associated with the interrupt to read the data.

To output data successfully, the microprocessor must make sure that the output device is not busy. There are also two ways to do this:

1. *The polling method.* The interface chip has a data register that holds data that are to be output to an output device. New data should be sent to the interface chip only when the data register of the interface chip is empty. In this method, the interface chip uses a status bit to indicate whether the output data register is empty. The microprocessor keeps checking the status flag until it indicates that the output data register is empty and then writes the data into it.

2. *The interrupt-driven method.* The interface chip asserts an interrupt signal to the microprocessor when the output data register is empty and can accept new data. The microprocessor then executes the service routine associated with the interrupt and outputs the data.

7.4.2   Synchronizing the Interface Chip and I/O Device

The interface chip is responsible for making sure that data are properly transferred to and from I/O devices. The following methods have been used to synchronize data transfer between the interface chip and I/O devices:

1. *The brute-force method.* Nothing special is done in this method. For input, the interface chip returns the voltage levels on the input port pins to the microprocessor. For output, the interface chip makes the data written by the microprocessor directly available on output port pins. This method is useful in situations in which the timing of data is unimportant. It can be used to test the voltage level of a signal, set the voltage of an output pin to high or low, or drive LEDs. All I/O ports of the 68HC11 can perform brute force I/O.

2. *The strobe method.* This method uses strobe signals to indicate that data are stable on input or output port pins. During input, the input device asserts a strobe signal when the data are stable on the input port pins. The interface chip latches data into the data register using the strobe signal. For output, the interface chip first places data on the output port pins. When the data become stable, the interface chip asserts a strobe signal to inform the output device to latch the data on the output port pins. This method can be used if the interface chip and I/O device can keep up with each other. Port B of the 68HC11 supports the strobed output mode, while port C supports the strobed input method.

3. *The handshake method.* The previous two methods cannot guarantee correct data transfer between an interface chip and an I/O

device when the timing of data is critical. For example, it takes a much longer time to print a character than it does to send a character to the printer electronics, so data shouldn't be sent to the printer if it is still printing. The solution is to use a handshake protocol between the interface chip and the printer electronics. There are two handshake methods: the *interlocked handshake* and the *pulse-mode handshake*. Whichever handshake protocol is used, two handshake signals are needed—one (call it H1) is asserted by the interface chip and the other (call it H2) is asserted by the I/O device. The handshake signal transactions for input and output are described in the following subsections. Note that the handshake operations of some interface chips may differ slightly from what we describe here. Port C of the 68HC11 supports the handshake I/O method.

INPUT HANDSHAKE PROTOCOL

The signal transaction of the input handshake protocol is illustrated in Figure 7.3.

Step 1

The interface chip asserts (or pulses) H1 to indicate its intention to input new data.

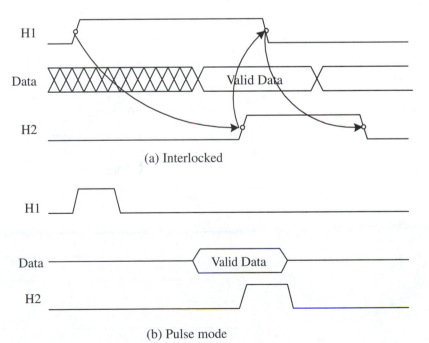

(a) Interlocked

(b) Pulse mode

Figure 7.3 ■ Input handshakes

Step 2
The input device puts valid data on the data port and also asserts (or pulses) the handshake signal H2.

Step 3
The interface chip latches the data and de-asserts H1. After some delay, the input device also de-asserts H2.

The whole process will be repeated if the interface chip wants more data.

OUTPUT HANDSHAKE PROTOCOL

The signal transaction of the output handshake protocol is shown in Figure 7.4. It also takes place in three steps:

Step 1
The interface chip places data on the data port and asserts (or pulses) H1 to indicate it has data to be output.

Step 2
The output device latches the data and asserts (or pulses) H2 to acknowledge the receipt of data.

Step 3
The interface chip de-asserts H1 following the assertion of H2. The output device then de-asserts H2.

The whole process will be repeated if the microprocessor has more data to be output.

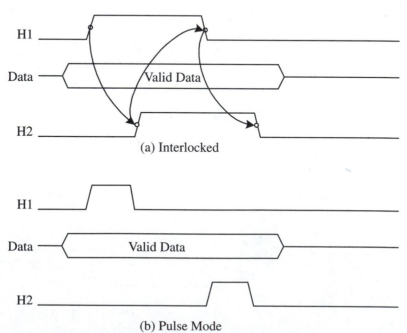

(a) Interlocked

(b) Pulse Mode

Figure 7.4 ■ Output handshaking

7.5   68HC11 Parallel I/O Overview

The 68HC11A8 has forty I/O pins arranged in five I/O ports. All of these pins serve multiple functions, depending on the operation mode and data in the control registers. Ports C and D are used as general-purpose I/O pins under the control of their associated data direction registers. Ports A, B, and E, with the exception of port A pin 7, are fixed-direction inputs or outputs and therefore do not have data direction registers. Port B, Port C, and STRA pin, and the STRB pin are used for strobed and handshake parallel I/O, as well as for general-purpose I/O.

Other series of 68HC11 microcontrollers have different numbers of I/O ports. Interested readers should refer to the appropriate technical data book.

7.5.1   Bidirectional I/O (Ports C and D)

Each pin of ports C and D has an associated bit in a specific port data register and another in a data direction register. The data direction register bits are used to specify the primary direction of data for each I/O pin. When an output pin is read, the value at the input to the pin driver is returned. When a pin is configured as input, that pin becomes a high-impedance input. If a write is executed to an input pin, its value does not affect the I/O pin but is stored in an internal latch. When the pin becomes an output, this value appears at the I/O pin. Data direction register bits are cleared by reset to configure I/O pins as inputs.

The AS and R/$\overline{\text{W}}$ pins are used for bus control in the expanded mode and as parallel I/O strobe signals (STRA and STRB) in the single-chip mode.

Other series of the 68HC11 have more bidirectional I/O ports than the 68HC11A8. A partial list is shown in Table 7.1. A data direction register is associated with each general-purpose bidirectional I/O port.

To configure a bidirectional I/O pin for input, set the associated direction bit in the data direction register to 0. To configure a bidirectional I/O pin for output, set the associated direction bit in the data direction register to 1.

Series	General I/O port names
A and E	8-bit port: C; 6-bit port; D
D	8-bit port: B, C, and D
F	8-bit port: A, C, and G; 6-bit port: D
G	8-bit port: A, C, G, H, and J; 6-bit port: D
K	8-bit port: A, B, C, F, G, and H; 6-bit port: D
L	8-bit port: C and G; 6-bit port: D
N	8-bit port: A, B, C, F, and H; 6-bit port: D

Table 7.1 ■ 68HC11 bidirectional I/O ports

7.5.2   Fixed Direction I/O (Ports A, B, and E)

In the 68HC11A8, the pins of ports A, B, and E (except for port A bit 7) have fixed data directions. When port A is being used for general-purpose I/O, bits 0, 1, and 2 are configured for input only, and writes to these pins have no effect. Bits 3, 4, 5, and 6 of port A are configured for output only, and reads of these pins return the levels sensed at the input to the pin drivers. Port A pin 7 can be configured as a general-purpose input or output using the DDRD7 bit (bit 7) of the pulse accumulator control register (PACTL). Port B is configured as output port only, and reads of these pins will return the levels sensed at the input of the pin drivers. Port E contains the eight A/D channel inputs, but these pins can also be used as general-purpose digital inputs. Writes to port E have no effect. The directions of port A pins vary with the chip version, as shown in Table 7.2.

7.5.3   Port Registers and I/O Operation

Each I/O port (with the exception of port C) has a data register associated with it. Port C also has a latched data register (which will be discussed shortly). The addresses of these data registers are listed in Table 7.3. The PORTCL register is part of the handshake I/O subsystem and will be explained in Section 7.7.

Chip version	Pin directions
A series (A0, A1, A7, A8)	1 bidirectional (PA7) 7 fixed directional, PA0-PA2 are input and PA3-PA6 are output
D series (D0, D3, 711D3)	2 bidirectional (PA7 and PA3), 4 fixed directional, PA0-PA2 are input and PA5 is output
E series (E0, E1, E9, E20, 711E9, 711E20, E8, 811E2)	2 bidirectional (PA7 and PA3), 6 fixed directional, PA0-PA2 are input and PA4-PA6 are output
F, G, K, L, M, and N series (see Table 4.1)	8 bidirectional (pin directions are to be set by programming the DDRA register)

1. The directions of PA7 and PA3 in the D and E series are set by programming the DDRA7 and DDRA3 bits of the PACTL register.
2. The direction of PA7 in A series is set by programming the DDRA7 bit of the PACTL register.

Table 7.2 ■ 68HC11 chip versions and port A pin directions

Port Register	Address
PORTA	$1000
PORTB	$1004
PORTC	$1003
PORTCL	$1005
PORTD	$1008
PORTE	$100A

Table 7.3 ■ Address of port data register

For a fixed direction I/O port, simply use a LOAD or STORE instruction to perform the I/O operation. For example,

```
LDAA    $100A
```

will load a byte from port E pins into accumulator A.

The instruction sequence

```
LDAA    #$36
STAA    $1004
```

will output the hex value $36 to port B pins.

So far, we have used the extended address to access I/O registers. Another way to access I/O registers is to use index-addressing mode. For example, to output the hex value $36 to port B, we can also use the following instruction sequence:

```
REGBAS    equ    $1000      ; I/O register block base address
PORTB     equ    $04        ; offset of PORTB data register from REGBAS


          LDX    #REGBAS
          LDAA   #$36
          STAA   PORTB,X
```

The advantage of this method is that the resulting code will be shorter when several I/O operations need to be performed.

To input from and output to a bidirectional port, we need to configure the port direction before the I/O operation is performed. The following example illustrates how to output a value to port D.

Example 7.1

▼ ───

Write an instruction sequence to set port D as an output, output the value $CD to it.

Solution: Since port D is bi-directional, we need to configure it for output.

```
REGBAS    equ      $1000
PORTD     equ      $08
DDRD      equ      $09


          LDX      #REGBAS       ; base address of I/O register block
          LDAA     #$3F          ; configure port D as an output port
          STAA     DDRD,X        ;       "
          LDAA     #$CD          ; output the value $CD to port D
          STAA     PORTD,X       ;       "
          END
```

Most C compilers (including ImageCraft C compiler) have defined the I/O port registers in an include file (for example, ImageCraft C defines them in hc11.h) so that users can reference them by using their names.

In C language, the operation in Example 7.1 can be implemented in two statements:

```
DDRD = 0x3F;      /* configure port D for output */
PORTD = 0xCD;     /* output data to port D */
```

▲

7.6   68HC11 Strobed I/O

Port B of the 68HC11 supports strobed output. Port C support strobed input as well as handshake input and output. The strobe mode and handshake mode I/O are controlled by the parallel I/O control register (PIOC), whose contents are shown in Figure 7.5.

When port C pins are configured to wired-or mode, they become open-drain I/O pins. In this configuration, external pull-up resistors are needed to pull port C pins to high. Wired-or mode allows port C pins to be directly connected to other open-drain–type pins. In this configuration, there is no danger of destructive conflicts if two output drivers try to drive the same node at the same time. The port C pin driver is illustrated in Figure 7.6.

When the CWOM bit is set to 1, the voltage applied to the gate of the P-channel transistor is high and hence disables (turns off the channel of) the P-transistor. The PIN will be pulled to low if the N-transistor gate input is high. Without a pull-up resistor, the drain voltage of the N-transistor (i.e., the pin voltage) will be indeterminate if the gate voltage of the N-transistor is low.

The parallel I/O strobe mode can be selected by setting the handshake bit (HNDS) in the PIOC register to 0. Port C becomes a strobed input port with the STRA pin as the edge-detecting latch command input. Port B becomes a strobed output port with the STRB pin as an output strobe signal. The EGA bit and the INVB bit define the active edges of STRA and STRB.

7	6	5	4	3	2	1	0	
STAF	STAI	CWOM	HNDS	OIN	PLS	EGA	INVB	PIOC at $1002

value after reset

0	0	0	0	0	U	1	1

STAF: *Strobe A flag.* This bit is set when a selected edge occurs on the STRA signal.

STAI: *Strobe A interrupt enable.* When the STAF and STAI bits are both equal to 1, a hardware interrupt request will be made to the CPU.

CWOM: *Port C wired-or mode.*
    0: All port C outputs are normal CMOS outputs.
    1: All port C outputs act as open-drain outputs.

HNDS: *Handshake/simple strobe mode select.*
    0: simple strobe mode
    1: handshake mode

OIN: *Output/input handshake.*
    0: input handshake
    1: output handshake

PLS: *Pulse/interlocked handshake operation.*
    0: interlocked handshake selected
    1: pulse handshake selected

EGA: *Active edge for STRA.*
    0: falling edge
    1: rising edge

INVB: *Invert STRB.*
    0: STRB active low
    1: STRB active high

Figure 7.5 ■ PIOC register

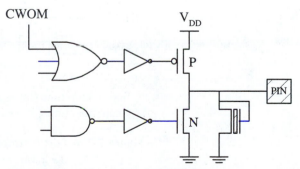

Figure 7.6 ■ Port C pin driver circuit

7.6.1   Strobed Input Port C

In this mode, there are two addresses where port C can be read, the PORTC data register and the latched port C register (PORTCL). Reading the PORTC register returns the current values on the port C pins. Reading the PORTCL register returns the contents of the latch PORTCL. The STRA pin signal edge loads the pin values into the PORTCL register. The EGA bit of the PIOC register selects the active edge of the STRA. The STRA pin interrupt may be generated to inform CPU the arrival of new data if both the STAI bit and the STAF flag are set to 1. The STRA interrupt shares the interrupt vector with the $\overline{\text{IRQ}}$ pin interrupt. The STAF flag will be cleared by reading the PIOC register and then reading the PORTCL register. The signal transaction for the port C strobed input (where the active edge of STRA is the rising edge) is shown in Figure 7.7. The parameter t_{IS} is the data setup time while t_{IH} is the data hold time required by port C strobed input.

7.6.2   Strobed Output Port B

In this mode, the STRB pin is a strobed output that is pulsed for two E clock periods each time there is a write to port B. The INVB bit in the PIOC register controls the polarity of the pulse on the STRB pin. The active level is low when the INVB bit is 0. The signal transaction for the port B strobe output is shown in Figure 7.8. The STRB signal is active high in the diagram. Port B data become valid t_{PWD} ns after the rising edge of the E clock. The STRB signal is then asserted (high in Figure 7.8) t_{DEB} ns after the falling edge of the E clock. The STRB signal will go high for two E clock cycles. An external output device will use the STRB signal to latch the port B data.

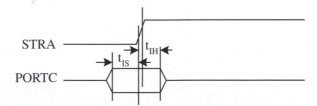

t_{IS}: input setup time (60 ns at 2 MHz)
t_{IH}: input hold time (100 ns at 2 MHz)

Figure 7.7 ■ Port C strobed input timing
(Redrawn with permission of Motorola)

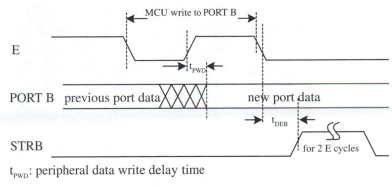

t_{PWD}: peripheral data write delay time

t_{DEB}: E fall to STRB valid delay time

Figure 7.8 ■ Port B strobe output timing
(Redrawn with permission of Motorola)

7.7   Port C Handshake I/O

The parallel I/O handshake modes involve port C and the STRA and STRB pins. There are two basic modes (input and output) and two variations on the handshake signals (pulsed and interlocked). In all handshake modes, STRA is an edge-detecting input and STRB is a handshake output pin.

7.7.1   Port C Input Handshake Protocol

In the input handshake protocol, port C is a latching input port, STRA is an edge-sensitive latch command from external system that is driving port C, and STRB is a ready output pin controlled by logic in the 68HC11.

When it is ready for new data, the 68HC11 asserts (or pulses) the STRB signal, then the external device places new data on the port C pins and asserts (or pulses) the STRA signal. The active edge of the STRA signal latches port C data into the PORTCL register, sets the STAF flag, and de-asserts the STRB signal. De-assertion of the STRB pin automatically inhibits the external device from strobing new data into port C. Reading the PORTCL register asserts the STRB signal, indicating that new data can now be applied to port C. Note that reading the PORTC data register does not assert the STRB signal. The signal transaction for port C handshake signals is shown in Figure 7.9. The STRB signal is active high in the diagram.

The STRB pin can be configured (by setting the PLS control bit) to be a pulse output (pulse mode) or a static output (interlocked mode). The port C data direction register bits should be cleared for each pin that is to be used as a latched input pin. However, some port C pins can be used as latched inputs with the handshake protocol while, at the same time, using some port C pins

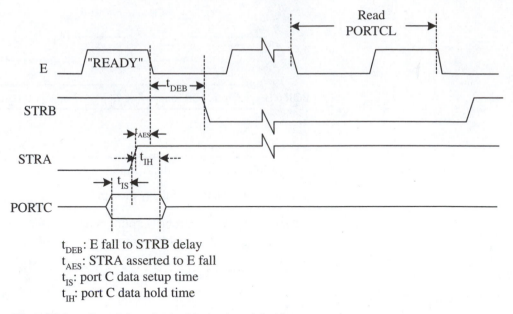

t_DEB: E fall to STRB delay
t_AES: STRA asserted to E fall
t_IS: port C data setup time
t_IH: port C data hold time

Figure 7.9 ■ Port C Interlocked input handshaking
(Redrawn with permission of Motorola)

as static inputs (read them from PORTC register), and some port C lines as static outputs. Reads of the PORTC register always returns the static logic level at port C pins (for pins configured as inputs). Writes to either the PORTC or PORTCL send information to the same port C output register without affecting the input handshake strobes.

7.7.2   Port C Output Handshake Protocol

In the output handshake protocol, port C is an output port, STRB is used to indicate that port C has new data ready, and STRA is used to indicate that the output data has been accepted by the external device. In a variation of this output handshake protocol, STRA can also be used as an output-enable input and as an edge-sensitive acknowledge input.

The 68HC11 places data on the port C output lines and then asserts STRB to indicate that data is already available. The external device then latches the data and pulses the STRA signal to indicate that new data may be placed on the port C output pins. The active edge on the STRA pin causes the STRB pin to be de-asserted and the STAF status flag to be set. In response to the setting of the STAF flag, our program transfers new data (if we want to) out of port C as required. Writing data to the PORTCL register causes data to appear on port

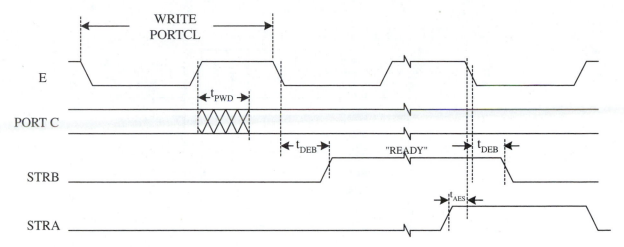

t_{PWD}: Peripheral data write delay time, 150 ns max (at 2 MHz)
t_{DEB}: E fall to STRB delay, 225 ns (at 2 MHz)
t_{AES}: STRA asserted to E fall setup time, 0 ns (at 2 MHz)

Figure 7.10 ■ Port C interlocked output handshake timing
(Redrawn with permission of Motorola)

C pins and asserts the STRB pin. The signal transaction for the port C hand-shake output is shown in Figure 7.10.

In Figure 7.10, both handshake signals are active high. When the 68HC11 writes a byte into the PORTCL register, the data becomes valid t_{PWD} ns after the rising edge of the E clock. The STRB handshake output is asserted t_{DEB} ns after the falling edge of the E clock signal, and the STRB signal is asserted by the output device to latch data. After latching the port C data, the output device asserts the STRA signal to inform the 68HC11 that the data have been latched. The STRB output is then dropped to tell the output device that the data on port C pins are not new. The output device also drops the STRA signal to complete the handshake cycle. When the STRA input pin is active, all port C pins obey the data direction register so that pins that are config-ured as inputs are high impedance. When the STRA pin is activated, all port C pins are forced to output regardless of the data in the data direction regis-ter. Port C pins intended as static outputs or normal handshake outputs should have their corresponding data direction register bits set, and pins intended as tristate handshake outputs should have their corresponding data direction register bits clear. Writing data to PORTC will change the non-handshake pins in port C, whereas writing data to PORTCL will change full-handshake port C pins.

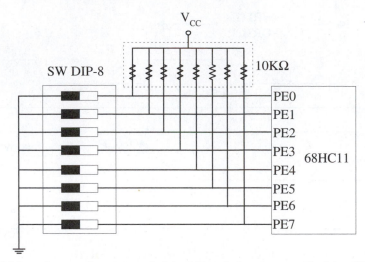

Figure 7.11 ■ Connecting a set of eight DIP switches to port E of the 68HC11

7.8   Simple Input Devices

An input device is any device that sends a binary number to the microprocessor. Examples of input devices include switches, analog-to-digital converters, keyboard, and so on. A set of eight DIP switches is probably the simplest input device that we can find. Such switches can be connected directly to port E or port C of the 68HC11, as shown in Figure 7.11. When a switch is closed, the associated port E input is 0. Otherwise, the associated port E pin has an input of 1. Each port E pin is pulled up to high via a 10KΩ resistor when the associated switch is open.

Example 7.2
▼

Write a sequence of instructions to read the value from an eight-DIP switch connected to port E of the 68HC11.

Solution:

```
REGBAS    EQU    $1000
PORTE     EQU    $0A

          LDX    #REGBAS      ; place the base address of the I/O register block in X
          LDAA   PORTE,X      ; read a byte from port E
```

In C language, this operation can be achieved as follows:

```
#include <hc11.h>
main ()
{
    char          xx;
    ...
    xx = PORTE;    /* read a byte from DIP switch */
    ...
}
```

7.9   Interfacing Parallel Ports to the Keyboard

A keyboard is arranged as an array of switches, which can be mechanical, membrane, capacitive, or Hall-effect in construction. In mechanical switches, two metal contacts are brought together to complete an electrical circuit. In membrane switches, a plastic or rubber membrane presses one conductor onto another; this type of switch can be made very thin. Capacitive switches internally comprise two plates of a parallel plate capacitor; pressing the key cap effectively increases the capacitance between the two plates. Special circuitry is needed to detect this change in capacitance. In a Hall-effect key switch, the motion of the magnetic flux lines of a permanent magnet perpendicular to a crystal is detected as a voltage appearing between the two faces of the crystal—it is this voltage that registers a switch closure.

Because of their construction, mechanical switches have a problem called *contact bounce*. Instead of producing a single, clean pulse output, closing a mechanical switch generates a series of pulses because the switch contacts do not come to rest immediately. This phenomenon is illustrated in Figure 7.12, where it can be seen that a single physical push of the button results in multiple electrical signals being generated and sent to the computer. (In Figure 7.12, when the key is not pressed, the voltage output to the computer rises to 5 V.) The response time of the switch is several orders of magnitude slower than that of a computer, so the computer could read a single switch closure

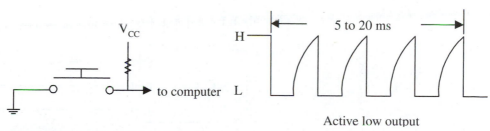

Figure 7.12 ■ Contact bounce

many times during the time the switch is operated, interpreting each low signal as a new input when in fact only one input is being sent.

Because of the contact bounce in the keyboard, a debouncing process is needed. A keyboard input program can be divided into three stages, which will be discussed in the following subsections:

1. Keyboard scanning to find out which key was pressed.
2. Key debouncing to make sure a key was indeed pressed.
3. Table lookup to find the ASCII code of the key that was pressed.

7.9.1 Keyboard Scanning Techniques

A keyboard can be arranged as a linear array of switches if it has only a few keys. A keyboard with more than a few keys is often arranged as a matrix of switches that uses two decoding and selecting devices to determine which key was pressed by coincident recognition of the row and column of the key.

As shown in Figure 7.13, a CMOS analog multiplexor MC14051 and a 3-to-8 decoder are used to scan the keyboard to determine which key is pressed. In this discussion, we will assume either no key or only one key is pressed. A scanning program is invoked to search for the pressed key, and then the debouncing routine is executed to verify that the key is closed. For a 64-key

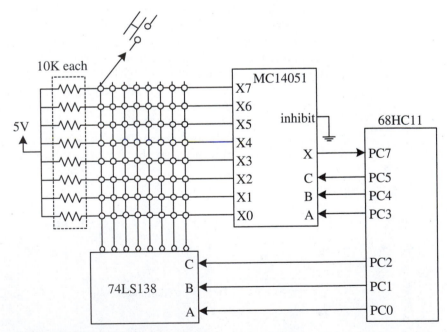

Figure 7.13 ■ Keyboard structure

keyboard, six bits are needed to select the rows and columns (three bits for each). The MC14051 has three select inputs (A, B, and C), eight data inputs (X0-X7), one inhibit input, and one common output (X). The common output is active high. When a data input is selected by the select signals, its voltage will be passed to the common output.

The MC14051 is used to select the row, and a 74LS138 3-to-8 decoder is used to select the column. The data inputs (X7-X0) of the MC14051 are pulled up to high. When a key is pressed, the row and the column where the keyboard is pressed are shorted together by the switch. Since the 3-to-8 decoder has low output, the common output of the MC14051 will be driven to low and can be detected by the microcontroller.

Port C is used to handle the keyboard scanning in the following program. Since port C is a bidirectional port, it must be configured properly. In our example, we will use the most significant bit of port C to detect whether a key has been pressed. Therefore, pin 7 of port C should be configured for input and the lower six bits should be configured for output. Bit 6 is not used. The following two instructions will configure port C as desired:

```
DDRC    EQU     $07
REGBAS  EQU     $1000
        LDAA    #$3F
        LDX     #REGBAS
        STAA    DDRC,X
```

If two adjacent keys in the same matrix row are pressed at the same time, the output pins of the 74LS138 will be shorted together and may possibly damage the decoder. Practical keyboard circuits include diodes to prevent such damage to the interface circuitry.

The ideas behind the keyboard scanning program are illustrated in the flowchart in Figure 7.14.

Since the lowest six bits of the PORTC register are used to specify the row number and column number, these two indices can be incremented by a single INC instruction. If the keyboard is not pressed and the lowest six bits of the PORTC registers are all 1s, the row number and column number must be reset to 0. The scanning program is as follows:

```
REGBAS  EQU     $1000
KEYBD   EQU     $03                          ; use port C for keyboard

        LDX     #REGBAS
resetc  CLR     KEYBD,X                      ; start from row 0 and column 0
scan    BRCLR   KEYBD,X $80 debnce           ; detect a pressed key?
        BRSET   KEYBD,X $3F resetc           ; need to reset row & column count?
        INC     KEYBD,X                      ; check the next row or column
        BRA     scan
        ...
```

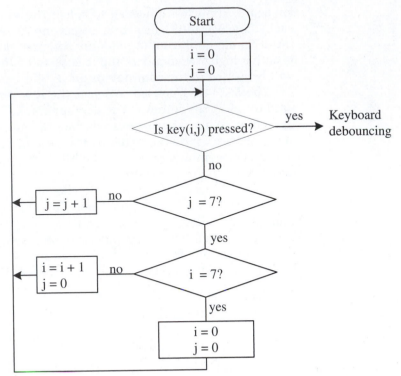

Figure 7.14 ■ Flowchart for keyboard scanning for the circuit in Figure 7.13

Keyboard decoding can be done strictly in software without using any hardware decoder or multiplexor. One such example is given in problem E7.16. A membrane-type keyboard is often used in products that require the user to make the selection. However, programs become much longer when a hardware decoder is not used.

7.9.2 Keyboard Debouncing

Contact bounce is due to the dynamics of a closing contact. The signal falls and rises a few times within a period of about 5 ms as a contact bounces. Since a human being cannot press and release a switch in less than 20 ms, a debouncer will recognize that the switch is closed after the voltage is low for about 10 ms and will recognize that the switch is open after the voltage is high for about 10 ms.

Both hardware and software solutions to the key bounce problem are available. Using a good switch can reduce keyboard bouncing. A mercury switch is much faster, and optical and Hall-effect switches are free of bounce. Hardware solutions to contact bounce include an analog circuit that uses a resistor and a capacitor to smooth the voltage and two digital solutions that use set-reset flip-flops or CMOS buffers and double-throw switches.

In practice, hardware and software debouncing techniques are both used (but not at the same time). Dedicated hardware scanner chips are also used frequently—typical ones are the National Semiconductor 74C922 and 74C923. These two chips perform keyboard scanning and debouncing.

HARDWARE DEBOUNCING TECHNIQUES

- *Set-reset flip-flops.* Before being pressed, the key is touching the set input. When pressed, the key moves toward the reset position. Before it has settled down, the key bounces. When the key touches the reset position, the voltage at Q goes low. When the key is bouncing, it touches neither the set terminal nor the reset terminal. In this situation, both the set and reset inputs are pulled down to low by the pull-down resistor. Since both set and reset inputs are grounded, the output Q will not change. Therefore, the key will be recognized as closed. This solution is shown in Figure 7.15a.

- *Noninverting CMOS gates with high-impedance.* When the switch is pressed, the input of the buffer 4050 is grounded and hence V_{OUT} is forced to the ground level. When the switch is bouncing, the feedback resistor R keeps the output voltage stay at low. This is due to the high input impedance of 4050, which causes a negligible voltage drop on the feedback resistor. Thus the output is debounced. This solution is shown in Figure 7.15b.

- *Integrating debouncers.* The RC constant of the integrator (smoothing filter) determines the rate at which the capacitor charges up toward the supply voltage once the ground connection via the switch has been removed. As long as the capacitor voltage does not exceed the logic zero threshold value, the V_{OUT} signal will continue to be recognized as a logic zero. This solution is shown in Figure 7.15c.

SOFTWARE DEBOUNCING TECHNIQUES

An easy software solution to the key bounce problem is the wait-and-see technique. When the input drops, indicating that the switch might be closed, the program waits 10 ms and looks at the input again. If it is high, the program decides that the input signal was noise or that the input is bouncing—if it is bouncing, it will certainly be cleared later if the key was actually pressed. In either case, the program returns to wait for the input to drop. The following sample program uses this technique to do the debouncing:

```
REGBAS     EQU      $1000
KEYBD      EQU      $03
TEN_MS     EQU      2000

debnce     LDY      #REGBAS
           LDX      #TEN_MS
wait       NOP                      ; wait for 10 ms
           NOP                      ;    "
           DEX                      ;    "
           BNE      wait            ;    "
           LDAA     KEYBD,Y         ; recheck the same key
           BMI      scan
           JMP      getcode         ; the key is indeed pressed
           ...
```

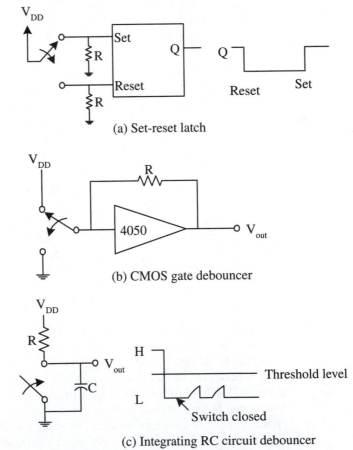

(a) Set-reset latch

(b) CMOS gate debouncer

(c) Integrating RC circuit debouncer

Figure 7.15 ■ Hardware debouncing techniques

7.9.3 ASCII Code Table Lookup

After the key has been debounced, the keyboard should reference the ASCII table and send the corresponding ASCII code to the CPU. The code segment for looking up the ASCII code in a table is:

```
keytab    FCC    "0123456789"
          FCC    ....

          ...
getcode   LDX    #REGBAS
          LDAB   KEYBD,X
          ANDB   #$3F         ; compute the address of the ASCII code
          LDX    #keytab      ; of the pressed key
          ABX
          LDAA   0,X

          ...
```

Example 7.3

Write a C routine to read a character from the keyboard. This routine will perform keyboard scanning, debouncing, and ASCII code lookup and return the ASCII code to the caller.

Solution: The C program follows the same logic flow. When detecting a pressed key, the function *read_kb* calls the function *delay10ms* to wait for 10 ms and recheck the same key. If the key is really pressed, *read_kb* calls the function *get_ascii* to read the ASCII code from the ASCII code table tab[64]. It is difficult to create an exact time delay using program loops in high-level language. Instead, we will use the output compare function to create a 10 ms delay. Output compare function will be discussed in chapter 8.

```
char get_ascii();
void delay10ms();
char tab[64] = {......};            /* ASCII code table */

char read_kb()
{
    char scanned;
    scanned = 0;
    PORTC = 0;
    while (1) {
        while (!scanned) {
            if (PORTC & 0x80) {       /* if key not pressed */
                if ((PORTC & 0x3F) == 0x3F) /* reach row 7 & column 7 */
```

```
                                PORTC = 0x00;       /* reset to row 0 & column 0 */
                        else
                                PORTC++;            /* scan the next key */
                }
                else
                        scanned = 1;                /* detect a pressed key */
        }
        delay10ms();                                /* wait 10 ms to recheck the same key */
        if (!(PORTC & 0x80))
                return (get_ascii());               /* the key is really pressed */
        else {
                scanned = 0;
        }
    }
}

/* the following routine use OC2 function to create 10 ms delay */
void delay10ms()
{
    TFLG1 = 0x40;                           /* clear OC2F flag */
    TOC2 = TCNT + 20000;                    /* start an OC2 operation to create 10 ms delay */
    while (!(TFLG1 & 0x40))                 /* wait until OC2F is set */
            ;
}
char get_ascii()
{
    char i;
    i = PORTC & 0x3F;                       /* obtain the row and column number for table */
    return tab[i];                          /* lookup */
}
```

▲

7.10   Interfacing 68HC11 to a Keypad

A 64-key keyboard is too much for many applications. Instead, people are using a 12- to 24-key keypad for many applications that need only very simple input capability. A membrane keypad with no more than 16 keys can easily be interfaced with an 8-bit bidirectional I/O port such as port C. An example is shown in Figure 7.16. In this example, the rows and columns of the keypad are simply conductors. Port C pins PC0-PC3 are pulled up to high by the pull-up resistors. Whenever a keypad is pressed, the corresponding row and column will be shorted together just like the case of a keyboard. To scan a row, we will set that row to low. The row selection of the 16-key keypad is shown in Table 7.4.

PC7	PC6	PC5	PC4	Selected keys
1	1	1	0	0, 1, 2, and 3
1	1	0	1	4, 5, 6, and 7
1	0	1	1	8, 9, A, and B
0	1	1	1	C, D, E, and F

Table 7.4 ■ Sixteen-key keypad row selections

For the circuit shown in Figure 7.16, we need to use software-only technique to perform keypad scanning and debouncing. We scan the keypad from the row controlled by PC4 toward the row controlled by PC7 and repeat.

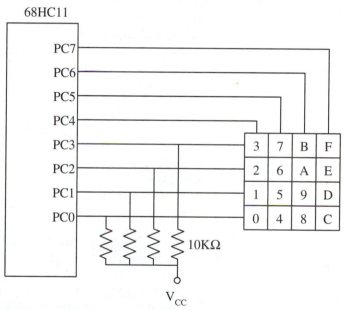

Figure 7.16 ■ Sixteen-key keypad connected to 68HC11

Example 7.4

Write a C program to perform keypad scanning and debouncing for the circuit shown in Figure 7.16. This program will continue to perform scanning until a key is pressed and return the ASCII code to the caller.

Solution: The following routine will continue to scan the keypad until it detects a key is pressed. The routine will then call a subroutine to wait for 10 ms and reexamine the same key. If the key is still pressed, then the ASCII code of the pressed key will be returned to the caller.

```c
void wait_10ms ( );

char get_key( )
{
    DDRC = 0xF0;                    /* configure upper 4 port C pins for output lower 4 pins
                                       for input */

    while (1) {
        PORTC = 0xE0;               /* set PC4 to low to scan the first row */
        if (!(PORTC & 0x01)) {
            wait_10ms( );
            if (!(PORTC & 0x01))
            return 0x30;            /* return the ASCII code of 0 */
        }
        if (!(PORTC & 0x02)) {
            wait_10ms( );
            if (!(PORTC & 0x02))
                return 0x31;        /* return the ASCII code of 1 */
        }
        if (!(PORTC & 0x04)) {
            wait_10ms( );
            if (!(PORTC & 0x04))
                return 0x32;        /* return the ASCII code of 2 */
        }
        if (!(PORTC & 0x08)) {
            wait_10ms( );
            if (!(PORTC & 0x08))
                return 0x33;        /* return the ASCII code of 3 */
        }
        PORTC = 0xD0;               /* set PC5 to low to scan the second row */
        if (!(PORTC & 0x01)) {
            wait_10ms( );
            if (!(PORTC & 0x01))
                return 0x34;        /* return the ASCII code of 4 */
        }
        if (!(PORTC & 0x02)) {
            wait_10ms( );
            if (!(PORTC & 0x02))
                return 0x35;        /* return the ASCII code of 5 */
        }
        if (!(PORTC & 0x04)) {
            wait_10ms( );
            if (!(PORTC & 0x04))
```

```
                return 0x36;        /* return the ASCII code of 6 */
        }
        if (!(PORTC & 0x08)) {
            wait_10ms( );
            if (!(PORTC & 0x08))
                return 0x37;        /* return the ASCII code of 7 */
        }
        PORTC = 0xB0;               /* set PC6 to low to scan the third row */
        if (!(PORTC & 0x01)) {
            wait_10ms( );
            if (!(PORTC & 0x01))
                return 0x38;        /* return the ASCII code of 8 */
        }
        if (!(PORTC & 0x02)) {
            wait_10ms( );
            if (!(PORTC & 0x02))
                return 0x39;        /* return the ASCII code of 9 */
        }
        if (!(PORTC & 0x04)) {
            wait_10ms( );
            if (!(PORTC & 0x04))
                return 0x41;        /* return the ASCII code of A */
        }
        if (!(PORTC & 0x08)) {
            wait_10ms( );
            if (!(PORTC & 0x08))
                return 0x42;        /* return the ASCII code of B */
        }
        PORTC = 0x70;               /* set PC7 to low to scan the fourth row */
        if (!(PORTC & 0x01)) {
            wait_10ms( );
            if (!(PORTC & 0x01))
                return 0x43;        /* return the ASCII code of C */
        }
        if (!(PORTC & 0x02)) {
            wait_10ms( );
            if (!(PORTC & 0x02))
                return 0x44;        /* return the ASCII code of D */
        }
        if (!(PORTC & 0x04)) {
            wait_10ms( );
            if (!(PORTC & 0x04))
                return 0x45;        /* return the ASCII code of E */
        }
```

```
        if (!(PORTC & 0x08)) {
            wait_10ms( );
            if (!(PORTC & 0x08))
                return 0x46;        /* return the ASCII code of F */
        }
    }
}
```

This program can be easily translated into assembly language and hence will be left as an exercise problem.

7.11   Simple Output Devices

A microprocessor is commonly connected to many more types of output devices than input devices. These output devices include light-emitting diode (LED) indicators, segmented displays constructed from LEDs, liquid crystal displays (LCDs), motors, relays, DACs (digital to analog converters), and even vacuum-tube devices such as fluorescent displays.

7.11.1   A Single LED

An LED will illuminate if it is forward-biased and has enough current flowing through it. Depending on the manufacturer, the current required to light an LED brightly may range from a few mA to more than 10 mA. The voltage drop across the LED when it is forward-biased can range from about 1.7 V to 2.3V.

LED indicators are easy to interface to the microcontroller if the interface circuit has at least 10 mA of current available to drive the LED. In the following discussion, we shall assume that the LED requires about 10 mA to light and that the forward voltage drop is about 2 V. Figure 7.17 illustrates a single

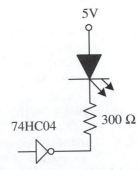

Figure 7.17 ■ A simple LED connected to a CMOS inverter through a current-limiting resistor

LED indicator connected so that a logic 1 causes it to illuminate. The 74HC04 inverter has a low output voltage of about 0.1 V and a high output voltage of about 4.9 V. The current flow through the LED is about 10 mA when the input to the inverter is 1. When the LED is lit, it glows red, green, or yellow. Notice in this interface that an inverter provides the required 10 mA of current through a series current-limiting resistor. The inverter selected here can sink up to 20 mA of current.

Example 7.5

Use the 68HC11 port B pins PB3, PB2, PB1, and PB0 to drive blue, green, red, and yellow LEDs. Light the blue LED for 2 seconds, then the green LED for 4 seconds, then the red LED for 8 seconds, and finally the yellow LED for 16 seconds. Repeat this operation forever.

Solution: The circuit connection is shown in Figure 7.18. As we have explained, sufficient current must be provided to illuminate the LEDs. The 68HC11 itself does not have the current capability to light the LEDs, so the 74HC04s are used to provide the driving capability required by the LED. The 74HC04 inverter has the capability to sink 20 mA of current.

Port B of the 68HC11 must output the appropriate values to turn the LEDs on and off. A two-level loop is used to create the desired delay. The inner loop creates a delay of 1/10 second and must be repeated the appropriate number of times. To create 2-, 4-, 8-, and 16-second delays, the inner loop must be repeated 20, 40, 80, and 160 times. The values $08, $04, $02, and $01 should be written into port B in order to turn on the blue, green, red, and yellow lights, respectively. The light pattern and the repetition count are placed in a table. The program reads the loop count and the light pattern until the end of the table is reached and then repeats. The assembly program is as follows:

```
REGBAS   equ     $1000
PORTB    EQU     $04

         ORG     $C000
loop     LDY     #lt_tab
next     LDAB    0,Y              ; get the repetition count
         LDAA    1,Y              ; select the port B pin to be pulled up to high
         INY                      ; move to the next entry
         INY                      ;         "
         LDX     #REGBAS
         STAA    PORTB,X
pt_lp    LDX     #20000
tenth_s  NOP                      ; 1/10 second delay loop
         NOP                      ;         "
```

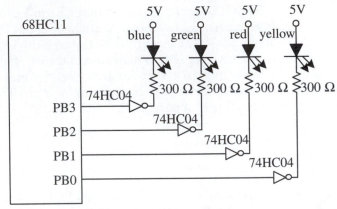

Figure 7.18 ■ LEDs circuit connection

```
          DEX                    ;     "
          BNE      tenth_s       ;     "
          DECB
          BNE      pt_lp
          CPY      #lt_tab+8     ; is this the end of lt_tab?
          BNE      next
          BRA      loop

lt_tab    FCB      20,$08        ; LED light pattern table
          FCB      40,$04        ;     "
          FCB      80,$02        ;     "
          FCB      160,$01       ;     "
          END
```

7.11.2   The Seven-Segment Display

As shown in Figure 7.19a, a seven-segment display consists of seven LED segments (a, b, c, d, e, f, and g) along with an optional decimal point (segment h). There are two types of seven-segment displays. In a *common-cathode seven-segment display*, the cathodes of all seven LEDs are tied together (must be connected to low voltage), and a segment will be lit whenever a high voltage is applied at the corresponding segment input. In a *common-anode seven-segment display*, the anodes of all seven LEDs are tied in common (must be connected to high voltage), and a segment will be lit whenever a low voltage is applied at the corresponding segment input. Diagrams of a common-cathode and common-anode seven-segment display are shown in Figure 7.19. The current required to light a segment is similar to that required to light an LED.

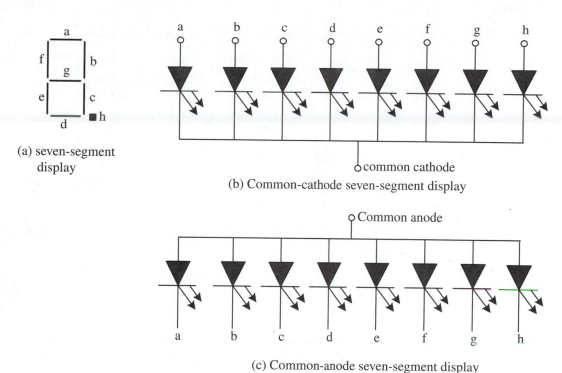

(a) seven-segment display

(b) Common-cathode seven-segment display

(c) Common-anode seven-segment display

Figure 7.19 ■ Seven-segment displays

Seven-segment displays are mainly used to display binary-coded-decimal (BCD) digits. Any output port of the 68HC11 can be used to drive seven-segment displays. Either port B or port C (port D does not have enough pins) can be used to output the segment pattern. One port will be adequate if there is only one display to be driven. The total supply current of the 68HC11 (shown in Table 7.5) is not enough to drive the seven-segment display directly, but this problem can be solved by adding a buffer chip such as 74ALS244. In Figure 7.20, a common-cathode seven-segment display is driven by port B.

Operation mode	Operating frequency	Total supply current
single-chip	DC ~ 2 MHz	15 mA
	3 MHz	27 mA
expanded	DC ~ 2 MHz	27 mA
	3 MHz	35 mA

Total supply current is the total current supplied by the power supply.

Table 7.5 ■ Total supply current of the 68HC11A8

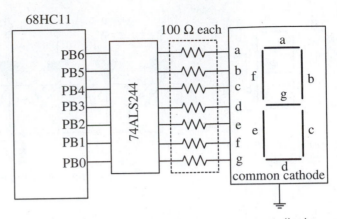

Figure 7.20 ■ Driving a single seven-segment display

The output high voltage (V_{OH}) of the 74ALS244 is about 3 V. Since the voltage drop of one LED segment is about 2.0V, a 100-Ω-current-limiting resistor could limit the current to about 10 mA, which would be adequate to light an LED segment. The light patterns corresponding to the ten BCD digits are shown in Table 7.6. The numbers in Table 7.6 requires that segments a, ..., g be connected from the most significant pin to the least significant pin of the output port.

We often have a need to display multiple BCD digits at the same time, and this can be accomplished by using the time-multiplexing technique, which will be discussed shortly. An example of circuit that displays six BCD digits is shown in Figure 7.21. In Figure 7.21, the common cathode of a seven-segment display is connected to the collector of an NPN transistor. When a port D pin voltage is high, the connected NPN transistor will be driven into saturation

| BCD | Segments | | | | | | | Corresponding |
digit	a	b	c	d	e	f	g	hex number
0	1	1	1	1	1	1	0	$7E
1	0	1	1	0	0	0	0	$30
2	1	1	0	1	1	0	1	$6D
3	1	1	1	1	0	0	1	$79
4	0	1	1	0	0	1	1	$33
5	1	0	1	1	0	1	1	$5B
6	1	0	1	1	1	1	1	$5F
7	1	1	1	0	0	0	0	$70
8	1	1	1	1	1	1	1	$7F
9	1	1	1	1	0	1	1	$7B

Table 7.6 ■ BCD to seven-segment decoder

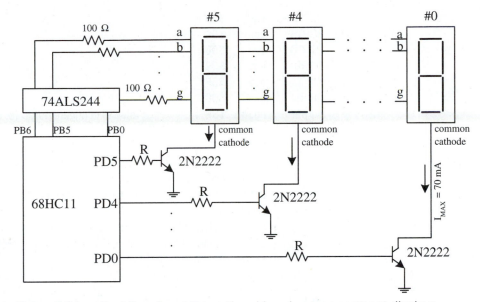

Figure 7.21 ■ Port B and port D together drive six seven-segment displays

region. The common cathode of the display will then be pulled down to low (about 0.2V), allowing the display to be lit. By turning the six NPN transistors on and off in turn many times in one second, multiple digits can be displayed. A 2N2222 transistor can sink from 100 mA to 300 mA of current. The maximum current that flows into the common cathode is about 70 mA (7×10 mA) and hence can be sunk by a 2N2222 transistor. The resistor R should be selected so that the transistor 2N2222 can be driven into saturation region and at the same time won't draw too much current from the 68HC11. A value of several hundred to 1K ohms will be OK.

Example 7.6

▼

Write a sequence of assembly instructions and C language statements to display 6 on the #3 seven-segment display.

Solution: To display a 6 on the #3 display, the numbers $5F and $08 should be written into the port B and port D data registers, respectively. The number $5F is the segment pattern of digit 6. The number $08 sets the port D pin PD3 to 1 and other port D pins to 0, allowing only the #3 display to light. The instruction sequence is as follows:

```
REGBAS    equ    $1000
PORTB     equ    $04
PORTD     equ    $08
DDRD      equ    $09
SIX       equ    $5F
third     equ    $08
output    equ    $3F              ; value to configure port D for output


...
LDX       #REGBAS
LDAA      #OUTPUT
STAA      DDRD,X            ; configure port D for output
LDAA      #SIX
STAA      PORTB,X           ; send out the light pattern of 6
LDAA      #third            ; select display #3 to be lighted
STAA      PORTD,X           ;        "
...
```

In C language, this can be achieved by the following statements:

```
DDRD = 0x3F;     /* configure port D for output */
PORTB = 0x5F;    /* output the segment of 6 */
PORTD = 0x08;    /* enable #3 display to light */
```

The circuit in Figure 7.21 can display six digits simultaneously by using the time-multiplexing technique, in which each seven-segment display is lighted in turn for a short period of time and then turned off. When one display is lighted, all the other displays are turned off. Within one second, each seven-segment display is lighted and then turned off many times. Because of the *persistence of vision*, the six displays will appear to be lighted simultaneously and continuously. In the following example, each display will be lighted in turn for 1 ms and turned off.

Example 7.7

Display 123456 on the six seven-segment displays shown in Figure 7.21.

Solution: We will display the digit 1, 2, 3, 4, 5, and 6 on display #5, #4, ..., and #0, respectively. To light these digits, we need to write the numbers $20, $10, $08, $04, $02, and $01 to port D respectively. A table lookup technique is best suited to this problem. The display patterns are shown in Table 7.7.

Seven-segment display	Displayed BCD digit	Port B	Port D
#5	1	$30	$20
#4	2	$6D	$10
#3	3	$79	$08
#2	4	$33	$04
#1	5	$5B	$02
#0	6	$5F	$01

Table 7.7 ■ Table of display patterns for example 7.6

This table can be created by using the following assembler directives:

```
display   FCB   $30,$20
          FCB   $6D,$10
          FCB   $79,$08
          FCB   $33,$04
          FCB   $5B,$02
          FCB   $5F,$01
```

The table lookup algorithm for this program is illustrated in Figure 7.22. If we use *display* to represent the starting address of the table, then *display+12* is the address of the byte immediately following the end of the table. The assembly program that implements this algorithm is as follows:

```
REGBAS   equ   $1000
PORTB    equ   $04
PORTD    equ   $08
DDRD     equ   $09
OUTPUT   equ   $3F

         ORG   $C000        ; starting address of the program
         LDX   #REGBAS
         LDAA  #OUTPUT
         STAA  DDRD,X       ; configure port D for output
forever  LDY   #display
next     LDAA  0,Y
         LDX   #REGBAS
         STAA  PORTB,X      ; send out the display pattern
         LDAA  1,Y
         STAA  PORTD,X      ; enable appropriate display to light
         INY
         INY
* the following 5 instructions create a delay of 1 ms
```

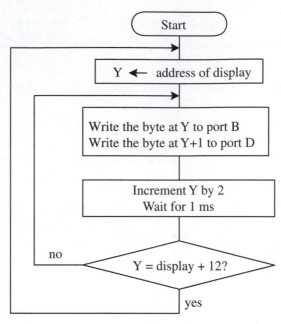

Note: display is the label of the display table

Figure 7.22 ■ Time-multiplexed seven-segment display algorithm

```
            LDX     #200
again       NOP
            NOP
            DEX
            BNE     again

            CPY     #display+12
            BEQ     forever
            BRA     next
display     FCB     $30,$20
            FCB     $6D,$10
            FCB     $79,$08
            FCB     $33,$04
            FCB     $5B,$02
            FCB     $5F,$01
            END
```

The C language version of this program is as follows:

```
#include              <hc11.h>
char display[6][2] =  {{0x30, 0x20}, {0x6D, 0x10}, {0x79, 0x08}, { 0x33, 0x04},
                      {0x5B, 0x02}, {0x5F, 0x01}};
```

```
void delay_1ms ( );
main ( )
{
    int i;
    while (1) {
        for (i = 0; i < 6; i++) {
            PORTB = display[i][0];
            PORTD = display[i][1];
            delay_1ms ( );
        }
    }
}


/* the following subroutine uses OC2 function to create 1 ms delay */
void delay_1ms ( )
{
    TFLG1    = 0x40;              /* clear OC2F flag */
    TOC2     = TCNT + 2000;
    while (!(TFLG1 & 0x40));
}
```

▲

7.11.3   Liquid Crystal Displays (LCDs)

The liquid crystal displays (LCDs) found in many digital watches and most calculators are among the most familiar output devices. The main advantage of the LCD is that it has very high contrast, unlike the LED display, so it can be seen extremely well in very bright light. The main problem with the LCD display is that it requires a light source in dim or dark areas because it produces no light of its own. Light sources are often included at the edge of the glass to flood the LCD display with light in dimly lit environments. However, the light consumes more power than the entire display.

Figure 7.23 illustrates the basic construction of an LCD display. The LCD type of display that is most common today allows light to pass through it whenever it is activated. Earlier LCD displays absorbed light. Like LEDs, LCDs are often constructed to display segmented digits. Activation of a segment requires a low-frequency bipolar excitation voltage of 30 to 1000 Hz. The polarity of this voltage must change, or else the LCD will not be able to change very quickly.

The LCD functions in the following manner. When a voltage is placed across the segment, it sets up an electrostatic field that aligns the crystals in the liquid. This alignment allows light to pass through the segment. If no voltage is applied across a segment, the crystals appear to be opaque because they are randomly aligned. Random alignment is assured by the AC excitation voltage applied to each segment. In a digital watch, the segments appear to darken

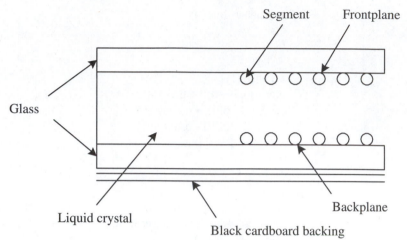

Figure 7.23 ■ A liquid crystal display (LCD)

when they are activated because light passes through the segment to a black cardboard backing that absorbs all light. The area surrounding the activated segment appears brighter in color because the randomly aligned crystals reflect much of the light. In a backlit computer display, the segment appears to grow brighter because of a light placed behind the display; the light is allowed to pass through the segment when it is activated.

A few manufacturers have developed color LCD displays for use with computers and small color televisions. These displays use three segments (dots) for each picture element (pixel), and the three dots are filtered so that they pass red, blue, and green light. By varying the number of dots and the amount of time each dot is active, just about any color and intensity can be displayed. White light consists of 59% green, 30% red, and 11% blue light. Secondary colors are magenta (red and blue), cyan (blue and green), and yellow (red and green). Any other colors, not just secondary colors, can be obtained by mixing red, green, and blue light.

LCDs are often sold in a module with LCDs and controller unit built in. The Hitachi HD44780 is one of the most popular LCD display controllers being used today. In the following section we will describe an LCD module that uses this controller.

7.11.4   The Optrex DMC-20434 LCD Kit

The DMC-20434 manufactured by Optrex is a 4 × 20 LCD kit that uses the HD44780 as its display controller. This LCD kit can plug into the LCD port provided by the CMD-11A8 evaluation board via a 14-pin cable. The block diagram and a photograph of the DMC-20434 are shown in Figure 7.24 and 7.25.

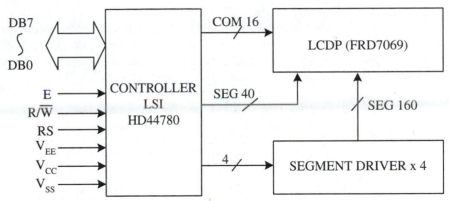

Figure 7.24 ■ Block diagram of the DMC-20434 LCD kit

The DB7~DB0 pins are used to exchange data with the microcontroller. The E pin is an enable signal to the kit. The $R/\overline{W}$ pin determines the direction of data transfer. The RS pin selects the register to be accessed. The value 1 selects the data register, whereas the value 0 selects the control register. The V_{cc} and V_{ss} pins should be connected to the power supply and the ground, respectively. The V_{EE} pin is used to control the brightness of the display and is often connected to a variable resistor.

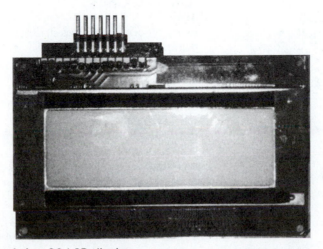

Figure 7.25 ■ A 4 × 20 LCD display

The DMC-20434 can be used as a memory mapped device and enabled by an address decoder. The signal E should be connected to one of the address decoder output in this configuration. The DMC-20434 can also be interfaced directly with an I/O port. In this configuration, we will need to use I/O pins to controller the E, R/$\overline{W}$, and RS pins.

Two addresses are assigned to the DMC-20434 in the CMD-11A8 demo board. The address $B5F0 is assigned to the control register, whereas the address $B5F1 is assigned to the data register.

THE SETUP OF THE DMC-20434 LCD KIT

All LCD kits need to be set up before they can be used. To setup the DMC-20434, we need to write commands to the command register. Commands that are used to configure the DMC-20434 are listed in Table 7.8. An LCD command takes much longer time to execute than a 68HC11 instruction does. Before a command is completed, we cannot send a new command to the LCD kit. We need to check the most significant bit (bit 7) of the LCD kit command register to find out if the LCD is ready for a new operation. A value 1 indicates that the LCD kit is still busy.

Command	Data	Time delay
Clear display, set cursor to home	$01	1.65 ms
Set cursor to home	$02	1.65 ms
Entry mode:		
cursor decrement, shift off	$04	40 µs
cursor decrement, shift on	$05	40 µs
cursor increment, shift off	$06	40 µs
cursor increment, shift on	$07	40 µs
Display control:		
display, cursor, and cursor blink off	$08	40 µs
display on, cursor and cursor blink off	$0C	40 µs
display and cursor on, cursor blink off	$0E	40 µs
display, cursor, and cursor blink on	$0F	40 µs
Cursor/Display shift:		
cursor shift left	$10	40 µs
cursor shift right	$14	40 µs
display shift left	$18	40 µs
display shift left	$1C	40 µs
Display function (default 4 x 20)	$3C	40 µs
Character generator RAM address set	$40-$7F	40 µs
Display RAM address set	$80-$FF	40 µs
(4 x 20 display = $80-$CF)		

Table 7.8 ■ Commands for DMC-20434 LCD kit

Example 7.8

Write a subroutine to initialize the DMC-20434 LCD kit.

Solution: The subroutine for LCD initialization is as follows:

```
LCD_CMD    EQU     $B5F0             ; address of the LCD control register
lcd_init   PSHX
           LDX     #LCD_CMD
           BRSET   0,X $80 *         ; wait until the LCD is not busy
           LDAA    #$3C
           STAA    0,X               ; set 4x20 display
           BRSET   0,X $80 *
           LDAA    #$01              ;
           STAA    0,X               ; clear display and move cursor to home
           BRSET   0,X $80 *
           LDAA    #$0F
           STAA    0,X               ; turn on display
           BRSET   0,X $80 *
           LDAA    #$06
           STAA    0,X               ; set cursor increment, shift off
           BRSET   0,X $80 *
           LDAA    #$14
           STAA    0,X               ; set cursor shift right
           BRSET   0,X $80 *
           LDAA    #$02
           STAA    0,X               ; move cursor to home
           BRSET   0,X $80 *
           PULX
           RTS
```

The addresses $B5F0 and $B5F1 are not I/O register addresses of the 68HC11. We will add the following two statements to the hc11.h file of the ImageCraft C compiler (hc11.h is an include file of the ImageCraft C compiler) so that we can use the symbols LCD_CMD and LCD_DAT to access the registers of the LCD kit:

```
#define LCD_CMD    *(unsigned char volatile *)(0xB5F0)
#define LCD_DAT    *(unsigned char volatile *)(0xB5F1)
```

The C language version of this routine is as follows:

```
void lcd_init()
{
    while(LCD_CMD & 0x80);    /* wait until LCD is ready */
    LCD_CMD = 0x3C;           /* set 4 × 20 display */
```

```
          while(LCD_CMD & 0x80);
          LCD_CMD = 0x01;              /* clear display and move cursor to home */
          while(LCD_CMD & 0x80);
          LCD_CMD = 0x0F;              /* turn on display */
          while(LCD_CMD & 0x80);
          LCD_CMD = 0x06;              /* turn display on, cursor and blink off */
          while(LCD_CMD & 0x80);
          LCD_CMD = 0x14;              /* set cursor increment, turn shift off*/
          while(LCD_CMD & 0x80);
          LCD_CMD = 0x02;              /* move cursor to home */
          while(LCD_CMD & 0x80);


      }
```

OUTPUT DATA ON THE DMC-20434 LCD KIT

The DMC-20434 can display four rows of alphanumeric characters. Each row can hold up to 20 characters. The DMC-20434 LCD kit does not display alphanumeric character in sequential order due to its internal line wrapping. If we did not adjust its line wrapping, it will display the first 20 characters in the first line, the second 20 characters in the third line, the third 20 characters in the second line, and the fourth 20 characters in the fourth line. If the string is longer than 80 characters, the fifth line will overwrite the first line, the sixth line will overwrite the third line, etc. Axiom Manufacturing provides the following assembly routine to output a character to the LCD kit, which performs the line wrapping adjustment:

```
LCD_CMD     EQU    $B5F0          ; LCD control register address
LCD_DAT     EQU    $B5F1          ; LCD data register address


lcdputch2lcd

            STAA   LCD_DAT        ; display character
lcdlp       LDAA   LCD_CMD        ; read next character position
            BMI    lcdlp          ; test if busy and wait if true

            CMPA   #$13           ; test for line 1 wrap
            BEQ    lcd1           ; if match, correct line wrap
            CMPA   #$53           ; test for line 2 wrap
            BEQ    lcd2           ; if match, correct line wrap
            CMPA   #$27           ; test for line 3 wrap
            BEQ    lcd3           ; if match, correct line wrap
            RTS
* correct line 1 wrap from line 3 to line 2
lcd1        LDAA   #$40           ; load line 2 start position
            ORAA   #$80           ; set command bit
            STAA   LCD_CMD        ; write to display
```

```
                RTS
* correct line 2 wrap from line 4 to line 3
lcd2            LDAA   #$14          ; load line 3 start position
                ORAA   #$80          ; set command bit
                STAA   LCD_CMD       ; write to display
                RTS
* correct line 3 wrap from line 2 to line 4
lcd3            LDAA   #$54          ; load line 4 start position
                ORAA   #$80          ; set command bit
                STAA   LCD_CMD       ; write to display
                RTS
```

We can write an assembly subroutine to output a string to the LCD kit by calling the previous subroutine. The following routine outputs a string pointed to by index register Y:

```
putstr2lcd      LDAA   0,Y
                BEQ    done
                JSR    putch2lcd
                INY
                BRA    putstr2lcd
done            RTS
```

The C language versions of these two routines are as follows:

```c
void putch2lcd (char ch)
{
    LCD_DAT = ch;
    while (LCD_CMD & 0x80);

    if (LCD_CMD == 0x13) {
        LCD_CMD = 0x40 | 0x80;      /* correct line 1 wrap from line 3 to line 2 */
        while (LCD_CMD & 0x80);
    }
    if (LCD_CMD == 0x53) {          /* correct line 2 wrap from line 4 to line 3 */
        LCD_CMD = 0x14 | 0x80;
        while (LCD_CMD & 0x80);
    }
    if (LCD_CMD == 0x27) {          /* correct line 3 wrap from line 2 to line 4 */
        LCD_CMD = 0x54 | 0x80;
        while (LCD_CMD & 0x80);
    }
}

void putstr2lcd (char *ptr)
{
    while (*ptr) {
```

```
        putch2lcd (*ptr);
        ptr++;
    }
}
```

7.11.5   Interfacing with a D/A Converter Using the 68HC11 Output Ports

Digital-to-analog conversion is required when a digital code must be converted to an analog signal. The analog signal can be used to control the output level of another system, for example, the flow level of a fluid system or the voltage level of a stereo system. A general D/A converter consists of a network of precision resistors, input switches, and level shifters that activate the switches that convert a digital code to an analog voltage or current. A D/A converter may also contain input or output buffers, amplifiers, and internal references.

D/A converters commonly have a *fixed* or *variable reference* voltage level. The reference voltage level can be generated internally or externally. It determines the switching threshold of the precision switches that form the controlled impedance network that controls the value of the output signal.

Fixed reference D/A converters have current or voltage reference output values that are proportional to the digital input. *Multiplying D/A converters* produce an output signal that is proportional to the product of a varying reference level and a digital code.

The AD557 by Analog Devices is an 8-bit D/A converter that produces an output voltage proportional to the digital input code. The AD557 can operate with a +5 V power supply, the reference voltage is generated internally, and input latches are available for microprocessor interfacing. The pin configuration of the AD557 is shown in Figure 7.26.

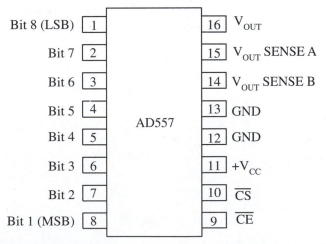

Figure 7.26 ■ AD557 Pin Layout

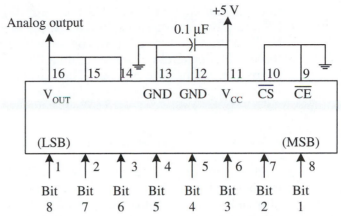

Figure 7.27 ■ AD557 unipolar output configuration, 0 V to 2.55 V operation

The AD557 has a unipolar 0 V to +2.55 V output range when operated as shown in Figure 7.27. The output voltage can be calculated using the following equation:

$$V_{OUT} = \text{(decimal equivalent of input code)} \times 0.01\ V$$

The output of this D/A converter can be amplified to any value needed in your application. The scaling circuit will be discussed in chapter 11.

The AD557 has input latches to simplify interfacing synchronous circuits and microprocessor systems to the AD557 for D/A conversion. The input latches are controlled by the chip select ($\overline{CS}$) and the chip enable ($\overline{CE}$) inputs. If the latches are not required, the chip select and chip enable inputs are grounded. In this mode, the output voltage settles within 1 μs after the digital value is applied at the input pins.

Table 7.9 shows the latched input operation of the AD557. Data is latched into the input latches on the rising edge of either $\overline{CS}$ or $\overline{CE}$. The data are held

Input data	$\overline{CE}$	$\overline{CS}$	D/A converter data	Latch condition
0	0	0	0	"transparent"
1	0	0	1	"transparent"
0	↑	0	0	latching
1	↑	1	1	latching
0	0	↑	0	latching
1	0	↑	1	latching
x	1	x	previous data	latching
x	x	1	previous data	latching

Table 7.9 ■ AD557 input latch operation

in the input latch until both $\overline{CS}$ and $\overline{CE}$ return to a logic low level. When both $\overline{CS}$ and $\overline{CE}$ are at the logic low level, the data are transferred from the input latch to the D/A converter for conversion to an analog voltage. By combining the use of a parallel output port and a D/A converter, a wide variety of waveforms can easily be generated. The following example generates a sawtooth waveform using port B of the 68HC11 and the AD557 D/A converter.

Example 7.9

▼

Use the 68HC11 port B and an AD557 D/A converter to generate a sawtooth waveform. The circuit connection is shown in Figure 7.28. At the beginning, a 0 is written into the PORTB register, causing the AD557 to output a minimum value close to 0 V. After a short delay, the PORTB register is incremented by 1 and the AD557 output is incremented by one step. This process continues until the hex value in the PORTB register reaches its maximum value of $FF and the AD557 output reaches its maximum analog output voltage. Incrementing the PORTB register one more time causes the hex number in the PORTB register to roll over to $00, so the AD557 output falls from its maximum value to its minimum value in one step. The process will repeat until the 68HC11 is reset or the power is turned off.

The following assembly program generates the sawtooth waveform:

```
REGBAS   EQU   $1000
PORTB    EQU   $04
```

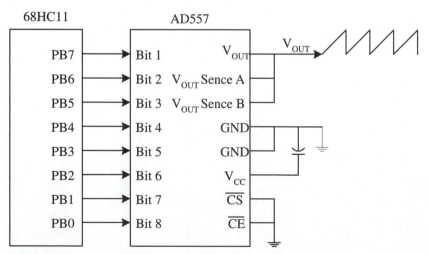

Figure 7.28 ■ Using an AD557 to generate a sawtooth waveform

```
          ORG    $C000
          LDX    #REGBAS
          CLR    PORTB,X
again     INC    PORTB,X        ; increment the output by one step
          BRA    again
          END
```

The C language version of this program is as follows:

```
#include    <hc11.h>
main ()
{
    PORTB = 0;
    while (1) {
        PORTB ++;
    }
    return 0;
}
```

Example 7.10

Using the same circuit as in Example 7.8, write a program to generate the waveform shown in Figure 7.29.

Solution: The solution to this problem is fairly straightforward and is outlined in the flowchart shown in Figure 7.30. The assembly program that generates this waveform is as follows:

```
REGBAS    EQU    $1000
PORTB     EQU    $04
          ORG    $C000
          LDY    #REGBAS
start     CLR    PORTB,Y    ; output 0 V
```

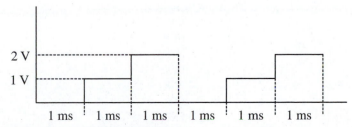

Figure 7.29 ■ Another periodic waveform

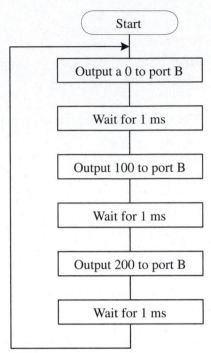

Figure 7.30 ■ Algorithm for generating
the waveform in Example 7.10

```
            JSR     delay_1ms    ; wait for 1 ms
            LDAA    #100
            STAA    PORTB,Y      ; output 1 V
            JSR     delay_1ms
            LDAA    #200
            STAA    PORTB,Y      ; output 2 V
            JSR     delay_1ms
            BRA     start
* the following subroutine creates a delay of 1 ms
delay_1ms   LDX     #200
again       NOP
            NOP
            DEX
            BNE     again
            RTS
            END
```

The C language version of the program is as follows:

```
#include    <hc11.h>
void delay_1ms ( );
main ( )
{
    while (1) {
        PORTB = 0;         /* output 0 V */
        delay_1ms ( );     /* wait for 1 ms */
        PORTB = 100;       /* output 1 V */
        delay_1ms ( );
        PORTB = 200;       /* output 2 V */
        delay_1ms ( );
    }
    return 0;
}
void delay_1ms ( )
{
    TFLG1   = 0x40;             /* clear OC2F flag */
    TOC2    = TCNT + 2000;      /* start an OC2 operation with 1 ms delay */
    While (!(TFLG1 & 0x40));    /* wait for 1 ms */
}
```

▲

7.12 Centronics Printer Interface

The standard Centronics printer interface allows the transfer of byte-wide, parallel data under the control of two handshake lines, $\overline{\text{DATA STROBE}}$ and $\overline{\text{ACKNLG}}$ (acknowledge). Short data setup (50 ns) and hold (100 ns) times with respect to the active edge of the $\overline{\text{DATA STROBE}}$ input are required. The Centronics interface provides the following signals:

- D7-D1: Data pins
- BUSY: Printer busy signal (output from printer electronics)
- PE: printer error
- SLCT: Printer on line (output from printer)
- $\overline{\text{DATA STROBE}}$: Input to the printer; active low. Data is latched into the printer by the falling edge of this signal.
- $\overline{\text{ACKNLG}}$: Acknowledge is active low and is an output from the printer.

PE, SLCT, and BUSY are status signals and are not all needed in any particular printer design. The Centronics interface timing is shown in Figure 7.31.

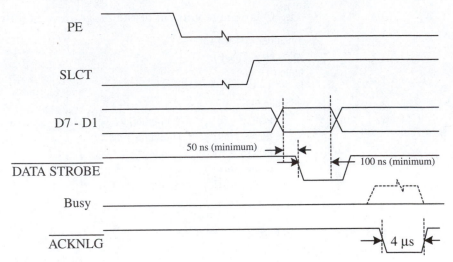

Figure 7.31 ■ Centronics printer interface timing requirements

Example 7.11

Interface a Centronics printer with 68HC11 port C.
You are expected to

1. synchronize data transfer between port C and the printer using the pulse-mode output handshake protocol.
2. use the polling method to determine whether the printer is ready for another character.
3. write routines to initialize the printer, output a character to the printer, and output a string that is terminated by a null character.

The BUSY signal is not needed in this example. The hardware connections are shown in Figure 7.32.

Solution: First, we need to verify that the data setup and hold time requirements with respect to the active edge of $\overline{\text{DATA STROBE}}$ are satisfied. A printer using the Centronics interface requires a data setup time of 50 ns and a data hold time of 100 ns. The data setup time (see Figure 7.8) is equal to

Period of E ÷ 2 − t_{PWD} + t_{DEB} = 250 ns − 150 ns + 225 ns = 325 ns > 50 ns (required)

The data hold time (see Figure 7.8) with respect to the active edge of $\overline{\text{DATA STROBE}}$ is also satisfied because port C data does not change during

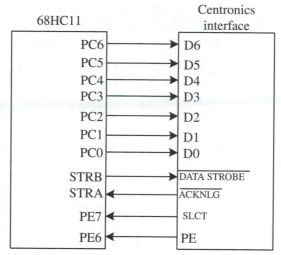

Figure 7.32 ■ Interfacing a Centronics printer
to port C

the period when STRB is active, which is two E cycles (1 μs at 2 MHz) in the pulse-mode output handshake.

Port C should be configured as follows:

1. Configure port C for output, i.e., write the value $FF into the DDRC register.

2. Program the PIOC register so that the following parameters are selected:

 ■ Output pulse mode handshake

 ■ STRA interrupt disabled

 ■ STRA falling edge as the active edge because $\overline{ACKNLG}$ is active low

 ■ STRB active low so that the falling edge of STRB can be used to latch data into the printer interface

 ■ Normal port C pins (not wire-ORcd)

With this setting, the value $1C should be written into the PIOC register.

The routines to initialize port C, output a character, and output a string using the polling method follow. A string is terminated by the null character (ASCII code 0). The printer initialization routine *prt_init* requires no incoming parameter; its role is to make sure that the printer is on-line and in working condition before it initializes the port C output function. Printer errors can be caused by paper jams, the printer being out of paper, and so on. The printer error condition is indicated by a logic high on the PE6 pin. The PE7 pin indicates whether the printer is on-line. Usually a printer is on-line when its power is on and there is a parallel connector between the computer and the printer.

```
REGBAS   EQU     $1000              ; base address of I/O register block
DDRC     EQU     $07                ; offset of DDRC from REGBAS
PIOC     EQU     $02                ; offset of PIOC from REGBAS
PORTCL   EQU     $05                ; offset of PORTCL from REGBAS
PORTE    EQU     $0A                ; offset of PORTE from REGBAS
OUTPUT   EQU     $FF                ; value to configure port C for output

prt_init PSHA
         PSHX
         LDX     #REGBAS
         BRSET   PORTE,X $40 *      ; don't go if printer error
         BRCLR   PORTE,X $80 *      ; don't go if printer is not on-line
         LDAA    #OUTPUT
         STAA    DDRC,X             ; configure port C for output
         LDAA    #$1C
         STAA    PIOC,X             ; initialize PIOC
         PULX
         PULA
         RTS
```

The C language version of this subroutine is as follows:

```
void prt_init ( )
{
    while ((PORTE & 0x40) || (!(PORTE & 0x80)));
    DDRC = 0xFF;
    PIOC = 0x1C;
}
```

The routine *prt_char*, which outputs the character in accumulator A, has two incoming parameters:

    1. the base address of the I/O register block, which is passed in index register X.

    2. the character to be output, which is passed in accumulator A.

The subroutine *prt_char* uses polling method to output a character.

```
prt_char BRCLR   PIOC,X $80 *      ; poll the STAF flag
         STAA    PORTCL,X
         RTS
```

The C language version of this routine is as follows:

```
void prt_char (char ch)
{
    while (!(PIOC & 0x80));
    PORTCL = ch;
}
```

The subroutine prt_str, which outputs a string, has two incoming parameters:

1. the base address of the I/O register block, which is passed in index register X.
2. the starting address of the string to be output, which is passed in index register Y.

This routine gets a character from the string and then calls the subroutine *prt_char* to output it; it repeats this operation until the character is a NULL character, which indicates the end of the string. After outputting a character, the string pointer Y is incremented by one.

```
prt_str    PSHA
next       LDAA    0,Y            ; get the next character
           BEQ     quit           ; is it a null character?
           JSR     prt_char
           INY                    ; move the string pointer
           BRA     next
quit       PULA
           RTS
```

The C language version of this routine is as follows:

```
void prt_str (char *ptr)
{
    while (*ptr) {
        prt_char (*ptr);
        ptr ++;
    }
}
```

Example 7.12

Rewrite the *prt_char* and *prt_str* routines in Example 7.10 using the interrupt-driven approach.

Solution: In the interrupt-driven output method, the *prt_str* routine is called with the same parameters as in Example 7.10. This routine enables the interrupt, gets the next character to be output, and waits for the printer interrupt. When there is an interrupt, the interrupt service routine *prt_char* is executed and a character is printed. This process is repeated until the whole string is printed and the routine *prt_str* disables the interrupt and returns to the caller. The printer initialization routine is identical to its counterpart in the polling method.

The routine *prt_str* has two incoming parameters:

1. the I/O register block base address, which is passed in index register X.
2. the starting address of the string to be output, which is passed in index register Y.

The string pointer will be incremented by the printer interrupt service.

```
prt_str   BSET    PIOC,X $40      ; enable the STRA interrupt
          CLI
again     LDAA    0,Y             ; wait for the interrupt
          BNE     again           ; at the end of the string?
          BCLR    PIOC,X $40      ; disable STRA interrupt when done
          SEI
          RTS
```

The routine prt_char is implemented as an interrupt service routine and the contents of Y are incremented during the output process. Because all CPU registers are pushed into the system stack and then restored when the interrupt is exited, the interrupt service routine must also update the copy of the Y register in the stack. The new value of Y will be lost if this is not done. The old copy of the index register Y is located five bytes below the top of the stack. The interrupt-driven *prt_char* routine is as follows:

```
prt_char  STAA    PORTCL,X
          INY
          TSX                     ; X points to the top byte of the stack
          STY     5,X             ; update the Y value in the stack
          RTI
```

The interrupt vector table should also be set up properly. The starting address of the STRA interrupt is located at $FFF2.

```
ORG   $FFF2
FDB   prt_char
```

For EVB, EVBU, and CMD-11A8 evaluation boards, we need to set up the interrupt vector jump table as follows:

```
ORG   $00EE
JMP   prt_char
```

In C language, we need to use global variable to point to the string to be output so that it becomes accessible to the interrupt service routine because it cannot receive any incoming parameter. Assume we add the following statement as a global statement:

```
char *ptr;    /* global character pointer */
```

We need to assign the starting address of the string to be output to this pointer before calling the *prt_str* routine.

The interrupt-driven version of the *prt_str* routine is as follows:

```
void prt_str ( )
{
    PIOC |= 0x40;     /* enable STRA interrupt */
    INTR_ON ( );
    while (*ptr);     /* wait for the interrupt to occur */
    PIOC &= 0xBF      /* disable STRA interrupt */
    INTR_OFF ( );
}
```

The STRA interrupt service routine is as follows:

```
#pragma interrupt_handler prt_char ( )
void prt_char ( )
{
    PORTCL = *ptr++;
}
```

▲

7.13   The i8255 Programmable Peripheral Interface

The 68HC11 has very limited number of I/O ports. If external memory is needed, then both port B and C cannot be used for I/O. If more I/O ports are needed, there are three solutions:

- Add parallel interface chip(s) to expand the number of parallel ports. The Intel Programmable Peripheral Interface (PPI) chip i8255 is a parallel interface chip that can be added to the 68HC11 to expand its number of parallel ports. The CMD-11A8 evaluation board incorporates an i8255 chip.
- Use the serial peripheral interface (SPI) and its supporting chips. This alternative will be discussed in chapter 11.
- Choose a member that has more I/O ports.

The Intel i8255 PPI has three 8-bit ports: Port A, B, and C. Port A and B can be programmed as either input or output, port A can also be programmed as bidirectional, and port C can be programmed as input, output, or a pair of 4-bit bidirectional control ports (one for port A and one for port B).

Furthermore, port C lines can be individually set or reset in order to generate strobe signals for controlling external devices.

The signals and internal organization of the i8255 are shown in Figure 7.33. In Figure 7.33, signals A1-A0 select the registers within the PPI as follows:

A1	A0	port selected
0	0	Port A
0	1	Port B
1	0	Port C
1	1	control (write only)

7.13.1   Operation of the i8255

Address inputs A1 and A0 select one of the four i8255 registers: port A data, port B data, port C data, and control registers. The control register controls the overall operation of the i8255. Depending on the setting of its most significant bit, the control register serves two functions: when set to 1, mode definition; when cleared to 0, bit set/reset. The mode definition format and bit set/reset format of the control register are illustrated in Figure 7.34 and Figure 7.35. Any of the eight bits of port C can be set or reset by writing a value into the port C control register as shown in Figure 7.35.

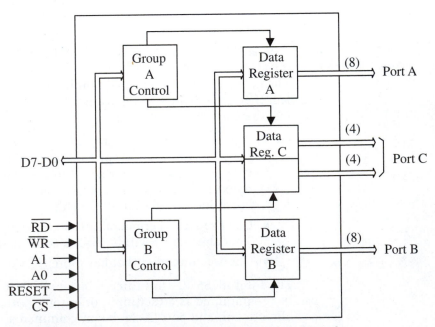

Figure 7.33 ■ Intel i8255 programmable peripheral interface

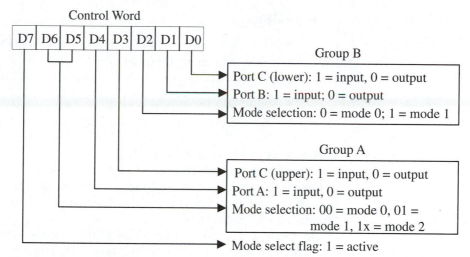

Figure 7.34 ■ i8255 control register mode definition format

7.13.2   i8255 Operation Modes

The i8255 has three operation modes: mode 0, mode 1, and mode 2.

MODE 0 (BASIC INPUT AND OUTPUT)

This functional configuration provides simple input and output operations for each of the three ports. No "handshaking" is required, data are simply written onto or read from a specified port.

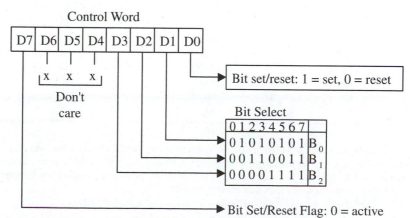

Figure 7.35 ■ i8255 control register bit set/reset format

In mode 0,

- There are two 8-bit ports and two 4-bit ports.
- Any port can be input or output.
- Outputs are latched.
- Inputs are not latched.
- 16 different input/output configurations are possible in this mode.

MODE 1 (STROBED INPUT/OUTPUT)

This functional configuration provides a means for transferring I/O data to or from a specified port in conjunction with strobes or "handshaking" signals. In mode 1, port A and port B use the lines on port C to generate or accept these "handshaking" signals.

In mode 1, three ports are divided into two groups. Each group contains one 8-bit port and one 4-bit control/data port. The 8-bit data port can be either input or output. Both input and output are latched. The 4-bit port is used for control and status of the 8-bit data port. The functions of Port C pins are illustrated in Table 7.10.

The function of a port C pin depends on whether its associated port is input or output. For example, when port A is configured for input, the pins PC7 and PC6 are used as general-purpose I/O whose directions are set by bit 3 of the control register. The signal transactions for mode 1 input and output are shown in Figure 7.36 and Figure 7.37, respectively.

Pin	Pin Function			
	Port A as input	Port A as output	Port B as input	Port B as output
PC7	I/O	$\overline{OBF}_A$	NA	NA
PC6	I/O	$\overline{ACK}_A$	NA	NA
PC5	IBF_A	I/O	NA	NA
PC4	$\overline{STB}_A$	I/O	NA	NA
PC3	$INTR_A$	$INTR_A$	NA	NA
PC2	NA	NA	$\overline{STB}_B$	$\overline{ACK}_B$
PC1	NA	NA	IBF_B	$\overline{OBF}_B$
PC0	NA	NA	$INTR_B$	$INTR_B$

Note:
$\overline{OBF}_x$ (X = A or B): Output buffer full flip-flop. The OBF will go low to indicate that the CPU has written data out to the specified port.
$\overline{ACK}_x$: Acknowledge input. A low on this pin informs the i8255 that the data from port A or B have been accepted.
$INTR_x$: Interrupt request. A high on this pin can be used to interrupt the CPU when an output device has accepted data transmitted by the CPU.
$\overline{STB}_x$: Strobe Input. A low on this input loads data into the input latch.
IBF_x: Input buffer full. A high on this pin indicates that data have been loaded into the input latch.

Table 7.10 ■ Port C pin functions in mode 1

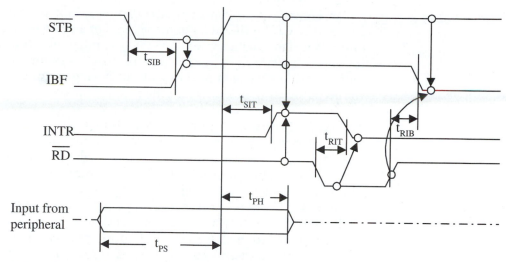

Figure 7.36 ■ Mode 1 input timing diagram

In Figure 7.36, when the input buffer is not full, the input device places data on input port pins and pulses the STB signal. The falling edge of the STB signal latches data into the input port data register and sets the input buffer full flip-flop. The i8255 asserts the interrupt request signal INTR and the CPU reads the data in the interrupt service routine. After the input data have been read, the input buffer full flag is cleared. This process will be repeated if there are more data to be input.

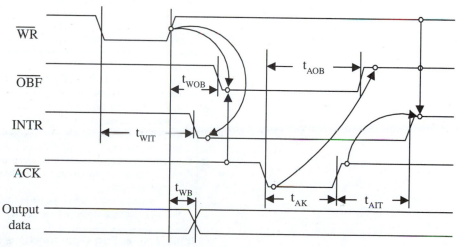

Figure 7.37 ■ Mode 1 output timing diagram

In Figure 7.37, when the output buffer is not full and the CPU is interrupted, the CPU writes data to the i8255. For a delay of t_{WB} after the rising edge of the WR signal, data appear on i8255 port pins. The $\overline{OBF}$ signal, which stands for *output buffer full*, is asserted by the write operation. The falling edge of the $\overline{OBF}$ signal latches data into the output device and the output device asserts the $\overline{ACK}$ signal to acknowledge the receipt of data.

MODE 2 (STROBED BIDIRECTIONAL BUS I/O)

This configuration provides a means for communicating with a peripheral device or structure on a single 8-bit bus for both transmitting and receiving data. "Handshaking" signals are provided to maintain proper bus flow discipline in a manner similar to mode 1.

In mode 2, only port A is used. Port A becomes an 8-bit bidirectional bus port whereas five bits (PC3~PC7) of port C are used as control port. Both input and output are latched. Functions of port C pins are shown in Table 7.11.

The signal transactions for mode 2 operation are shown in Figure 7.38. Like mode 1 operation, data I/O can be either hardware (interrupt) driven or software (polling) driven.

Pin	Pin Function
PC7	$\overline{OBF}_A$
PC6	$\overline{ACK}_A$
PC5	IBF_A
PC4	$\overline{STB}_A$
PC3	$INTR_A$
PC2	I/O
PC1	I/O
PC0	I/O

Note:
$\overline{OBF}_A$: Output buffer full. The OBF will go low to indicate that the CPU has written data out to port A.
$\overline{ACK}_A$: Acknowledge input. A low on this pin enables the tristate output buffer of port A to send out data.
$INTR_A$: Interrupt request. A high on this pin can be used to interrupt the CPU for both input or output.
$\overline{STB}_A$: Strobe Input. A low on this pin loads data into the input latch.
IBF_A: Input buffer full. A high on this pin indicates that data have been loaded into the input latch.

Table 7.11 ■ Port C pin functions in mode 2

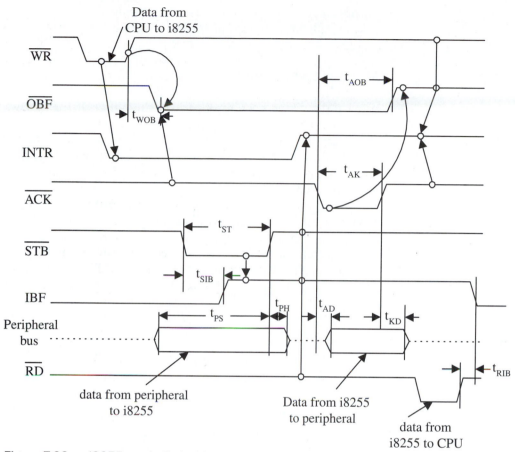

Figure 7.38 ■ i8255 mode 2 signal transactions

MODE 2 DATA OUTPUT

In the software-driven approach, the software tests the $\overline{\text{OBF}}$ signal to determine whether the output buffer is empty. If it is, then data are written out via the STORE instruction. The external circuitry also monitors the $\overline{\text{OBF}}$ signal to decide if the microprocessor has sent new data to the i8255. As soon as the output circuitry sees logic 0 on $\overline{\text{OBF}}$, it sends back the $\overline{\text{ACK}}$ signal to remove data from the output buffer. The low level of the $\overline{\text{ACK}}$ signal causes the $\overline{\text{OBF}}$ signal to go high and also enables the tristate output buffer so that data may be read.

In the interrupt-driven approach, the software first enables the i8255 to interrupt the microprocessor. When the output buffer is empty (indicated by the high level of $\overline{\text{OBF}}$ pin), it generates an interrupt request to the CPU. The CPU outputs a byte to the i8255 in the interrupt service routine. The byte written into the i8255 causes the $\overline{\text{OBF}}$ signal to go low, which informs the output device to latch the data. The output device asserts the $\overline{ACK}$ signal to acknowledge the receipt of the data, which causes the $\overline{OBF}$ signal to become inactive.

MODE 2 DATA INPUT

In the software-driven approach, the program tests the IBF signal to determine if data have been strobed into the buffer. If IBF is low, the input device places data on port A pins and asserts the $\overline{STB}$ signal to strobe data into the input buffer. After this, the IBF signal goes high. When the user program detects that the IBF signal is high, it reads the data from the input buffer. The IBF signal goes low after the data is read. This process will be repeated as long as there are data to be input.

In the interrupt-driven approach, the software enables the i8255 to interrupt the microprocessor. Whenever the IBF signal is low, the input device places data on the port pins and asserts the $\overline{STB}$ signal to strobe data into the input buffer. After data are strobed into the input buffer, the IBF signal is asserted and an interrupt is generated. The CPU reads the data from the input buffer when executing the interrupt service routine. Reading data brings the IBF signal to low, which further causes the $\overline{STB}$ signal to go high.

Four bytes are assigned to the i8255 on the CMD-11A8 evaluation board, where

1. $B5F4 is assigned to port A register
2. $B5F5 is assigned to port B register
3. $B5F6 is assigned to port C register
4. $B5F7 is assigned to control register

Example 7.13

▼

Write down an instruction sequence to configure the i8255 on the CMD-11A8 evaluation board to operate in mode 0:

- Port A for input
- Port B for output
- Upper port C for input
- Lower port C for output

Solution: The setting of the control register is as follows:

- bit 7: set to 1 to choose mode select
- bits 6 & 5: set to 00 to configure port A for mode 0 operation
- bit 4: set to 1 to configure port A for input

- bit 3: set to 1 to configure upper port C pins for input
- bit 2: set to 0 to configure port B for mode 0 operation
- bit 1: set to 0 to configure port B for output
- bit 0: set to 0 to configure lower port C pins for output

We need to write the binary value 10011000_2 into the control register. The following instructions will configure the i8255 as desired:

```
LDX    #$B5F4
LDAA   #$98        ; place the control word into accumulator A
STAA   3,X         ; write out the contents of accumulator A
```

Example 7.14

Write down an instruction sequence to configure port A for output in mode 1 and port B for input also in mode 1 for the i8255 in the CMD-11A8 evaluation board. Configure the unused port C pins (PC4 and PC5) as general input.

Solution: To configure pins PC4 and PC5 for input, set the bit 3 of the control register to 1. The value to be written into the control register is 10101110_2. The following instructions will configure the i8255 as desired:

```
LDX    #$B5F4
LDAA   #$AE        ; place the control word in accumulator A
STAA   3,X         ; write out the control word
```

Example 7.15

Use the auxiliary port of the CMD-11A8 evaluation board to drive four seven-segment displays. Use the port A of the i8255 to drive the segment pattern and use the upper four bits of the port B (i8255) to drive the display select signals. The circuit connection is shown in Figure 7.39. Write an instruction sequence to display 1999 on these displays.

Solution: We need to configure i8255 in mode 0 for this operation because we are not using the handshake capability of the i8255. Because both port A and B must be configured for output, we need to write the value of $80 into the control register. We will light the four seven-segment displays in turn every 1 ms. The resistance value R should be in the range from several hundred to

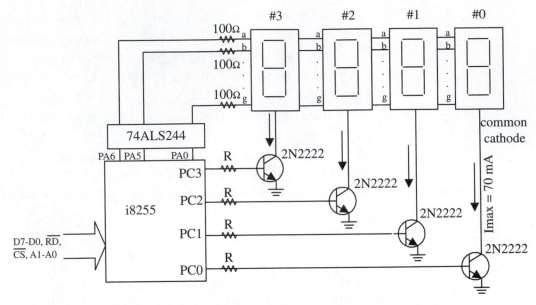

Figure 7.39 ■ Use port A and C of i8255 to drive four seven-segment displays

1 KΩ so that the NPN transistor is driven into saturation region when a high voltage is output from a port B pin.

The following program will achieve what we want:

```
PPI_BAS   EQU   $B5F4         ; i8255 port A data register address
INIT_VAL  EQU   $80           ; value to initialize the i8255 control register
PA        EQU   0             ; offset of port A from PPI_BAS
PB        EQU   1             ; offset of port B from PPI_BAS


          LDAA  #INIT_VAL
          LDX   #PPI_BAS
          STAA  3,X           ; initialize the i8255 control register
forever   LDY   #distab       ; place the base address of the display table in Y
next      LDX   #PPI_BAS      ; place the base address of i8255 in X
          LDAA  0,Y
          STAA  PA,X          ; send out the digit pattern
          LDAB  1,Y
          STAB  PB,X          ; send out the select value
          INY
          INY
          LDX   #200
```

```
again      NOP                          ; 1 ms delay loop
           NOP                          ;     "
           DEX                          ;     "
           BNE      again               ;     "
           CPY      #distab+8
           BEQ      forever
           BRA      next
distab     FCB      $30,$08
           FCB      $7B,$04
           FCB      $7B,$02
           FCB      $7B,$01
           END
```

We need to add the following statements to the hc11.h file so that we can use labels to reference i8255 registers:

```
#define PPI_PA       *(unsigned char volatile *)(0xB5F4)    /* port A data register */
#define PPI_PB       *(unsigned char volatile *)(0xB5F5)    /* port B data register */
#define PPI_PC       *(unsigned char volatile *)(0xB5F6)    /* port C data register */
#define PPI_CMD      *(unsigned char volatile *)(0xB5F7)    /* i8255 control register */
```

After adding these statements we are ready to write the C language version of this program.

```
#include    <hc11.h>
char distab [4][2] = {{0x30, 0x08}, {0x7B, 0x04}, {0x7B, 0x02}, {0x7B, 0x01}};
void wait_1ms ( );
main ( )
{
    PPI_CMD = 0x80;                 /* initialize the i8255 control register */
    while (1) {
        for (i = 0; i < 4; i++) {
            PPI_PA = distab[i][0];   /* output segment pattern */
            PPI_PB = distab[i][1];   /* output digit select */
            wait_1ms ( );            /* wait for 1 ms */
        }
    }
    return 0;
}
/* the following routine use OC2 function to create 1 ms time delay */
void wait_1ms ( )
{
    TFLG1 = 0x40;               /* clear OC2F flag */
    TOC2 = TCNT + 2000;
    while (!(TFLG1 & 0x40));
}
```

7.14   Summary

Peripheral devices are pieces of equipment that exchange data with a computer. Examples of peripheral devices include switches, light-emitting diodes, cathode-ray tube (CRT) screens, printers, modems, keypads, keyboards, and disk drives. The speed and characteristics of these devices are very different from those of the processor, so they are not connected to the processor directly. Instead, interface chips are used to resolve the differences between the microprocessor and these peripheral devices.

The major function of an interface chip is to synchronize the data transfer between the processor and an I/O device. An interface chip consists of control registers, status registers, data registers, data direction registers, and control circuitry. There are two aspects in the I/O transfer synchronization: the synchronization between the CPU and the interface chip and the synchronization between the interface chip and the I/O device. There are two methods to synchronize the CPU and the interface chip: polling method and interrupt method. There are three methods for synchronizing the interface chip and the I/O device: brute force, strobe, and handshake methods.

Different processors deal with I/O differently. Some processors have dedicated instructions and addressing modes for performing I/O operations. In these processors, I/O devices occupy a separate memory space from the main memory. Other processors use the same instructions and addressing modes to access I/O devices. I/O devices and main memory share the same memory space.

The 68HC11A8 has 40 I/O pins arranged in five I/O ports. All of these I/O pins serve multiple functions, depending on the operation mode and data in the control registers. Ports C and D are bidirectional I/O ports. Ports A, B, and E are fixed-direction I/O ports. The pin 7 of port A is an exception because it is bidirectional. Each bidirectional I/O port has an associated data direction register for configuring data transfer direction.

The 68HC11 supports strobed I/O. Port B can be configured to perform strobed output, and port C can be configured to perform strobed input. Port C also supports handshake I/O. Parameters related to handshake I/O are programmed via the PIOC register.

Input devices such as DIP switches, keyboards, and keypads are discussed. Both assembly and C languages are used to perform I/O programming. The key bouncing problem and its solution are explored. The characteristics and interfacing of LEDs and LCDs are explained and their programming is demonstrated.

The Centronics printer interface standard is discussed and its interfacing with the 68HC11 port C is illustrated. Digital-to-analog conversion is required when a digital code must be converted to an analog signal. The analog signal can be used to control the output of another system, for example, the flow level of a fluid system or the voltage level of a stereo system. The D/A converter AD557 from Analog Devices is used as an example to demonstrate the generation of waveforms.

When in expanded mode, ports B and C are not available for general-purpose I/O operation. We have three solutions for this issue: (1) use an interface chip like the Intel i8255 to expand I/O ports for the microcontroller, (2) use the serial peripheral interface (SPI) and its supporting devices, and (3) choose a 68HC11 member with more I/O ports.

The function, operation, and programming of the i8255 are illustrated in detail.

7.15   Exercises

E7.1. What is isolated I/O? What is memory-mapped I/O?

E7.2. Describe interlocked input handshaking.

E7.3. Write an instruction sequence to configure port C to operate with the following parameters:

- enable STRA interrupt
- no port C wired-or
- input handshaking
- pulse mode handshake protocol
- STRA falling edge as active edge
- STRB active low

E7.4. Write a sequence of instructions and C statements to read in the current port C pin logic levels.

E7.5. *Traffic Light Controller.* Use the EVB, EVBU, or CMD-11A8 evaluation board and green, yellow, and red LEDs to simulate a traffic-light controller. The traffic-light patterns and durations for the east-west and north-southbound traffic are given in Table 7E.1. Write an assembly and a C program to control the light patterns, and connect the circuit to demonstrate the changes in the lights.

East-west			North-south			
green	red	yellow	green	red	yellow	Duration (seconds)
1	0	0	0	1	0	20
0	0	1	0	1	0	4
0	1	0	1	0	0	15
0	1	0	0	0	1	3

Table 7E.1 ■ Traffic light pattern

E7.6. Compute the average maximum current that flows into the collector of the 2N2222 transistor in Figure 7.21. Note that each seven-segment display is illuminated for only $\frac{1}{6}$ second per second.

E7.7. Calculate the period of the sawtooth waveform generated by the circuit in Example 7.8. How would you double the period of the sawtooth waveform?

E7.8. Operate the AD557 D/A converter in the transparent mode (the chip enable and chip select are tied to low). Set V_{ref} to 5V. What will the output voltage be when the digital input is 0, 16, 32, 64, 128, and 192?

E7.9. Write a program to control the circuit shown in Figure 7.27 to generate a *triangular waveform*. Observe the waveform using an oscilloscope. Answer the following questions:

  a. What is the highest frequency of the waveform that can be achieved?
  b. How can you double and quadruple the period of the generated triangular waveform?

E7.10. Use the DMC-20434 LCD kit to display the message "Microcontroller is very fun to learn!"

E7.11. Write an instruction sequence to read a byte from the port A of the i8255 in the CMD-11A8 evaluation board. The byte must be left in accumulator A.

E7.12. Suppose you are assigned to add a Centronics printer to the port A of the CMD-11A8 evaluation board. Write the routines to perform printer initialization, output one character, and output a string. Provide both the polling and the interrupt-driven versions of the routines. Write these routines in assembly and C languages.

E7.13. Write assembly routines to perform keypad scanning and debouncing for the keypad circuit shown in Figure 7.16.

E7.14. Use the 68HC11 port B to drive the D/A converter AD557 and write a program to generate the waveform shown in Figure 7E.1. The circuit is shown in Figure 7.28.

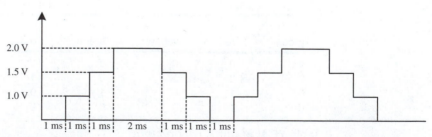

Figure 7E.1 ■ Periodic waveform to be generated

7.16 Lab Experiments and Assignments

L7.1. *Interrupt-driven input experiment.* The purpose of this lab is to practice interrupt-driven input. You will apply three interrupt pulses to the $\overline{IRQ}$ pin. Each interrupt pulse will cause one byte to be read from an 8-DIP switch. Connect an 8-DIP switch to the port E of the demo board as shown in Figure 7.11. The procedure for the experiment is as follows:

Step 1

Write a main program that performs the following operations:
  a. initializes an $\overline{IRQ}$ interrupt count *ircnt* to 3.
  b. initializes the buffer pointer *buf_ptr* to $00.
  c. stays in a wait loop and keeps checking the interrupt count *ircnt;* the main program will disable the $\overline{IRQ}$ interrupt and jump to the Buffalo monitor when *ircnt* is 0.

Step 2

Write an $\overline{IRQ}$ service routine that performs the following operations:
  a. reads the 8-DIP switch and saves the value at the memory location pointed to by *buf_ptr.*
  b. increments the buffer pointer *buf_ptr* by 1.
  c. decrements the $\overline{IRQ}$ interrupt count *ircnt* by 1.

Step 3

Set up the 8-DIP switch properly so that it represents the value that the user wants to input to the evaluation board.

Step 4

Apply a debounced pulse to the $\overline{IRQ}$ pin of the evaluation board so that the $\overline{IRQ}$ service routine reads the 8-DIP switch value.

Step 5

Repeat steps 3 and 4 three times and then stop.

Display the contents of the value that you just entered on the monitor screen and check if the values have been read correctly.

L7.2. *Seven-segment display experiment.* Use ports B and D to drive two seven-segment displays. Port B pins drive the segment inputs and port D pins PD1 and PD0 drive the bases of two 2N2222 NPN transistors. The collectors of these two 2N2222 transistors are connected to the common cathodes of the seven-segment displays. Write a program to perform the following operations:

Step 1

Display the BCD digits 00 on these two seven-segment displays.

Step 2

Use program delay loops to create delays and increment the seven-segment displays every second.

Step 3

Display the decimal values from 00-59 and then roll back to 00.

Step 4

Repeat Step 3 three times and then return to the monitor program of the evaluation board.

L7.3. *Time-of-day input and display.* Use port E to read time-of-day from an 8-DIP switch and use port B and port D to drive six seven-segment displays. Use the circuits in Figure 7.11 and 7.21 and follow this procedure:

Step 1

Write a main program that performs the following operations:

      a. initializes an $\overline{\text{IRQ}}$ interrupt count *ircnt* to 3.

      b. initializes the buffer pointer *buf_ptr* to $00.

      c. stays in a wait loop and keeps checking the interrupt count *ircnt*; the main program will disable the $\overline{\text{IRQ}}$ interrupt and jump to the Buffalo monitor when *ircnt* is 0.

Step 2

Write an $\overline{\text{IRQ}}$ interrupt service routine that performs the following operations:

      a. reads the 8-DIP switch and saves the value at the memory location pointed to by *buf_ptr*.

      b. increments the buffer pointer *buf_ptr* by 1.

      c. decrements the $\overline{\text{IRQ}}$ interrupt count *ircnt* by 1.

Step 3

Set up the 8-DIP switch properly so that it represents the value that the user wants to input to the demo board.

Step 4

Apply a debounced pulse to the $\overline{\text{IRQ}}$ pin of the evaluation board so that the $\overline{\text{IRQ}}$ service routine reads the 8-DIP switch value.

Step 5

Repeat steps 3 and 4 three times to input the current hours, minutes, and seconds digits.

Step 6

The main program should call a delay routine to update the time-of-day. The time-of-day is displayed in 24-hour format. The main program does not request the user to reenter the current time-of-day if the user enters incorrect values. However, it will reset the time to a valid value when the time is to be updated. For example, if out-of-range seconds digits (i.e., ≥ 60) are entered, they will be reset to 00 in the next second; if out-of-range minutes digits (i.e., ≥ 60) are entered, they will be reset to 00 in the next update of the minutes digits; if out-of-range hours digits are entered, they will be reset to 00 in the next update of the hours digits.

8

68HC11 Timer Functions

8.1 Objectives

After completing this chapter you should be able to

- use the input-capture function to measure the duration of a pulse or the period of a square wave

- use the input-capture function to measure the duty cycle of a periodic waveform and the phase difference between two signals

- use the output-compare function to create a time delay

- use the output-compare function to generate a pulse or a square waveform

- use two output-compare functions to control the same pin

- use the forced output-compare function

- use the input-capture function and pulse accumulator as edge-sensitive interrupt sources

- use the pulse accumulator to count the number of events occurred within a time interval

- use the real-time interrupt function to generate periodic interrupts

8.2 Introduction

There are many applications that require a dedicated timer system, including

- delay creation and measurement
- period measurement
- event counting
- time-of-day tracking
- periodic interrupt generation to remind the processor to perform routine tasks
- waveform generation

Without a dedicated timer system, some of these applications will be very difficult to implement. The 68HC11 includes a sophisticated timer system to support these applications. At the heart of the 68HC11 timer functions is the 16-bit free-running main timer. One of the 68HC11 timer functions is called *input capture*, which can latch the main timer value into a register when the rising or falling edge (selected by the user program) of a signal arrives. This ability allows the user to measure the width or period of an unknown signal, and it can also be used as a time reference for triggering other operations. There are three input-capture functions in the 68HC11.

Another 68HC11 timer function is *output compare,* which compares the value of the main timer with that of an output-compare register once every E clock cycle and performs the following operations when they are equal:

1. (optionally) triggers an action on a pin (set to high, set to low, or toggle its value).
2. sets a flag in a register.
3. (optionally) generates an interrupt to the microcontroller.

The output-compare function is typically used to create a delay or generate a digital waveform with a specified frequency and duty cycle. The key to using the output-compare function is to make a copy of the main timer, add a delay to this copy, and then save this sum into an *output-compare register.* The delay that is added determines when the value of the main timer will be equal to that of the output-compare register. The 68HC11 has five output-compare functions.

The third 68HC11 timer function is the *real time interrupt* (RTI) function. There are some routine tasks that must be performed periodically and cannot tolerate two successive performances being separated by more than some predetermined time limit; checking the temperature and pressure in some process control plants is an example of such a task. These applications can be implemented as interrupt service routines to be invoked by the timer interrupt. The RTI function can generate periodic interrupts and hence is ideal for these applications. Because of the nature of the RTI interrupt, the required task will be performed within a specified time limit (hence the name *real-time*).

The fourth function is the *computer operating properly* (COP) subsystem. Once enabled, this function can be used to reset the computer system if a software program has bugs in it and fails to take care of the COP system within a time limit. The COP system has been discussed in detail in section 6.6.5 of chapter 6.

The fifth 68HC11 timer function is the 8-bit *pulse accumulator*. This function is often used to count events occurring within some time limit, measure the frequency of an unknown signal, or to measure the duration of a pulse.

The timer system involves more registers and control bits than any other subsystem in the 68HC11. Each of the three input-capture functions has its own 16-bit time-capture latch (input-capture register), and each of the five output-compare functions has a 16-bit compare register. All timer functions, including the timer overflow and RTI, have their own interrupt controls and separate vectors.

8.3   Free-Running Main Timer

The main timer (TCNT), shown in Figure 8.1, is clocked by the output of a four-stage prescaler (divided by 1, 4, 8, and 16), which in turn is driven by the E clock signal. The main timer is cleared to 0 during a reset and is a read-only

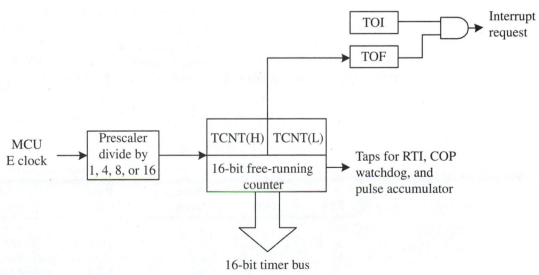

Figure 8.1 ■ 68HC11 main timer system

	7	6	5	4	3	2	1	0	
	TOI	RTII	PAOII	PAII	0	0	**PR1**	**PR0**	TMSK2 at $1024
value after reset	0	0	0	0	0	0	0	0	

	7	6	5	4	3	2	1	0	
	TOF	RTIF	PAOVF	PAIF					TFLG2 at $1025
value after reset	0	0	0	0	0	0	0	0	

Figure 8.2 ■ TMSK2 and TFLG2 registers

register except in test mode. When the count changes from $FFFF to $0000, the timer overflow flag (TOF) bit in *timer flag register 2* (TFLG2) is set. The timer overflow flag can be cleared by writing a 1 to it. An interrupt can be enabled by setting the timer overflow interrupt enable bit (TOI) of *timer interrupt mask register 2* (TMSK2). The contents of TFLG2 and TMSK2 are shown in Figure 8.2. Bits that are related to the free-running timer are in boldface.

The prescale factor for the main timer is selected by bits 1 and 0 of TMSK2, as shown in Table 8.1. The prescale factor can be changed only once within the first 64 E clock cycles after a reset, and the resultant count rate stays in effect until the next reset.

The *timer counter* (TCNT) register is meant to be read using a double-byte read instruction such as LDD or LDX. During a 16-bit read, the upper byte is accessed in the first bus cycle, while the lower byte is returned in the next bus cycle. The low-order half of the counter is momentarily frozen when the upper byte is accessed. This procedure assures that the two bytes read from TCNT belong with each other. If the user accesses TCNT with two 8-bit reads, the result might not be correct because the lower byte of TCNT would be incremented when the upper byte is accessed.

			Overflow period	
PR1	PR0	Prescale factor	2 MHz E clock	1 MHz E clock
0	0	1	32.77 ms	65.54 ms
0	1	4	131.1 ms	262.1 ms
1	0	8	262.1 ms	524.3 ms
1	1	16	524.3 ms	1.049 ms

Table 8.1 ■ Main timer clock frequency vs. PR1 and PR0

Example 8.1

▼
What values will be in accumulator A and B after the execution of the following three instructions if TCNT contains $5EFE when the upper byte is accessed? Assume the PR1 and PR0 bits of the TMSK1 register are 00.

```
REGBAS   EQU   $1000        ; base address of I/O register block
TCNTH    EQU   $0E          ; offset of TCNTH from REGBAS
TCNTL    EQU   $0F          ; offset of TCNTL from REGBAS


         LDX   #REGBAS
         LDAA  TCNTH,X
         LDAB  TCNTL,X
```

Solution: Since the bits of PR1 and PR0 (in TMSK2) are set to 00, the prescale factor of TCNT is 1. The E clock is used as the clock input to the main timer. The instruction "LDAA TCNTH,X" accesses the upper byte of TCNT. Therefore, accumulator A will get the value of $5E. The instruction "LDAB TCNTL,X" accesses the lower byte four E clock cycles later. At that time TCNT will be incremented to $5F02. Therefore, accumulator B will get the value of $02.

If the instruction "LDD TCNT,X" is used to access TCNT, then accumulators A and B will receive $5E and $FE, respectively. This example shows that the value returned from two 8-bit accesses can be very different from the actual value in TCNT.

▲

The TCNT register can be used to keep track of time as well as to create time delays. However, implementing time delays using TCNT may not be easy. For example, to create a delay of 10 ms at 2 MHz clock rate, a very naive approach will be to write a program as follows:

```
REGBAS   EQU   $1000
TCNT     EQU   $0E
```

```
* A 10-ms delay at the 2-MHz clock rate is equivalent to 20,000 clock
* cycles. By adding 20000 to the current value of TCNT and waiting
* until this value is smaller than the incrementing TCNT, a 10 ms delay
* is created.
```

```
(1)         LDD   TCNT,X
(2)         ADDD  #20000
(3) LOOP    CPD   TCNT,X
(4)         BGT   LOOP
            ...
```

There are several potential problems in this approach:

1. Adding 20000 to the current value of TCNT could cause the value to go from positive to negative, causing the instruction 4 to fall through because the BGT instruction performs a signed comparison.

2. The first problem can be solved by using the BHI instruction, which performs an unsigned comparison.

3. However, for some other delay, the resultant sum may be smaller than the current value in TCNT because of the limit of 16-bit addition. In this situation, overflow occurs. For example, adding a delay count of $9000 to the current value of TCNT ($8000) will result in a 16-bit sum of $1000. This situation makes the previous solution unusable because the BHI instruction will also fall through. This problem can be solved by checking the carry flag.

4. The longest delay that can be generated by this approach is limited to about 32.7 ms ($(2^{16} - 1)$ E clock cycles) at 2 MHz. The user program must keep track of the number of timer overflows in order to generate a longer delay.

The solution of this problem will be left to the reader as an exercise.

8.4   Input-Capture Functions

Some applications need to know the arrival time of events. In a computer, *physical time* is often represented by the count in a counter, while the occurrence of an event is represented by a signal edge (either the rising or falling edge). The time when an event occurs can be recorded by latching the count when a signal edge arrives, as illustrated in Figure 8.3.

The 68HC11 timer system has three input capture channels to implement this operation. Each input-capture channel includes a 16-bit input-capture register, input edge-detection logic, and interrupt-generation logic as shown in Figure 8.4. In the 68HC11, physical time is represented by the count in the 16-bit free-running timer counter (TCNT). An input-capture function is used to record the time when an external event occurs. When a selected edge is detected at the input pin, the contents of TCNT are latched into a specified

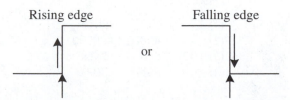

Figure 8.3 ■ Events represented by signal edges

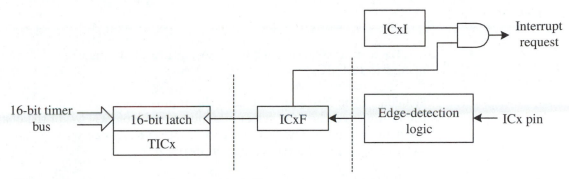

Figure 8.4 ■ Input-capture function block diagram

input-capture register. The user can select the signal edge to be captured and may optionally generate an interrupt to the microcontroller when the selected edge is detected. The captured 16-bit timer value is stored in the input-capture register. There are three input-capture registers:

1. TIC1 (located at $1010 and $1011)
2. TIC2 (located at $1012 and $1013)
3. TIC3 (located at $1014 and $1015)

The TICx (x = 1, 2, or 3) registers are not affected by reset and cannot be written by software. Pins PA2, PA1, PA0 are input only and are used as input-capture channels 1, 2, and 3, respectively, when they are enabled. If an input-capture channel is disabled, the corresponding pin becomes a general-purpose input.

The user can select the signal edge to capture by programming the TCTL2 register. The TCTL2 register is 6-bit, and its contents are shown in Figure 8.5. The signal edge to be captured is specified by two bits. The user can select to capture the rising or falling or both edges. When the edge-select bits are 00, the corresponding input-capture channel is disabled.

EDG1B	EDG1A	EDG2B	EDG2A	EDG3B	EDG3A	TCTL2 at $1021
0	0	0	0	0	0	

value after reset

EDGxB	EDGxA	
0	0	capture disabled
0	1	capture on rising edge
1	0	capture on falling edge
1	1	capture on both edges

x = 1,...,3

Figure 8.5 ■ Contents of TCTL2

Example 8.2

Write an instruction sequence to capture the rising edge of the signal connected to PA0 (IC3).

Solution: The PA0 pin is used as the input-capture channel IC3. To capture the rising edge, the edge select bits (bits 1 and 0 of TCTL2) of the channel IC3 must be set to 01. The following instruction sequence will set up TCTL2 to capture the time of a rising edge on PA0.

```
TCTL2     EQU    $21
REGBAS    EQU    $1000

          LDX    #REGBAS
          BCLR   TCTL2,X $02      ; clear bit 1 to 0 without affecting other bits
          BSET   TCTL2,X $01      ; set bit 0 to 1 without affecting other bits
```

The input-capture status register bit in the TFLG1 register (at $1023) will be set when a selected edge is detected at the corresponding input-capture pin. The contents of the TFLG1 register are shown in Figure 8.6. Bits 7 to 3 are output-compare status bits, and bits 2 to 0 are input-capture status bits.

A status flag can be cleared by writing a 1 to it. Two methods are most often used:

- Load an accumulator with a mask that has a 1 (or 1s) in the bit(s) corresponding to the flag(s) to be cleared; then write this value to TFLG1 or TFLG2.
- Use a bit clear (BCLR) instruction with a mask having zeros in the positions corresponding to the flags to be cleared and ones in all other bits. For example, the instruction "BCLR TFLG1,X %01111111" will clear the OC1F flag.

It is not appropriate to use the bit set (BSET) instruction to clear the flags in the timer flag registers because this could inadvertently clear one or more of the other flags in the register. The BSET instruction is a read-modify-write instruction that reads the operand, ORs this value with a mask having one(s)

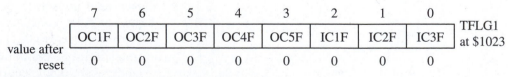

	7	6	5	4	3	2	1	0	
	OC1F	OC2F	OC3F	OC4F	OC5F	IC1F	IC2F	IC3F	TFLG1 at $1023
value after reset	0	0	0	0	0	0	0	0	

Figure 8.6 ■ The contents of the TFLG1 register

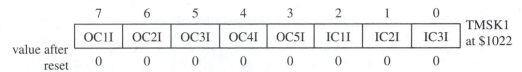

	7	6	5	4	3	2	1	0	
	OC1I	OC2I	OC3I	OC4I	OC5I	IC1I	IC2I	IC3I	TMSK1 at $1022
value after reset	0	0	0	0	0	0	0	0	

Figure 8.7 ■ The contents of the TMSK1 register

in the bit(s) to be set, and writes the resultant value back to the operand address. Using this method, all flags that are set at the time the flag register is read will be cleared.

The input-capture subsystem can optionally generate an interrupt request when a selected edge is detected. A *mask register* (TMSK1 at $1022) is provided to enable and disable the interrupt requests. When a bit is set, the corresponding interrupt request can be generated. The bit assignment of the TMSK1 register is as follows: bits 2 to 0 enable or disable interrupts from input-capture channels 1 to 3, and bits 7 to 3 enable or disable interrupt requests from output-compare channels 1 to 5. After reset, all bits are cleared to 0, which disables all output-compare and input-capture interrupts. The contents of TMSK1 are shown in Figure 8.7. The memory locations that holds interrupt vectors of IC1, IC2, and IC3 are listed in Table 8.2.

Channel	Pin	68HC11 vector location	EVB pseudo vector location
IC1	PA2	$FFEE-$FFEF	$E8-$EA
IC2	PA1	$FFEC-$FFED	$E5-$E7
IC3	PA0	$FFEA-$FFEB	$E2-$E4

Table 8.2 ■ Locations that hold input-capture interrupt vectors

8.5 Applications of the Input-Capture Function

The input-capture function has many applications. Examples include:

- *Event arrival time recording.* Some applications may need to compare the arrival times of several different events. The input-capture function is very suitable for this. The number of events that can be compared is limited by the number of input-capture channels.
- *Period measurement.* To measure the period of an unknown signal, the input-capture function should be configured to capture the main timer values corresponding to two consecutive rising or falling edges, as illustrated in Figure 8.8.

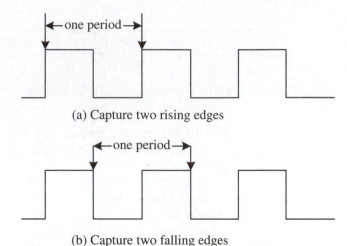

(a) Capture two rising edges

(b) Capture two falling edges

Figure 8.8 ■ Period measurement by capturing two consecutive edges

- *Pulse-width measurement.* To measure the width of a pulse, the rising and falling edges are captured, as shown in Figure 8.9. Since the free-running timer is 16-bit, it can only measure a signal period (or pulse width) no longer than 2^{16} E clock cycles. If the period of a slower signal needs to be measured, the number of timer overflows that occur between the two edges must be taken into account.

- *Interrupt generation.* The three input-capture inputs can serve as three edge-sensitive interrupt sources. Once enabled, interrupts will be generated on the selected edge(s).

- *Event counting.* An event can be represented by a signal edge. An input-capture channel can be used in combination with an output-compare function to count the number of events that occur during an interval. Whenever an edge arrives, an interrupt is generated. An event counter can be set up and incremented by the input-capture interrupt service routine. This application is illustrated in Figure 8.10.

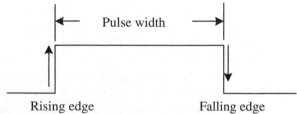

Figure 8.9 ■ Pulse-width measurement using input capture

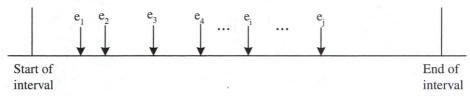

Figure 8.10 ■ Using an input-capture function for event counting

- *Time reference.* In this application, an input-capture function is used in conjunction with an output-compare function. For example, if the user wishes to activate an output signal a certain number of clock cycles after detecting an input event (a rising or falling edge), the input-capture function would be used to record the time at which the edge is detected. A number corresponding to the desired delay would be added to this captured value and stored to an output-compare register. This application is illustrated in Figure 8.11.
- *Duty cycle measurement.* Duty cycle is the percent of time that the signal is high within a period in a periodic digital waveform. The measurement of duty cycle is illustrated in Figure 8.12.

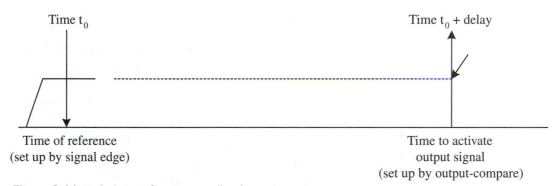

Figure 8.11 ■ A time reference application

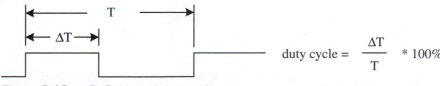

Figure 8.12 ■ Definition of duty cycle

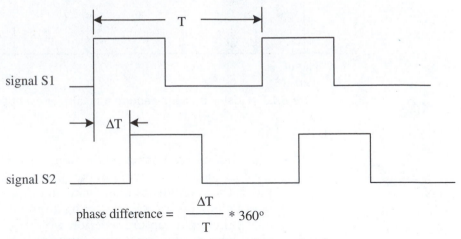

$$\text{phase difference} = \frac{\Delta T}{T} * 360°$$

Figure 8.13 ■ Phase difference definition for two signals

- *Phase difference measurement.* Phase difference is defined as two signals that have the same frequency but do not coincide in their rising and falling edges. The definition of phase difference is illustrated in Figure 8.13.

Example 8.3

▼

Use the input-capture function IC1 of the 68HC11 to measure the period of an unknown signal. The period is known to be shorter than 32 ms, and the signal is connected to the IC1 pin. Write a program to set up the input-capture function to perform the measurement and store the result in memory.

Solution: The circuit connection is shown in Figure 8.14.

Because the period of the unknown signal is less than 32 ms, we don't need to take the main timer overflow into account. To capture the rising edge of IC1, we need to set the value of the register TCTL2 to %00010000. Either the

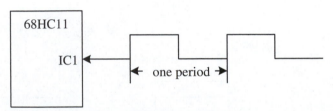

Figure 8.14 ■ Period measurement

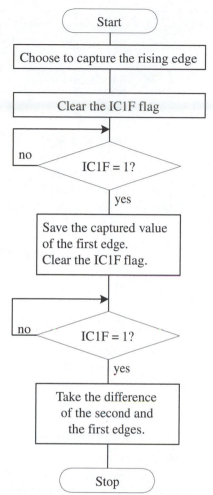

Figure 8.15 ■ Logic flow of the period-measurement program (polling method)

polling or interrupt method can be used to measure the signal frequency. The algorithm of the polling version of the measurement program is shown in Figure 8.15. In this method, the user program checks whether the IC1F flag of TFLG1 is set. When the IC1F flag (bit 2 of the TFLG1 register) is set to 1, a signal edge is captured.

The polling version of the period measurement program is as follows:

```
REGBAS    EQU    $1000        ; I/O register block base address
TFLG1     EQU    $23          ; offset of TFLG1 from REGBAS
TIC1      EQU    $10          ; offset of TIC1 from REGBAS
```

```
TCTL2       EQU     $21                 ; offset of TCTL2 from REGBAS
IC1rise     EQU     $10                 ; value to select the IC1 rising edge
CLEAR       EQU     $04                 ; value to be written into TFLG1 to clear IC1F flag


            ORG     $0000
EDGE1       RMB     2                   ; the captured first edge value
period      RMB     2                   ; period in number of E clock cycles


            ORG     $C000               ; starting address of the program
            LDX     #REGBAS
            LDAA    #CLEAR              ; clear the IC1F flag
            STAA    TFLG1,X             ;          "
* configure TCTL2 to capture the rising edge of IC1
            LDAA    #IC1rise            ; configure to capture the rising edge of IC1
            STAA    TCTL2,X             ;          "
            BRCLR   TFLG1,X $04 *       ; wait for the arrival of the first rising edge
            LDD     TIC1,X              ; save the captured first edge
            STD     EDGE1               ;          "
            LDAA    #CLEAR              ; clear the IC1F flag
            STAA    TFLG1,X             ;          "
            BRCLR   TFLG1,X $04 *       ; wait for the arrival of the next rising edge
            LDD     TIC1,X              ; compute the difference of two edges
            SUBD    EDGE1               ;          "
            STD     period              ; saved the period
            SWI                         ; break to monitor
            END
```

The C language version of this program is as follows:

```
#include <hc11.h>
#include <stdio.h>
main ( )
{
    unsigned int edge1, period;

    TFLG1 = 0x04;              /* clear the IC1F flag */
    TCTL2 = 0x10;              /* capture the rising edge of IC1 */
    while (!(TFLG1 & 0x04));   /* wait for the arrival of the first edge */
    edge1 = TIC1;             /* save the first edge */
    TFLG1 = 0x04;              /* clear the IC1F flag */
    while (!(TFLG1 & 0x04));   /* wait for the arrival of the second edge */
    period = TIC1 - edge1;
    printf("\n The period of the signal is %d E clock cycles. \n", period);
    return 0;
}
```

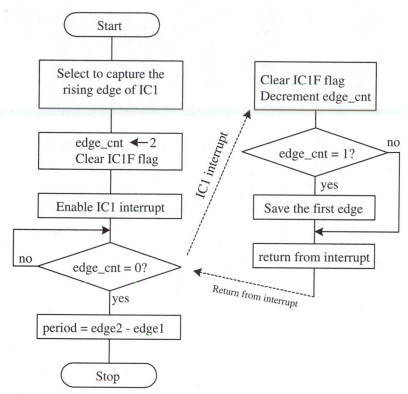

Figure 8.16 ■ Interrupt-driven method for period measurement

To use the interrupt-driven method, we need to enable the IC1 interrupt. By setting the IC1I bit of the TMSK1 register and clearing the I flag of the CCR register, IC1 is enabled to interrupt the CPU. The logic flow of the interrupt-driven version of the period measurement program is shown in Figure 8.16. The dotted lines in Figure 8.16 indicate program control transfers caused by interrupts and returns from interrupts.

The interrupt-driven version of the period-measurement program using the input-capture function is as follows:

```
REGBAS    EQU    $1000         ; base address of the I/O register block
TFLG1     EQU    $23           ; offset of TFLG1 from REGBAS
TMSK1     EQU    $22           ; offset of TMSK1 from REGBAS
TIC1      EQU    $10           ; offset of TIC1 from REGBAS
TCTL2     EQU    $21           ; offset of TCTL2 from REGBAS
IC1rise   EQU    $10           ; value to select the rising edge of IC1 to capture
IC1I      EQU    $04           ; mask to select the IC1I bit of the TMSK1 register
```

```
          IC1FM      EQU     $FB                       ; mask to clear IC1F flag using the BCLR instruction

                     ORG     $00
          edge_cnt   RMB     1                         ; edge count
          EDGE1      RMB     2                         ; captured first edge value
          period     RMB     2                         ; period in number of E clock cycles

                     ORG     $FFE8
                     FDB     IC1_ISR                   ; set up IC1 interrupt vector


          **************************************************************************
          * Use the following assembler directives to replace the previous two directives if
          * this program is to be run on the EVB, EVBU, or the CMD-11A8
          *
          *          ORG     $E8
          *          JMP     IC1_ISR
          **************************************************************************
                     ORG     $C000                     ; starting address of the main program
                     LDS     #$DFFF                    ; set up stack pointer (for EVB only)
                     LDX     #REGBAS
                     LDAA    #IC1rise                  ; prepare to capture the rising edge of IC1
                     STAA    TCTL2                     ;       "
                     BCLR    TFLG1,X IC1FM             ; clear the IC1F flag of the TFLG1 register
                     LDAA    #2                        ; initialize edge count to 2
                     STAA    edge_cnt                  ;       "
                     BSET    TMSK1,X IC1I              ; set the IC1 interrupt enable bit
                     CLI                               ; enable interrupt to 68HC11

          * The following two instructions wait until the variable edge_cnt has been decremented to 0

          WAIT       TST     edge_cnt                  ; check if two edges have been captured
                     BNE     WAIT
                     LDD     TIC1,X                    ; get the second edge time
                     SUBD    EDGE1                     ; take the difference of edge1 and edge 2
                     STD     period
                     ....                              ; do something else

          * IC1 interrupt service routine is as follows
          IC1_ISR    LDX     #REGBAS
                     BCLR    TFLG1,X IC1FM             ; clear the IC1F flag
                     DEC     edge_cnt
                     BEQ     SKIP                      ; Is this the second edge?
                     LDD     TIC1,X
                     STD     EDGE1                     ; save the arrival time of the first edge
          SKIP       RTI
                     END
```

The C language version of this program is as follows:

```c
#include <hc11.h>
#include <stdio.h>
int edge_cnt;
unsigned int edge1, period;

void IC1_ISR( );
main ( )
{
    *(unsigned char *)0xe8 = 0x7E;      /* 7E is "jmp" */
    *(void (**)())0xe9 = IC1_ISR;       /* set up pseudo vector for Buffalo monitor */
    TFLG1 = 0x04;                        /* clear IC1F flag */
    edge_cnt = 2;
    TMSK1 |= 0x04;                       /* enable IC1 interrupt locally */
    TCTL2 = 0x10;                        /* configure to capture the rising edge of IC1 */
    INTR_ON ( );                         /* enable IC1 interrupt globally */
    while (edge_cnt);                    /* wait until two edges have been captured */
    period = TIC1 - edge1;
    printf("\n The period is %d E clock cycles \n", period);
    return 0;
}

#pragma interrupt_handler IC1_ISR ( )
void IC1_ISR ( )
{
    TFLG1 = 0x04;           /* clear IC1F flag */
    if (edge_cnt == 2)
        edge1 = TIC1;       /* save the first edge */
    —edge_cnt;
}
```

Example 8.4

Write a subroutine to measure the pulse width of an unknown signal that is connected to the IC1 pin. Return the pulse width in D. Assume the main timer prescale factor is 1 and the pulse width is shorter than 32 ms. The E clock frequency is 2 MHz.

Solution: To measure the pulse width, first the rising and then the falling edge must be captured. The program first sets up IC1 to capture the rising edge. After the arrival of the first edge, IC1 is reconfigured to capture the falling edge. The pulse width is the difference of these two edges. The polling approach is easier, and the corresponding program is as follows:

```
REGBAS      EQU     $1000               ; base address of the I/O register block
TFLG1       EQU     $23                 ; offset of TFLG1 from REGBAS
TIC1        EQU     $10                 ; offset of TIC1 from REGBAS
TCTL2       EQU     $21                 ; offset of TCTL2 from REGBAS
IC1rise     EQU     $10                 ; value to select the rising edge of IC1
IC1fall     EQU     $20                 ; value to select the falling edge of IC1
IC1F        EQU     $04                 ; mask to select the IC1F flag
temp        EQU     0                   ; offset of temp from the top of the stack


pul_width   PSHX
            PSHY
            DES                         ; allocate two bytes for the local variable temp
            DES
            LDX     #REGBAS
            TSY
            LDAA    #IC1rise            ; prepare to capture the rising edge
            STAA    TCTL2,X             ;            "
            LDAA    #IC1F               ; clear the IC1F flag
            STAA    TFLG1,X             ;            "
rise        BRCLR   TFLG1,X IC1F rise   ; wait for the arrival of the rising edge
            LDD     TIC1,X              ; save the rising edge
            STD     temp,Y              ;            "
```

* The following two instructions configure TCTL2 to capture the falling edge of IC1

```
            LDAB    #IC1fall
            STAB    TCTL2,X             ; select to capture the falling edge of IC1
            LDAA    #IC1F               ; clear the IC1F flag
            STAA    TFLG1,X             ;            "
fall        BRCLR   TFLG1,X IC1F fall   ; wait for the arrival of the falling edge
            LDD     TIC1,X              ; get the captured second edge
            SUBD    temp,Y              ; leave the pulse width in D
            INS                         ; deallocate the local variable temp
            INS
            PULY
            PULX
            RTS
```

The C language version of the subroutine is as follows:

```
unsigned int pul_width ( )
{
    unsigned temp;
    TCTL2 = 0x10;                /* prepare to capture the rising edge */
    TFLG1 = 0x04;               /* clear IC1F flag */
    while (!(TFLG1 & 0x04));     /* wait for the arrival of the rising edge */
    TFLG1 = 0x04;
```

```
    temp = TIC1;
    TCTL2 = 0x20;               /* prepare to capture the falling edge */
    while (!(TFLG1 & 0x04));    /* wait for the arrival of the falling edge */
    return (TIC1 - temp);
}
```

The programs in Examples 8.3 and 8.4 work for any signal period and pulse width less than $(2^{16} - 1)$ E clock cycles. If the period or pulse width is longer, the TCNT register will completely roll over at least once. We must then count the number of times TCNT rolls around by counting the number of times TOF sets. Let

ovcnt	= the main timer overflow count
diff	= the difference of two edges
edge1	= the captured time of the first edge
edge2	= the captured time of the second edge

Then the period, or pulse width, can be calculated by the following equations:

Case 1
edge2 $\geq$ edge1

   period (or pulse width) = ovcnt $\times$ 2^{16} + diff

Case 2
edge2 $<$ edge1

   period (or pulse width) = (ovcnt - 1) $\times$ 2^{16} + diff

In case 2, the timer overflows at least once even if the period or pulse width is shorter than $(2^{16} - 1)$ E clock cycles. Therefore, we need to subtract 1 from the timer overflow count in order to get the correct result.

Example 8.5

Write a program that uses input-capture channel IC1 to measure a period longer than 2^{16} E clock cycles (32.77 ms for 2MHz E clock).

Solution: We will use the polling method to capture two consecutive edges and include a timer overflow interrupt-handling routine to keep track of the number of main timer overflows. Two bytes are reserved for this purpose so that we can measure a period or pulse width as long as 35.8 minutes (65536 $\times$ 32.77 $\div$ 1000 $\div$ 60 $\approx$ 35.8). The period is obtained by appending the difference of the two captured edges to the overflow count. The idea of the program logic is illustrated in Figure 8.17.

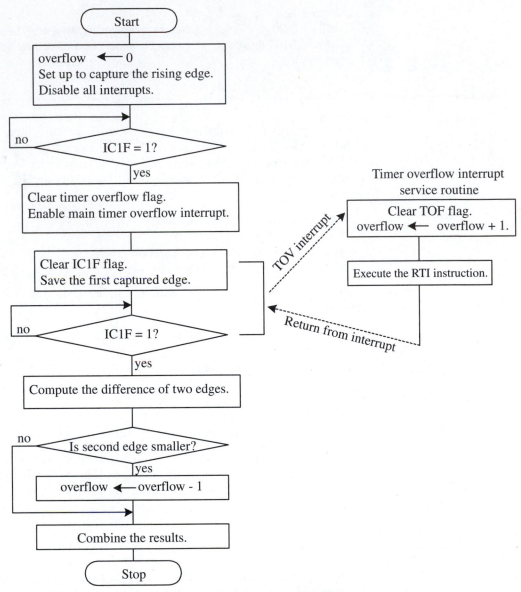

Figure 8.17 ■ Logic flow for measuring period of slow signals

The program is:

```
REGBAS   EQU   $1000          ; base address of I/O register block
TFLG1    EQU   $23            ; offset of TFLG1 from REGBAS
TIC1     EQU   $10            ; offset of TIC1 from REGBAS
TCTL2    EQU   $21            ; offset of TCTL2 from REGBAS
```

```
TMSK1      EQU     $22                    ; offset of TMSK1 from REGBAS
TMSK2      EQU     $24                    ; offset of TMSK2 from REGBAS
TOF        EQU     $80                    ; mask to select timer overflow flag
IC1rise    EQU     $10                    ; value to select the rising edge of IC1
IC1F       EQU     $04                    ; mask to select the IC1F flag
TOI        EQU     $80                    ; mask to select the timer overflow interrupt


           ORG     $0000
edge1      RMB     2                      ; the captured first edge value
overflow   RMB     2                      ; number of main timer overflows
period     RMB     2                      ; period value


           ORG     $FFDE                  ; timer overflow interrupt vector
           FDB     TOV_ISR                ;        "
```

```
******************************************************************
* Use the following assembler directives to set up pseudo interrupt vector so that this program
* can be run on EVB, EVBU, or CMD-11A8
*
*      ORG     $00D0
*      JMP     TOV_ISR
******************************************************************
```

```
           ORG     $C000
           LDS     #$DFFF                 ; set stack pointer (for EVB only)
           SEI                            ; disable all interrupts before initialization
           CLR     overflow               ; initialize main timer overflow count to 0
           CLR     overflow+1             ;        "
           LDX     #REGBAS
           LDAA    #IC1rise               ; select to capture the rising edge
           STAA    TCTL2,X                ;        "
           LDAA    #IC1F                  ; clear the IC1F flag in TFLG1 register
           STAA    TFLG1,X                ;        "
           BCLR    TMSK1,X $FF            ; disable all input-capture and output-compare
*                                         ; interrupts
           BRCLR   TFLG1,X IC1F *         ; wait for the arrival of the first rising edge
```

```
* The timer-overflow flag must be cleared before enabling the timer-overflow interrupt
```

```
           LDAA    #TOF
           STAA    TFLG2,X                ; clear the TOF flag in TFLG2 register
           BSET    TMSK2,X TOI            ; enable the main-timer-overflow interrupt
           CLI                            ; enable interrupt to the 68HC11
           LDD     TIC1,X                 ; save the captured rising edge
           STD     EDGE1                  ;        "
```

```
        LDAA    #IC1F                   ; clear the IC1F flag
        STAA    TFLG1,X                 :        "
        BRCLR   TFLG1,X IC1F *          ; wait for the second rising edge
        LDD     TIC1,X                  ; take the difference of two edges
        SUBD    edge1                   ;        "
        STD     period                  ; save the period
        BCC     next                    ; check if the captured first edge is larger
```

* When the second edge is smaller, we need to subtract the timer overflow count by 1 using the
* following three instructions

```
        LDD     overflow
        SUBD    #1
        STD     overflow
next    ......                          ; do something else
```

* The timer overflow interrupt service is in the following

```
TOV_ISR LDX     #REGBAS
        LDAA    #TOF                    ; clear the TOF flag
        STAA    TFLG2,X                 ;        "
```

* The following three instructions increment the timer overflow count by 1

```
        LDD     overflow
        ADDD    #1
        STD     overflow
        RTI
        END
```

The C language version of this program is as follows:

```c
#include <stdio.h>
#include <hc11.h>
unsigned edge1, overflow;
unsigned long period;

void TOV_ISR ( );
main ( )
{
    *(unsigned char *)0xd0 = 0x7E;      /* 7E is "jmp" */
    *(void (**)())0xd1 = TOV_ISR;       /* set up TOV pseudo vector entry */

    INTR_OFF ( );
    overflow = 0;
    TFLG1 = 0xFF;                       /* clear all output-compare and input-capture flags */
```

```
    TFLG2 = 0x80;                      /* clear TOF flag */
    TCTL2 = 0x10;                      /* configure to capture IC1's rising edge */
    TMSK1 = 0x00;                      /* disable all input-capture and output-compare
                                          interrupts */

    while (!(TFLG1 & 0x04));           /* wait for the arrival of the first rising edge on IC1 */
    TFLG1 = 0x04;                      /* clear IC1F flag */
    edge1 = TIC1;                      /* save the first rising edge */
    TMSK2 = 0x80;                      /* enable timer overflow interrupt */
    INTR_ON ( );                       /*          "          */
    while (!(TFLG1 & 0x04));           /* wait for the arrival of the second edge */
    if (TIC1 < edge1)                  /* if the second edge is smaller */
        overflow --;                   /* then decrement the overflow count */
    period = overflow *65536;          /* combine the result */
    period += TIC1 - edge1;            /*          "          */
    printf("\n The period of the unknown signal is %d E clock cycles. \n", period);
    return 0;
}


#pragma interrupt_handler TOV_ISR ( )
void TOV_ISR ( )
{
    TFLG2 = 0x80;                      /* clear TOF flag */
    overflow ++;
}
```

8.6   Output-Compare Functions

The 68HC11A8 has five output-compare functions. Output-compare functions can be used to inform the 68HC11 that an event has occurred and to control the level on output pins PA7-PA3. As shown in Figure 8.18, each output-compare function consists of

1. a 16-bit comparator
2. a 16-bit compare register (TOCx)
3. an output action pin (OCx—can be pulled down to 0, pulled up to 1, or toggled)
4. an interrupt request circuit
5. a forced-compare function (FOCx)
6. control logic

The five output-compare registers occupy memory locations from $1016 through $101F, as shown in Table 8.3.

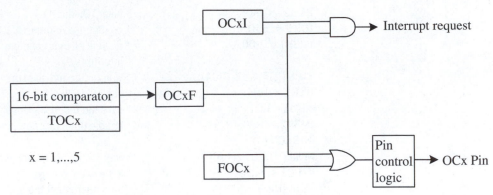

Figure 8.18 ■ Output-compare function block diagram

8.6.1 Operation of the Output-Compare Function

The main application of an output-compare function is performing an action at a specific time in the future (when the 16-bit free-running counter reaches a specific value). The action might be to toggle a signal, turn on a switch, and so on. To use an output-compare function, the user

1. makes a copy of the current contents of the free-running main timer (TCNT).
2. adds to this copy a value equal to the desired delay.
3. stores the sum into an output-compare register.

The user has the option of specifying the action to be activated on the selected output-compare pin by programming the TCTL1 register. The comparator compares the value of TCNT and that of the specified output-compare register in every E clock cycle. If they are equal, the specified action on the output-compare pin is activated and the associated status bit in TFLG1 will be

Register	Address
TOC1	$1016 - $1017
TOC2	$1018 - $1019
TOC3	$101A - $101B
TOC4	$101C - $101D
TOC5	$101E - $101F

Table 8.3 ■ The output-compare registers in the 68HC11A8

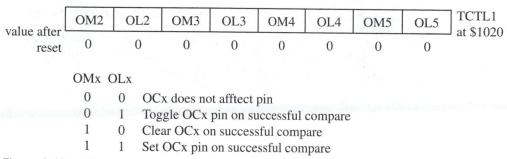

OM2	OL2	OM3	OL3	OM4	OL4	OM5	OL5	TCTL1
								at $1020

value after reset

| 0 | 0 | 0 | 0 | 0 | 0 | 0 | 0 |

OMx	OLx	
0	0	OCx does not afftect pin
0	1	Toggle OCx pin on successful compare
1	0	Clear OCx on successful compare
1	1	Set OCx pin on successful compare

Figure 8.19 ■ The contents of the TCTL1 register (Reprinted with permission of Motorola)

set to 1. An interrupt request will be generated if it is enabled. The 16-bit output-compare register for each output-compare function can be read and written at any time. The TOCx registers are forced to $FFFF during a reset.

The following actions can be activated on an output-compare pin:

- pull up to high
- pull down to low
- toggle

The action of an OC pin can be selected by programming the TCTL1 register (at $1020) as shown in Figure 8.19.

A successful compare will set the corresponding flag bit in the TFLG1 register. An interrupt will be generated if it is enabled. An output-compare interrupt is enabled by setting the corresponding bit in the TMSK1 register. The memory locations that hold interrupt vectors for output-compare interrupts are listed in Table 8.4. Output-compare functions are not disabled at any time. Therefore, the output-compare flags in the TFLG1 register will be set most of the time. It is very important to clear those flags before using output-compare functions. The output-compare 1 (OC1) function is different from other output-compare functions and will be discussed later.

Output compare	Pin	68HC11 vector location	Buffalo pseudo vector location
OC1	PA7	$FFE8 - $FFE9	$00DF - $00E1
OC2	PA6	$FFE6 - $FFE7	$00DC - $00DE
OC3	PA5	$FFE4 - $FFE5	$00D9 - $00DB
OC4	PA4	$FFE2 - $FFE3	$00D6 - $00D8
OC5	PA3	$FFE0 - $FFE1	$00D3 - $00D5

Table 8.4 ■ Interrupt vector locations for output-compare functions

8.6.2   Applications of the Output-Compare Function

An output-compare function can be programmed to perform a variety of functions. Generation of a single pulse, a square-wave, and a specific delay are among the most popular applications.

Example 8.6

Generate an active high 1 KHz digital waveform with 40 percent duty cycle from output-compare pin OC2. Use the polling method to check the success of the output compare. The frequency of the E clock is 2 MHz, and the prescale factor to the free-running timer is 1.

Solution: An active high 1KHz waveform with 40 percent duty cycle looks like this:

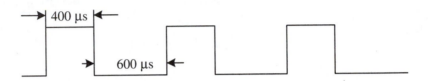

The logic flow of this problem is illustrated in the flowchart shown in Figure 8.20.

For the 2MHz E clock, 800 and 1200 clock cycles correspond to delays of 400 µs and 600 µs, respectively. The following program implements this algorithm using the OC2 function:

```
REGBAS   EQU    $1000           ; base address of I/O register block
PORTA    EQU    $00             ; offset of PORTA from REGBAS
TOC2     EQU    $18             ; offset of TOC2 from REGBAS
TCNT     EQU    $0E             ; offset of TCNT from REGBAS
TCTL1    EQU    $20             ; offset of TCTL1 from REGBAS
TFLG1    EQU    $23             ; offset of TFLG1 from REGBAS
LOTIME   EQU    1200            ; value to set low time to 600 µs
HITIME   EQU    800             ; value to set high time to 400 µs
TOGGLE   EQU    $40             ; value to select the toggle action on pin OC2
OC2      EQU    $40             ; mask to select OC2 pin & OC2F flag
CLEAR    EQU    $40             ; value to clear OC2F flag
sethigh  EQU    $40             ; value to set OC2 pin to high

         ORG    $C000           ; starting address of the program
         LDX    #REGBAS
         BSET   PORTA,X OC2      ; set OC2 pin to high
```

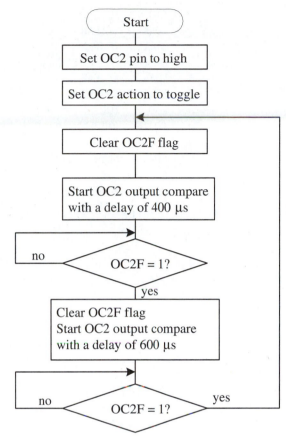

Figure 8.20 ■ The program logic flow for digital waveform generation

```
        LDAA    #TOGGLE              ; select toggle as the output compare action
        STAA    TCTL1,X             ;        "
```

* The following three instructions start the OC2 output-compare with a delay of 400 µs

```
        LDD     TCNT,X              ; start OC2 with a delay of 400 µs
        ADDD    #HITIME             ;        "
        STD     TOC2,X              ;        "
        LDAA    #CLEAR              ; clear OC2F flag
        STAA    TFLG1,X             ;        "

high    BRCLR   TFLG1,X OC2 high    ; wait until OC2F flag set to 1
        LDD     TOC2,X              ; toggle OC2 pin (pulled to high) after
*                                   ; 1200 cycles
        ADDD    #LOTIME             ;        "
```

```
                    STD     TOC2,X              ;       "
                    LDAA    #CLEAR              ; clear the OC2F flag
                    STAA    TFLG1,X             ;       "
        low         BRCLR   TFLG1,X OC2 low     ; wait until OC2F set to 1
                    LDAA    #CLEAR              ; clear OC2F flag
                    STAA    TFLG1,X             ;       "
                    LDD     TOC2,X              ; start the next OC2 compare operation
                    ADDD    #HITIME             ; which will toggle OC2 pin 400 µs later
                    STD     TOC2,X              ;       "
                    BRA     high
                    END
```

The C language version of the program is as follows:

```c
#include   <hc11.h>
main ()
{
    PORTA | = 0x40;                 /* set OC2 pin to high */
    TCTL1 = 0x40;                   /* select "toggle" as the OC2 pin action */
    TOC2 = TCNT + 800;              /* start OC2 compare with 800 cycles as the delay */
    TFLG1 = 0x40;                   /* clear the OC2F flag */
    while (1) {
        while (!(TFLG1 & 0x40));    /* wait until 800 E clock cycles is over */
        TFLG1 = 0x40;               /* clear OC2F flag */
        TOC2 = TOC2 + 1200;         /* start the next OC2 operation */
        while (!(TFLG1 & 0x40));    /* wait until 1200 E cycles is over */
        TFLG1 = 0x40;
        TOC2 = TOC2 + 800;
    }
    return 0;
}
```

Example 8.7

Write a function to generate a delay of 1 second using the OC2 function. The E clock frequency is 2 MHz, and the prescale factor to the main timer is 1.

Solution: Because the output-compare register is only 16-bit, the longest delay that can be generated in one output-compare operation is only 32.7 ms at 2 MHz E clock frequency. In order to generate a delay of 1 second, we will need to perform multiple output-compare operations. For example, we can perform forty such output-compare operations with each output-compare operation creating a delay of 25 ms. The following program implements this idea with the OC2 function. A memory location is used to keep track of the number of output-compare operations remained to be performed.

```
        REGBAS    EQU    $1000          ; base address of I/O register block
        TOC2      EQU    $18            ; offset of TOC2 from REGBAS
        TCNT      EQU    $0E            ; offset of TCNT from REGBAS
        TFLG1     EQU    $23            ; offset of TFLG1 from REGBAS
        OC2       EQU    $40            ; mask to select OC2 pin & OC2F flag
        CLEAR     EQU    $40            ; value to clear OC2F flag
        DLY25ms   EQU    50000          ; the number of E clock cycles to generate
*                                       ; 25 ms delay
        onesec    EQU    40             ; the number of output-compare operations to
*                                       ; be performed
        oc2_cnt   EQU    0              ; offset of oc2_cnt from the top of the stack
        delay_1s  PSHX                  ; save register X in the stack
                  PSHY
                  DES                   ; allocate one byte as the OC2 count
                  TSY                   ; Y points to the top byte of the stack
                  LDX    #REGBAS
                  LDAA   #onesec        ; initialize output-compare count
                  STAA   oc2_cnt,Y      ;          "
                  LDD    TCNT,X         ; prepare to start OC2 operation
        wait      ADDD   #DLY25ms       ; add 25 ms delay
                  STD    TOC2,X         ; start the OC2 compare operation
                  LDAA   #OC2           ; clear OC2F flag
                  STAA   TFLG1,X        ;          "
                  BRCLR  TFLG1,X OC2 *  ; wait until OC2F is set
                  LDAA   #OC2           ; clear the OC2F flag
                  STAA   TFLG1,X        ;          "
                  DEC    oc2_cnt,Y      ; decrement the output compare counter
                  BEQ    exit           ; Is one second expired?
                  LDD    TOC2,X         ; prepare to start the next OC2 compare operation
                  BRA    wait
        exit      INS                   ; deallocate local variable
                  PULY
                  PULX
                  RTS
```

The C language version of the function is as follows:

```c
void delay_1s ( )
{
    unsigned char oc2_cnt;
    oc2_cnt = 100;              /* initialize the total number of OC2 operations to be
                                   performed */
    TFLG1 = 0x40;              /* clear OC2F flag */
    TOC2 = TCNT + 20000;       /* start an OC2 operation with 10 ms as the delay */
    while (oc2_cnt) {
        while (!(TFLG1 & 0x40));    /* wait for 10 ms */
```

```
              TFLG1 = 0x40;
              -- oc2_cnt;
              TOC2 = TOC2 + 20000;     /* start the next output compare operation */
        }
  }
```

Example 8.8

▼

Suppose an alarm device is already connected properly and the subroutine to turn on the alarm is also available. Write an assembly program to implement the alarm timer—it should call the given alarm subroutine when the alarm time is reached.

Solution: We will use the OC2 function to update the current time. Each OC2 output-compare operation generates a delay of 20 ms. In order to generate a delay of 1 minute, 3000 such output-compare operations (call each such operation a *tick*, with one tick equal to 20 ms) must be performed. With a 2 MHz E clock frequency, the delay is equal to the duration of 40000 E clock cycles. An interrupt will be requested when the OC2 output-compare succeeds. The OC2 interrupt service routine decrements the ticks by one and updates the current time. The interrupt service routine also compares the current time with the predefined alarm time. When they are equal, a subroutine is called to invoke the alarm device.

```
        REGBAS   EQU   $1000        ; base address of I/O register block
        TOC2     EQU   $18          ; offset of TOC2 from REGBAS
        TCNT     EQU   $0E          ; offset of TCNT from REGBAS
        TFLG1    EQU   $23          ; offset of TFLG1 from REGBAS
        TMSK1    EQU   $22          ; offset of TMSK1 from REGBAS
        CLEAR    EQU   $40          ; value to clear OC2F flag
        DLY20ms  EQU   40000        ; the number of E clock cycles needed to generate a
*                                   ; 20 ms delay
        one_min  EQU   3000         ; number of output compares to be performed
*                                   ; to generate a delay of 1 minute
                 ORG   $0000
HOURS    RMB   1
MINUTES  RMB   1
TICKS    RMB   2
ALARM    RMB   2
ROUTINE  FDB   ALHNDER      ; the starting address of the alarm handler
                 ORG   $FFE6        ; interrupt vector location for OC2
                 FDB   OC2_ISR      ; set up OC2 interrupt vector
```

```
*****************************************************************
* Use the following assembler directives to replace the previous two directives if the program
* is to be run on the EVB, EVBU, or CMD-11A8
*
*              ORG      $00DC
*              JMP      OC2_ISR
*****************************************************************
```

```
              ORG      $C000         ; starting address of the program
              LDS      #$00FF        ; initialize the stack pointer (set to $DFFF for EVB)
              SEI                    ; disable interrupt to 68HC11 at the beginning
              LDD      #one_min      ; initialize the output compare count to generate
              STD      TICKS         ; 1 minute delay
              LDX      #REGBAS
              LDAA     #CLEAR        ; clear the OC2F flag
              STAA     TFLG1,X       ;          "
              STAA     TMSK1,X       ; enable the OC2 interrupt
              LDD      TCNT,X        ; start an OC2 output compare operation with a
              ADDD     #DLY20ms      ; delay equal to 20 ms
              STD      TOC2,X        ;          "
              CLI                    ; enable interrupt to 68HC11
loop          BRA      loop          ; loop forever and wait for OC2 interrupts
```

* the interrupt service routine of OC2 is in the following

```
OC2_ISR       LDX      #REGBAS
              LDAA     #CLEAR        ; clear OC2F flag
              STAA     TFLG1,X       ;          "
              LDD      TOC2,X
              ADDD     #DLY20ms      ; start the next 20 ms delay
              STD      TOC2,X        ;          "
              LDY      TICKS         ; decrement the minute ticks
              DEY                    ;          "
              STY      TICKS         ;          "
              BNE      CASE2         ; Is 1 minute expired?
              LDD      #one_min      ; Re-initialize the one minute counter
              STD      TICKS         ;          "
              LDD      HOURS         ; load the hours and minutes
              INCB                   ; increment the minute
              CMPB     #60           ; Is it time to increment the hour?
              BNE      CASE1         ;          "
              CLRB                   ; clear minutes to 0
              INCA                   ; increment the hour
              CMPA     #24           ; Is it time to reset hours to 0?
```

```
                      BNE    CASE1        ;        "
                      CLRA                ; reset hours to 00
          CASE1       STD    HOURS        ; save the current time in memory
                      CPD    ALARM        ; compare to alarm time
                      BNE    CASE2
                      LDX    ROUTINE      ; If match then get the subroutine address
                      JSR    0,X          ; call the alarm subroutine
          CASE2       RTI
                      END
```

▲

Example 8.9

▼

LED Flashing. Flashing the LEDs is a common operation during the booting process of many single-board computers. This operation can be achieved easily by using an output port (say port B of the 68HC11A8) and an output-compare function. The circuit connection is shown in Figure 8.21. The flashing pattern can be anything you like. For example, assume we use eight LEDs and we want the LEDs to flash in the following manner:

1. Light all LEDs for ¼ seconds and off for ¼ seconds—this pattern is repeated four times.
2. Light one LED at a time for one second—from the LED controlled by the most significant output port pin to the LED controlled by the least significant output port pin.
3. Reverse the order of display in step 2.
4. Turn off all of the LEDs.

Write a C program to perform this operation.

Solution: We will write two subroutines to implement this operation. The first routine outputs the LED pattern and calls the delay routine for the desired amount of delay. We will use the table lookup method to implement the operation. The delay routine will perform a specified number of OC2 operations (each operation creates 10 ms delay) to create the desired delay.

```
unsigned char flash_tab [25][2] = {{0xFF, 25},{0x00, 25}, {0xFF, 25},{0x00, 25}, {0xFF,
25},{0x00, 25}, {0xFF, 25},{0x00, 25}, {0x80, 100}, {0x40, 100}, {0x20, 100}, {0x10, 100},
{0x08, 100}, {0x04, 100}, {0x02, 100}, {0x01, 100}, {0x01, 100}, {0x02, 100}, {0x04, 100},
{0x08, 100}, {0x10, 100}, {0x20, 100}, {0x40, 100}, {0x80, 100}, {0x00, 100}};
void delay (char k);
void flash ( )
{
    int i;
    for (i = 0; i < 25; i++) {
        PORTB = flash_tab[i][0];
```

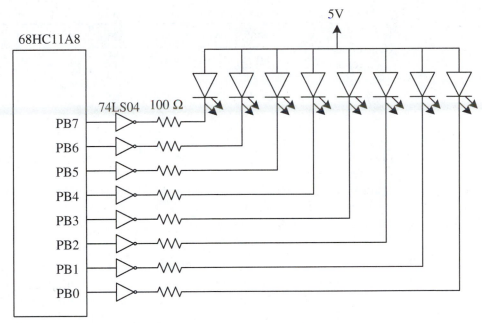

Figure 8.21 ■ An LED-flashing circuit driven by port B

```
            delay(flash_tab[i][1]);
       }
}

void delay (char k)              /* ¼ seconds delay is created when k = 25 and a second
                                    delay is created when k = 100 */
{
    TFLG1 = 0x40;                /* clear OC2F flag */
    TOC2 = TCNT + 20000;         /* start an OC2 operation with 20000 E cycles of delay */
    while (k) {
        while (!(TFLG1 & 0x40)); /* wait for 20000 E cycles to expired */
        TFLG1 = 0x40;
        -- k;
        TOC2 += 20000;           /* start the next OC2 operation */
    }
}
```

8.7   Use OC1 to Control Multiple Output-Compare Functions

The output-compare function OC1 is special because it can control up to five output-compare functions at the same time. The register OC1M is used to specify the output-compare functions to be controlled by OC1. The value

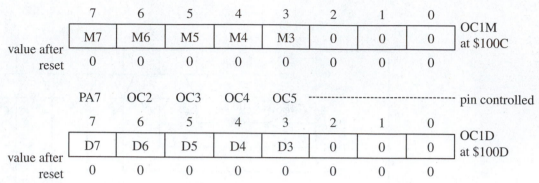

Figure 8.22 ■ Contents of the OC1M and OC1D registers (Reprinted with permission of Motorola)

that any OCx (x = 1, ... ,5) pin is to assume when the value of TOC1 equals TCNT is specified by the OC1D register. To control an output-compare pin using OC1, the user sets the corresponding bit of OC1M. When a successful OC1 compare is made, each affected pin assumes the value of the corresponding bit of OC1D. The contents of the OC1M and OC1D registers and the pin controlled by each bit are shown in Figure 8.22.

Example 8.10

▼

What values should be written into OC1M and OC1D if we want pins OC2 and OC3 to assume the values of 0 and 1 when OC1 compare succeeds?

Solution: Bits 6 and 5 of OC1M must be set to 1, and bits 6 and 5 of OC1D should be set to 0 and 1, respectively. The following instruction sequence will set up these values:

```
REGBAS  EQU   $1000        ; base address of I/O register block
OC1M    EQU   $0C          ; offset of OC1M from REGBAS
OC1D    EQU   $0D          ; offset of OC1D from REGBAS

        LDX   #REGBAS
        LDAA  #$60
        STAA  OC1M,X       ; specify pins to be controlled by OC1
        LDAA  #$20
        STAA  OC1D,X       ; specify the value to assume on a successful OC1 compare
```

The following C statements will achieve the same goal:

```
OC1M = 0x60;
OC1D = 0x20;
```

The OC1 function is especially suitable when several functions are to be controlled by one signal. The PA7 pin is a bidirectional pin that can be controlled by OC1. In order for PA7 to be controlled by OC1, it must be configured for output by setting the DDRA7 bit of the PACTL register to 1. OC1 can also be used in conjunction with one or more other output-compare functions to achieve even more timing flexibility. OC1 can control a timer output even when one of the other output-compare functions is already controlling the same pin, thus allowing the user to schedule two successive edges of each output signal at once. Example 8.11 illustrates this application.

Example 8.11

Using OC1 function to control multiple OC pins. An application requires control of five valves. The first, third, and fifth valves should be opened for 5 seconds and then closed for 5 seconds. When these three valves are closed, the second and fourth valves must be opened and vice versa. This process is repeated forever. The requirements of this application can be satisfied by using the OC1 function. Pins OC1 to OC5 are used to control valves 1 to 5, respectively. When the OCx (x = 1,…,5) pin is high, the corresponding valve will be opened. Write a program to perform the specified operations.

Solution: For this application, all output-compare pins must be controlled by OC1. The PA7 pin must be set for output. The value %11111000 should be written into OC1M. In order to open valves 1, 3, and 5, the value %10101000 should be written into OC1D. In order to open valves 2 and 4, the value %01010000 should be written into OC1D. To create a delay of 5 seconds, 200 output-compare operations will be performed, with each output compare creating a delay of 25 ms. The following program implements these ideas:

```
REGBAS    EQU    $1000        ; base address of I/O register block
PACTL     EQU    $26          ; offset of PACTL from REGBAS
OC1D      EQU    $0D          ; offset of OC1D from REGBAS
OC1M      EQU    $0C          ; offset of OC1M from REGBAS
TOC1      EQU    $16          ; offset of TOC1 from REGBAS
TCNT      EQU    $0E          ; offset of TCNT from REGBAS
TFLG1     EQU    $23          ; offset of TFLG1 from REGBAS
oc1m_in   EQU    $F8          ; value to initialize the OC1M register
oc1d_in1  EQU    $A8          ; value to initialize the OC1D register to open
*                             ; valves 1,3,5
oc1d_in2  EQU    $50          ; value to initialize the OC1D register to open
*                             ; valves 2,4
five_sec  EQU    200          ; number of OC1 compares to be performed
*                             ; to create 5 seconds delay
```

```
CLEAR       EQU     $40             ; value to clear OC2F flag
DLY25ms     EQU     50000           ; the number of E clock cycles to generate
*                                   ; 25 ms delay
PA7         EQU     $80             ; mask to select PA7 pin
OC1         EQU     $80             ; mask to specify OC1
OC1F        EQU     $7F             ; value to select the OC1F flag


            ORG     $0000
oc1_cnt     RMB     1               ; keep track of the number of OC1 operations
*                                   ; remained to be performed


            ORG     $C000           ; starting address of the program
            LDX     #REGBAS
            BSET    PACTL,X PA7     ; set the PA7 pin for output
            LDAA    #oc1m_in        ; allow OC1 to control five pins
            STAA    OC1M,X          ;           "
            BCLR    TFLG1,X OC1F    ; clear OC1F flag
            LDAA    #five_sec       ; initialize the OC1 count
            STAA    oc1_cnt         ;           "
            LDAA    #oc1d_in1       ; set pins OC1, OC3, & OC5 to high after 25 ms
            STAA    OC1D,X          ;           "
            LDD     TCNT,X          ; start an OC1 output compare
repeat1     ADDD    #DLY25ms        ; to create a delay of 25 ms
            STD     TOC1,X          ;           "
            BRCLR   TFLG1,X OC1 *   ; wait until OC1F flag is set to 1
            BCLR    TFLG1,X OC1F    ; clear the OC1F flag
            DEC     oc1_cnt         ; decrement the output-compare count
            BEQ     change          ; At the end of 5 seconds change the valve setting
            LDD     TOC1,X          ; prepare to start the next OC1 compare
            BRA     repeat1
```

* The following two instructions change the valve setting

```
change      LDAA    #oc1d_in2       ; set to open OC2, OC4
            STAA    OC1D,X          ;           "
            LDAA    #five_sec       ; re-initialize output-compare count
            STAA    oc1_cnt         ;           "
repeat2     LDD     TOC1,X          ; start the next OC1 operation
            ADDD    #DLY25ms        ;           "
            STD     TOC1,X          ;           "
            BRCLR   TFLG1,X OC1 *   ; wait unit 25 ms expired
            BCLR    TFLG1,X OC1F    ; clear OC1F flag
            DEC     oc1_cnt         ; decrement the number of OC1 compares
*                                   ; to be performed
            BEQ     switch          ; switch the valve setting
            BRA     repeat2
```

```
switch       LDAA    #five_sec         ; re-initialize the output compare count
             STAA    oc1_cnt           ;      "
             LDAA    #oc1d_in1         ; change the valve setting
             STAA    OC1D,X            ;      "
             LDD     TOC1,X            ; prepare to start another OC1 operation
             BRA     repeat1           ;      "
             END
```

The C language version of the program is as follows:

```c
#include <hc11.h>
main ( )
{
    unsigned int oc1_cnt;
    PACTL |= 0x80;               /* configure PA7 pin for output */
    OC1M = 0xF8;                 /* allow OC1 to control OC1–OC5 pins */
    TFLG1 = 0x80;                /* clear OC1F flag */
    TOC1 = TCNT + 20000;         /* start an OC1 operation */
    while (1) {
        OC1D = 0xA8;             /* value to set PA7, PA5, PA3 to high */
        oc1_cnt = 500;           /* number of OC1 operations needed to create
                                    5 seconds delay */

        while (oc1_cnt) {
            while (!(TFLG1 & 0x80));
            TFLG1 = 0x80;
            TOC1 += 20000;       /* start the next OC1 operation */
            oc1_cnt --;
        }
        OC1D = 0x50;             /* value to set PA6 and PA4 to high */
        oc1_cnt = 500;
        while (oc1_cnt) {        /* stay in this loop for 5 seconds */
            while (!(TFLG1 & 0x80));
            TFLG1 = 0x80;
            TOC1 += 20000;
            oc1_cnt --;
        }
    }
    return 0;
}
```

Two output-compare functions can control the same pin simultaneously. Thus OC1 can be used in conjunction with one or more other output-compare functions to achieve even more timing flexibility. We can generate a digital waveform with a given duty cycle by using OC1 and any other output-compare function.

Example 8.12

Use OC1 and OC2 together to generate a 5 KHz digital waveform with 40 percent duty cycle (active high).

Solution: The idea in using OC1 and OC2 together to generate the specified waveform is to use the OC1 function to pull the OC2 pin to high every 200 μs and use the OC2 function to pull the OC2 pin to low 80 μs later. The waveform is shown in Figure 8.23. Output-compare interrupts for OC1 and OC2 will be enabled. Each interrupt-service routine clears the interrupt flag and then starts another output-compare operation with a 200-μs delay.

```
REGBAS    EQU    $1000          ; base address of I/O register block
TMSK1     EQU    $22            ; offset of TMSK1 from REGBAS
PORTA     EQU    $00            ; offset of PORTA from REGBAS
OC1D      EQU    $0D            ; offset of OC1D from REGBAS
OC1M      EQU    $0C            ; offset of OC1M from REGBAS
TOC1      EQU    $16            ; offset of TOC1 from REGBAS
TOC2      EQU    $18            ; offset of TOC2 from REGBAS
TCTL1     EQU    $20            ; offset of TCTL1 from REGBAS
TCNT      EQU    $0E            ; offset of TCNT from REGBAS
TFLG1     EQU    $23            ; offset of TFLG1 from REGBAS
tctl1_in  EQU    $80            ; value to set the OC2 action to pull the OC2 pin to low
oc1m_in   EQU    $40            ; value to allow OC1 to control the OC2 pin
oc1d_in   EQU    $40            ; value to be written into OC1D to pull OC2 pin to high
fiveKHz   EQU    400            ; timer count for 5 KHz
diff      EQU    160            ; the count difference between two OC functions

          ORG    $FFE6
          FDB    OC2_ISR        ; set up OC1 interrupt vector
          FDB    OC1_ISR        ; set up OC2 interrupt vector
```

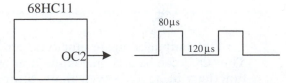

Figure 8.23 ■ Using OC1 and OC2 to generate a 40 percent duty cycle waveform

```
****************************************************************
* Replace the previous three assembler directives with the following directives if the program is
* to be run on the EVB, EVBU, or CMD-11A8
*
*            ORG      $00DC
*            JMP      OC2_ISR        ; pseudo interrupt vector for OC2
*            JMP      OC1_ISR        ; pseudo interrupt vector for OC1
****************************************************************

             ORG      $C000          ; set to $2000 for CMD-11A8
             LDS      #$DFFF         ; set up stack pointer (for EVB only)
             LDX      #REGBAS
             LDAA     #tctl1_in      ; define OC2 action to pull OC2 pin to low
             STAA     TCTL1,X        ;          "
             LDAA     #oc1d_in       ; define OC1 action to pull OC2 pin to high
             STAA     OC1D,X         ;          "
             LDAA     #oc1m_in       ; allow OC1 to control OC2
             STAA     OC1M,X         ;          "
             BSET     TMSK1,X $C0    ; enable OC1 and OC2 interrupts
             BCLR     PORTA,X $40    ; pull OC2 pin to low
             LDD      TCNT,X
             ADDD     #fiveKHz       ; add 200 µs delay
             STD      TOC1,X         ; start OC1 compare operation
             ADDD     #diff
             STD      TOC2,X         ; start OC2 compare operation
             BCLR     TFLG1,X $3F    ; clear OC1F and OC2F flags
             CLI                     ; enable interrupt to 68HC11
loop         BRA      loop           ; infinite loop to wait for interrupts from OC1 & OC2

* The service routines of OC1 and OC2 are in the following

OC1_ISR      BCLR     TFLG1,X $7F    ; clear OC1F flag
             LDD      TOC1,X
             ADDD     #fiveKHz       ; start the next OC1 operation
             STD      TOC1,X         ;          "
             RTI

OC2_ISR      BCLR     TFLG1,X $BF    ; clear OC2F flag
             LDD      TOC2,X
             ADDD     #fiveKHz       ; start the next OC2 operation
             STD      TOC2,X         ;          "
             RTI
             END
```

The C language version of this program is as follows:

```
#include <hc11.h>

void OC1_ISR ( );
void OC2_ISR ( );

main ( )
{
    *(unsigned char *)0xdf = 0x7E;        /* 7E is "jmp" */
    *(void (**)())0xe0 = OC1_ISR;         /* set up OC1 pseudo vector entry */
    *(unsigned char *)0xdc = 0x7E;
    *(void (**)())0xdd = OC2_ISR;
    OC1M = 0x40;                          /* allow OC1 to control OC2 pin */
    OC1D = 0x40;                          /* configure OC1 to pull OC2 pin to high */
    TCTL1 = 0x80;                         /* configure OC2 to pull OC2 pin to low */
    PORTA &= 0xBF;                        /* pull OC2 pin to low */
    TOC1 = TCNT + 400;                    /* start an OC1 operation with 400 E clock cycles
                                             delay */
    TOC2 = TOC1 + 160;                    /* start OC2 operation that will succeed 160 E cycles
                                             later */
    TMSK1 |= 0xC0;                        /* enable OC1 and OC2 interrupts */
    INTR_ON ( );                          /*              "          */
    while (1);                            /* infinite loop */
    return 0;
}

#pragma interrupt_handler OC1_ISR ( )
void OC1_ISR ( )
{
    TFLG1 = 0x80;                         /* clear OC1F flag */
    TOC1 += 400;                          /* start the next OC1 operation */
}
#pragma interrupt_handler OC2_ISR ( )
void OC2_ISR ( )
{
    TFLG1 = 0x40;                         /* clear OC2F flag */
    TOC2 += 400;
}
```

▲

8.8   Forced Output-Compare

There may be applications in which the user requires an output-compare to occur immediately instead of waiting for a match between TCNT and the proper TOC register. This situation arises in spark-timing control in some

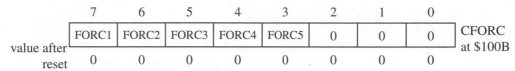

	7	6	5	4	3	2	1	0	
	FORC1	FORC2	FORC3	FORC4	FORC5	0	0	0	CFORC at $100B
value after reset	0	0	0	0	0	0	0	0	

Figure 8.24 ■ Contents of the CFORC register
(Redrawn with permission of Motorola)

automotive engine control applications. To use the forced output-compare mechanism, the user would write to the CFORC register with ones in the bit positions corresponding to the output-compare channels to be forced. At the next timer count after the write to CFORC, the forced channels will trigger their programmed pin actions to occur. The forced actions are synchronized to the timer counter clock, which is slower than E clock signal if a prescale factor has been specified. The forced output-compare signal causes pin actions but does not affect the OCxF bit or generate interrupts. Normally, the force mechanism would not be used in conjunction with the automatic pin action that toggles the corresponding output-compare pin. The contents of CFORC (at $100B) are shown in Figure 8.24. CFORC always reads as all zeroes.

Example 8.13

▼

Suppose that the contents of the TCTL1 register are %10011000. What would occur on pins PA6-PA3 on the next clock cycle if the value %01111000 is written into the CFORC register?

Solution: The contents of the TCTL1 configure the output-compare actions shown in Table 8.5.

Because the CFORC register specifies that the output compare functions 2 to 5 are to be forced immediately, the actions specified in the third column in Table 8.5 will occur immediately.

Bit positions	Value	Action to be triggered
7 6	1 0	clear OC2 pin
5 4	0 1	toggle OC3 pin
3 2	1 0	clear OC4 pin
1 0	0 0	no effect on OC5 pin

Table 8.5 ■ Pin actions on OC2-OC5 pins

▲

8.9   Real-Time Interrupt (RTI)

When enabled, the real-time interrupt (RTI) function of the 68HC11 can be used to generate periodic interrupts, and these periodic interrupts can be used to perform program switching in a multitasking operating system (see chapter 6).

In a multitasking operating system, a *task* is the execution image of a program. Therefore, two tasks can run the same program. The operating system places all the tasks waiting for execution in a queue data structure. Recall that the queue data structure has a head and a tail. Elements can be appended only to the tail and removed only from the head; the queue is therefore a first-in-first-out data structure. When a timer interrupt occurs, the running task is suspended and the operating system takes over the CPU control. The operating system appends the suspended task to the tail of the queue and then removes the task from the head of the queue for execution.

The period of the real-time interrupt is programmable. Its setting is controlled by the lowest two bits of the PACTL register (located at $1026), as shown in Table 8.6. If the divide factor is 1, the real time interrupt will be generated every 4.1 ms with a 2 MHz E clock. Because the RTI generates interrupt periodically, it is suitable for any application that requires periodic processing, such as updating the time-of-day or generating a delay. However, the user should know that the RTI is not as good as the output-compare function in terms of the resolution of the delays it generates.

RTR1 bit 1	RTR0 bit 0	$(E/2^{13})$ divided by
0	0	1
0	1	2
1	0	4
1	1	8

Table 8.6 ■ RTI clock source prescale factor

Example 8.14

▼

Use RTI to generate a delay of 10 seconds.

Solution: To generate a delay of 10 seconds using the RTI function, we need to deal with three registers:

1. We need to enable RTI interrupt, so we need to set the RTII bit (bit 6) of TMSK2 to 1.

2. Because the interrupt interval of RTI is much shorter than 10 seconds, many RTI interrupts are required to generate a delay of 10 seconds. The RTIF flag (bit 6) of the TFLG2 register will need to be cleared many times.

3. We choose a divide factor of 8, so a delay of about 32.77 ms is generated by each RTI interrupt. The total number of RTI interrupts required to generate a delay of 10 seconds is $1000 \times 10 \div 32.77 = 305.15 \approx 305$. The value to be written into the lowest two bits of the PACTL register is 11_2.

The key ideas of the solution to this problem are illustrated in the flowchart in Figure 8.25.

```
REGBAS     EQU     $1000          ; base address of I/O register block
TMSK2      EQU     $24            ; offset of TMSK2 from REGBAS
TFLG2      EQU     $25            ; offset of TFLG2 from REGBAS
PACTL      EQU     $26            ; offset of PACTL from REGBAS
TENSEC     EQU     305            ; total RTI interrupts in 10 seconds
RTIF       EQU     $40            ; mask to select the RTIF flag

           ORG     $0000
rti_cnt    RMB     2              ; reserve two bytes for RTI interrupt count
           ORG     $FFF0          ; set up RTI interrupt vector
           FDB     RTI_ISR        ;        "
```

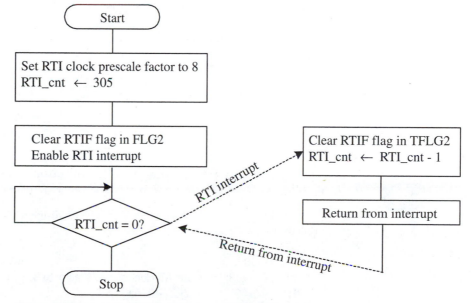

Figure 8.25 ■ Using RTI to create a 10-second delay

```
*****************************************************************
* The following directives must be used to replace the previous two directives if this program
* is to be executed on the EVB, EVBU, or CMD-11A8
*
*            ORG     $00EB
*            JMP     RTI_ISR
*****************************************************************

             ORG     $C000           ; starting address of the program
             LDS     #$00FF          ; set up stack pointer (set to $DFFF for the EVB)
             LDX     #REGBAS
             LDD     #TENSEC         ; initialize the RTI interrupt count
             STD     rti_cnt         ;          "
             BSET    PACTL,X $03     ; set the RTI clock source prescale factor to 8
             LDAA    #RTIF
             STAA    TFLG2,X         ; clear RTI flag
             STAA    TMSK2,X         ; enable RTI interrupt
             CLI                     ; enable interrupt to the 68HC11

* The following two instructions wait until the rti_cnt is decremented to 0

loop         LDD     rti_cnt
             BNE     loop
             ....                    ; do something else

* The RTI interrupt service routine is in the following

RTI_ISR      LDX     #REGBAS
             LDAA    #RTIF
             STAA    TFLG2,X
             LDX     rti_cnt
             DEX
             STX     rti_cnt
             RTI
             END
```

The C language routine that uses RTI function to create a 10-second delay is as follows:

```
int rti_cnt;                /* total number of RTI interrupts remained */
void delay_10s ( )
{
    rti_cnt = 305;
    PACTL |= 0x03;       /* set RTI clock prescale factor to 8 */
    TFLG2 = 0x40;        /* clear RTIF flag */
```

```
        TMSK2 |= 0x40;    /* enable RTI interrupt */
        INTR_ON ( );       /*           "        */
        while (rti_cnt);
        TMSK2 &= 0xBF;    /* disable RTI interrupt */
    }
    #pragma interrupt_handler RTI_ISR ( )
    void RTI_ISR ( )
    {
        TFLG2 = 0x40;       /* clear RTIF flag */
        rti_cnt --;
    }
```

8.10   Pulse Accumulator

The 68HC11 has an 8-bit pulse accumulator (PACNT) that can be configured to operate in either of two basic modes:

1. *Event counting mode.* The 8-bit counter increments at each active edge of the PA7 (PAI) pin. Many microcontroller applications involve counting things. These things are called *events*, but in real applications they might be anything: pieces on an assembly line, cycles of an incoming signal, or units of time. To be counted by the pulse accumulator, these things must be translated into rising or falling edges on the PAI pin. A trivial example of event counting might be to count pieces on an assembly line: a light emitter/detector pair could be placed across a conveyor so that as each piece passes the sensor, the light beam is interrupted and a logic-level signal that can be connected to the PAI pin is produced.

2. *Gated accumulation mode.* The 8-bit counter is clocked by a free-running E-divided-by-64 clock, subject to the PAI signal being active. One common use of this mode is to measure the duration of a single pulse. The counter is set to 0 before the pulse starts, and the resultant pulse time is read directly when the pulse is finished.

The PAI pin must be configured for input and the PA function must be enabled before the PA system can be used. Interrupts are often used in pulse accumulator applications. There are two independent interrupt sources for the pulse accumulator:

1. detection of a selected edge on the PAI pin
2. rollover (overflow) of the 8-bit counter, i.e., from $FF to $00

8.10.1   Pulse Accumulator–Related Registers

The four registers described below are related to the pulse accumulator. The contents of these registers are shown in Figure 8.26, where the bits related to the pulse accumulator are in boldface. These bits are also discussed below:

1. PACNT: an 8-bit pulse accumulation counter.

2. TMSK2: timer mask register 2. Bits 5 and 4 are related to the operation of the pulse accumulator.
   PAOVI: pulse accumulator overflow interrupt enable. When this bit is set to 1, the pulse accumulator overflow interrupt is enabled.
   PAII: pulse accumulator input interrupt enable. When this bit is set to 1, the pulse accumulator input edge interrupt is enabled.

3. TFLG2: timer flag register 2. Bits 5 and 4 are related to the operation of the pulse accumulator.
   PAOVF: pulse accumulator overflow flag. This bit is set to 1 whenever PACNT rolls over from $FF to $00.
   PAIF: pulse-accumulator-input-edge-detect flag. This bit is set to 1 each time a selected edge is detected at the PA7/PAI pin.

4. PACTL: pulse accumulator control register. Bits 7 to 4 are related to the operation of the pulse accumulator.

	7	6	5	4	3	2	1	0	
	TOI	RTII	**PAOVI**	**PAII**	0	0	PR1	PR0	TMSK2 at $1024
Value after reset	0	0	0	0	0	0	0	0	
	TOF	RTIF	**PAOVF**	**PAIF**	0	0	0	0	TFLG2 at $1025
Value after reset	0	0	0	0	0	0	0	0	
	DDRA7	**PAEN**	**PAMOD**	**PEDGE**	0	0	RTR1	RTR0	PACTL at $1026
Value after reset	0	0	0	0	0	0	0	0	
	bit 7	bit 6	bit 5	bit 4	bit 3	bit 2	bit 1	bit 0	PACNT at $1027
Value after reset	-	-	-	-	-	-	-	-	

Figure 8.26 ■ Registers related to the pulse accumulator function
(Redrawn with permission of Motorola)

PAMOD	PEDGE	Action on clock
0	0	PAI falling edge increments the counter
0	1	PAI rising edge increments the counter
1	0	A 0 on PAI inhibits counting
1	1	A 1 on PAI inhibits counting

Table 8.7 ■ Combinations of PAMOD and PEDGE bits

DDRA7: data direction control for port A pin 7. This bit must be set to 0 when the pulse accumulator is enabled.

PAEN: pulse accumulator enable. When this bit is set to 1, the pulse accumulator is enabled.

PAMOD: pulse accumulator mode select. This bit sets the operation mode of the pulse accumulator:

0: external event-counting mode.

1: gated time accumulation mode.

PEDGE: pulse accumulator edge select. This bit has different meanings depending on the state of the PAMOD bit as shown in Table 8.7.

8.10.2 Pulse Accumulator Applications

The 68HC11 pulse accumulator has several interesting applications, such as interrupting after N events, pulse duration measurement, and frequency measurement.

Example 8.15

Interrupt After N Events. In this application, external events are converted into signals and connected to the PAI pin. N must be smaller than 255. Write a program to interrupt the 68HC11 after N events.

Solution: By writing the twos complement of N into PACNT, the counter will overflow after N events and generate an interrupt. The program is as follows:

```
REGBAS   EQU   $1000            ; base address of I/O register block
TMSK2    EQU   $24              ; offset of TMSK2 from REGBAS
TFLG2    EQU   $25              ; offset of TFLG2 from REGBAS
PACTL    EQU   $26              ; offset of PACTL from REGBAS
PACNT    EQU   $27              ; offset of PACNT from REGBAS
PA_INI   EQU   $50              ; value to be written into PACTL to enable PA & select
*                               ; event counting mode, rising edge as active edge
N        EQU   ...              ; event count
```

```
ORG    $C000              ; starting address of the program
LDX    #REGBAS
BCLR   TFLG2,X $DF        ; clear the PAOVF flag to make sure it is not set
LDAA   #N
NEGA
STAA   PACNT,X            ; place the 2's complement of N in PACNT
LDAA   #PA_INI
STAA   PACTL,X            ; initialize the pulse accumulator
BSET   TMSK2,X $20        ; enable PACNT overflow interrupt
CLI                       ; enable interrupt to the 68HC11
END
```

The C language version of the program is as follows:

```c
#include <hc11.h>
void PAOV_ISR ( );
main ( )
{
    *(unsigned char *)0xcd = 0x7E;
    *(void (**)())0xce = PAOV_ISR;
    PACNT = ~N + 1;                    /* place -N in PACNT */
    PACTL = 0x50;                      /* enable PA function */
    TMSK2 |= 0x20;                     /* enable PAOV interrupt */
    INTR_ON ( );
    ...
}
#pragma interrupt_handler PAOV_ISR ( )
void PAOV_ISR ( )
{
    ...
}
```

The pulse accumulator can also be used to measure the frequency of an unknown signal. The procedure is as follows:

1. Set up the pulse accumulator to operate in event-counting mode.
2. Connect the unknown signal to the PAI pin.
3. Select the active edge (rising or falling).
4. Use one of the output-compare functions to create a delay of 1 second.
5. Use a memory location to keep track of the number of active edges arriving during 1 second.

6. Enable the PAI interrupt. The interrupt-handling routine increments the edge-count by one.

7. Disable the pulse accumulator interrupt at the end of 1 second.

Example 8.16

▼

Write a program to measure the frequency of an unknown signal connected to PAI pin.

Solution: We will use the OC2 function to perform 40 output-compare operations, with each operation creating a delay of 25 ms. This will create a time base of 1 second. The PAI interrupt is enabled, and a service routine is written that will increment the frequency count by 1 whenever it is invoked. The memory location used as the frequency counter must be initialized to 0. Whenever an active edge arrives at the PAI pin, an interrupt that causes the frequency counter to be incremented by 1 is requested. The number of active edges arrived in 1 second is the frequency of the unknown signal.

```
REGBAS   EQU    $1000            ; base address of I/O register block
TCNT     EQU    $0E              ; offset of TCNT from REGBAS
TOC2     EQU    $18              ; offset of TOC2 from REGBAS
TFLG1    EQU    $23              ; offset of TFLG1 from REGBAS
TMSK2    EQU    $24              ; offset of TMSK2 from REGBAS
TFLG2    EQU    $25              ; offset of TFLG2 from REGBAS
PACTL    EQU    $26              ; offset of PACTL from REGBAS
PACNT    EQU    $27              ; offset of PACNT from REGBAS
CLEAR    EQU    $40              ; mask to clear OC2F flag
OC2      EQU    $40              ; mask to select the OC2F flag
OC2DLY   EQU    50000            ; output compare delay count to generate
*                                ; 25 ms delay
PA_IN    EQU    $50              ; value to be written in PACTL to enable PA function,
*                                ; select event-counting mode, rising edge as
*                                ; active edge, and set PA7 pin for input
onesec   EQU    40               ; number of output compares to be performed
STOP     EQU    $10              ; value to disable PAI interrupt

         ORG    $00
oc2_cnt  RMB    1
FREQCY   RMB    2                ; memory locations to keep track of the number
*                                ; of active edges arrived in one second
         ORG    $FFDA
         FDB    PAI_ISR          ; set up PAI interrupt vector
```

```
        *********************************************************************
        * Use the following directives to replace the above two directives if this program is to be
        * executed on the EVB, EVBU, or CMD-11A8
        *
        *        ORG     $00CA
        *        JMP     PAI_ISR              ; set up PAI pseudo interrupt vector on EVB
        *********************************************************************

                 ORG     $C000               ; starting address of the program
                 LDS     #$DFFF              ; set up stack pointer (for EVB only)
                 LDX     #REGBAS
                 LDAA    #onesec
                 STAA    oc2_cnt             ; set up the number of OC2 operations to be
        *                                    ; performed
                 LDD     #0
                 STD     FREQCY              ; initialize the frequency counter to 0
                 LDAA    #PA_IN
                 STAA    PACTL,X             ; initialize the PA function
                 BCLR    TFLG2,X $EF         ; clear the PAIF flag
                 BSET    TMSK2,X $10         ; enable the PAI interrupt
                 CLI                         ; enable interrupt to the 68HC11
                 LDD     TCNT,X
sec_loop         ADDD    #OC2DLY             ; start OC2 operation to create 25 ms delay
                 STD     TOC2,X              ;        "
                 LDAA    #CLEAR
                 STAA    TFLG2,X             ; clear the OC2F flag
                 BRCLR   TFLG1,X OC2 *       ; wait until the OC2F flag is set
                 LDD     TOC2,X
                 DEC     oc2_cnt
                 BNE     sec_loop            ; Repeat if one second is not expired yet
                 LDAA    #STOP
                 STAA    TMSK2,X             ; disable PAI interrupt
                 .                           ; some other operations may follow here
                 .
                 .
```

* The PAI interrupt service routine is in the following

```
PAI_ISR          LDX     #REGBAS
                 BCLR    TFLG2,X $EF         ; clear the PAIF flag
                 LDD     FREQCY
                 ADDD    #1                  ; increase the frequency count
                 STD     FREQCY              ;        "
                 RTI
                 END
```

The C language version of the program is as follows:

```c
#include <stdio.h>
#include <hc11.h>
void PAI_ISR ( );
unsigned int frequency;

main ( )
{
    unsigned int oc2_cnt;

    *(unsigned char *)0xca = 0x7E;
    *(void (**)())0xcb = PAI_ISR;
    frequency = 0;
    PACTL = 0x50;                        /* enable PA, rising edge, choose event counting
                                            mode */
    TFLG2 = 0x10;                        /* clear PAIF flag */
    oc2_cnt = 100;                       /* total OC2 operations to be performed */
    TOC2 = TCNT + 20000;                 /* start an OC2 operation with 20000 E cycles as
                                            delay */
    TFLG1 = 0x40;                        /* clear OC2F flag */
    TMSK2 |= 0x10;                       /* enable PAI interrupt */
    INTR_ON ( );                         /*           "        */
    while (oc2_cnt) {
        while (!(TFLG1 & 0x40));
        TFLG1 = 0x40;
        TOC2 += 20000;                   /* start the next OC2 operation */
        oc2_cnt --;
    }
    TMSK2 &= 0xEF;                        /* disable PAI interrupt */
    INTR_OFF ( );
    printf("\n The frequency of the unknown signal is %d \n", frequency);
    return 0;
}

#pragma interrupt_handler PAI_ISR ( )
void PAI_ISR ( )
{
    TFLG2 = 0x10;                        /* clear PAIF flag */
    frequency ++;
}
```

Using the PAI edge interrupt to measure the frequency has fairly high interrupt processing overhead, which will prevent us from measuring frequencies slightly higher than 43 KHz. This problem can be avoided by using the PAOV interrupt, and you will be able to measure frequencies higher than 600 KHz with 2 MHz E clock. This method will be left as a lab exercise problem.

▲

The pulse accumulator can be set up to measure the duration of an unknown signal using the gated pulse accumulation mode. Remember that the pulse accumulator can count only if the PAI input is asserted, i.e., when the pulse is present. By counting the number of clock cycles that occur when the pulse is present, the pulse duration can be measured. The clock input to the pulse accumulator is E ÷ 64. The procedure for measuring the pulse width of an unknown signal (positive pulse) is as follows:

Step 1
Set up the pulse accumulator to operate in the gated time accumulation mode, and initialize the pulse accumulator counter (PACNT) to 0.

Step 2
Select the falling edge as the active edge. In this setting, the pulse accumulator counter will increment when the signal connected to the PAI pin is high and generate an interrupt to the 68HC11 on the falling edge.

Step 3
Enable the PAI active edge interrupt and wait for the arrival of the active edge of PAI.

Step 4
Stop the pulse accumulator counter when the interrupt arrives.

Because the pulse accumulator counter is 8-bit, it can only measure the width of a pulse with duration no longer than $2^8 \times 64 = 16386$ E clock cycles. If you need to measure the duration of a very slow pulse, you can keep track of the number of pulse accumulator counter overflows. Let n be the number of PA overflows and m be the contents of the PACNT when the PAI falling edge arrives. Then the duration of the unknown signal can be calculated using the following expression:

$$\text{pulse width} = (2^8 \times n + m) \times 64 \times T_E$$

where T_E is the period of the E clock.

Example 8.17

▼

Example 8.17   Write a program to measure the duration of an unknown signal connected to the PAI pin using the idea just described.

Solution:

```
REGBAS    EQU    $1000           ; base address of I/O register block
TMSK2     EQU    $24             ; offset of TMSK2 from REGBAS
```

```
        TFLG2     EQU    $25           ; offset of TFLG2 from REGBAS
        PACTL     EQU    $26           ; offset of PACTL from REGBAS
        PACNT     EQU    $27           ; offset of PACNT from REGBAS
        STOP      EQU    $00           ; value to stop pulse accumulator
        PA_IN     EQU    $60           ; value to be written into PACTL to enable PA, select
        *                              ; gated accumulation mode, and select active high
                  ORG    $00
ov_cnt            RMB    2             ; two bytes to keep track of the PACNT overflow count
pa_cnt            RMB    1             ; a byte to hold the count of PACNT
edge              RMB    1             ; PAI edge interrupt count

                  ORG    $FFDA
                  FDB    PAI_ISR       ; set up PAI edge interrupt vector
                  FDB    PAOV_ISR      ; set up PAOV interrupt vector
```

```
****************************************************************
* Use the following directives to replace the previous three directives if the program is to be
* run on the EVB, EVBU, or CMD-11A8
*
*                 ORG    $CA
*                 JMP    PAI_ISR       ; set up PAI pseudo vector
*                 JMP    PAOV_ISR      ; set up PAOV pseudo vector
****************************************************************
```

```
                  ORG    $C000         ; starting address of the program
                  LDS    #$DFFF        ; set up the stack pointer (for EVB)
                  LDX    #REGBAS
                  LDD    #0
                  STD    ov_cnt        ; initialize the PA overflow count to 0
                  LDAA   #1
                  STAA   edge          ; initialize the number of PAI edge interrupt count to 1
                  BCLR   TFLG2,X $CF   ; clear PAOVF and PAIF flags to 0
                  LDAA   #PA_IN
                  STAA   PACTL,X       ; initialize the pulse accumulator
                  CLR    PACNT,X       ; initialize PACNT to 0
                  BSET   TMSK2,X $30   ; enable PAI and PAOV interrupts
                  CLI                  ; enable interrupt to the 68HC11
wait              TST    edge          ; wait for the falling edge of PAI
                  BNE    wait          ;        "
                  BCLR   PACTL,X $40   ; disable pulse accumulator
                  LDAA   PACNT,X
                  STAA   pa_cnt        ; append the contents of PACNT to the overflow count
                  .                    ; do something else
                  .
                  .
```

* The PAI and PAOV interrupt service routines are in the following

```
PAI_ISR     BCLR    TFLG2,X $EF     ; clear the PAIF flag
            DEC     edge            ; reset the edge flag to 0
            RTI


PAOV_ISR    BCLR    TFLG2,X $DF     ; clear the PAOVF flag
            LDD     ov_cnt
            ADDD    #1              ; increment the PACNT overflow count
            STD     ov_cnt          ;        "
            RTI
            END
```

The C language version of the program is as follows:

```c
#include <stdio.h>
#include <hc11.h>
void PAI_ISR ( );
void PAOV_ISR ( );
unsigned int paov_cnt, pai_cnt, edge;

main ( )
{
    unsigned long pulse_width;
    *(unsigned char *)0xca = 0x7E;
    *(void (**)( ))0xcb = PAI_ISR;
    *(unsigned char *)0xcd = 0x7E;
    *(void (**)( ))0xce = PAOV_ISR;
    paov_cnt = 0;
    edge = 1;
    TFLG2 = 0x30;                    /* clear PAIF and PAOVF flags */
    PACTL = 0x60;                    /* initialize PA function */
    PACNT = 0;
    TMSK2 |= 0x30;                   /* enable PAI and PAOV interrupts */
    INTR_ON ( );                     /*            "          */
    while (edge);
    PACTL &= 0xBF;                   /* disable PA function */
    pulse_width = paov_cnt << 8 + PACNT;
    printf("\n The pulse width of the signal is %d \n",pulse_width);
    return 0;
}

#pragma interrupt_handler PAI_ISR ( )
void PAI_ISR ( )
{
    TFLG2 = 0x10;                    /* clear PAIF flag */
```

```
        edge --;
    }

    #pragma interrupt_handler PAOV_ISR ( )
    void PAOV_ISR ( )
    {
        TFLG2 = 0x20;                    /* clear PAOVF flag */
        paov_cnt ++;
    }
```

▲

8.11   Summary

Port A pins PA2-PA0 can be used as general-purpose input pins or input-capture channels IC1-IC3. Port A pins PA6-PA3 can be used as general-purpose output pins or output-compare pins OC2-OC5. The PA7 pin is bidirectional; when configured for input, it can be used as a general-purpose input pin or as the pulse accumulator input.

Physical time for the 68HC11 is the value in the 16-bit free-running timer. When the selected edge is detected on an input-capture pin, the value of the free-running timer is latched into a 16-bit input-capture register. This feature is often used to measure the pulse width and the signal frequency. Input-capture channels can also be used as additional interrupt sources, or time references.

Each output-compare function has a 16-bit register, a 16-bit comparator, and an output-compare action pin. Output-compare functions are often used to generate a delay, to create a time base for measurements, or to generate digital waveforms. The steps in using an output-compare function include making a copy of the current contents of TCNT, adding a value equal to the desired delay to the copy of TCNT, and storing the sum back to an output-compare register. An output-compare pin can activate any of the following actions: pull up to high, pull down to low, and toggle.

An output-compare operation can be forced to take effect immediately without setting any output-compare flag. Output compare 1 can control up to five OC pins simultaneously. Each of the output-compare pins OC2-OC5 can be controlled by two output-compare functions simultaneously.

The RTI function can be used to generate periodic interrupts so that the microcontroller can be reminded to perform routine works.

The 8-bit pulse accumulator can be enabled to operate either in the event-counting or the gated time accumulation mode. The pulse accumulator can be used to measure the frequency and pulse width of an unknown signal and as an edge-sensitive interrupt source.

The memory locations that hold the timer interrupt vectors are listed in Table 8.8.

Interrupt source	68HC11 vector location	EVB pseudovector location
pulse accumulator input edge	$FFDA - $FFDB	$CA - $CC
pulse accumulator overflow	$FFDC - $FFDD	$CD - $CF
timer overflow	$FFDE - $FFDF	$D0 - $D2
timer output compare 5	$FFE0 - $FFE1	$D3 - $D5
timer output compare 4	$FFE2 - $FFE3	$D6 - $D8
timer output compare 3	$FFE4 - $FFE5	$D9 - $DB
timer output compare 2	$FFE6 - $FFE7	$DC - $DE
timer output compare 1	$FFE8 - $FFE9	$DF - $E1
timer input capture 3	$FFEA - $FFEB	$E2 - $E4
timer input capture 2	$FFEC - $FFED	$E5 - $E7
timer input capture 1	$FFEE - $FFEF	$E8 - $EA
real time interrupt	$FFF0 - $FFF1	$EB - $ED

Table 8.8 ■ 68HC11 timer interrupt vector locations

8.12   Exercises

A 2MHz E clock is used in all of the following problems.

E8.1 What value will be loaded into accumulator D if the value of TCNT is $1000 when the instruction LDD $100E is being fetched?

E8.2 What value should be written into TCTL1 to toggle the voltage on the PA5 pin on successful output compares?

E8.3 Rewrite the program in Example 8.6 and use interrupt-driven method to generate the waveform.

E8.4 Write a program to use the OC2 function to generate a 5 KHz square waveform with 50 percent duty cycle.

E8.5 Write a program to use the OC2 function to generate a 2 KHz square waveform with 30 percent duty cycle.

E8.6 Use OC1 and OC2 function together to generate a 1 KHz digital waveform with 20 percent duty cycle.

E8.7 Write a program to generate ten pulses from the OC3 pin. Each pulse has 1 ms low time and 0.5 ms high time.

E8.8 Write a program to turn a light on for 5 seconds and then off for 5 seconds, repeating this cycle continuously. The light is controlled by pin PB0. The light is turned on when PB0 is high and off when it is low. Use the RTI interrupts to implement this operation.

E8.9 Write a program to interrupt 10 ms after a positive edge has been detected on the PA2 pin.

E8.10 How often will the pulse accumulator overflow if PA7 is constantly high in the gated counting mode? Assume the pulse accumulator counter has been programmed to count when the PA7 pin is high.

E8.11 Write a program that uses the OC2 function to create a delay of 20 ms after the rising edge has been detected on the IC1 pin.

E8.12 What will happen at the end of the execution of the following instructions?

```
TCTL1    EQU    $20
CFORC    EQU    $08
TFLG1    EQU    $23
TOC2     EQU    $12
TOC3     EQU    $14
TOC4     EQU    $16
TOC5     EQU    $18
TCNT     EQU    $0E

         LDX    #$1000
         LDD    TCNT,X
         ADDD   #40000
         STD    TOC2,X
         STD    TOC3,X
         STD    TOC4,X
         STD    TOC5,X
         LDAA   #%01011011
         STAA   TCTL1,X
         LDAA   #$FF
         STAA   TFLG1,X
         LDAA   #%01111000
         STAA   CFORC,X
```

E8.13 Write a program to output a 10-ms positive pulse on the PA3 pin every time a rising edge is detected on PA1.

E8.14 Write a subroutine to measure the duty cycle of a digital waveform. The channel to which the signal is connected is specified in accumulator A. This routine will return the duty cycle as a percentage in accumulator A. For example, the duty cycle of 60 percent will cause the decimal number (BCD) 60 to be returned in A.

E8.15 Write a subroutine to measure the phase difference of two digital waveforms. Your program will need to detect the channels to which the signals are connected. The phase difference in degree will be returned in accumulator D.

E8.16 Write a subroutine to generate a pulse on the OC5 pin 10 ms after a rising edge has been detected on the IC2 pin. The pulse width is 1 ms.

8.13   Lab Exercises and Assignments

L8.1 *Frequency Measurement.* Use the pulse accumulator function to measure the frequency of an unknown signal. The procedure is as follows:

1. Connect the signal to the PAI (PA7) pin.
2. Also connect the signal to an oscilloscope or a frequency counter.

3. Set up a 1 second time base using one of the output-compare functions for frequency measurement.

4. Allocate three bytes of on-chip SRAM to accumulate the frequency counts. The frequency counter must be reset to 0 before the program is executed.

5. Enable PAOV interrupt to keep track of the number of times that the PACNT overflows in one second.

6. Increment the PAOV count when there is a PACNT-overflow interrupt.

7. At the end of the 1-second period append the contents of PACNT to the PAOV count to obtain the frequency of the unknown signal.

8. Convert the binary frequency count to decimal and display it on the monitor screen.

9. Change the signal frequency from about 1 Hz to 650KHz and repeat the same measurement. The frequency range to be measured is as follows:

   (1) 1 to 10 Hz
   (2) 10 to 100 Hz
   (3) 100 to 1000 Hz
   (4) 1 KHz to 10 KHz
   (5) 10 KHz to 100 KHz
   (6) 100 KHz to 650 KHz

Make two measurements in each range. What is the highest frequency that you can measure?

L8.2 *Waveform generation.* Use an output-compare function pin to generate an active high digital waveform. The duty cycle and the frequency must be programmable. The procedure is as follows:

1. Connect one of the output-compare function pins to the oscilloscope input.

2. Enter the frequency (as a multiple of hundred) in the on-chip SRAM at location $00. For example, the number 20 stands for the frequency $20 \times 100 = 2$ KHz. The frequency is no higher than 20 KHz, for which the number 200 would be entered.

3. Enter the duty cycle in the on-chip SRAM at $01. For example, the number 30 stands for duty cycle 30 percent.

4. Write a program to

   ■ read the frequency and duty cycle from the SRAM
   ■ use the OC2 pin to generate the specified waveform
   ■ set a time base of one minute using the OC3 function
   ■ jump back to the monitor at the end of 1 minute

5. Download the program into the 68HC11 evaluation board and execute the program. The waveform should appear on the oscilloscope screen.

6. Repeat the same experiment with several different frequencies.

L8.3 *Time-of-Day Display.* Use an output-compare function, an input capture function, two parallel ports, and seven-segment displays to implement the time-of-day clock. The procedure is as follows:

Step 1
Write a main program that invokes the Buffalo library routines to put out a prompt that asks the user to enter the current time in the format of *hh mm ss*. The main program then invokes the Buffalo input routine to read the time entered by the user. The main program will perform input checking until a valid time is entered.

Step 2
Use one of the output-compare functions to create delays, for example, 25 ms. Each successful output compare operation will generate an interrupt request to the microcontroller. The output-compare interrupt service routine will check to see if 1 second has passed. If so, it will update the current time.

Step 3
Use port B to drive the segment pattern inputs for seven-segment displays. Six seven-segment displays are required to display the time-of-day. Use port D to drive the display select signals as shown in Figure 7.21.

L8.4 *LEDs flashing experiment.* Use port B to drive eight LEDs and generate the following flashing patterns:

1. Light one LED at a time for ½ second from the LED controlled by PB7 down to the LED controller by PB0.

2. Repeat step 1 but in the reverse order.

3. Light all the LEDs for ½ second and off for ½ second for three times.

9

68HC11 Serial Communication Interface

9.1 Objectives

After completing this chapter you should be able to

- define different types of data links

- explain the four aspects of the RS232 interface standard

- explain the data transmission process for the RS232 interface standard

- explain the types of data transmission errors that occur in serial data communication

- establish a null modem connection

- explain the operation of the SCI subsystem

- wire the SCI pins to an RS232 connector

- program the SCI subsystem to perform terminal I/O

- explain the operation of the Motorola 6850 ACIA chip

- interface the 6850 ACIA to the 68HC11 and an RS232 connector

- program the 6850 ACIA to do data transfer

9.2   Introduction

Asynchronous serial data transfer is widely used for data communication between two computers or between a computer and a terminal (or printer), either with or without a modem. A dedicated or a public phone line can be used as a medium for data transmission.

A basic data communication link is shown in Figure 9.1. A terminal at one end of the link communicates with a computer at the other end. The communication link consists of a data terminal equipment (DTE) and an associated modem (data communication equipment [DCE]) at each end. The term *modem* is derived from the process of accepting digital bits, changing them into a form suitable for analog transmission (*modulation*), receiving the signal at the other station, and transforming it back to its original digital representation (*demodulation*).

There are two types of communication links, point-to-point and multidrop, as illustrated in Figure 9.2. In the former, the two end stations communicate as peers, while in the later, one device is designated as the master (primary) and the others as slaves (secondaries). Each multidrop station has its own unique *address*, with the primary station controlling all data transfers over the communication link.

A communication link can consist of either two or four wires. A two-wire link provides a signal line and a ground, and a four-wire link provides two such pairs. A communication link can be used in three different ways:

1. *Simplex link.* The line is dedicated to either transmission or reception, but not both.
2. *Half-duplex link.* The communication link can be used for either transmission or reception, but only in one direction at a time.
3. *Full-duplex link.* Both transmission and reception can proceed simultaneously. This link requires four wires.

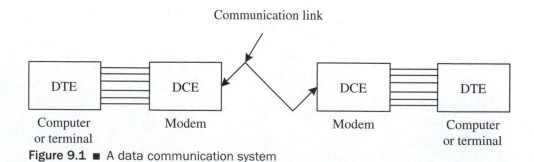

Figure 9.1 ■ A data communication system

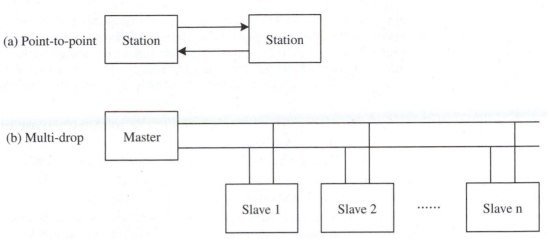

Figure 9.2 ■ Point-to-point and multi-drop communication links

Most data communication takes place over the telephone line or over lines that are engineered to the telephone-line specifications. The various standards organizations have therefore published many recommendations defining how these connections and communications should be made. The International Telegraph & Telephone Consultative Committee (CCITT) V series define these connections. The International Standard Organization (ISO) also publishes many standards, some of which describe the mechanical connectors used by computers, terminals, modems, and other devices. Readers who are interested in these standards should refer to books on data communication and computer networking.

Data communication engineers and technicians use the term *mark* to indicate a binary 1 and *space* to indicate a binary 0. The terms *high* and *low* are confusing when applied to serial data communication because the logic levels used on a serial communication circuit may not be those used in standard TTL or CMOS digital circuits. The RS232 standard, for example, represents a binary 0, or space, with a positive voltage and a binary 1, or mark, with a negative voltage.

9.3   RS232 Standard

The RS232 standard published by the Electronic Industry Association (EIA) is one of the most widely used physical level interfaces in the world and is most prevalent in North America. It specifies twenty-five interchange circuits for DTE/DCE use. In 1960 EIA established the RS232 standard for interfacing between a computer and a modem. It has experienced several revisions since then. The latest revision, RS-232-E, was published in July 1991. In this revision, the EIA has replaced the prefix RS (recommended standard) with the prefix EIA. The change represents no change to the standard, but was made to allow a user to identify more easily the source of the standard. We will mix the uses of RS-232 and EIA-232 throughout this chapter. There are four aspects of the RS232 standard:

1. The electrical aspect specifies the voltage level, rise time, and fall time of each signal.
2. The functional specifications specify the function of each signal.
3. The mechanical specifications specify the number of pins and the shape and dimensions of the connector.
4. The procedural specifications specify the sequence of events for transmitting data, based on the functional specifications of the interface.

9.3.1   EIA-232-E Electrical Specification

Most of the EIA-232-E electrical specifications are not important to the end user. However, we do need to know the following:

- The interface is rated at a signal rate of <20 kbps.
- The signal can transfer correctly within 15 meters. Greater distance and higher data rate are possible with better design.
- Driver maximum output voltage (when open-circuit) is −25 V to +25 V.
- Driver minimum output voltage (loaded output) is −25 V to −5 V or +5 V to +25 V.
- Receiver input voltage range is −25 V to +25 V.
- A voltage more negative than −3 V at receiver's input is interpreted as a logic 1
- A voltage more positive than +3 V at receiver's input is interpreted as a logic 0.

9.3.2   EIA-232-E Functional Specification

The EIA-232-E interface provides one primary channel and one secondary channel. Table 9.1 summarizes the functional specification of each circuit. Because the secondary channel is rarely used, only the functions of signals in the primary channel will be discussed. There is one data circuit in each direction, so full-duplex operation is possible. One ground lead is for protective isolation; the other serves as the return circuit for both data leads. Hence transmission is unbalanced, with only one active wire. The timing signals provide clock pulses for synchronous transmission. When the DCE is sending data over circuit BB, it

Pin No.	Circuit	Description
1	–	Shield
2	BA	Transmitted data
3	BB	Received data
4	CA/CJ	Request to send/ready for receiving[1]
5	CB	Clear to send
6	CC	DCE ready
7	AB	Signal common
8	CF	Received line signal detector
9	–	(reserved for testing)
10	–	(reserved for testing)
11	–	unassigned[3]
12	SCF/CI	Secondary received line signal detection/data rate selector (DCE source)[2]
13	SCB	Secondary clear to send
14	SBA	Secondary transmitted data
15	DB	Transmitter signal element timing (DCE source)
16	SBB	Secondary received data
17	DD	Receiver signal element timing
18	LL	Local loopback
19	SCA	Secondary request to send
20	CD	DTE ready
21	RL/CG	Remote loopback/signal quality detector
22	CE	Ring indicator
23	CH/CI	Data signal rate selector (DTE/DCE source)[2]
24	DA	Transmitter signal element timing (DTE source)
25	TM	Test mode

1. When hardware flow control is required, circuit CA may take on the functionality of circuit CJ. This is one change from the former EIA-232.
2. For designs using interchange circuit SCF, interchange circuits CH and CI are assigned to pin 23. If SCF is not used, CI is assigned to pin 12.
3. Pin 11 is unassigned. It will not be assigned in future versions of EIA-232. However, in international standard ISO 2110, this pin is assigned to select transmit frequency.

Table 9.1 ■ EIA-232-E signals

also sends 1-0 and 0-1 transitions on circuit DD, with transitions timed to the middle of each BB signal element. When the DTE is sending data, either the DTE or DCE can provide timing pulses, depending on the circumstances. The control signals are explained in the procedural specifications.

9.3.3 EIA-232-E Mechanical Specification

The EIA-232-E uses a 25-pin D-type connector as shown in Figure 9.3. The 25-pin connector is available from many vendors. The exact dimension can be found in the standard document published by EIA.

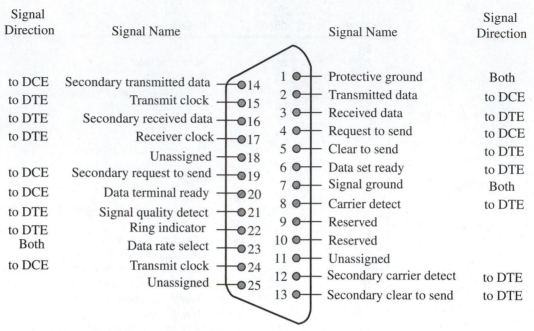

Figure 9.3 ■ EIA-232-E connector and pin assignment

9.3.4   EIA-232-E Procedural Specification

The sequence of events that occurs during data transmission using the EIA-232-E is easier to understand by studying examples. We will use two examples to explain the procedure.

In the first example, two DTEs are connected with a point-to-point link using an asynchronous dedicated-line modem. The modem requires only the following circuits to operate:

- Signal ground (AB)
- Transmitted data (BA)
- Received data (BB)
- Request to send (CA)
- Clear to send (CB)
- Data set ready (CC)
- Carrier detect (CF)

Before the DTE can transmit any data, the data-set-ready (DSR) circuit must be asserted to indicate that the modem is ready to operate. This lead should be asserted before the DTE attempts the request-to-send. The DSR pin can simply be connected to the power supply of the DCE to indicate that it is switched on and ready to go. When the DTE is ready to send data, it asserts request-to-send. The modem responds, when ready, with clear-to-send, indicating that data may be transmitted over circuit BA. If the arrangement is half-duplex, then request-to-send also inhibits the receive mode. The DTE sends data to the local modem bit-serially. The local modem modulates the data into the carrier signal and transmits the resultant signal over the dedicated communication lines. Before sending out modulated data, the local modem sends out a carrier signal to the remote modem so that the remote modem is ready to receive the data. The remote modem detects the carrier and asserts the carrier-detect signal. The assertion of the carrier-detect signal tells the remote DTE that the local modem is transmitting. The remote modem receives the modulated signal from the local modem, demodulates it to recover the data, and sends it to the remote DTE over the received-data pin. The circuit connections are illustrated in Figure 9.4.

The second example involves two DTEs exchanging data through a public telephone line. Now one of the DTEs (initiator) must dial the phone (automatically or manually) to establish the connection. Two additional leads are required for this application:

- Data terminal ready (CD)
- Ring indicator (CE)

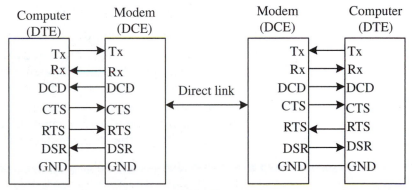

The following acronyms are used in Figure 9.4:

Tx: transmit data   CTS: clear to send
Rx: receive data   RTS: request to send
DCD: data carrier detect   DSR: data set ready

Figure 9.4 ■ Point-to-point asynchronous connection

The data transmission in this configuration can be divided into three steps:

Step 1

Establishing the connection. The following events occurred during this step:

1. The transmit DTE asserts the data-terminal-ready (DTR) signal to indicate to the local modem that it is ready to make a call.

2. The local modem opens the phone circuit and dials the destination number. The number can be stored in the modem or transmitted to the modem by the computer via the transmit-data pin.

3. The remote modem detects a ring on the phone line and asserts ring indicator (RI) to indicate to the remote DTE that a call has arrived.

4. The remote DTE asserts the DTR signal to accept the call.

5. The remote modem answers the call by sending a carrier signal to the local modem via the phone line. It also asserts the DSR signal to inform the remote DTE that it is ready for data transmission.

6. The local modem asserts both DSR and DCD signals to indicate that the connection is established and it is ready for data communication.

7. For full-duplex data communication, the local modem also sends a carrier signal to the remote modem. The remote modem then asserts the DCD signal.

Step 2

Data transmission. The following events occur during this step:

1. The local DTE asserts the RTS signal when it is ready to send data.

2. The local modem responds by asserting the CTS signal.

3. The local DTE sends data bit-serially to the local modem over the transmit-data pin. The local modem then modulates its carrier signal to transmit the data to the remote modem.

4. The remote modem receives the modulated signal from the local modem, demodulates it to recover the data, and sends it to the remote DTE over the received-data pin.

Step 3

Disconnection. Disconnection requires only two steps:

1. When the local DTE has finished the data transmission, it drops the RTS signal.

2. The local modem then de-asserts the CTS signal and drops the carrier (equivalent to hanging up the phone).

The circuit connection for this example is shown in Figure 9.5.

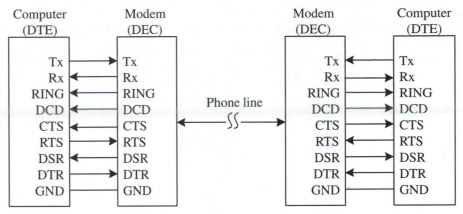

Figure 9.5 ■ Asynchronous connection over public phone line

A timing signal is not needed in an asynchronous transmission, but the following rules must be observed to guarantee that data are received correctly:

1. Data must be transferred character by character.

2. A character must consist of one start bit, seven or eight data bits, an optional parity bit, and one, one and a half, or two stop bits.

3. The start bit must be low.

4. The least significant bit (LSB) must be transferred first, and the most significant bit (MSB) must be transferred last.

5. The stop bit must be high.

6. A clock with a frequency equal to sixteen (or sixty-four) times the data bit rate must be used to detect the arrival of the start bit and determine the value of each data bit. A clock with sixteen times the bit rate can tolerate a difference of about 5% in the clocks at the transmitter and receiver. The methods for detecting the start bit and determining the values of data bits vary with different communication chips. This issue will be addressed in later sections.

With this protocol, the format of a character is shown in Figure 9.6. The start bit and stop bit identify the beginning and end of a character. They also permit a receiver to resynchronize a local clock to each new character.

Start bit	0	1	2	3	4	5	6	7	Stop bit 1	Stop bit 2

Figure 9.6 ■ The format of a character

Example 9.1

▼

The letter A is to be transmitted in the format with eight data bits, no parity, and one stop bit. Sketch the output.

Solution: Letters are transmitted in ASCII code. The ASCII code for the letter A is $41 or %01000001, which will be followed by a stop bit. Since the least significant bit is transmitted first, the data pattern for letter A in the EIA-232-E format will be binary value 0100000101. Because the EIA-232-E interface standard uses negative voltage for logic 1 and uses positive voltage for logic 0, the waveform on the EIA-232 interface will be opposite to that from the output of the microcontroller (in CMOS technology). The output waveforms are shown in Figure 9.7.

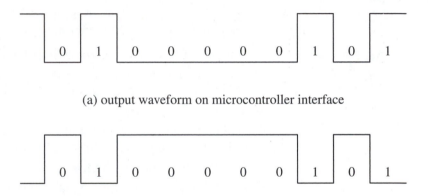

(a) output waveform on microcontroller interface

(b) output waveform on EIA-232-E interface

Figure 9.7 ■ Data format for letter A

▲

The EIA-232-E standard also supports synchronous transmission. Transmitter timing and receiver timing signals will be needed in a synchronous transmission. However, the synchronous EIA-232 serial communication is beyond the scope of this text and will not be discussed.

Example 9.2

▼

How long does it take to transmit one character at a speed of 9600 bauds? Each character is transmitted using a format with seven data bits, even parity, and two stop bits.

Solution: Each character consists of eleven bits (one start bit, seven data bits, one parity bit, and two stop bits). Each bit requires 104 µs (= 1 sec ÷ 9600). Thus each character will require 11 × 104 µs = 1.145 ms to transmit.

9.3.5   Data Transmission Errors

The following types of errors may occur during data transfer using asynchronous serial transmission:

- *Framing errors.* A framing error occurs when a received character is improperly framed by the start and stop bits; it is detected by the absence of the first stop bit. This error indicates a synchronization error, faulty transmission, or a break condition.

- *Receiver overrun.* One or more characters in the data stream were lost. That is, a character or a number of characters were received but were not read from the buffer before subsequent characters were received.

- *Parity errors.* A parity error occurs when an odd number of bits change value. It can be detected by a parity error detecting circuit.

9.3.6   The Null Modem Connection

Probably one of the most popular applications of EIA-232 interface in today's microprocessor laboratories is to connect a terminal (PC running a terminal program) to a single-board computer. Both the computer and the terminal are DTEs, but the EIA-232 standard is not for a direct connection between two DTEs. In order to make this scheme work, a null modem is needed—the null modem interconnects leads in such a way as to fool both DTEs into thinking that they are connected to modems. The null modem connection is shown in Figure 9.8.

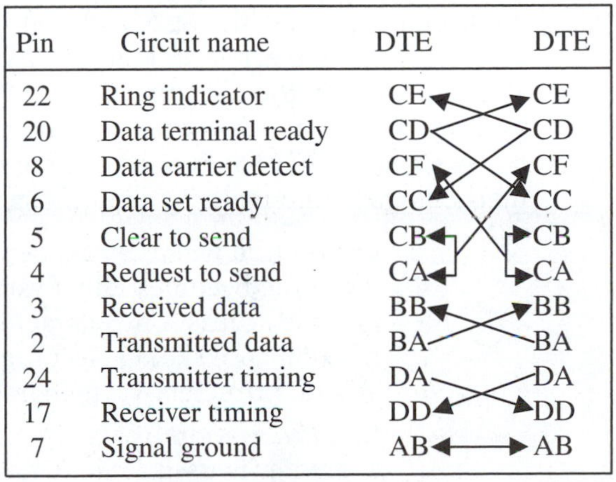

Pin	Circuit name	DTE	DTE
22	Ring indicator	CE	CE
20	Data terminal ready	CD	CD
8	Data carrier detect	CF	CF
6	Data set ready	CC	CC
5	Clear to send	CB	CB
4	Request to send	CA	CA
3	Received data	BB	BB
2	Transmitted data	BA	BA
24	Transmitter timing	DA	DA
17	Receiver timing	DD	DD
7	Signal ground	AB	AB

Figure 9.8 ■ Null modem connection

The transmitter-timing and receiver-timing signals are not needed in asynchronous data transmission. The ring-indicator is not needed, either, because the data transmission is not through a phone line.

9.4 The 68HC11 Serial Communication Interface (SCI)

A full-duplex asynchronous serial communication interface (SCI) is incorporated on the chip of the 68HC11 microcontroller. The SCI transmitter and receiver are functionally independent but use the same data format and bit rate.

The 68HC11 SCI subsystem can be used to communicate with multiple pieces of data communication equipment or just one DTE. The first situation is called *multidrop* environment. In a multidrop environment, an SCI receiver can be put to sleep when a message is not being sent to it. The sleeping SCI can be configured to wake up using either the *idle line* or an *address mark*. When the correct address is detected, the SCI receiver will start to receive the data message until the end of the message, which will be indicated by the idle line.

9.4.1 Data Format

The SCI subsystem can handle a character that consists of one start bit, eight or nine data bits, and one stop bit. A *break* is defined as the transmission or reception of a low for at least one complete character frame time. An *idle line* is defined as a continuous logic high on the RxD line for at least a complete character time.

9.4.2 Wake-Up Feature

In a multidrop environment, the receiver wake-up feature reduces SCI service overhead for the multiple-receiver system. Software in each SCI receiver evaluates the first character(s) of each message. If the message is intended for a different receiver, the SCI is placed in a sleep mode so that the rest of the message will not generate requests for service. Whenever a new message is started, logic in the sleeping receivers causes them to wake up so they can evaluate the initial character(s) of the new message.

A sleeping SCI receiver can be configured to wake up using either of two methods: *idle line wake-up* or *address mark wake-up:*

1. In *idle line wake-up*, a sleeping receiver wakes up as soon as the RxD line becomes idle. Systems using this type of wake up must provide at least one character time of idleness between messages in order to wake up sleeping receivers, but they must not allow any idle time between characters within a message.

2. In *address mark wake-up*, the most significant bit (MSB) of the character is used to indicate whether the character is an address

(1) or a data (0) character. Sleeping receivers will wake up whenever an address mark character is received. Systems using this method to wake up set the MSB of the first character in each message and leave it clear for all other characters in the message. Idle periods may occur within messages, and no idle time is required between messages for this wake-up method.

9.4.3 Start Bit Detection

The SCI uses a clock signal that is sixteen times the bit rate to detect the arrival of a valid start bit or decide the logic level of a data bit. When RxD is high for at least three sampling clock cycles and is then followed by a low voltage, the SCI will look at the third, fifth, and seventh samples after the first low sample (these are called verification samples) to determine if a valid start bit has arrived. This process is illustrated in Figure 9.9. If the majority of these three samples are low, then a valid start bit is detected. Otherwise, the SCI will repeat the process. After detecting the start bit, the SCI will begin to shift in the data bit. A noise flag is set if all three verification samples do not detect a logic 0. A valid start bit could be assumed with a set noise flag present.

9.4.4 Determining the Logic Value of Data Bits

The SCI also uses a clock that is sixteen times the bit rate to decide the logic value of any data bit and the stop bit. If the majority of the eighth, ninth, and tenth samples are 1, then the data bit is determined to be 1. Otherwise,

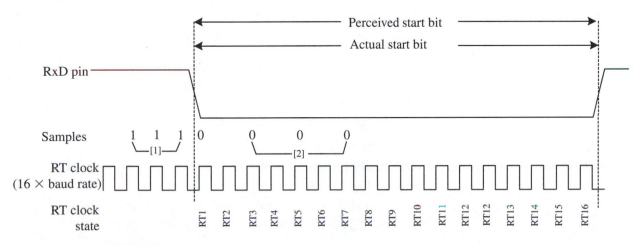

1. A 0 following three 1s.
2. Majority of samples 3, 5, and 7 are 0s

Figure 9.9 ■ Detection of start bit (ideal case)

the data bit is determined to be 0. If one sample differs from the other two samples, the noise flag will also be set.

9.4.5 Receive the Stop Bit

The stop bit is a 1. The noise flag will be set during the sampling of the stop bit if those three samples are not all identical.

9.4.6 Receive Errors

The 68HC11 SCI detects the following receiving errors:

- *Framing errors (FE):* A framing error occurs when the expected stop bit is not detected during the process of receiving a character. If a framing error occurs without a break (ten zeros for an 8-bit format or eleven zeros for a 9-bit format) being detected, the circuit will continue to operate as if there actually were a stop bit and the start edge will be placed artificially. The FE flag in the SCSR register will also be set.

- *Receiver overrun (OR):* The overrun bit is set when the next byte is ready to be transferred from the receive-shift register to the receive-data register (SCDR) but this register is already full (RDRF bit of SCSR is set). When an overrun error occurs, the data that caused the overrun are lost and the data that was already in SCDR are not disturbed. The OR bit will be set.

- *Noise error (NE):* The noise flag bit is set if there is noise on any of the received bits, including the start and stop bits. The NE bit is not set until the RDRF flag is set.

All error flags will be cleared when the SCSR is read and then the SCDR is read.

9.4.7 The Receive Process of a Character

The receive process in an SCI includes the following steps:

1. Detect the start bit.
2. Shift in eight (or nine) data bits.
3. Shift in the stop bit.
4. When a correct stop bit is detected, the data bits will be loaded into the receive data register from the receive shift register.

9.4.8 The Transmitting Process of a Character

The SCI transmitter uses an internally generated bit-rate clock to serially shift data out of the TxD pin. A normal transmission is initiated by enabling the transmitter and then writing data into the transmit data register (TDR). The SCI is double-buffered, so a new character can be written into the transmitter queue whenever the TDRE status flag is set to 1.

The transmitter logic adds a start bit (0) and a stop bit (1) to the data characters presented by the CPU for transmission. The transmitter can be configured to send characters with eight or nine data bits. When the TDR register is able to accept a new data character, the TDRE status flag (bit 7 of the SCSR register) is set and an interrupt can be optionally generated. The transmit-complete-status flag (bit 6 of the SCSR register) is set, and an optional interrupt is produced when the transmitter has finished sending everything in its queue. The SCI transmitter is also capable of sending idle-line characters and break characters, which are useful in a multi-drop SCI network.

9.4.9   SCI Registers

The SCI receiver has a receive shift register and a receive data register (RDR). The SCI transmitter has one transmit-data register (TDR) and one transmit-shift register. The SCI status register (SCSR) indicates the receive status, the transmit status, and errors that have occurred. Although RDR and TDR are separate registers, they share one address ($102F). Two control registers set up all SCI operation parameters. The data rate of SCI subsystem is controlled by the BAUD register. The addresses and contents of these registers are described in the following paragraphs.

SERIAL COMMUNICATION DATA REGISTER (SCDR)
RDR and TDR together are referred to as the SCDR register. The SCDR register is 8-bit and is located at $102F.

SERIAL COMMUNICATION STATUS REGISTER (SCSR)
The contents of the SCSR register are as follows:

	7	6	5	4	3	2	1	0	
	TDRE	TC	RDRF	IDLE	OR	NF	FE	0	SCSR at $102E
value after reset	1	1	0	0	0	0	0	0	

TDRE: *transmit data register empty.* When set to 1, this bit indicates that the transmit data register is empty.

TC: *transmit complete.* When set to 1, this bit indicates that the transmitter has completed sending and has reached an idle state. This bit is useful in systems where the SCI is driving a modem. When TC is set at the end of a transmission, the modem can be disabled without any delay.

RDRF: *receive data register full.* When set to 1, this bit indicates that the receive data register is full.

IDLE: *idle-line condition.* An idle-line condition is detected when this bit is set to 1. When this bit is 0, the RxD line is either active now or has never been active since IDLE was cleared.

OR: *receiver-overrun flag.* When set to 1, this bit indicates the receiver-overrun condition.

NF: *noise flag.* When set to 1, this flag indicates that the communication line is noisy.

FE: *framing error.* When set to 1, this bit indicates that a framing error occurred.

The condition flags related to transmission (TDRE & TC) can be cleared by reading the SCSR register followed by writing a byte into the SCDR register. Condition flags related to reception (RDRF, IDLE, OR, NF, FE) can be cleared by reading the SCSR register followed by reading the SCDR register.

SERIAL COMMUNICATION CONTROL REGISTER 1 (SCCR1)

The contents of the SCCR1 register are as follows:

	7	6	5	4	3	2	1	0	
	R8	T8	0	M	WAKE	0	0	0	SCCR1 at $102C
value after reset	u	u	0	0	0	0	0	0	

R8:   *receive data bit 8.* When the SCI system is configured for 9-bit data characters, this bit acts as an extra bit of the RDR. The MSB of the received character is transferred into this bit at the same time that the remaining eight bits are transferred from the serial receive shifter to the SCDR register.

T8:   *transmit data bit 8.* If the M bit is set, this bit provides a storage location for the ninth bit in the transmit data character. It is not necessary to write to this bit for every character transmitted, only when the value of this bit is to be different from that of the previous character.

M: *data format bit*

        0 = 1 start bit, 8 data bits, 1 stop bit
        1 = 1 start bit, 9 data bits, 1 stop bit

WAKE: *wake up method select*

    0 = idle line is selected as wake up method

    1 = address mark is selected as wake up method

Bits 5, 2, 1, and 0 are not implemented.

SERIAL COMMUNICATION CONTROL REGISTER 2 (SCCR2)

The contents of the SCCR2 register are as follows:

	7	6	5	4	3	2	1	0	
	TIE	TCIE	RIE	ILIE	TE	RE	RWU	SBK	SCCR2 at $102D
Value after reset	0	0	0	0	0	0	0	0	

TIE: *transmit interrupt enable.* When this bit is set to 1, an SCI interrupt is generated if the TDRE flag in the SCSR register is set to 1.

TCIE: *transmit complete interrupt enable.* When TCIE is set to 1, an SCI interrupt is requested if the TC flag in the SCSR register is set to 1.

RIE: *receive interrupt enable.* When RIE is set to 1, an SCI interrupt is requested if either the RDRF or the OR flag in the SCSR register is set to 1.

ILIE: *idle line interrupt enable.* When ILIE set to 1, an SCI interrupt is requested if the IDLE flag in the SCSR register is set to 1.

TE: *transmit enable.* When this bit is set, the transmit shift register output is applied to the TxD pin. Depending on the state of control bit M in SCCR1, a preamble of ten (M = 0) or eleven (M = 1) consecutive ones (representing the idle line) is transmitted when the software sets the TE bit from a cleared state.

RE: *receive enable.* The SCI receiver is enabled when this bit is set to 1. While SCI receiver is disabled, the RDRF, IDLE, OR, NF, and FE status flags cannot become set. If these bits were set, turning off RE bit does not cause them to be cleared.

RWU: *receiver wake up.* When the receiver-wake-up bit is set, it puts the receiver to sleep and enables the "wake up" function.

SBK: *send break.* If the send break bit is set and then cleared, the transmitter sends ten (M = 0) or eleven (M = 1) zeroes and then reverts to idle or to sending data. If SBK remains set, the transmitter will continuously send whole blocks of zeroes until cleared. At the completion of break code, the transmitter sends at least one high bit to guarantee recognition of a valid start bit.

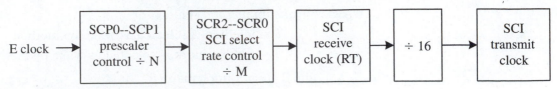

Figure 9.10 ■ SCI rate generator division
(Redrawn with permission of Motorola)

BAUD RATE REGISTER (BAUD)

The baud rate of the SCI system is derived by dividing the E clock signal by two factors specified in the BAUD register, as illustrated in Figure 9.10. The contents of the BAUD register are:

	7	6	5	4	3	2	1	0	
	TCLR	0	SCP1	SCP0	RCKB	SCR2	SCR1	SCR0	BAUD at $102B
value after reset	0	0	0	0	0	u	u	u	

TCLR: *clear baud rate counter.* This bit is used to clear the baud rate counter chain during factory testing. TCLR is zero and cannot be set while in normal operating modes.

SCP1 - SCP0: *SCI baud rate prescale selects.* The E clock is divided by the factors shown in Table 9.2, which are determined by the settings of SCP0 and SCP1. This prescaled output provides an input to a divider that is also controlled by SCR2-SCR0.

RCKB: *SCI baud rate clock check.* This bit is used during factory testing to enable the exclusive-OR of the receiver clock and transmitter clock to be driven out the TxD pin. RCKB bit is 0 and cannot be set during normal operation.

SCR2 - SCR0: *SCI baud rate selects.* The output of prescaler is further divided by a number selected by these three bits, as shown in Table 9.3.

SCP1	SCP0	Divide processor clock by
0	0	1
0	1	3
1	0	4
1	1	13

Table 9.2 ■ Baud rate prescale factor

SCR2	SCR1	SCR0	Divide processor clock by
0	0	0	1
0	0	1	2
0	1	0	4
0	1	1	8
1	0	0	16
1	0	1	32
1	1	0	64
1	1	1	128

Table 9.3 ■ SCI baud rate select

Example 9.3

▼

Determine the baud rate prescale and baud rate select factors needed to set the baud rate to 9600 with a 2 MHz E clock.

Solution: The frequency used to determine the bit value is $16 \times 9600 = 153,600$. Multiplying this value by 13, we get $153,600 \times 13 = 1996800 \approx 2$ MHz. Therefore we need a prescale factor of 13 and a baud rate select value of 0. To set these two parameters, the value of $30 should be written into the BAUD register.

▲

9.5 SCI Interfacing

Because the SCI circuit uses 0 V and 5 V to represent logic 0 and 1, respectively, it cannot be connected directly to the EIA-232 interface. A voltage translation circuit is needed to translate the voltage levels of the SCI signals (RxD & TxD) to and from those of the corresponding EIA-232 signals. In this section, we will discuss the use of the MAX232 chip from MAXIM to do the voltage translation. A MAX232 requires only a 5 V power supply to convert a 0 and 5 V input to a +10 V and −10 V output. The pin assignments and the use of each pin of the MAX232 are shown in Figure 9.11. Adding an RS-232 driver/receiver chip will then allow the 68HC11 SCI subsystem to

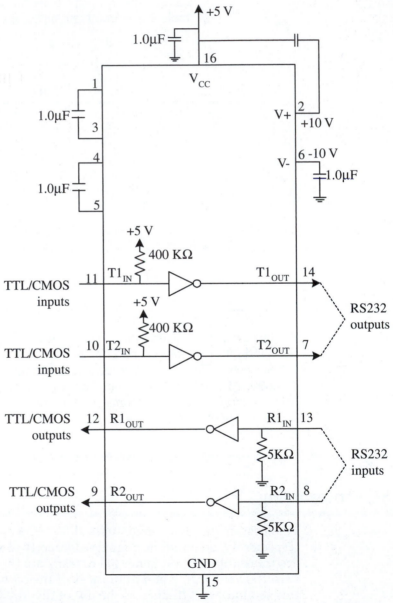

Figure 9.11 ■ Pin assignments and connections of the MAX232

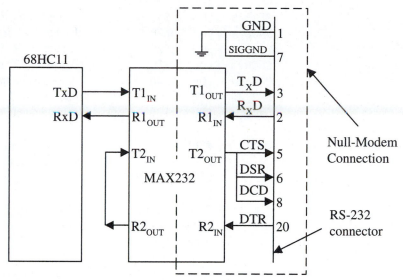

Figure 9.12 ■ Diagram of the SCI's EIA-232 side circuit connection

communicate to the RS-232 interface circuit. An example of such a circuit is shown in Figure 9.12.

Example 9.4

Write a subroutine to initialize the 68HC11 SCI subsystem (E clock frequency is 2 MHz) to operate with the following parameters:

- 9600 baud
- one start bit, eight data bits, and one stop bit
- no interrupt for receive and transmit
- enable receive and transmit
- idle line wake up
- do not send break

Solution: According to Example 9.3, the value of $30 should be written into the BAUD register to set the specified prescale and baud rate select factors. The data format and wake-up method are set by the SCCR1 register, so the value $00 should be written into the SCCR1 register. For the remaining parameters, the value $0C should be written into the SCCR2 register. The initialization of the SCI system is performed by the routine ONSCI.

```
REGBAS    EQU    $1000         ; I/O register base address
BAUD      EQU    $2B           ; offset of BAUD from REGBAS
SCCR1     EQU    $2C           ; offset of SCCR1 from REGBAS
SCCR2     EQU    $2D           ; offset of SCCR2 from REGBAS
CR1_INI   EQU    $00           ; 8 data bits, address mark wake up
CR2_INI   EQU    $0C           ; enable transmitter and receiver but disable all interrupts
baud_ini  EQU    $30           ; set baud rate to be 9600

ONSCI     PSHX                 ; save X on the stack
          PSHA                 ; save A on the stack
          LDX    #REGBAS
          LDAA   #baud_ini
          STAA   BAUD,X        ; initialize BAUD register
          LDAA   #CR1_INI
          STAA   SCCR1,X       ; initialize SCCR1
          LDAA   #CR2_INI
          STAA   SCCR2,X       ; initialize SCCR2
          PULA                 ; restore A from stack
          PULX                 ; restore X from stack
          RTS
```

The C language version of the routine is as follows:

```
void sci_on ( )
{
    BAUD = 0x30;      /* set up appropriate baud rate */
    SCCR1 = 0x00;
    SCCR2 = 0x0C;
}
```

Example 9.5

▼

Write a subroutine to send a break to the communication port controlled by the SCI subsystem. The duration of the transmitted break is approximately 200,000 E clock cycles, or 100 ms at 2.0 MHz.

Solution: A break character is represented by ten or eleven consecutive zeroes and can be sent out by setting the bit 0 of the SCCR2 register. As long as bit 0 of the SCCR2 register remains set, the SCI will keep sending out the break character.

```
REGBAS    EQU    $1000         ; I/O register base address
SCCR2     EQU    $2D           ; offset of SCCR2 from REGBAS

sendbrk   PSHX                 ; save register
          LDX    #$1000
```

```
                    BSET    SCCR2,X $01       ; set send break bit
```

* The following three instructions create a delay of about 100 ms.

```
                    LDY     #28571            ; delay for 100 ms
       txwait1      DEY                       ; seven clock cycles each loop
                    BNE     txwait1
                    BCLR    SCCR2,X $01       ; clear send-break bit
                    PULX                      ; restore X
                    RTS
```

The C language version of the routine is as follows:

```
void send_break ( )
{
    char    i;
    SCCR2 |= 0x01;              /* set the bit that triggers send break */
    TFLG1 = 0x40;              /* clear OC2F flag */
    TOC2 = TCNT + 20000;       /* start an OC2 operation with 10 ms delay */
    for (i = 0; i < 10; i++) {
            while (!(TFLG1 & 0x40));
            TFLG1 = 0x40;
            TOC2 += 20000;
    }
    SCCR2 &= 0xFE;             /* clear the bit that sends break */
}
```

Example 9.6

Write a subroutine to output a character from the SCI subsystem using the polling method. The character to be output is passed in accumulator A.

Solution: A new character should be sent out only if the transmit data register is empty. When the polling method is used, the subroutine will wait until the TDRE bit (bit 7) of the SCSR register is set before sending out the new character. The ASCII code of the character is sent out, and since the ASCII code has only seven bits, bit 7 of accumulator A is cleared to 0 before the code is sent out. The subroutine is as follows:

```
REGBAS     equ     $1000          ; I/O register block base address
SCSR       equ     $2E            ; offset of SCSR from REGBAS
SCDR       equ     $2F            ; offset of SCDR from REGBAS
TDRE       equ     $80            ; mask to check transmit data register empty

SCIputch   PSHX                   ; save X
```

```
            LDX        #REGBAS
            BRCLR      SCSR,X TDRE *        ; is transmit data register empty?
```

* The following instructions clear bit 7 to make sure the most significant bit of the ASCII
* code is a 0

```
            ANDA       #$7F                 ; mask parity bit
            STAA       SCDR,X               ; send character
            PULX                            ; restore X
            RTS
```

The C language version of the routine is as follows:

```
#define    TDRE    0x80
void sci_putch (char xch)
{
    while (!(SCSR & TDRE));
    SCDR = xch & 0x7F;          /* clear bit 7 before sending out */
}
```

Example 9.7

Write a subroutine to input a character from the SCI subsystem using the polling method. The character will be returned in accumulator A.

Solution: Using the polling method, this subroutine will wait until the RDRF bit (bit 5) of the SCSR register is set and then read the character held in the SCDR register.

```
REGBAS      equ      $1000         ; I/O register block base address
SCSR        equ      $2E           ; offset of SCSR from REGBAS
SCDR        equ      $2F           ; offset of SCDR from REGBAS
RDRF        equ      $20           ; mask to check receive data register full

SCIgetch    PSHX
            LDX      #REGBAS
            BRCLR    SCSR,X RDRF *  ; is receive data register full?
            LDAA     SCDR,X         ; read the SCI data register
            PULX
            RTS
```

The C language version of the routine is as follows:

```
#define    RDRF    0x20
char sci_getch ( )
```

```
    {
        while (!(SCSR & RDRF)); /* wait until the RDRF flag is set */
        return SCDR;
    }
```

▲

9.6 Serial Communication Support Chips

Unlike the 68HC11, a general-purpose microprocessor does not have a serial communication interface to allow communication to another computer using a phone line or modem, so a peripheral chip is required to interface the microprocessor to a DCE. This peripheral chip is called a UART (universal asynchronous receiver transmitter). The function of the UART chip is to perform serial-to-parallel and parallel-to-serial conversions and also to handle the serial communication protocol such as the RS-232. The SCI subsystem is the on-chip UART for the 68HC11. The Motorola MC6850, Rockwell R6551, Intel i8251 and Zilog Z8530 are among the most popular UART chips. These chips operate on a 5 V power supply. However, the RS-232 signal levels are different from the voltage levels of these chips. Thus, when interfacing a serial interface chip to a DCE, we need some form of voltage level translation or buffering. Two such widely used line drivers (receivers) are the Motorola MC1448 and MC1449. The MC1448 requires $\pm$12 V power supply to operate, and the MC1489 requires only a 5 V power supply. Most earlier single board microcomputers require a power supply that can supply -12, $+5$ and $+12$ V outputs simply because the line drivers require $\pm$12 V power supply to operate. However, several companies have introduced RS-232 line drivers (for example, the MAX232 from MAXIM and the MC145403 from Motorola) that can operate with a single 5 V power supply and generate RS-232 compatible outputs. These line drivers thus eliminate the need for a $\pm$ 12 V power supply.

In some applications, the user may need more than one serial communication interface—a requirement that cannot be satisfied by the 68HC11 SCI. One possible solution is to use an external UART chip, as is done in EVB and CMD-11A8. The EVB board has a terminal and a host port. The terminal port is controlled by an MC6850 chip (ACIA), and the host port is controlled by the serial communication interface (SCI) on the 68HC11. The CMD-11A8 also has two communication ports: *com1* and *com2*. The com1 port is controlled by the SCI function, whereas the com2 port is controlled by a R6551 chip. In the following section we will discuss the operation of the MC6850 and the terminal port design of the EVB board.

9.7 The MC6850 ACIA

The MC6850 is a UART chip designed to work with the 8-bit MC6800 family microprocessors, and it can also work with the 68000 microprocessor

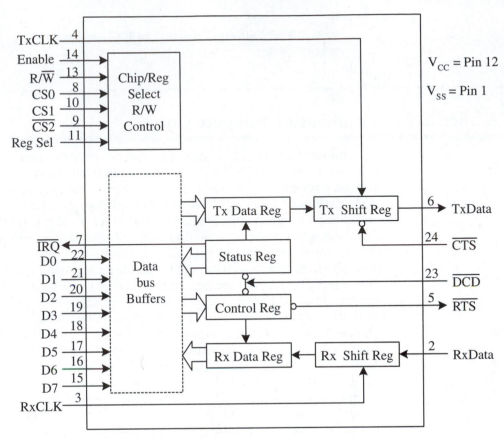

Figure 9.13 ■ MC6850 block diagram
(Redrawn with permission of Motorola)

and the 68HC11 microcontroller. The block diagram for the MC6850 is shown in Figure 9.13. The MC6850 has two 8-bit data registers, one each for receiving and transmitting. It also has a programmable control register and a read-only status register.

There are three chip select inputs, two active high (CS0 and CS1) and one active low ($\overline{CS2}$). The register select input, which is usually connected to the address line A0 in MC6800 systems, is used in conjunction with the R/$\overline{W}$ input to select one of the four on-chip registers, as illustrated in Table 9.4.

9.7.1 ACIA Registers

The operation of the ACIA is controlled by four 8-bit registers: control register, status register, transmit-data register, and receive-data register. The function of these registers are discussed in the following subsections:

Register Select Input	R/$\overline{W}$	Register
1	0	Tx data register
1	1	Rx data register
0	0	control register
0	1	status register

Table 9.4 ■ MC6850 register selection

THE CONTROL REGISTER

The 8-bit control register is divided into four fields. The meaning of each field is illustrated in Figure 9.14.

The *counter divide field* selects the divide ratio used in both the transmit and the receive sections of the ACIA. The receive bit rate is set equal to the frequency of RxCLK divided by the factor selected by this field. The transmit bit rate is set equal to the frequency of TxCLK divided by the factor selected by this field.

The *format select field* selects the data character length, number of stop bits, and the parity. The *transmit control field* provides control over the interrupt resulting from the transmit data register empty condition, the RTS output, and the transmission of a break level. When this field is 10, the RTS signal will not be asserted. When it is 01, the transmit interrupt will be enabled, and when it is 11, a break will be transmitted by the TxData pin. A *break* is a condition in which the transmitter output is held at the active level (low) continuously. A break can be used to force an interrupt at a distant receiver, because

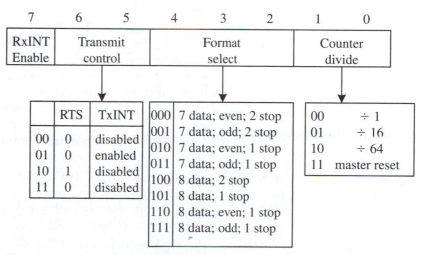

Figure 9.14 ■ MC6850 control register

the asynchronous serial format precludes the existence of a space level (low) for longer than about 10-bit periods. The *receive interrupt enable bit* will enable all receive interrupts when set to 1. Interrupts will be requested only when the conditions for interrupt are satisfied.

THE STATUS REGISTER

The contents of this register are illustrated in Figure 9.15.

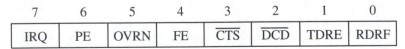

7	6	5	4	3	2	1	0
IRQ	PE	OVRN	FE	$\overline{\text{CTS}}$	$\overline{\text{DCD}}$	TDRE	RDRF

Figure 9.15 ■ The MC6850 status register

IRQ: *interrupt request flag.* The IRQ flag is set to 1 when the $\overline{\text{IRQ}}$ pin is active low to indicate that an interrupt has been requested. This bit is cleared by a read operation to the receive data register or a write operation to the transmit data register.

PE: *parity error.* This flag indicates that the number of ones in the character does not agree with the preselected odd or even parity.

OVRN: *overrun flag.* This flag indicates that a character or a number of characters in the data were received but not read from the receive data register before the subsequent characters were received. This situation occurs primarily at higher baud rates. The overrun condition is cleared by reading data from the receive data register or by a master reset.

FE: *framing error flag.* The FE bit is set to 1 when a framing error occurs. A framing error indicates that the received character is improperly framed by the start and the stop bits; it is detected by the absence of the first stop bit. This error indicates a synchronization error, faulty transmission, or a break condition normally caused by pressing the Break key on a terminal. The Break key causes the transmission of a continuous stream of null characters for a period of 250 to 500 ms. A framing error can occur only during receive time.

$\overline{\text{CTS}}$:   *clear to send flag.* The $\overline{\text{CTS}}$ bit reflects the current level of the $\overline{\text{CTS}}$ input (active low) from a modem. A high level on the $\overline{\text{CTS}}$ pin will also inhibit the TDRE bit from going to 1. A master reset to the ACIA does not affect the $\overline{\text{CTS}}$ bit.

$\overline{\text{DCD}}$:   *data carrier detect flag.* The $\overline{\text{DCD}}$ bit will go high when the $\overline{\text{DCD}}$ input from the modem has gone high to indicate that a carrier is not present. The $\overline{DCD}$ bit's being high causes an interrupt request to be generated when the receive-interrupt-enable bit is set. After the $\overline{DCD}$ input is returned low, the $\overline{DCD}$ input remains high until it is cleared by first reading the status register and then the data register or until a master reset occurs. If the $\overline{DCD}$ input remains high after read status and read data or master reset has occurred, the interrupt is cleared and the $\overline{DCD}$ status bit will remain high and will follow the $\overline{DCD}$ input.

TDRE: *transmit data register empty.* The TDRE bit is set to 1 when the character in the transmit data register has been shifted out from the TxData pin. This bit will be cleared when there is a new character written into the transmit data register.

RDRF: *receive data register full.* The RDRF bit is set to 1 when there is a character in the receive data register. It is cleared when the CPU reads the receive data register, and it is also cleared by a master reset. The cleared or 0 value of this bit indicates that the contents of the receive data register are not current. The data-carrier-detect bit being high also causes RDRF to indicate empty.

THE TRANSMIT DATA REGISTER

Data is written into the transmit data register during the negative transition of the E(nable) signal when the ACIA has been addressed with RS pin high and R/$\overline{W}$ pin low. Writing data into the register causes the TDRE bit in the status register to be cleared. If the transmitter is idle, the data will be loaded into the transmit shift register (not accessible to the programmer) and then shifted out from the TxData pin. The TDRE bit in the status register will be set at the moment when the data are loaded into the transmit shift register.

THE RECEIVE DATA REGISTER

Data are automatically transferred to the empty receive data register (RDR) from the receive shift register upon receiving a complete character. This event causes the RDRF bit in the status register to be set to 1. Data can be read through the data bus by addressing the ACIA and setting the RS and R/$\overline{W}$ pins to high. The RDRF bit will be reset to 0 when the receive data register is read. When the receive data register is full, the automatic transfer of data from the receive shift register to the receive data register is inhibited.

9.7.2 The ACIA Operation

At the bus interface, the ACIA appears as two addressable memory locations. Internally, there are four addressable registers: two read-only and two write-only registers. The read-only registers are status and receive data registers; the write-only registers are control and transmit data registers.

The master reset (counter control field) in the control register must be set during system initialization to insure the reset condition and to prepare for programming the ACIA functional configuration when the communication is required. During the first master reset, the $\overline{\text{IRQ}}$ and $\overline{\text{RTS}}$ outputs are held at the logic 1 level. On all other master resets, $\overline{\text{RTS}}$ output can be programmed high or low with the $\overline{\text{IRQ}}$ output high.

THE TRANSMIT OPERATION

A typical transmitting sequence consists of reading the ACIA status register either as a result of an interrupt or in a polling sequence. A character may be written into the transmit data register if the TDRE bit in the status register is 1. This character is transferred to the transmit shift register, where it is serialized and transmitted to the TxData pin, preceded by a start bit and followed by one or two stop bits. A parity bit can be optionally added to the character and will occur between the last data bit and the first stop bit. As long as the character in the transmit data register is transferred to the shift register, next character can be written into the transmit data register.

THE RECEIVE OPERATION

The receiving sequence starts with the detection of the start bit of a character. The RxCLK clock, which is 1 or 16 or 64 times the bit rate, is used to detect the arrival of a start bit. A divide-by-one clock is used when the receive data is externally synchronized with the RxCLK input. For the divide-by-16 or divide-by-64 counter divide factor, the start bit is detected by 8 or 32 consecutive low samples from the RxData pin. A character will be shifted in after the detection of the start bit. As the character is received, parity is checked and the error is available in the status register along with any framing error, overrun error, and receive data register full condition. The receiver is also double-buffered so that a character can be read from the data register as another character is being received in the shift register.

ACIA INTERRUPTS

Only one *transmit interrupt* can be requested by the ACIA. A transmit interrupt will be requested when it is enabled (when bits 6 and 5 of control register are 01) and when the transmit data register is empty. When the $\overline{\text{RTS}}$ pin is high, a transmit interrupt cannot be requested.

A *receive interrupt* will be requested when the receive interrupt is enabled and the receive data register is full, or a receiver overrun or data carrier detect goes high (signifying a loss of carrier from a modem). An interrupt caused by the receive data register full can be cleared by reading the receive data register. An interrupt caused by overrun or loss of carrier can be cleared by reading the status register and then reading the receive data register. Note that bit 7 of the ACIA control register enables all receive interrupts at once.

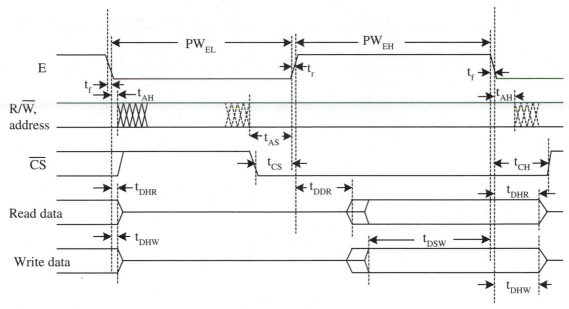

Figure 9.16 ■ Microprocessor side bus timing diagram for the MC6850
(Redrawn with permission of Motorola)

9.7.3   The Bus Timing of ACIA

The ACIA pins can be divided into two categories: processor side pins and RS232 side pins. Processor side pins allow the microprocessor to access and control the functioning of the ACIA. The processor side bus timing characteristics for a 2 MHz MC6850 are shown in Figure 9.16, and the timing parameter values are listed in Table 9.5.

The MC6850 requires the $\overline{CS}$ signal to be valid 40 ns before the rising edge of the E input (at 2 MHz). When read by the processor, the data from the ACIA will be available on the data bus 150 ns after the rising edge of the E input and will remain valid for at least 20 ns (possibly up to 50 ns). When being written to by the processor, the ACIA requires the input data to be valid 60 ns before the falling edge of the E input and remain valid for at least 10 ns after the falling edge of the E input. The address input (register select) to the ACIA must be valid 40 ns before the rising edge of the E input and remain valid 10 ns after the falling edge of the E input.

9.7.4   Interfacing an ACIA to the 68HC11

The MC6850 can be interfaced to the 68HC11 when the 68HC11 is configured in expanded mode. The data pins on the microprocessor side can be

Characteristic	Symbol	Minimum (ns)	Maximum (ns)
cycle time	t_{cyc}	500	10000
pulse width, E low	PW_{EL}	210	9500
pulse width, E high	PW_{EH}	220	9500
clock rise and fall time	t_r, t_f	–	20
address hold time	t_{AH}	10	–
address setup time before E	t_{AS}	40	–
chip select time before E	t_{CS}	40	–
chip select hold time	t_{CH}	10	–
read data hold time	t_{DHR}	20	50
write data hold time	t_{DHW}	10	–
output data delay time	t_{DDR}	–	150
input data setup time	t_{DSW}	60	–

Table 9.5 ■ Timing characteristics of 2-MHz MC6850

(Redrawn with permission of Motorola)

connected directly to the 68HC11 port C pins. The register select pin (reg sel) can be connected to the lowest address pin, A0.

An address space must be assigned to the MC6850 registers so that they can be accessed by the 68HC11, and this address space assignment will affect the address-decoding scheme. As we discussed in chapter 5, the partial address-decoding scheme can simplify the decoder design and will be used here. Before making the address assignment, we must also consider if we need to add other devices to the 68HC11. There are two sides for the interfacing of a MC6850: processor side and the RS-232 (DCE) side. The following two examples will demonstrate interfacing the MC6850 to the 68HC11.

Example 9.8

▼

Given the following address space assignment of a 68HC11-based single board computer, design a circuit to interface the MC6850 to the 68HC11A1 and verify that all the timing requirements are met.

$0000-$1FFF: part of this space is occupied by on-chip memories and registers and the rest of this space is not assigned.
$2000-$3FFF: 6840 programmable timer chip
$4000-$5FFF: 8KB SRAM HM6264
$6000-$7FFF: i8255 PPA
$8000-$9FFF: ACIA
$A000-$BFFF: not assigned
$C000-$DFFF: not assigned
$E000-$FFFF: on-chip 8KB ROM

Solution: An address decoder is required to generate chip enable signals for these external components, and a latch is needed to latch the lowest eight address signals (A7-A0). The AS signal is used to latch these address signals, while address signals A12 and A11 are tied to the chip select signals CS1 and CS0, as shown in Figure 9.17, thus reducing the address space allocated to the ACIA to $9800-$9FFF.

We also need to compare the timing requirements for the MC6850 and the 68HC11:

> *E clock:* The E(nable) input to the 6850 is from the 68HC11. By comparing Table 9.5 and Table 5.5, it is easy to see that the rise/fall time and E clock low and high pulse-width requirements are satisfied. These comparisons are summarized in Table 9.6.

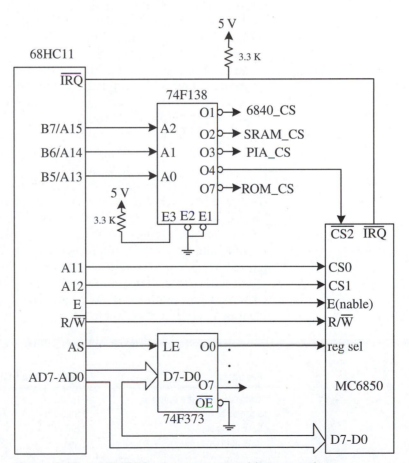

Figure 9.17 ■ MC6850 microprocessor side connections

Parameter	Required time (ns)	Actual time (ns)
cycle time	500 (min.)	500
rise and fall time	20 (max.)	20
clock pulse width (high)	210 (min.)	227
clock pulse width (low)	220 (min.)	222

Table 9.6 ■ Comparison of E clock timing requirements

Address (reg sel) and R/$\overline{W}$ Inputs: The multiplexed low address signals can be latched and propagated to the output of the 74F373 41.5 ns before the rising edge of the E(nable) input. This value (41.5 ns) is derived as follows: the delay time from AS-valid to E-Rise is 53 ns, and the propagation delay of 74F373 from LE-valid to output-stable is 11.5 ns; 53 ns − 11.5 ns = 41.5 ns, and hence the timing requirement t_{AS} of the MC6850 is satisfied. The 2MHz 6850 requires a 10-ns address hold time (t_{AH}) after the falling edge of the E clock. The 74F373 will not latch a new value earlier than 53 ns before the next E clock rising edge, thus providing at least 174 ns (227 ns − 53 ns) of address hold time. The address hold time requirement is therefore satisfied. The R/$\overline{W}$ output from the 68HC11 is valid 94 ns before the rising edge of the E clock, and the MC6850 requires this signal to be valid 40 ns before the rising edge of the E input, so this requirement is also satisfied. These comparisons are summarized in Table 9.7.

Chip select ($\overline{CS2}$, CS1, and CS0): High address signals (A15-A13) from the 68HC11 are stable 94 ns before the rising edge of E clock, and the address decoder output becomes stable 86 ns before the rising edge of the E clock. The MC6850 requires all three chip select signals to be valid 40 ns before the rising edge of the E clock. The 68HC11 holds these address signals for at least 33 ns after the falling edge of the E clock, and hence the chip select output $\overline{CS2}$ will not be changed until 41 ns after the falling edge of the E clock. The chip select signals CS1 and CS0 are connected to A12 and A11, respectively, and their timing requirements (t_{CS}) are easily satisfied. These timing comparisons are listed in Table 9.8.

Parameter	Required time (ns)	Actual time (ns)
address setup time before E	40 (min.)	41.5
address hold time	10 (min.)	>= 174
R/$\overline{W}$ setup time	40 (min.)	94
R/$\overline{W}$ hold time	10 (min.)	30

Table 9.7 ■ Address (reg sel) and R/$\overline{W}$ timing comparisons

Parameter	Required time (ns)	Actual time (ns)
chip select time before E for $\overline{CS2}$	40 (min.)	86
chip select hold time for $\overline{CS2}$	10 (min.)	41
chip select time before E for CS1	40 (min.)	94
chip select hold time for CS1	10 (min.)	33
chip select time before E for CS0	40 (min.)	94
chip select hold time for CS0	10 (min.)	33

Table 9.8 ■ Comparisons of chip select signals

Read data setup time and hold time (68HC11 requirements): The MC6850 presents the data 150 ns after the rising edge of the E clock (72 ns before the falling edge of the E clock), satisfying the read data setup time requirement of the 68HC11 (30 ns is required). The 6850 stops driving the microprocessor side data bus 20 to 50 ns after the falling edge of the E clock, but the 68HC11 requires a data hold time of 0-83 ns, so this timing requirement is violated. However, as we discussed in chapter 5, the capacitance on the data bus would hold the data for a longer length of time, and thus the data hold time requirement can be met. The timing comparisons are shown in Table 9.9.

Parameter	Required time (ns)	Actual time (ns)
68HC11 read data setup time	30 (min.)	72
68HC11 read data hold time	0-83 (min.)	20-50, but held by data bus capacitance

Table 9.9 ■ Timing comparisons for 68HC11 read data requirements

Write data set up time and *hold time (MC6850 requirements):* On a write bus cycle, the 68HC11 presents the data 128 ns after the rising edge of the E clock (or 94 ns before the falling edge of the E clock). The MC6850 requires a write data setup time of 60 ns, so this requirement is easily satisfied. The 68HC11 holds the write data for at least 33 ns, which also exceeds the requirement of the MC6850 (10 ns). These comparisons are summarized in Table 9.10.

Parameter	Required time (ns)	Actual time (ns)
6850 write data setup time	60	94
6850 write data hold time	10	33

Table 9.10 ■ Timing comparisons with MC6850 write data requirements

From this analysis, we conclude that the circuit connections shown in Figure 9.17 satisfy all the timing requirements of the 68HC11 and the MC6850.

Example 9.9

Implement the null modem connection to the ACIA so that the 68HC11 can talk to a PC via a straight-through RS-232 cable and connector. This implementation should allow the user to select the following baud rates: 300, 600, 1200, 2400, 4800, and 9600.

Solution: To provide the required range of baud rates, appropriate components must be combined with a selection of ACIA transmit- and receive-clock input divide factors. In this example, we will use a 2.4576MHz crystal oscillator to generate the transmit- and receive-clock signals. Since the output from a crystal oscillator is a sine wave, we need to convert it to a square wave. A Schmitt-Trigger inverter such as a 74HC14 is perfect for this job because it can "square up" slow input rise and fall times. The pin assignments of the 74HC14 are shown in Figure 9.18.

In order to allow the user to select the appropriate baud rate, a ripple counter such as the 74HC4040 can be used. This device consists of twelve master-slave

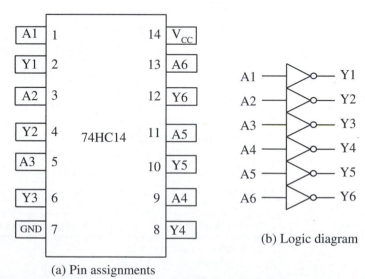

(a) Pin assignments

(b) Logic diagram

Figure 9.18 ■ 74HC14 pin assignments and logic diagram

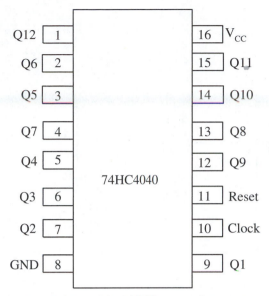

Figure 9.19 ■ 74HC4040 pin assignment

flip-flops. The output of each flip-flop feeds the next, and the frequency of each output is half that of the preceding one. The state of the counter advances on the falling edge of the clock input. The pin assignments of the 74HC4040 are shown in Figure 9.19. By feeding the output of the 74HC14 Schmitt-Trigger inverter to the clock input of the 74HC4040, the frequencies of Q1 through Q7 will be 1.2288MHz, 0.6144MHz, 0.3072MHz, 0.1536MHz, 0.0768MHz, 0.0384MHz, and 0.0192MHz, respectively. By dividing these frequencies by 64, baud rates of 9600, 4800, 2400, 1200, 600, and 300 can be derived. A jumper can be used to allow the user to select the baud rate.

A MAX232 is used to translate voltage from the normal CMOS level to the RS-232 level or vise versa. The null-modem connection as shown in Figure 9.8 should be followed to allow the direct connection of the ACIA and a PC using a straight through RS-232 cable. The RS-232 side connection of the ACIA is shown in Figure 9.20. Because both the $\overline{DCD}$ and $\overline{CTS}$ inputs to the ACIA are tied to low, the MC6850 will be ready to transmit or receive data immediately after power-on. Since the terminal port is to be connected to a terminal or a PC, the TxD output should be connected to the RxD input to the terminal, and the RxD input should be connected to the TxD output from the terminal. The RTS signal is not needed in this connection.

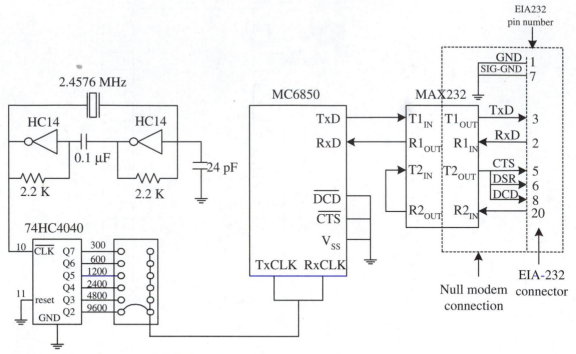

Figure 9.20 ■ Diagram of EIA232 side circuit connection

9.7.5   Configuring the ACIA

In Figure 9.20, the TxCLK and RxCLK inputs to the ACIA are tied together and derived by dividing a 2.4576 MHz clock signal by four. The 9600 baud rate is chosen (other baud rate can be chosen by changing the jumper), so the counter divide factor is 64. The configuration of the ACIA is fairly straightforward.

Example 9.10

▼

Write an instruction sequence to configure the ACIA in Figure 9.20 to operate with the following parameters:

1. disable receive and transmit interrupts

2. counter divide factor is 64

3. data format is eight data bits, one stop bit, and no parity bit

Solution: The user must write a byte into the ACIA control register to set up these parameters accordingly.

1. To disable receive interrupt, set bit 7 to 0.
2. To disable transmit interrupt, set bits 6 and 5 to 00 or 10.
3. To select the data format of eight data bits with one stop bit, set the format select field (bits 4, 3, and 2) to 101.
4. To select the counter divide factor of 64, set the counter divide field to 10.
5. The address $9800 & $9801 will be used to access the control/status and transmit/receive data registers, respectively.

To set up these parameters, a value of $16 or $56 should be written into the ACIA control register. The following instruction sequence will initialize the ACIA accordingly:

```
acia_ini    EQU     $16
            LDAA    #acia_ini
            STAA    $9800
```

9.7.6   A Terminal I/O Package for ACIA

A set of flexible I/O routines is useful during the process of software and hardware development, particularly for making the following operations possible:

- *Checking the results of program execution:* information can be displayed on the PC monitor screen to allow the user to examine the results of the program execution.
- *Interaction with the prototype:* the user may provide different data to the application via the PC keyboard to check the correctness of the program logic.

The following list includes some very useful I/O routines:

ONACIA: initializes the ACIA control register.

GETCH: returns a character in accumulator A from the ACIA receive data register.

PUTCH: places the contents of accumulator A in the ACIA transmit data register.

GETSTR: inputs a string that is terminated by a carriage return (CR) character. The input string is to be stored in a buffer pointed to by Y.

PUTSTR: outputs a null-terminated string that is pointed to by Y.

NEWLINE: outputs a CR/LF character to the terminal (LF stands for line feed character).

CHPRSNT: clears the Z bit of the CCR register if a character is found in the ACIA receive data register. Otherwise, it sets the Z bit.

PUTHEX: prints the 8-bit contents of A in two hex digits.

ECHOFF: eliminates the character echo on input from the terminal port.

ECHON: causes characters to be echoed on input from the terminal port.

LFON: expands a CR into CR/LF pair, and expands a LF into LF/CR pair.

LFOFF: turns off newline expansion.

The I/O package given in this section can be run on an EVB without modification. The constant definitions for the I/O package are as follows:

```
ACIA      EQU   $9800    ; base address of the ACIA
CR        EQU   $0D      ; ASCII code of carriage return
LF        EQU   $0A      ; ASCII code of line feed
TDRE      EQU   $02      ; position of the transmit data register empty bit
RDRF      EQU   $01      ; position of the receive data register full bit
control   EQU   $0       ; offset of the ACIA control register from the ACIA base
STATUS    EQU   $0       ; offset of the ACIA status register from the ACIA base
XMIT      EQU   $01      ; offset of the ACIA transmit data register from ACIA base
RCV       EQU   $01      ; offset of the ACIA receive data register from ACIA base
masterst  EQU   $03      ; value to master reset the ACIA
CTL_INI   EQU   $16      ; value to configure ACIA to eight data bits, one stop bit, no
*                        ; parity, no receive and transmit interrupt,
*                        ; counter divide factor 64
```

The subroutine ONACIA first resets the ACIA and then initializes the ACIA control register by writing $16 into it. This specifies the transmission timing (divide by 64), eight data bits with one stop bit and no parity, and all interrupts disabled.

```
ONACIA   PSHX                    ; save registers
         PSHA                    ;      "
         LDX     #ACIA
         LDAA    #masterst       ; reset ACIA
         STAA    control,X       ;      "
         LDAA    #CTL_INI        ; set up ACIA
         STAA    control,X       ;      "
         PULA                    ; restore registers
         PULX                    ;      "
         RTS
```

In C language, we add the following definition to the hc11.h file so we can use symbols to access ACIA registers:

```
#define ACIA_CTRL    *(unsigned char volatile *)( 0x9800)
#define ACIA_STAT    *(unsigned char volatile *)( 0x9800)
#define ACIA_XMIT    *(unsigned char volatile *)( 0x9801)
#define ACIA_RCV     *(unsigned char volatile *)( 0x9801)

void on_acia ( )
{
    ACIA_CTRL = 0x03;    /* master reset ACIA */
    ACIA_CTRL = 0x16;    /* configure ACIA parameters */
}
```

The logic flow of the subroutine GETCH is shown in the flowchart in Figure 9.21. Its instruction sequence is as follows:

```
GETCH:  PSHX                                ; save X
        PSHB                                ; save B
        LDX     #ACIA
retry   BRCLR   STATUS,X $70 noerr          ; are there parity, framing, and overrun errors?
        JSR     ONACIA                      ; if there is an error then reinitialize and try
*                                           ; again
        BRA     retry                       ;           "
noerr   BRCLR   STATUS,X RDRF retry         ; Is receive data register full ?
        LDAA    RCV,X                       ; read the character
        ANDA    #$7F                        ; clear the parity bit
        LDAB    ECHO                        ; test to see if echo flag is on
        BEQ     quit                        ; the echo flag is not set
        JSR     PUTCH                       ; echo the character to the screen
quit    PULB
        PULX
        RTS                                 ; return input character in A
```

The C language version of this routine is as follows:

```
char getchar ( )
{
    char xch;

    while (ACIA_STAT & 0x70) {       /* wait until error is eliminated */
        on_acia ( );
    }
    while (!(ACIA_STAT & 0x01));     /* wait until receive data register is full */
    xch = ACIA_RCV & 0x7F;           /* mask out parity bit */
    if (ECHO_ON) putchar (xch);      /* echo the character if echo flag is on */
    return xch;
}
```

The subroutine PUTCH needs to check whether the auto linefeed expansion flag is set. If this flag is set, then both the CR and LF characters will be

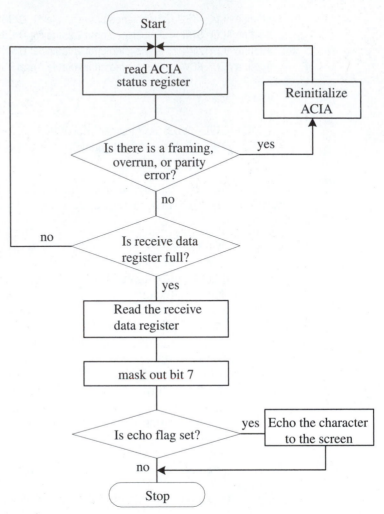

Figure 9.21 ■ Flowchart of the GETCH routine

expanded to the CR/LF pair. The effect is to move the screen cursor to the first column of the next line of the monitor screen. The flowchart of the PUTCH routine is shown in Figure 9.22. The PUTCH subroutine is as follows:

```
PUTCH   PSHB                    ; save B
        BSR     OUTCH           ; output the character
        TST     AUTOLF          ; check to see if auto line-feed flag is set
        BEQ     OUT2            ; prepare to return if not set
        CMPA    #CR             ; check to see if the output character is a carriage
*                               ; return
        BNE     OUT1
```

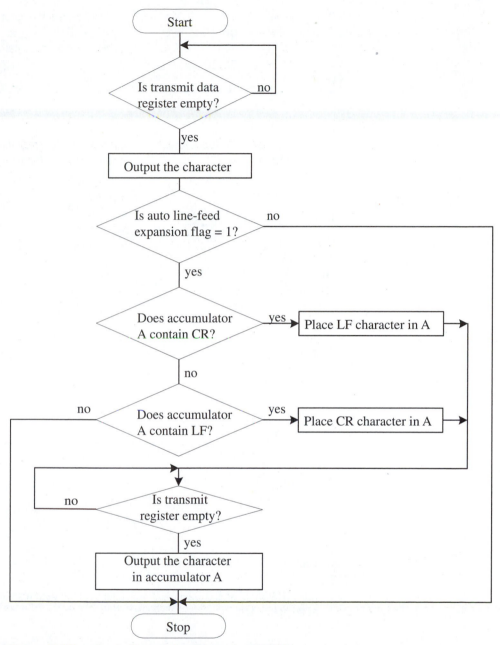

Figure 9.22 ■ Logic flow of the subroutine PUTCH

```
LDAA    #LF                 ; also output a line feed after a carriage return,
BSR     OUTCH               ;        "
BRA     OUT2
```

```
OUT1    CMPA    #LF                     ; check to see if the output character is a line feed
        BNE     OUT2
        LDAA    #CR                     ; If yes, also output a carriage return
        BSR     OUTCH                   ;       "
OUT2    PULB                            ; restore B
        RTS

OUTCH   LDX     #ACIA
        BRCLR   STATUS,X TDRE *         ; Is transmit data register empty?
        ANDA    #$7F                    ; mask parity bit before output
        STAA    XMIT,X                  ; output the character
        RTS
```

In C language,

```c
char auto_lf;
#define CR    0x0D;
#define LF    0x0A;
void outchar (char xch);
void putchar (char xch)
{
    outchar (xch);
    if (auto_lf) {
        switch (xch) {
            case CR:
                outchar(LF);
                break;
            case LF:
                outchar(CR);
                break;
            default:
                break;
        }
    }
}
void outchar (char xch)
{
    while (!(ACIA_STAT & 0x02));    /* wait until transmit data register is empty */
    ACIA_XMIT = xch;
}
```

The subroutine GETSTR inputs a string terminated by the CR character from the keyboard. The index register Y contains the address of the input buffer where the string is to be stored. The subroutine appends a null character to the end of the string.

```
GETSTR  PSHA
GSLOOP  JSR     GETCH
```

```
            CMPA    #CR             ; is it a carriage return?
            BEQ     GFINIS          ; end of get string
            STAA    0,Y             ; save the character in the buffer
            INY                     ; increment the buffer pointer
            BRA     GSLOOP          ; return to get string loop
GFINIS      LDAA    #$00            ; terminate the string with a NULL character
            STAA    0,Y             ;       "
            PULA                    ; restore A
            RTS
```

In C language,

```
/* assume enough space has been reserved to hold the string */
void getstr (char *ptr)
{
    char xch;
    while ((xch = getchar ( )) != '\n')    /* read keyboard input until CR is entered */
        *ptr++ = xch;
    *ptr = '\0';
}
```

The subroutine PUTSTR outputs the string pointed to by the index register Y. This subroutine outputs the string character by character by calling the PUTCH subroutine until the NULL character is encountered.

```
PUTSTR      PSHA                    ; save A
PSLOOP      LDAA    0,Y             ; get a character
            BEQ     PFINIS          ; is this the end of the string ?
            JSR     PUTCH           ; if not, output the character
            INY                     ; increment the pointer
            BRA     PSLOOP
PFINIS      PULA                    ; restore A
            RTS
```

In C language,

```
void putstr (char *ptr)
{
    while (*ptr) {
        putchar (*ptr);
        ptr ++;
    }
}
```

* The following subroutine outputs a CR/LF pair.

```
NEWLINE     PSHA                    ; save A
            LDAA    #$01
            STAA    AUTOLF          ; turn on auto line feed expansion
```

```
        LDAA    #CR         ; output a CR which will be expanded into CR/LF
        JSR     PUTCH       ; pair
        PULA                ; restore A
        CLR     AUTOLF      ; clear the auto line feed flag
        RTS
```

In C language,

```
void newline ( )
{
    auto_lf = 1;
    putchar(CR);
    auto_lf = 0;
}
```

The following subroutine checks to see if a character is present at the input terminal port. The Z bit of the CCR register is updated.

```
CHPRSNT PSHA                ; save A
        PSHX                ; save X
        LDX     #ACIA
        LDAA    STATUS,X    ; check the status register
        BITA    #RDRF       ; test the RDRF bit and update the Z bit
        PULX                ; restore X
        PULA                ; restore A
        RTS
```

In C language, a 1 is returned if a character is present at the terminal port. Otherwise, a 0 is returned.

```
int chprsnt ( )
{
    if (ACIA_STAT & 0x01)
        return 1;
    else return 0;
}
```

The following routine outputs the contents of A as two hex digits.

```
PUTHEX  PSHX                ; save registers
        PSHB                ;     "
        PSHA                ;     "
        TAB                 ; make a copy of A in B
        LSRB                ; move the upper hex digit into the lower
        LSRB                ; half of B
        LSRB                ;     "
        LSRB                ;     "
        LDX     #HEXDIG     ; load the hex digit Table address into X
        ABX                 ; X points the ASCII code of the upper hex digit
        LDAA    0,X         ; get the ASCII code
```

```
        JSR     PUTCH           ; output the upper hex digit
        PULB                    ; put A to B again
        PSHB                    ; restore the stack
        ANDB    #$0F            ; mask out the upper hex digit
        LDX     #HEXDIG         ; load the hex digit Table address into X
        ABX                     ; point X to the ASCII code of the LSB hex digit
        LDAA    0,X             ; get the ASCII code of the lower hex digit
        JSR     PUTCH           ; output the lower hex digit
        PULA                    ; restore registers
        PULB                    ;       "
        PULX                    ;       "
        RTS
HEXDIG  FCC     "0123456789ABCDEF" ; hex digits ASCII code table.
```

The C language version of the routine is as follows:

```c
char hex_tab [16] = {'0', '1', '2', '3', '4', '5', '6', '7', '8', '9', 'A', 'B', 'C', 'D', 'E', 'F'};
void puthex (xch)
{
    char xx;
    xx = xch & 0xF0;           /* mask out the lower four bits */
    xx = xx >> 4;
    putchar (hex_tab[xx]);
    xx = xch & 0x0F;           /* mask out the upper four bits */
    putchar (hex_tab[xx]);
}
```

The following routine turns off the echo on input:

```
ECHOFF  CLR     ECHO
        RTS
```

In C language,

```c
void echoff ( )
{
    echo = 0;
}
```

The following routine turns on the echo flag:

```
ECHON   LDAA #1
        STAA    ECHO        ; set the echo flag
        RTS
```

In C language,

```c
void echon ( )
{
    echo = 1;
}
```

The following routine turns on the newline expansion flag:

```
LFON    LDAA #1
        STAA        AUTOLF        ; set the auto line feed flag
        RTS
```

In C language,

```
void lfon ( )
{
   autolf = 1;
}
```

The following routine turns off the line feed expansion flag:

```
LFOFF    CLR    AUTOLF        ; clear the line feed expansion
         RTS


ECHO     FCB    0             ; input echo flag
AUTOLF   FCB    0             ; line feed expansion flag
```

In C language,

```
void lfoff ( )
{
   autolf = 0;
}
```

Example 9.11

▼

Write an instruction sequence to output the following message to the terminal screen. "Hello, world!"

```
        ORG    $C000         ; user RAM starting address
        LDS    #$DFFF        ; initialize the stack pointer
        LDY    #MSG          ; put the starting address into Y register
        JSR    PUTSTR        ; call put string routine
        ......
MSG     FCC    "Hello, world!"  ; message to be output
        END
```

▲

9.8   Summary

The EIA-232 standard is one of the most widely used physical level interfaces for data communications in the world. It specifies 25 interface circuits for DTE/DCE use. A computer or a terminal is a DTE, whereas a modem is

DCE. Since 1960, this standard has been revised several times and the latest revision EIA-232-E was done is 1991. There are four aspects of the EIA-232 standard: electrical specifications, functional specifications, mechanical specifications, and procedural specifications.

The EIA-232 interface is designed to facilitate the data communications between two computers or a computer and a terminal using a dedicated line or a public phone line. Data communications over the EIA-232 interface require the use of a modem. However, a modem is not needed when two DTEs are very close by using the null modem connection.

Data communication is proceeded character by character. Each character consists of one start bit, seven or eight data bits, an optional parity bit, and one, or one and one half, or two stop bits. The start bit is low and the stop bit is high. The least significant bit is transmitted first and the most significant bit is transmitted last.

Three types of data transmission errors may happen: framing errors, receiver overrun, and parity errors.

The 68HC11 has incorporated a full-duplex asynchronous serial communication interface (SCI) on the chip. The 68HC11 SCI can be used in either point-to-point or multipoint data communication. The SCI has a wake-up feature that can reduce the processing overhead in the multipoint environment.

The SCI uses a clock signal that is 16 times the bit rate to detect the arrival of a valid start bit or determine the logic level of a data bit. When RxD is high for at least three sampling clock cycles and is then followed by a low voltage, the SCI will look at the third, fifth, and seventh samples after the first low sample to determine whether a valid start bit has arrived. The SCI uses the majority function of the eighth, ninth, and tenth samples to determine the logic level of a bit.

Four registers are related to the operation of the SCI function: serial communication data register, serial communication control register 1, serial communication control register 2, serial communication status register, and baud rate register.

The voltage levels from the SCI are different from those of the EIA-232 interface. A voltage level translation is needed to interface the SCI to the EIA-232 interface. The MAX232 is a chip that can perform this translation.

The asynchronous serial communication interface is often implemented as a chip separate from the microprocessor. The Motorola MC6850, Intel 8251, Rockwell R6551 are examples. The EVB uses MC6850 to implement the terminal port. The CMD-11A8 uses SCI to control com1 port and the R6551 to control the com2 port.

9.9 Exercises

E9.1. What is a DTE? What is a DCE?

E9.2. What is multidrop serial communication?

E9.3. What is a simplex link? A half-duplex link? A full-duplex link?

E9.4. With regard to the voltage of the RS-232 driver output, what voltage is considered as a mark (1)? What voltage is considered as a space (0)?

E9.5. Sketch the TxD pin output of a UART chip for the letter K in RS-232 format. The letter K is to be transmitted using a format with eight data bits, even parity, and one stop bit.

E9.6. Compute the time required to transmit 100 characters using the data format of eight data bits, one stop bit, and no parity. The transmit speed is 4800 baud.

E9.7. How does the SCI system determine the logic level of data bits?

E9.8. Write a subroutine to output a null-terminated string using the polling method to the SCI. The string to be output is pointed to by index register Y.

E9.9. Write a subroutine to output a null-terminated string from the SCI using the interrupt-driven method. The string is pointed to by index register Y.

E9.10. Write a subroutine to input a string from the SCI using the polling method. The buffer to hold the string is pointed to by index register Y.

E9.11. Write a subroutine to input a string from the SCI using the interrupt-driven method. This routine will enable the receive interrupt and stay in a loop until the CR character is received. The starting address of the buffer to hold the string is in index register Y.

E9.12. What is a framing error? What is a receiver-overrun error?

E9.13. Write a routine so that it can input two hex digits from the ACIA, store the equivalent binary number in accumulator A, and echo the number to the screen. This program is to be run on the EVB using the terminal port.

E9.14. How many addresses are assigned to the ACIA registers in Figure 9.17?

E9.15. Write a byte into the BAUD register to select the baud rate of 4800 at 2 MHz E clock frequency.

9.10   Lab Exercises and Assignments

L9.1. Write a program to simulate the login session on any mini- or mainframe computer. Your program should perform the following steps:

1. The program begins by outputting the following message:

    BUFFALO Monitor Version 2.5
    login:

2. After putting out this message, the program waits for you to type in your user name, which should be terminated by the CR character.

3. After you enter your user name, the program outputs the following prompt:

    password:

4. The program then waits for you to enter the password. After the user name and password are both entered, the program starts to

search the username/password table to see if the user name and password match. The table contains fifteen pairs of user names and passwords. If the user name and password are correct, the program outputs the following message:

login successful;

If login is not successful, the program should output the following message and repeat the login process:

Invalid password or user ID.

L9.2. Write a program to be run on the evaluation board to compute the mean and median of an array with fifteen 8-bit numbers. Each number consists of two hex digits and will be entered from the keyboard. The program outputs the following prompt:

Please enter a two-digit hex number:

The user then enters the hex number. If the number contains an invalid digit, the program will ask the user to reenter a valid number. Each number is terminated by a carriage return.

Each prompt should appear on a separate line. After fifteen 8-bit numbers have been entered, the program computes the mean and median of the array and outputs them to the screen. The program also sorts the given array and also displays it on the screen. The following message should appear on the screen at the end of program execution:

The given array is: xx xx xx xx xx xx xx xx xx xx xx xx xx xx xx
The mean of the array is: yy
The median of the array is: zz

10

68HC11 Serial Peripheral Interface

10.1 Objectives

After completing this chapter you should be able to

- describe the operation of the 68HC11 SPI subsystem

- connect an SPI master with one or more SPI slave devices

- simulate the operation of the SPI

- use the SPI system and multiple HC589s to add parallel input ports to the 68HC11

- use the SPI system and multiple HC595s to add parallel output ports to the 68HC11

- use the SPI and M14489(s) to drive multiple seven-segment displays

- use the SPI, the MC145000, and the MC145001 to drive LCDs

- interface the SPI to the MC68HC68T1 time-of-day chip

10.2   Introduction

The number of pins on an 8-bit microcontroller is quite limited. However, it is desirable to implement as many I/O functions as possible on that limited number of pins. Many I/O devices have a low data rate, and using parallel interfaces can certainly satisfy data transfer needs. However, parallel data transfers require many data pins, and thus fewer functions can be implemented on the same number of signal pins. The solution is to use serial data transfer for low-speed I/O devices. The Motorola 68HC11 provides a serial peripheral interface (SPI) and a serial communication interface (SCI) for low-speed I/O devices. The SPI is a synchronous interface that requires the same clock signal to be used by the SPI subsystem and the external device. (The SCI is an asynchronous interface that allows different clock signals to be used by the SCI system and the external device. The SCI subsystem was discussed in chapter 9.)

The SPI subsystem allows several microcontrollers with the SPI function or SPI-type peripherals to be interconnected. In the SPI format, two types of devices are involved in data transfer: a *master* and one or more *slaves*. The master can initiate a data transfer, but a slave can only respond. A clock signal is required to synchronize the data transfer but is not included in the data stream and must be furnished as a separate signal by the master of the data transfer. The 68HC11 SPI subsystem may be configured as a master or a slave. The master SPI device initiates data transfers in the SPI format. The SPI subsystem is usually used for I/O port expansion and for interfacing with peripherals, but it can also be used for multiprocessor communication. The SPI is often used to interface with peripheral devices such as TTL shift registers, LED/LCD display drivers, phase-locked loop (PLL) chips, memory components with serial interface, or A/D converter systems that do not need a very high data rate.

10.3   Signal Pins

Four of the 68HC11 port D pins are associated with SPI transfers: $\overline{\text{SS}}$/PD5, SCK/PD4, MOSI/PD3, and MISO/PD2. Port D pins are bidirectional and must be configured for either input or output. All SPI output lines must have their corresponding data direction register bits set to 1. If one of these bits is 0, the line is disconnected from the SPI logic and becomes a general-purpose input line. All SPI input lines are forced to act as inputs regardless of what is in their corresponding data direction register bit.

1. MISO/PD2: *Master in slave out.* The MISO pin is configured as an input in a master device and as an output in a slave device. It is one of the two lines that transfer serial data in one direction with the most significant bit sent first. The MISO line of a slave device is placed in a high-impedance state if the slave device is not selected.

2. MOSI/PD3: *Master out slave in.* The MOSI pin is configured as an output in a master device and as an input in a slave device. It is the second of the two lines that transfer serial data in one direction with the most significant bit sent first.

3. SCK/PD4: *Serial clock.* The serial clock is used to synchronize data movement both in and out of the device through its MOSI and MISO lines. The master and slave devices are capable of exchanging one byte of information during a sequence of eight SCK clock cycles. Since SCK is generated by the master device, this line becomes an input on a slave device.

4. $\overline{SS}$/PD5: *Slave select.* The slave select (SS) input line is used to select a slave device. For a slave device, it has to be low prior to data transactions and must stay low for the duration of the transaction. The $\overline{SS}$ line on the master device must be tied high. If it goes low, a mode fault error flag (MODF) is set in the serial peripheral status register (SPSR). The $\overline{SS}$ pin can be configured as a general-purpose output by setting the bit 5 of the DDRD register to 1, thus disabling the mode fault circuit. The other three SPI lines are dedicated to the SPI whenever the SPI subsystem is enabled.

10.4   SPI-Related Registers

Three registers in the serial peripheral interface provide control, status, and data storage functions: the *serial peripheral control register* (SPCR), the *serial peripheral status register* (SPSR), and the *serial peripheral data I/O register* (SPDR). Because the directions of SPI pins must be set for proper operation, the DDRD register must also be programmed.

10.4.1   Serial Peripheral Control Register (SPCR)

The meaning and function of each bit in the SPCR register and the value of each bit after reset are as follows:

	7	6	5	4	3	2	1	0	SPCR
	SPIE	SPE	DWOM	MSTR	CPOL	CPHA	SPR1	SPR0	located at $1028
Value after reset	0	0	0	0	0	1	u	u	

u = undefined

SPIE: SPI *interrupt enable.* When this bit is set to 1, the SPI interrupt is enabled.

SPE: SPI *enable.* When this bit is set to 1, the SPI system is enabled.

DWOM: Port D *wired-or mode select.* When this bit is set to 1, all port D pins become open-drain. Otherwise, they are normal CMOS output pins.

MTSR: *Master mode select.* When this bit is set to 1, the SPI master mode is selected. Otherwise, the slave mode is selected.

CPOL: *Clock polarity.* When the clock polarity bit is 0 and data are not being transferred, a steady state low value is produced at the SCK pin of the master device. Conversely, if this bit is 1, the SCK pin will idle high. This bit is also used in conjunction with the clock phase control bit to produce the desired clock-data relationship between the master and the slave. The relationships between CPOL bit, CPHA bit, MOSI, and MISO are shown in Figure 10.1.

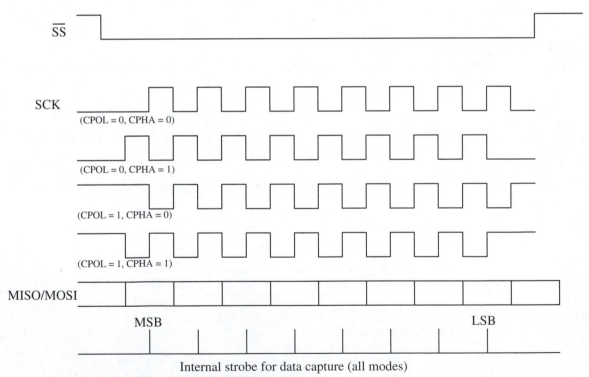

Figure 10.1 ■ SPI data clock timing diagram
(Redrawn with permission of Motorola)

Data are captured in the center of a bit time. Depending on the combination of the CPOL and CPHA setting, data bits can be captured on the rising (CPOL & CPHA = 00 or 11) or the falling (CPOL & CPHA = 01 or 10) edge of the SCK clock.

CPHA: *Clock phase.* The clock phase bit, in conjunction with the CPOL bit, controls the clock-data relationship between the master and the slave. The CPHA bit selects one of two fundamentally different clocking protocols. When CPHA = 0, the shift clock is the OR of SCK with $\overline{SS}$. As soon as $\overline{SS}$ goes low, the transaction begins and the first edge on the SCK signal invokes the first data sample. In this clock phase mode, $\overline{SS}$ must go high between successive characters in an SPI message. When CPHA = 1, the $\overline{SS}$ pin may be left low for several SPI characters. This setting requires fewer instructions during the data transfer phase.

SPR1 and SPR0: *SPI clock rate select.* These two bits control the frequency of the SCK clock output according to the scheme given by the following table:

SPR1	SPR0	E divided by
0	0	2
0	1	4
1	0	16
1	1	32

The highest data rate supported by the SPI is 1 Mbits per second at the 2 MHz E clock rate. The lowest data rate is 62.5 Kbits per second with a 2 MHz E clock frequency.

10.4.2   Serial Peripheral Status Register (SPSR)

The meaning and function of each bit in the SPSR register and the value of each bit after reset are as follows:

	7	6	5	4	3	2	1	0	SPSR located at $1029
	SPIF	WCOL	0	MODF	0	0	0	0	
Value after reset	0	0	0	0	0	0	0	0	

SPIF: *SPI transfer complete flag.* The SPIF flag bit is set on completion of data transfer between the processor and the external device. Clearing the SPIF bit is accomplished by reading the SPSR register followed by an access of the SPDR register. Unless the SPSR register is read first, attempts to write to the SPDR register are inhibited.

WCOL: *Write collision.* The write collision bit is set when an attempt is made to write to the SPDR register while data transfer is taking place. Clearing the WCOL bit is accomplished by reading the SPSR register followed by an access to the SPDR register.

MODF: *Mode fault.* The mode fault flag indicates that there may have been a multimaster conflict for system control. It allows a proper exit from system operation to a reset or default system state. The MODF bit is normally clear and is set only when the master device has its SS pin pulled low. Setting the MODF bit affects the internal serial peripheral interface system in the following ways:

1. An SPI interrupt is generated if SPIE = 1.
2. The SPE bit is cleared, disabling the SPI.
3. The MSTR bit is cleared, forcing the device into the slave mode.
4. The DDRD bits for the four SPI pins are forced to zeros.

Clearing the MODF bit is accomplished by reading the SPSR, followed by a write to the SPCR. Control bits SPE and MSTR can be restored by user software to their original state after the MODF bit has been cleared. The DDRD register must also be restored after a mode fault.

10.4.3   The Port D Data Direction Register (DDRD)

The meaning and function of each bit in the DDRD register and the value of each bit after reset are as follows:

	7	6	5	4	3	2	1	0	DDRD
	0	0	DDRD5	DDRD4	DDRD3	DDRD2	DDRD1	DDRD0	located at $1009
Controlled pin			$\overline{SS}$	SCK	MOSI	MISO	TxD	RxD	
Value after reset	0	0	0	0	0	0	0	0	

0 = input
1 = output

DDRD5: *Data direction control for port D bit 5 ($\overline{SS}$).* When the SPI system is enabled as a slave, the PD5/$\overline{SS}$ pin is the slave-select input, regardless of the value of DDRD5. When the SPI system is enabled as a master, the function of the PD5/$\overline{SS}$ pin depends on the value in DDRD5. If DDRD5 is set to 0, the $\overline{SS}$ pin is used as an input to detect mode-fault errors. A low on this pin indicates that some other device in a multiple-master system has become a master and is trying to select this MCU as a slave. When a low occurs on this pin, all SPI pins will be changed to high impedance to prevent damage to the SPI pins. When DDRD5 is set to 1, the $\overline{SS}$ pin is used as a general-purpose output pin and is not affected by the SPI system.

DDRD4: *Data direction control for port D bit 4 (SCK).* When the SPI system is enabled as a slave, the PD4/SCK pin acts as the SPI serial clock input, regardless of the state of DDRD4. When the SPI system is enabled as a master, the DDRD4 bit must be set to one to enable the SCK output.

DDRD3: *Data direction control for port D bit 3 (MOSI).* When the SPI system is enabled as a slave, the PD3/MOSI pin acts as the slave serial data input, regardless of the state of DDRD3. When the SPI system is enabled as a master, the DDRD3 bit must be set to 1 to enable the master serial data output. If a master device wants to initiate an SPI transfer to receive a byte of data from a slave without transmitting a byte, it might purposely leave the MOSI output disabled. SPI systems that tie MOSI and MISO together to form a single, bidirectional data line also need to selectively disable the MOSI output.

DDRD2: *Data direction control for port D bit 2 (MISO).* When the SPI system is enabled as a slave, the DDRD2 bit must be set to 1 to enable the slave serial data output. A master SPI device can simultaneously broadcast a message to several slaves as long as no more than one slave tries to drive the MISO pin. SPI systems that tie MOSI and MISO together to form a single, bidirectional data line also need to selectively disable the MISO output.

10.5 SPI Operation

The SPI circuit connection, data transfer timing, and the data transfer process are discussed in the following subsections.

10.5.1 SPI Circuit Connection

In a system that uses the SPI function, one device (normally the 68HC11) is configured as the master and the other devices are configured as slave devices. Either a peripheral chip or a 68HC11 can be configured as a slave device. The master SPI device controls the data transfer and can control one or more SPI slave devices simultaneously.

In a single-slave configuration, the circuit would be connected as shown in Figure 10.2. The $\overline{SS}$ pin of the master is tied to 5 V, and the $\overline{SS}$ pin on the slave SPI device is tied to ground.

In a multiple-slave configuration, all corresponding SPI pins except the $\overline{SS}$ pin can be tied together as shown in Figure 10.3. The master uses port pins to select a specific slave for data transfer. The $\overline{SS}$ input of the selected slave device should be set to low, but the $\overline{SS}$ input of a slave should be set to high when it is not selected.

Another way to connect multiple SPI slaves to one master is shown in Figure 10.4. Figure 10.4 differs from Figure 10.3 in three ways:

1. The MISO pin of each slave is wired to the MOSI pin of the slave device next to it. The MOSI pins of the master and slave 0 are still wired together.

2. The MISO pin of the master is wired to the MISO pin of the last slave device.

3. The $\overline{SS}$ inputs of all slaves are tied to ground to enable all slaves. No port pin from the master is needed to enable any individual slave device.

Thus the shift registers of the SPI master and slaves become a ring. The data of slave k are shifted to the master SPI, the data of the master are shifted to the slave 0, the data of slave 0 are shifted to the slave 1, etc. In this configuration, a minimal number of signal pins are used to control a large number of devices. However, the master does not have the freedom to select an arbitrary slave device for data transfer without going through other slave devices.

This type of configuration is often used to extend the capability of the SPI slave. For example, suppose there is an SPI-compatible seven-segment display driver/decoder that can drive only four digits. By using this configuration, up to $4 \times k$ digits can be displayed when k driver/decoders are cascaded together. An example of this configuration will be discussed in section 10.9.3.

Depending on the capability and role of the slave device, either the MISO or MOSI pin may not be used in the data transfer. Many SPI-compatible peripheral chips do not have the MISO pin.

10.5.2 SPI Data Transfer

An SPI transfer is initiated by writing data to the shift register in the master SPI device. During an SPI transfer, data are circulated eight bit positions;

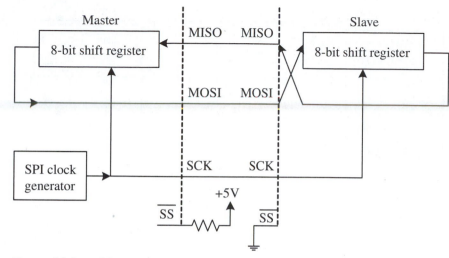

Figure 10.2 ■ SPI master-slave interconnection

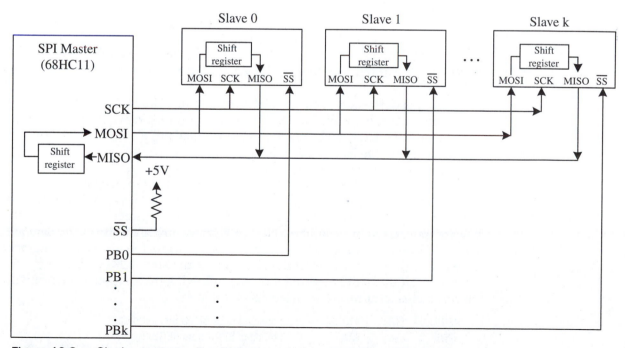

Figure 10.3 ■ Single-master and multiple-slave device connection (method 1)

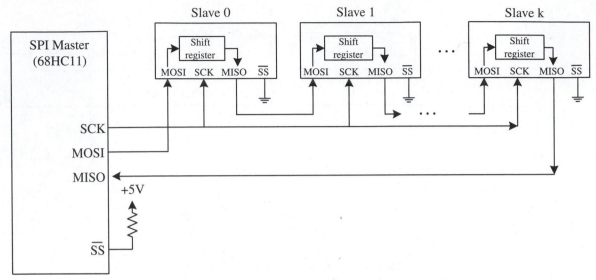

Figure 10.4 ■ Single-master and multiple-slave device connection (method 2)

thus data are exchanged between the master and the slave. Eight pulses are sent out from the master SCK pin to synchronize the data transfer between the master and the slave SPI devices.

The SPI is double-buffered on read but not on write. If a write is performed during a data transfer, the transfer occurs uninterrupted and the write will be unsuccessful. This condition sets the write collision (WCOL) status bit in the SPSR register. After a data byte is shifted, the SPIF flag of the SPSR register is set to 1.

In the master mode, the SCK pin is an output. The SCK clock idles high (or low) if the CPOL bit in the SPCR register is 1 (or 0). When data are written into the shift register, eight clock pulses are generated to shift eight bits of data, and then the SCK pin goes idle again.

In the slave mode, the slave start logic receives a logic low at the $\overline{SS}$ pin and a clock input at the SCK pin. Thus the slave is synchronized with the master. Data from the master are received serially at the slave MOSI line and loaded into the 8-bit shift register. After the 8-bit shift register is loaded, its data are transferred in parallel to the read buffer. During a write cycle, data are written into the shift register, and then the slave waits for a sequence of clock pulses from the master to shift the data out on the slave's MISO line.

In a single-slave SPI environment, the following instruction sequence will transfer a character from the master to the slave:

```
REGBAS   EQU    $1000        ; base address of I/O register block
SPDR     EQU    $2A          ; offset of SPDR from REGBAS
SPCR     EQU    $28          ; offset of SPCR from REGBAS
```

```
        SPSR     EQU    $29          ; offset of SPSR from REGBAS
        DDRD     EQU    $09          ; offset of DDRD from REGBAS
        SPI_DIR  EQU    $3A          ; value to set SPI pin directions, this value sets SS,
        *                            ; SCK, MOSI, and TxD pins for output. Other port D pins are
        *                            ; configured as input.
        SPI_INI  EQU    $54          ; value to initialize the SPI. This value enables SPI function,
        *                            ; disables the SPI interrupt, selects port D pins as normal
        *                            ; CMOS output, selects master mode, chooses the falling
        *                            ; edge to synchronize data transfer, and sets the data
        *                            ; rate to 1 Mbits/sec

                 ORG    $00
        data     RMB    $10
                  .
                  .
                  .
                 LDX    #REGBAS
                 LDAA   #SPI_DIR
                 STAA   DDRD,X       ; set SPI pin directions
                 LDAA   #SPI_INI
                 STAA   SPCR,X       ; initialize SPI system
                 LDAA   data         ; get data to be sent
                 STAA   SPDR,X       ; start SPI transfer
        WAIT     LDAB   SPSR,X       ; check bit 7 of SPSR to see if the SPI transfer if
                 BPL    WAIT         ; completed
                  .
                  .
                  .
```

The following C statements will output the variable *xx* to the output device via the MOSI pin:

```
DDRD = 0x3A;
SPCR = 0x54;
SPDR = xx;
while (!(SPSR & 0x80));     /* wait until 8 bits have been shifted out */
```

To read data from a slave SPI device, the master SPI device also writes a byte into the SPDR register. However, the byte written into the SPDR register is unimportant. The data from the slave are shifted into the SPDR register from the MISO pin. The following instruction sequence reads a byte from the slave SPI device:

```
        LDX    #REGBAS
        LDAA   #SPI_DIR
        STAA   DDRD,X        ; set SPI pin directions
        LDAA   #SPI_INI
```

```
        STAA    SPCR,X          ; initialize SPI system
        STAA    SPDR,X          ; start the SPI transfer
HERE    LDAB    SPSR,X          ; wait until a character has been shifted in
        BPL     HERE            ;      "
        LDAA    SPDR,X          ; place the byte in accumulator A
```

The following C statements will input a byte from the MISO pin:

```
DDRD = 0x3A;
SPCR = 0x54;
SPDR = 0x00;                /* write an arbitrary value to SPDR to trigger SCK pulses */
while (!(SPSR & 0x80));     /* wait until eight bits have been shifted in */
xx = SPDR;                  /* place the input data in variable xx */
```

10.5.3   Simulating the SPI

The SPI function is convenient to use when the number of data bits to be transferred is a multiple of eight. When the data to be transferred are not a multiple of eight bits, the SPI might be clumsy to use. In this situation, a software technique can be used to simulate the SPI operation. The peripheral device would use either the falling or the rising clock edge to shift data in/out. The following procedure can be used to shift data into (or out from) a peripheral device on the falling clock edge:

Step 1
Set the clock to high.

Step 2
Apply the data bit on the port pin that is connected to the serial data input pin of the peripheral device.

Step 3
Pull the clock to low.

Step 4
Repeat steps 1 to 3 as many times as needed.

If the rising edge is used to latch data, then the following changes should be made: set the clock to low in step 1, and pull the clock to high in step 3.

10.6   SPI-Compatible Peripheral Chips

The SPI is a protocol developed by Motorola to interface peripheral devices to a microcontroller. As long as a peripheral device supports the SPI interface protocol, it can be used with any microcontroller that implements the SPI function. Several manufacturers produce SPI-compatible peripheral chips. A partial list is given in Appendix G. The Motorola SPI protocol is compatible with the National Semiconductor Microwire protocol. Therefore, any peripheral device that can be interfaced with the SPI can also be interfaced with the Microwire protocol.

10.7   Interfacing the HC589 to SPI

One application of the HC589 is to expand the number of input ports to the 68HC11. The full name of the HC589 is 74HC589. This device is available from several manufacturers, each of whom may add a prefix for identification purpose; for example, Motorola adds the prefix MC. In this text, we will use the generic name HC589 to refer to this chip.

The HC589 is an 8-bit serial or parallel-input/serial-output shift register. Its block diagram and pin assignments are shown in Figure 10.5. The HC589 can accept serial or parallel data input and shift it out serially. The maximum shift clock rate for the HC589 is 6 MHz at room temperature.

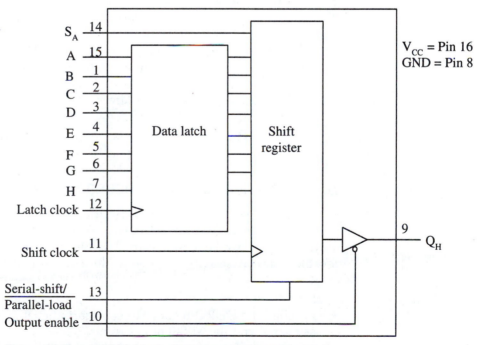

Figure 10.5 ■ HC589 block diagram and pin assignment

10.7.1   HC589 Pins

The functions of the HC589 signal pins are as follows:

A, B, C, D, E, F, G, and H:	*Parallel data inputs.* Data on these inputs are stored in the data latch on the rising edge of the latch clock input.
S$_A$:	*Serial data input.* Data on this pin are shifted into the shift register on the rising edge of the shift clock input if the serial-shift/parallel-load signal is high. Data on this pin are ignored when the serial-shift/parallel-load signal is low.
Serial-shift/ Parallel-load:	*Shift register mode control.* When a high level is applied to this pin, the shift register is allowed to shift data serially. When a low level is applied, the shift register accepts parallel data from the data latch.
Shift clock:	*Serial shift clock.* A low-to-high transition on this input shifts data on the serial data input into the shift register; data on stage H are shifted out from Q$_H$, where it is replaced by the data previously stored in stage G.
Latch clock:	*Data latch clock.* A low-to-high transition on this input loads the parallel data on inputs A-H into the data latch.
Output enable:	*Active-low output enable.* A high level applied to this pin forces the Q$_H$ output into high-impedance state. A low level enables the output. This control does not affect the state of the input latch or the shift register.
Q$_H$ output:	*Serial data output.* This is a three-state output from the last stage of the shift register.

10.7.2   Circuit Connection of HC589 and SPI

There are several different ways to interface HC589s to the SPI subsystem. One connection method is shown in Figure 10.6. With this method, multiple bytes can be loaded into HC589s at one time and then shifted into the 68HC11 bit-serially. The procedure for shifting in multiple bytes is as follows:

Step 1
Program the DDRD register to configure the SCK, TxD, and SS pins as output and the MISO pin as input.

Step 2
Program the SPCR register to enable the SPI function.

Step 3
Set the LC pin to low and then pull it to high. This will load the external data into the data latch in parallel.

Step 4
Set the $\overline{SS}$ pin to low to select the parallel load mode, which will load the contents of the data latch into the shift register.

Step 5
Set the $\overline{SS}$ pin to high to select serial shift mode.

Step 6
Write a byte into the SPDR to trigger eight SCK clock pulses to shift in eight bits.

Step 7
Repeat step 6 as many times as needed and save the data in a buffer.

Example 10.1

Write a program to input eight bytes from eight external HC589s connected as shown in Figure 10.6 and store the data at memory locations starting at $0000.

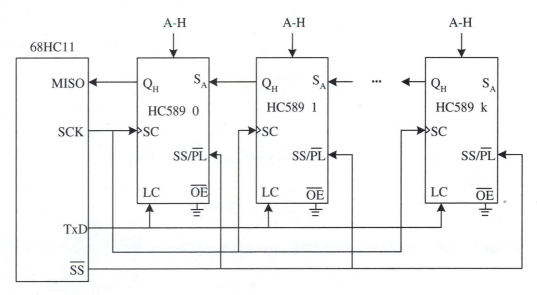

SC: shift clock
LC: latch clock
SS/$\overline{PL}$: serial shift/parallel load mode select
$\overline{OE}$: output enable

Figure 10.6 ■ Serial connection of multiple HC589s to an SPI

Solution: The program should

1. write the value \$3A into the DDRD register so that the $\overline{SS}$, SCK, MOSI, and TxD pins are configured for output and the MISO and RxD pins are configured for input.

2. write the value \$50 into the SPCR register to:

   a. disable the SPI interrupt

   b. enable the SPI

   c. set port D pins to be normal (not open-drain) CMOS output pins

   d. use the rising edge of the SCK signal to shift data

   e. select a 1 Mbit/sec data rate

The program is as follows:

```
REGBAS   EQU    $1000            ; base address of I/O register block
PORTD    EQU    $08              ; offset of PORTD from REGBAS
DDRD     EQU    $09              ; offset of DDRD from REGBAS
SPCR     EQU    $28              ; offset of SPCR from REGBAS
SPSR     EQU    $29              ; offset of SPSR from REGBAS
SPDR     EQU    $2A              ; offset of SPDR from REGBAS
SPCR_INI EQU    $50              ; value to initialize the SPCR register
SPI_DIR  EQU    $3A              ; value to initialize the DDRD register

         ORG    $C000            ; starting address of the program
         LDX    #REGBAS
         LDAA   #SPI_DIR
         STAA   DDRD,X           ; set port D pins directions
         LDAA   #SPCR_INI
         STAA   SPCR,X           ; initialize the SPI system
```

* The following two instructions latch data into HC589s in parallel

```
         BCLR   PORTD,X $02      ; create a rising edge to load data into data latch
         BSET   PORTD,X $02      ;          "
         BCLR   PORTD,X $20      ; select parallel load mode and load the contents
*                                ; of the data latch into the shift register
         BSET   PORTD,X $20      ; select serial shift mode
         LDAB   #8               ; loop count for transferring eight bytes
         LDY    #$0000           ; initialize buffer pointer Y
loop     STAA   SPDR,X           ; trigger an SPI data transfer
         BRCLR  SPDR,X $80 *     ; wait until the SPI transfer is done
         LDAA   SPDR,X           ; get one byte and
         STAA   0,Y              ; store it in the buffer
         INY                     ; move the buffer pointer
```

```
        DECB                        ; decrement loop count
        BNE     loop
        END
```

The C language version of the program is as follows:

```c
#include <hc11.h>
main ( )
{
    char buffer[8], i;
    DDRD = 0x3A;                    /* configure port D pin directions */
    SPCR = 0x50;                    /* configure SPI parameters */
    PORTD &= 0xFD;                  /* create a rising edge on the TxD pin */
    PORTD |= 0x02;                  /* to load data into data latch */
    PORTD &= 0xDF;                  /* transfer data from data latch into shift register */
    PORTD |= 0x20;                  /* select serial shift mode */
    for (i = 0; i < 8; i++) {
        SPDR = 0x00;                /* shift in data from the MISO pin */
        while (!(SPSR & 0x80));     /* wait until 8 bits have been shifted in */
        buffer[i] = SPDR;           /* save input data */
    }
    return 0;
}
```

Another way to connect the HC589s is to tie all the Q_H pins together and then use other port pins to selectively enable one of the HC589s to shift out data to the 68HC11. The circuit connection for this method is shown in Figure 10.7. This circuit configuration allows the user to selectively input data from a specific HC589.

The procedure for loading a byte from a specified HC589 into the SPDR register is as follows:

Step 1
Program the DDRD register to set the directions of the MISO, SCK, $\overline{SS}$, and TxD pins.

Step 2
Program the SPCR register to configure the SPI function.

Step 3
Set the TxD pin to low and then pull it to high to load external data into the data latch in parallel.

Step 4
Set the $\overline{SS}$ pin to low to select the parallel load mode, which will load the contents of the data latch into the shift register.

Step 5
Pull the $\overline{SS}$ pin to high to select the serial shift mode.

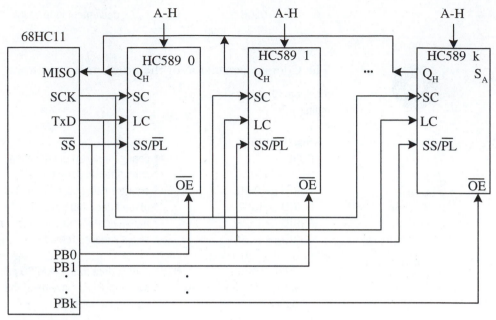

SC: shift clock
LC: latch clock
SS/$\overline{\text{PL}}$: serial shift /parallel load mode select
$\overline{\text{OE}}$: output enable

Figure 10.7 ■ Serial connection of multiple HC589s to an SPI

Step 6
Set the port B pin that controls the specified HC589 to low to enable
the shift register to output serial data. The remaining port B pins are set
to high.

Step 7
Write a byte into the SPDR register to trigger eight pulses from the SCK
pin to shift in the serial data. The external data are now in the SPDR
register ready for use.

Step 8
Repeat step 6 and 7 as many times as needed.

10.8   Interfacing the HC595 to SPI

The full name of the HC595 is 74HC595. Its block diagram and pin assignments are shown in Figure 10.8. The HC595 consists of an 8-bit shift register
and an 8-bit D-type latch with three-state parallel outputs. The shift register
accepts serial data and provides a serial output. The shift register also provides
parallel data to the 8-bit latch. The shift register and the latch have indepen-

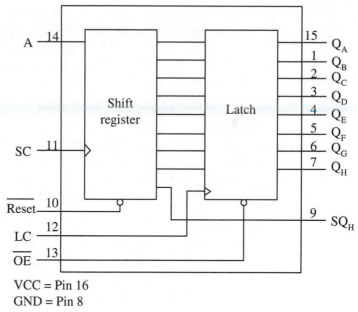

Figure 10.8 ■ HC595 block diagram and pin assignment

dent clock inputs. This device also has an asynchronous reset input. The maximum shift clock rate for the HC595 at room temperature is 6 MHz.

10.8.1   HC595 Pins

The function of each pin is as follows:

A: *Serial data input:* The data on this pin is shifted into the 8-bit serial shift register.

Shift clock: A low-to-high transition on this input causes the data at the serial input pin to be shifted into the 8-bit shift register.

Reset: Active low. A low on this pin resets the shift register portion of this device only. The 8-bit latch is not affected.

Latch clock: A low-to-high transition on this pin loads the shift register data into the output latch.

Output enable: Active-low output enable. A low on this pin allows the data from the latches to be presented at the outputs. A high on this input forces the outputs (Q_A to Q_H) into the high-impedance state. This pin does not affect the serial output.

Q_A *to* Q_H: noninverted, three-state, latch outputs.

SQ_H: *Serial data output:*   This is the output of the eighth stage of the 8-bit shift register. This output does not have tristate capability.

10.8.2   Circuit Connections of the HC595 and the SPI

One application of the HC595 is to expand the number of parallel output ports of the 68HC11. Many parallel ports can be added to the 68HC11 by using the SPI subsystem and multiple HC595s. Figure 10.9 shows a method of connecting multiple HC595s to the SPI subsystem.

The procedure for outputting multiple bytes using this connection is as follows:

Step 1
Program the DDRD register to configure each SPI pin and the TxD pin to appropriate directions. The $\overline{SS}$, MOSI, SCK, and TxD pins are configured for output, and the MISO and RxD pins for input. Write the value %00111010 into DDRD.

Step 2
Program the SPCR register to enable the SPI function, disable SPI interrupt, use the rising edge of the SCK signal to shift data in and out, select master mode, normal port D pins, and set the data rate to 1 Mbits/sec; to do this, write the value %01010000 into SPCR.

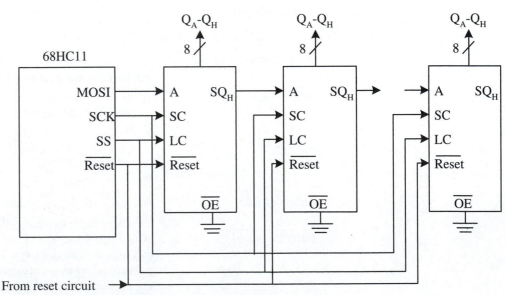

Figure 10.9 ■ Serial connection of multiple HC595s to the SPI

Step 3

Write a byte into the SPDR register to trigger eight output pulses from the SCK pin.

Step 4

Repeat step 3 as many times as needed.

Step 5

Set the $\overline{SS}$ pin to low and then pull it to high to load the data in the shift register of each HC595 into the output latch. After this step, the Q_A through Q_H pins of each HC595 contains valid data.

Example 10.2

Write a program to output three bytes to the first three HC595s in Figure 10.9.

Solution: The following program is based on the previous algorithm:

```
REGBAS      EQU     $1000           ; base address of the I/O register block
PORTD       EQU     $08             ; offset of PORTD from REGBAS
DDRD        EQU     $09             ; offset of DDRD from REGBAS
SPCR        EQU     $28             ; offset of SPCR from REGBAS
SPSR        EQU     $29             ; offset of SPSR from REGBAS
SPDR        EQU     $2A             ; offset of SPDR from REGBAS
SPI_DIR     EQU     $3A             ; value to be written into DDRD to set up SPI pin
*                                   ; directions
SPCR_INI    EQU     $50             ; value to initialize SPCR
K           EQU     $3

            ORG     $00
buffer      FCB     $11,$22,$33     ; values to be output from SPI

            ORG     $C000           ; starting address of the program
            LDX     #REGBAS
            LDAA    #SPI_DIR
            STAA    DDRD,X          ; set the directions of SPI pins
            LDAA    #SPCR_INI
            STAA    SPCR,X          ; initialize SPI system
            LDAB    #K              ; initialize loop count
            LDY     #buffer         ; load data buffer base address
ch_loop     LDAA    0,Y             ; get one character
            STAA    SPDR,X          ; output the character
wait_ch     LDAA    SPSR,X          ; wait until eight bits have been shifted out
            BPL     wait_ch
            INY                     ; move the buffer pointer
```

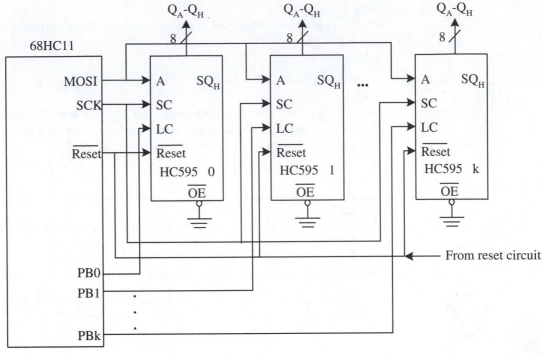

Figure 10.10 ■ Parallel connection of multiple HC595s to the SPI

```
        DECB                        ; decrement the loop count
        BNE     ch_loop
```

* The following two instructions create a rising edge from the $\overline{SS}$ pin to load the data into the
* output latch

```
        BCLR    PORTD,X $20
        BSET    PORTD,X $20
        END
```

Another method of expanding the number of parallel output ports is to connect multiple HC595s in parallel to the SPI subsystem, as shown in Figure 10.10.

With this method, the user can selectively output data to any HC595 by setting the corresponding port B pin after the data has been shifted to that HC595. The procedure for outputting data to I/O devices connected to the HC595s is as follows:

Step 1
Program the DDRD to set up the SPI pin directions.

Step 2
Program the SPCR register to enable the SPI subsystem, select the data rate, select the rising edge of the SCK signal to shift data, select master mode, and disable interrupt.

Step 3
Write a data byte into the SPDR register to trigger SPI data transfer.

Step 4
Set the PB_i pin to low and then pull it to high to load the byte from the shift register of the HC595 *i* into the output latch.

10.9   Interfacing the MC14489 to SPI

The Motorola MC14489 is a seven-segment display driver that can directly interface with individual lamps, seven-segment displays, or various combinations of both. LEDs wired with common cathodes are driven in a multiplexed-by-5 fashion. Communication with a microcontroller is via a SPI-compatible serial port. The MC14489 requires only a current-setting resistor to operate.

A single MC14489 can drive any one of the following: a 5-digit display plus decimals, a $4^1/_2$-digit display plus decimals and sign, or 25 lamps. A special technique allows driving $5^1/_2$ digits. A configuration register allows the drive capability to be partitioned to suit many additional applications. The on-chip decoder outputs seven-segment numbers 0 to 9, hexadecimal characters A to F, plus 15 letters and symbols.

10.9.1   The Signal Pins of MC14489

The diagram and the pin assignment are illustrated in Figure 10.11.
The functions of the MC14489 pins are as follows:

Data In: *Serial data input.* The bit stream begins with the most significant bit and is shifted in on the rising edges of Clock. When the device is not cascaded, the bit pattern is either one byte long to change the configuration register or three bytes long to update the display register. For two chips cascaded, the pattern is either four or six bytes, respectively. The display does not change or flicker during shifting, which allows slow serial data rate.

Clock: *Serial data clock input.* Low-to-high transitions on Clock shift bits available at Data In, while high-to-low transitions shift bits from Data Out. To guarantee proper operation of the power-on reset (POR) circuit, the Clock pin must not be floated or toggled during power-up. That is, the Clock pin must

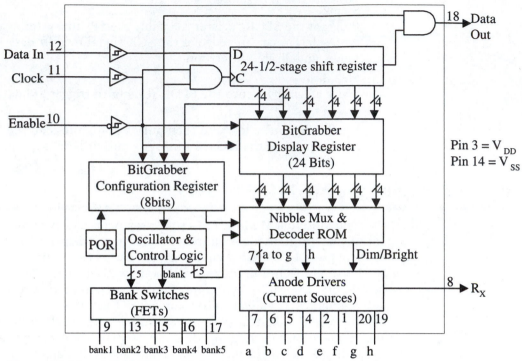

Figure 10.11 ■ MC14489 multi-character LED driver
(Redrawn with permission of Motorola)

be stable until the V_{DD} pin reaches at least 3 V. The highest clock frequency is 4 MHz.

Enable: *Active low enable input.* This pin allows the MC14489 to be used on a serial bus, sharing Data In and Clock with other peripherals. When $\overline{\text{Enable}}$ is in inactive state, Data Out is forced to the low state, shifting is inhibited. To transfer data to the device, this pin must be taken low. When the transfer is done, we must pull it to high. The low-to-high transition of $\overline{\text{Enable}}$ transfers data to either the configuration or display register, depending on the data stream length.

Data Out: *Serial data output.* Data are transferred out of the shift register through Data Out pin on the falling edges of the Clock signal. When cascaded, this pin is connected to the Data In pin of the next device. Data Out can be fed back to a microcontroller to perform a wrap-around test of serial data.

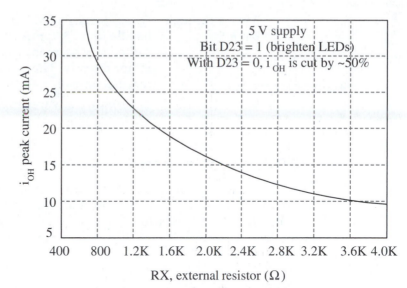

Figure 10.12 ■ **a** through **h** nominal current vs. R_x
(Redrawn with permission of Motorola)

R_x:

External current-setting resistor. A resistor tied between this pin and ground (V_{ss}) determines the peak segment drive current at pins **a** through **h.** This resistor ties to a current mirror with an approximate current gain of 10 when bit 23 of the shift register is a 1. With bit 23 set to 0, the peak current is reduced to half. Values of R_x range from 700 Ω to infinity. When $R_x = \infty$ (open circuit), the display is extinguished. The relationship between R_x and the segment current is shown in Figure 10.12.

a through h:

Anode-driver current sources. These outputs are closely-matched current sources that directly tie to the anodes of external discrete LEDs (lamps) or display segment LEDs. When used with lamps, outputs **a, b, c,** and **d** are used to independently control up to 20 lamps. Output **h** is used to control up to five lamps dependently. For lamps, the *No Decode* mode is selected via the configuration register, and **e, f,** and **g** are forced inactive. When used with segmented displays, outputs **a** through **g** drive segments **a** to **g,** respectively. Output **h** is used to drive the decimal point. If unused, **h** should be left open.

Bank 1 through Bank 5: *Diode-Bank FET switches.* These outputs can handle up to 320 mA and are tied to the common cathodes of segment displays or the cathodes of lamps directly. The display is refreshed at a minimal 1 KHz rate.

V_{DD}: *Most positive power supply.*

V_{SS}: *Most negative power supply.*

10.9.2   Operation of the MC14489

The configuration register controls the operation of the MC14489. The contents of the configuration register are shown in Figure 10.13.

Before sending data for display, we need to configure the operation of this device. The MC14489 has two modes: *hex decode* mode and *special decode* mode. In the hex decode mode, regular hex digits (from 0, 1,...9, A, B, C, D, E, and F) are displayed. In the special decode mode, the annunciators and some letters can be displayed. In either mode, displays can be made dimmer or brighter by clearing or setting the first bit of the display data sent to the MC14489. The MC14489 segment decoder function is illustrated in Table 10.1.

When there are no more than five digits to be displayed, one MC14489 is adequate. The circuit connection is shown in Figure 10.14. When there are only five digits to be displayed, the digit controlled by bank 5 is shifted out first, followed by the digit controlled by bank 4, and so on. Three bytes

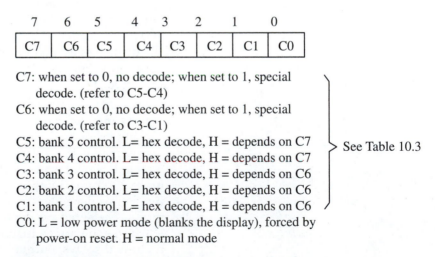

7	6	5	4	3	2	1	0
C7	C6	C5	C4	C3	C2	C1	C0

C7: when set to 0, no decode; when set to 1, special decode. (refer to C5-C4)

C6: when set to 0, no decode; when set to 1, special decode. (refer to C3-C1)

C5: bank 5 control. L= hex decode, H = depends on C7

C4: bank 4 control. L= hex decode, H = depends on C7   ⎱ See Table 10.3

C3: bank 3 control. L= hex decode, H = depends on C6

C2: bank 2 control. L= hex decode, H = depends on C6

C1: bank 1 control. L= hex decode, H = depends on C6

C0: L = low power mode (blanks the display), forced by power-on reset. H = normal mode

Figure 10.13 ■ The MC14489 Configuration register

Bank Nibble Value			7-segment Display Characters		Lamp Conditions No Decode ① (invoked via bits C1 to C7)			
Hex	Binary MSB LSB		Hex Decode (invoked via bits C1 to C5)	Special Decode (Invoked via bits C1 to C7)	d	c	b	a
$0	0 0 0 0		0					
$1	0 0 0 1		1	c				on
$2	0 0 1 0		2	H			on	
$3	0 0 1 1		3	h			on	on
$4	0 1 0 0		4	J		on		
$5	0 1 0 1		5 ②	L		on		on
$6	0 1 1 0		6	n		on	on	
$7	0 1 1 1		7	o		on	on	on
$8	1 0 0 0		8 ③	P	on			
$9	1 0 0 1		9 ④	r	on			on
$A	1 0 1 0		A	U	on		on	
$B	1 0 1 1		b	u	on		on	on
$C	1 1 0 0		C	y	on	on		
$D	1 1 0 1		d	–	on	on		on
$E	1 1 1 0		E	=	on	on	on	
$F	1 1 1 1		F	o	on	on	on	on

(1) In the *No Decode Mode,* outputs e, f, and g are unused and are all forced inactive. Output h's decoding is unaffected, i.e., unchanged from the other modes. The No Decode mode is used for three purposes:
    (a) Individually controlling lamps.
    (b) Controlling a half digit with sign.
    (c) Controlling annunciators—examples: AM, PM, UHF, kV, mm, Hg.
(2) Can be used as "cap S."
(3) Can be used as "cap B."
(4) Can be used as "small g."

Table 10.1 Triple-mode segment decoder function table
(Redrawn with permission of Motorola)

will be shifted out when there are five digits to be displayed. The first four bits of the display data are used to select the decimal point to be displayed and select whether to dim the displays or not. The use of the most significant four bits of the display data sent to the MC14489 is illustrated in Figure 10.15.

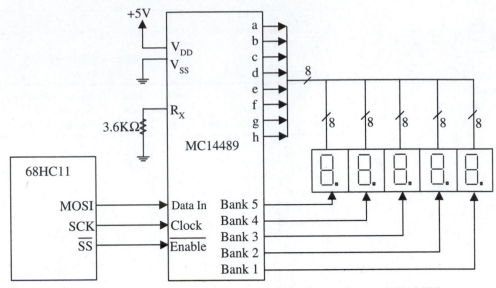

Figure 10.14 ■ 68HC11 driving five 7-segment displays using an MC14489

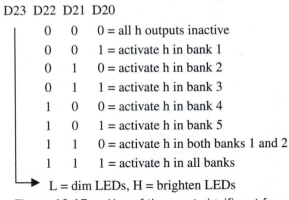

Figure 10.15 ■ Use of the most significant four
bits of the display data

Example 10.3

Write a program to display 997.04 from bank 5 to bank 1 in Figure 10.14.
Use the normal brightness to display these five digits.

Solution: The control byte to be written into the configuration register is as
follows:

bit 7: no decode, set to 0
bit 6: no decode, set to 0
bit 5: bank 5 hex decode, set to 0
bit 4: bank 4 hex decode, set to 0
bit 3: bank 3 hex decode, set to 0
bit 2: bank 2 hex decode, set to 0
bit 1: bank 1 hex decode, set to 0
bit 0: normal mode, set to 1

The display data are shown in Figure 10.16.
The SPCR register should be configured as follows:

bit 7 (SPIE): set to 0 to disable SPI interrupt
bit 6 (SPE): set to 1 to enable SPI function
bit 5 (DWOM): set to 0 to choose normal port D pins
bit 4 (MSTR): set to 1 to select master mode
bit 3 & 2 (CPOL & CPHA): set to 00 to use rising edges to shift data out
bit 1 & 0 (SPR1 & SPR0): set to 00 to choose 1MHz as the data rate

Write the value $50 to configure SPI.
The program is as follows:

```
REGBAS    EQU      $1000
SPCR      EQU      $28
SPSR      EQU      $29              ; offset of SPSR from REGBAS
SPDR      EQU      $2A
DDRD      EQU      $09
PORTD     EQU      $08
          ORG      $00
disp_dat  FCB      $B9,$97,$04
          ORG      $C000
          LDX      #$1000
          LDAA     #$50
          STAA     SPCR,X           ; configure SPI parameters
          LDAA     #$3A             ; configure SPI pin direction
          STAA     DDRD,X           ;        "
          BCLR     PORTD,X $20      ; enable SPI data transfer to MC14489
```

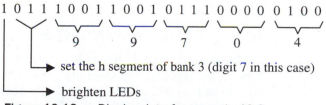

1 0 1 1 1 0 0 1 1 0 0 1 0 1 1 1 0 0 0 0 0 1 0 0
        9   9   7   0   4

set the h segment of bank 3 (digit 7 in this case)

brighten LEDs

Figure 10.16 ■ Display data for example 10.3

```
            LDAA    #$01
            STAA    SPDR,X          ; sent out the MC14489 configuration data
wait        LDAA    SPSR,X
            BPL     wait            ; wait until configuration data is sent out
            BSET    PORTD,X $20     ; transfer data into the configuration register
            BCLR    PORTD,X $20     ; enable SPI data transfer to MC14489
            LDY     #disp_dat
            LDAB    #3
loop        LDAA    0,Y
            STAA    SPDR,X
            BRCLR   SPSR,X $80 *    ; wait until a byte is shifted out
            INY
            DECB
            BNE     loop
            BSET    PORTD,X $20     ; transfer data into the display register
            END
```

In C language,

```c
#include <hc11.h>
main ( )
{
    int i;
    unsigned char disp_dat [3] = {0xB9, 0x97, 0x04};
    DDRD = 0x3A;                /* configure SPI pins directions */
    SPCR = 0x50;                /* configuration SPI parameters */
    PORTD &= 0xDF;              /* enable SPI data transfer */
    SPDR = 0x01;                /* send out MC14489 configuration data via SPI */
    while (!(SPSR & 0x80));     /* wait until data have been shifted out */
    PORTD |= 0x20;             /* load data into configuration register */
    PORTD &= 0xDF;              /* enable SPI data transfer */
    for (i = 0; i < 3; i++) {
            SPDR = disp_dat[i];
            while (!(SPSR & 0x80));
    }
    PORTD |= 0x20;             /* load data into display data register */
}
```

10.9.3 Cascading the MC14489s

Nothing special is needed to cascade multiple MC14489s together. It is done by connecting the Data Out pin of the current MC14489 to the Data In pin of the next MC14489. An example of cascading three MC14489s is shown in Figure 10.17.

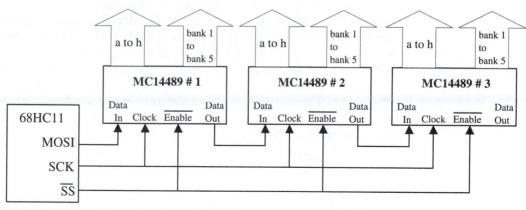

Figure 10.17 ■ Cascading three MC14489s

Before sending display data, these MC14489s must be configured. Seven bytes of data must be sent in. The first byte shifted in is used as the configuration byte for the rightmost MC14489; the fourth byte shifted in is used as the configuration byte of the middle MC14489; the last byte shifted in is used as to configure the leftmost MC14489. The remaining four bytes are not used for configuration purpose. The configuration information is shown in Figure 10.18.

1st byte	2nd byte	3rd byte	4th byte	5th byte	6th byte	7th byte (last)
Configuration register of device #3	don't care	don't care	Configuration register of device #2	don't care	don't care	Configuration register of device #1

⎯⎯⎯⎯⎯⎯⎯⎯➤ Time

Figure 10.18 ■ Configuration information for three cascaded MC14489s

Example 10.4

Write a program to display the following information (temperature at 12:00 of Aug. 2, 1999) on the 15 seven-segment displays driven by the three MC14489s shown in Figure 10.17:

25.5° C 12 00 08 02 99

Solution: The degree character can be represented by the special decode character of hex digit F (see Table 10.1). The letter *C* can be represented by hex digit C in hex decode mode. All the other characters can be represented by digits in regular hex decode mode. The leftmost five characters of the above information are to be displayed on displays driven by the leftmost MC14489. The middle five characters are to be displayed on the displays driven by the middle MC14489. The rightmost five characters are to be displayed on the displays driven by the rightmost MC14489. The configuration data for these three MC14489s are as follows:

The leftmost MC14489:
C7: set to 0 to select no decode
C6: set to 1 to select special decode (need to display degree character on bank 2)
C5: set to 0 to select hex decode
C4: set to 0 to select hex decode
C3: set to 0 to select hex decode
C2: set to 1 to select special decode
C1: set to 0 to select hex decode
C0: set to 1 to select normal mode

The middle and rightmost MC14489s:

C7 & C6: set to 0 to select no decode
C5 to C1: set to 0 to select hex decode
C0: set to 1 to select normal mode

The configuration data of MC14489 #3 must be sent out first and hence the complete configuration data are 01xxxx01xxxx45H. Here, x stands for don't care, and we will use 0 for it.

The rightmost five characters of the display data 80299 must be sent out first, followed by the middle five characters 12000, and then followed by the leftmost five characters 25.5°C.

The overall display data are as follows:
The leftmost MC14489 (Figure 10.19):

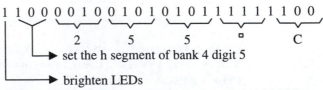

Figure 10.19 ■ Display data for leftmost MC14489

The middle MC14489 (Figure 10.20):

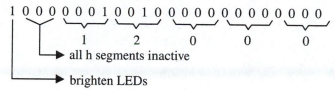

Figure 10.20 ■ Display data for the middle MC14489

The rightmost MC14489 (Figure 10.21):

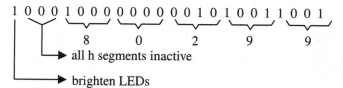

Figure 10.21 ■ Display data for the rightmost MC14489

The program that displays the specified information is as follows:

```
REGBAS    EQU     $1000
SPSR      EQU     $29
SPDR      EQU     $2A
SPCR      EQU     $28
PORTD     EQU     $08
DDRD      EQU     $09
          ORG     $C000
          LDX     #REGBAS
          LDAA    #$3A
          STAA    DDRD,X          ; configure SPI pin directions
          LDAA    #$50
          STAA    SPCR,X          ; configure SPI parameters
          LDAB    #7
          BCLR    PORTD,X $20     ; enable SPI data transfer
          LDY     #conf_dat
loop1     LDAA    0,Y             ; get a byte of configuration data
          STAA    SPDR,X
          BRCLR   SPSR,X $80 *    ; wait until a byte has been shifted out via SPI
          INY
          DECB
          BNE     loop1
          BSET    PORTD,X $20     ; load data into configuration registers
          BCLR    PORTD,X $20     ; enable SPI transfer to MC14489s
```

```
                    LDY      #disp_dat
                    LDAB     #9                      ; set up transfer byte count
        loop2       LDAA     0,Y
                    STAA     SPDR,X
                    BRCLR    SPSR,X $80 *
                    INY
                    DECB
                    BNE      loop2
                    BSET     PORTD,X $20             ; load data into display data register
                    SWI                              ; do something else
        conf_dat    FCB      01H,00H,00H,01H,00H,00H,45H
        disp_dat    FCB      88H,02H,99H,81H,20H,00H,C2H,55H,FCH
                    END
```

In C language,

```
#include <hc11.h>
main ( )
{
    int i;
    unsigned char conf_dat [] = {0x01, 0x00, 0x00, 0x01, 0x00, 0x00, 0x45};
    unsigned char disp_dat [] = {0x88, 0x02, 0x99, 0x81, 0x20, 0x00, 0xC2, 0x55, 0xFC};

    DDRD = 0x3A;                /* configure SPI pins directions */
    SPCR = 0x50;                /* configuration SPI parameters */
    PORTD &= 0xDF;              /* enable SPI data transfer */
    for (i = 0; i < 7; i++) {
        SPDR = conf_dat[i];
        while (!(SPSR & 0x80));
    }
    PORTD |= 0x20;              /* load data into configuration registers */
    PORTD &= 0xDF;              /* enable SPI data transfer */
    for (i = 0; i < 9; i++) {
        SPDR = disp_dat[i];
        while (!(SPSR & 0x80));
    }
    PORTD |= 0x20;
    return 0;
}
```

▲

10.10   Interfacing LCDs with SPI

Liquid crystal displays (LCDs) are finding broad use in instrument and lap-
top computer designs. LCDs can easily handle the display of numeric vari-

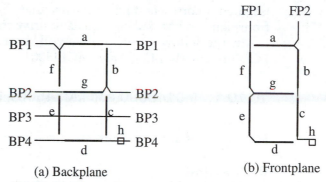

(a) Backplane   (b) Frontplane

Figure 10.22 ■ LCD backplane and frontplane connections

ables, units (for example, MHz and DM), and annunciators (for example, FREQ and AMPTD).

LCDs are often multiplexed to save connection pins. Several multiplex formats have become popular during the last few years. The Motorola LCD drivers that will be discussed here employ 1/4 multiplexing. In a 1/4 multiplexing LCD, each character is represented by a multiple of four segments. Like LED displays, an LCD digit consists of seven segments and an optional decimal point. Each LCD segment is turned on or off by controlling the backplane and frontplane voltages. Each digit is controlled by two frontplane and four backplane signals. These four backplane signals are multiplexed. The frontplane and backplane connections to a seven-segment LCD are shown in Figure 10.22. The segment truth table is given in Table 10.2. Note that there is no standard for backplane and frontplane connections on multiplexed LCD displays. Readers should refer to the manufacturer's literature for actual connections.

LCD drivers are available from many manufacturers. An LCD driver accepts the display pattern that specifies the segments to be turned on

	FP1	FP1
BP1	f	a
BP2	g	b
BP3	e	c
BP4	d	h

*Because there is no standard for backplane and frontplane connections on multiplexed displays, this truth table can be used only for the connection shown in Figure 10.22.

Table 10.2   LCD segment truth table*

(segments corresponding to a particular digit) and generates the appropriate frontplane and backplane signals to drive the LCD.

In the following section we will discuss the Motorola 1/4-multiplexing LCD drivers MC145000 and MC145001.

10.10.1 MC145000 and MC145001 LCD Drivers

Motorola provides two LCD drivers for the multiplexed LCDs. The MC145000 is a master unit, and the MC145001 is a slave unit.

OPERATIONS

The block diagrams of the MC145000 and MC145001 are shown in Figure 10.23 and 10.24. The driver is composed of two independent circuits: the data input circuit with its associated data clock and the LCD driver circuit with its associated system clock.

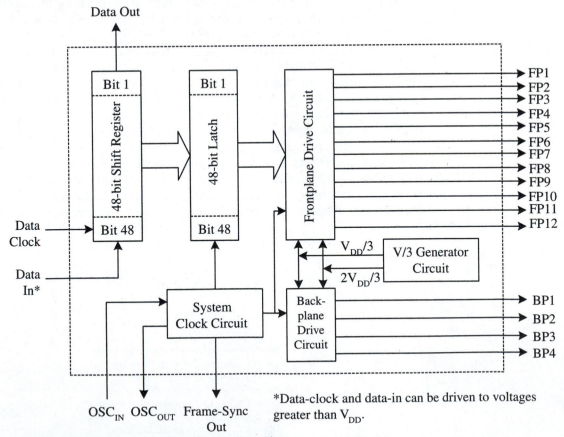

*Data-clock and data-in can be driven to voltages greater than V_{DD}.

Figure 10.23 ■ Block diagram of the MC145000 (master) LCD driver
(Redrawn with permission of Motorola)

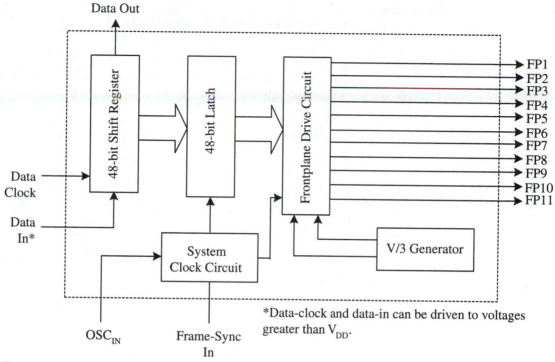

Figure 10.24 ■ Block diagram of the MC145001 (slave) LCD driver
(Redrawn with permission of Motorola)

In the MC145000, 48 bits of data are serially clocked into the shift register on the falling edge of the external data clock. Data in the shift register are latched into the 48-bit latch at the beginning of each frame period. The frame period, t_{frame}, is the time during which all the LCD segments are set to the desired on or off states. The relationship between the frame-sync pulse and the system clock (OSC_{out}) is shown in Figure 10.25. The data in the shift register of the LCD driver is loaded into the latch that drives the frontplane outputs when the frame-sync pulse is high. If new data are shifted in during this period, the display will flicker. To avoid this problem, new data should be shifted in when the frame-sync pulse is low. This can be easily achieved in the 68HC11 by

1. using frame-sync pulse as the $\overline{IRQ}$ input to the 68HC11
2. selecting edge sensitive for the $\overline{IRQ}$ input to the 68HC11
3. updating the display data in the interrupt-handling routine

The binary data present in the latch determine the appropriate waveform signal to be generated by the frontplane drive circuits, whereas the backplane waveforms are invariant. The frontplane and backplane waveforms, FPn and

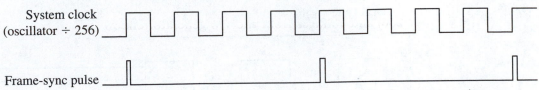

Figure 10.25 ■ The relationship between the system clock and the frame-sync pulse

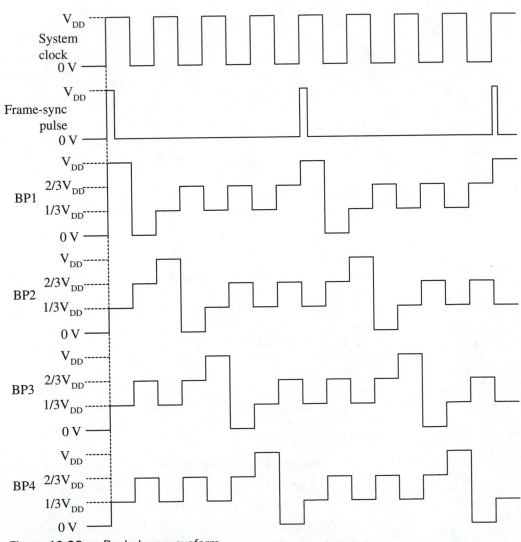

Figure 10.26 ■ Backplane waveform

BPn, are generated using the system clock (the oscillator output divided by 256) and voltages from the V/3 generator circuit (which divides V_{DD} into one-third increments). The backplane signal waveforms are shown in Figure 10.26, where you can see that the waveform of BPn+1 is shifted from the waveform of BPn by one fourth of a clock cycle. Each frontplane signal controls four segments. Because of the 1/4 multiplexing, either one segment or no segment is turned on at any time. The frontplane waveform of an on segment is shown in Figure 10.27a. The voltage across a segment is the voltage difference between the backplane and frontplane signals; an example of an on segment voltage waveform is shown in Figure 10.27b. The frontplane waveform of an

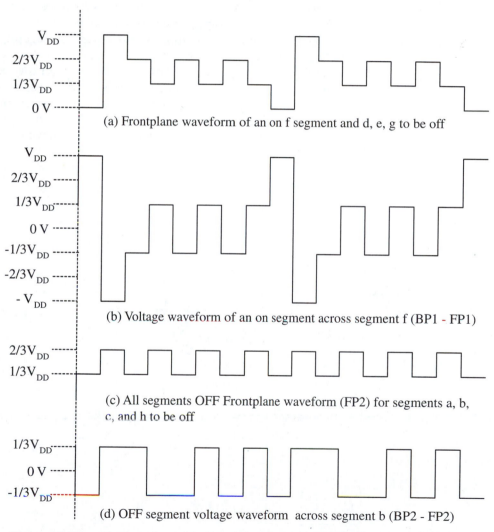

(a) Frontplane waveform of an on f segment and d, e, g to be off

(b) Voltage waveform of an on segment across segment f (BP1 - FP1)

(c) All segments OFF Frontplane waveform (FP2) for segments a, b, c, and h to be off

(d) OFF segment voltage waveform across segment b (BP2 - FP2)

Figure 10.27 ■ Examples of frontplane waveforms and segment voltage waveforms

off segment is shown in Figure 10.27c, and an example of an off segment voltage waveform is shown in Figure 10.27d. Twelve frontplane and four backplane drivers are available from the master unit. Since each digit needs two frontplane signals, a master LCD driver can drive six digits.

The slave unit consists of the same circuitry as the master unit, with two exceptions: it has no backplane driver circuitry, and its shift register and latch hold 44 bits. Eleven frontplane and no backplane drivers are available from the slave unit. One slave unit can drive 5½ digits.

Pin Descriptions

The pin assignments of the MC145000 and MC145001 are shown in Figure 10.28. The functions and connections for most pins are self-explanatory. Only pins OSC_{in} and OSC_{out} need to be explained.

The OSC_{in} signal is the input to the system clock circuit. The oscillator frequency is either obtained from an external oscillator or generated in the master unit by connecting an external resistor between the OSC_{in} and OSC_{out} pins. Figure 10.29 shows the relationship between the resistor value and frequency.

The OSC_{out} signal is the system clock output generated by the master unit. This signal is connected to the OSC_{in} input of each slave unit to synchronize the updating of display data for the master and slave driver units.

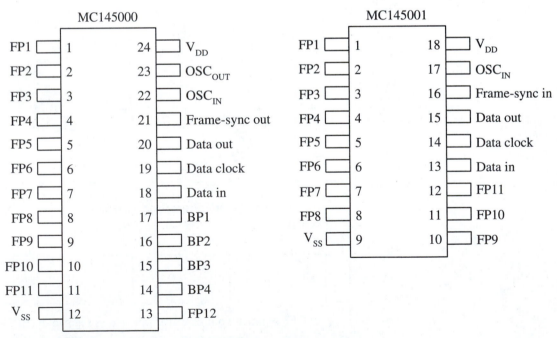

Figure 10.28 ■ MC145000 and MC145001 pin assignments

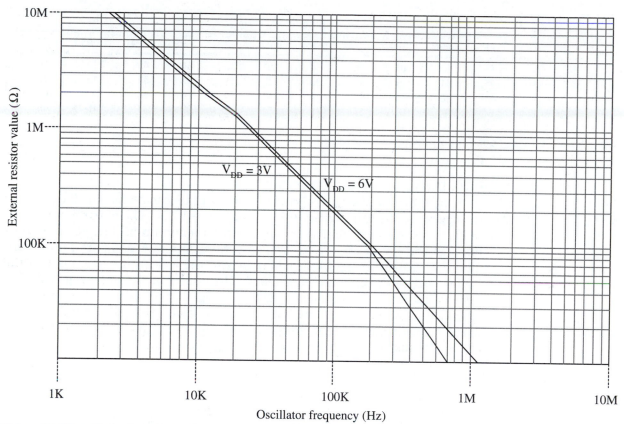

Figure 10.29 ■ Typical oscillator frequency vs. external resistance value

DISPLAY PATTERNS

The data patterns to be displayed must be written into the LCD drivers in the right order. The LCD controller multiplexes four bits to drive the same frontplane output pin. The master unit also activates the corresponding backplane output to turn segments on or off. To turn on a segment, a 1 must be applied to it; to turn off a segment, a 0 must be applied. The bit locations (in the latch) that control the master unit LCD segments located at each frontplane-backplane intersection are shown in Table 10.2. To display a hexadecimal digit, eight bits must be shifted into the LCD driver. Because segment a is at location 8, it must be shifted into the LCD driver first. The order for shifting the segment patterns into the LCD driver is a, b, c, h, f, g, e, and d. For a slave LCD driver, column FP12 should be deleted. By combining Tables 10.2 and 10.3, segment pattern for each BCD digit can be derived; these patterns are shown in Tables 10.4a and 10.4b.

Backplanes		Frontplanes											
		FP1	FP2	FP3	FP4	FP5	FP6	FP7	FP8	FP9	FP10	FP11	FP12
	BP1	4	8	12	16	20	24	28	32	36	40	44	48
	BP2	3	7	11	15	19	23	27	31	35	39	43	47
	BP3	2	6	10	14	18	22	26	30	34	38	42	46
	BP4	1	5	9	13	17	21	25	29	33	37	41	45

Table 10.3   Bit locations for controlling the LCD segments

LCD DRIVER SYSTEM CONFIGURATION

One master and several slave LCD driver units can be cascaded together to display more than six digits, as shown in Figure 10.30. The maximum number of slave units in a system is dictated by the maximum backplane drive capability of the device and by the system data update rate. Data are shifted serially first into the master unit and then into the following slave units on the falling edge of the common data clock. The oscillator is common to the master unit and each of the slave units. At the beginning of each frame period, the master unit generates a frame-sync pulse to ensure that all slave unit frontplane drive circuits are synchronized to the master unit's backplane drive circuits. The master unit generates the backplane signals for all the LCD digits in the system.

Digit	Segment								Hexdecimal representation
	a	b	c	h	f	g	e	d	
0	1	1	1	0	1	0	1	1	$EB
1	0	1	1	0	0	0	0	0	$60
2	1	1	0	0	0	1	1	1	$C7
3	1	1	1	0	0	1	0	1	$E5
4	0	1	1	0	1	1	0	0	$6C
5	1	0	1	0	1	1	0	1	$AD
6	0	0	1	0	1	1	1	1	$2F
7	1	1	1	0	0	0	0	0	$E0
8	1	1	1	0	1	1	1	1	$EF
9	1	1	1	0	1	1	0	0	$EC
A	1	1	1	0	1	1	1	0	$EE
B	0	0	0	0	1	0	1	0	$0A
C	0	1	1	0	1	0	1	0	$6A
D	0	1	1	0	1	0	1	1	$6B
E	0	0	0	0	0	1	0	0	$04
F	0	0	0	0	0	0	0	0	$00

Table 10.4a   BCD digits display patterns
(without decimal point)

Digit	Segment								Hexdecimal representation
	a	b	c	h	f	g	e	d	
0	1	1	1	1	1	0	1	1	$FB
1	0	1	1	1	0	0	0	0	$70
2	1	1	0	1	0	1	1	1	$D7
3	1	1	1	1	0	1	0	1	$F5
4	0	1	1	1	1	1	0	0	$7C
5	1	0	1	1	1	1	0	1	$BD
6	0	0	1	1	1	1	1	1	$3F
7	1	1	1	1	0	0	0	0	$F0
8	1	1	1	1	1	1	1	1	$FF
9	1	1	1	1	1	1	0	0	$FC
A	1	1	1	1	1	1	1	0	$FE
B	0	0	0	1	1	0	1	0	$1A
C	0	1	1	1	1	0	1	0	$7A
D	0	1	1	1	1	0	1	1	$7B
E	0	0	0	1	0	1	0	0	$14
F	0	0	0	1	0	0	0	0	$10

Table 10.4b BCD digits display patterns (with decimal point)

10.10.2 Interfacing the MC145000 and MC145001 to the 68HC11

Many LCDs can be driven by the MC145000 and MC145001 directly. One example is the eight-digit multiplexed LCD display (part #69) from LXD. Its block diagram and pin assignments are shown in Figure 10.31 and Figure 10.32, respectively. This LCD is compatible with Motorola LCD drivers MC145000 and MC145001, and it is a relatively large package in which many pins are not used. Figure 10.32 can be interpreted as follows:

- Backplane BP1 is assigned to pin 60 and 59.
- Backplane BP2 is assigned to pin 32 and 31.
- Backplane BP3 is assigned to pin 1 and 2.
- Backplane BP4 is assigned to pin 29 and 30.
- The characters A_i, B_i, C_i, D_i, E_i, F_i, G_i, and DP stand for the segments and decimal point of the ith digit.
- A pin associated with segment letters is a frontplane pin.

All frontplanes and backplanes have two pins assigned to them, i.e., the user can use either pin (or both pins tied together) to drive a given frontplane or backplane pin.

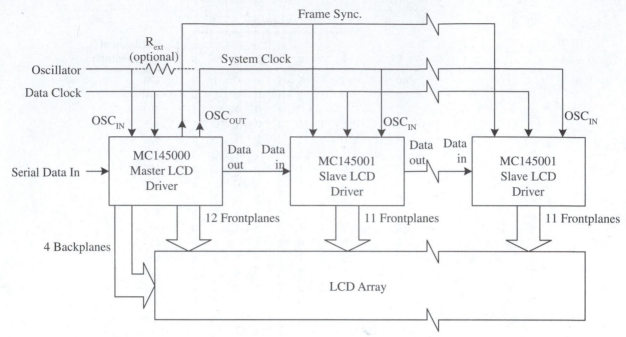

Figure 10.30 ■ Frontplane and backplane connections to a multiplexed-by-four seven-segment (plus decimal point) LCD

(Redrawn with permission of Motorola)

The user can display from one to eight BCD digits with this LCD package. Since the MC145000 can drive only six digits, one slave LCD driver unit will be needed if more than six digits are to be displayed. The circuit connections between the 68HC11 SPI system, the MC145000, and an eight-digit multiplexed LCD display are shown in Figure 10.33. Frame-sync pulse is not used in this example. The $\overline{SS}$ pin should be configured as an output to avoid any

Figure 10.31 ■ Block diagram of the eight-digit multiplexed LCD display from LXD Inc.

PAD	BP1	BP2	BP3	BP4	BP1	BP2	BP3	BP4	PAD
31	--	BP2	--	--	--	--	--	BP4	30
32	--	BP2	--	--	--	--	--	BP4	29
33	--	--	--	--	F1	G1	E1	D1	28
34	A1	B1	C1	DP	--	--	--	--	27
35	--	--	--	--	--	--	--	--	26
36	--	--	--	--	F2	G2	E2	D2	25
37	A2	B2	C2	DP	--	--	--	--	24
38	--	--	--	--	--	--	--	--	23
39	--	--	--	--	--	--	--	--	22
40	--	--	--	--	F3	G3	E3	D3	21
41	A3	B3	C3	DP	--	--	--	--	20
42	--	--	--	--	--	--	--	--	19
43	--	--	--	--	--	--	--	--	18
44	--	--	--	--	F4	G4	E4	D4	17
45	A4	B4	C4	DP	--	--	--	--	16
46	--	--	--	--	--	--	--	--	15
47	--	--	--	--	--	--	--	--	14
48	--	--	--	--	F5	G5	E5	D5	13
49	A5	B5	C5	DP	--	--	--	--	12
50	--	--	--	--	--	--	--	--	11
51	--	--	--	--	F6	G6	E6	D6	10
52	A6	B6	C6	DP	--	--	--	--	9
53	--	--	--	--	--	--	--	--	8
54	--	--	--	--	F7	G7	E7	D7	7
55	A7	B7	C7	DP	--	--	--	--	6
56	--	--	--	--	F8	G8	E8	D8	5
57	A8	B8	C8	DP	--	--	--	--	4
58	--	--	--	--	--	--	--	--	3
59	BP1	--	--	--	--	--	BP3	--	2
60	BP1	--	--	--	--	--	BP3	--	1

Figure 10.32 ■ Pin assignment of the eight-digit multiplexed LCD from LXD Inc.

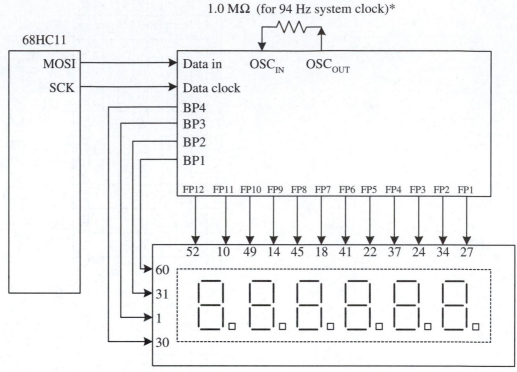

* The oscillator frequency is 256 times the system clock frequency or 24 KHz

Figure 10.33 ■ 68HC11 SPI drives six-digit LCD display

problem. A 1 MΩ resistor is connected between the OSC_{in} and OSC_{out} pins to set the oscillator frequency to about 24 KHz (see Figure 10.29). The system clock is derived by dividing this frequency by 256 and is about 94 Hz. The LCD display is updated at this rate and should be fairly stable.

Example 10.5

Write a small program to display 123456 on the LCD in Figure 10.33.

Solution: The rated data clock frequency is higher than the highest operating frequency of the 68HC11, so we choose the highest SPI frequency—1 MHz. To select this frequency, the SPR1 and SPR0 bits of the SPCR register must be set to 00. The SPI should be configured to operate in the master mode. Because data are clocked into the shift register on the falling edge of the data clock, the CPOL & CPHA bits in the SPCR register should be set to 01 (or 10). Port D pins should be set to normal CMOS output instead of open-drain. An interrupt is not needed for the SPI data transfer because the interrupt overhead is higher than polling for the SPI subsystem at 1Mbits/sec. With these parameters, the value to be written into the SPCR register is $54.

The port D data direction register must also be set properly. Configure the SS, SCK, MOSI, and TxD pins for output. The value $3A should be written into the DDRD register. The segment patterns are placed in a table in the order of the hexadecimal digits so that they can be accessed by using the index-addressing mode. The program is:

```
REGBAS   EQU    $1000           ; base address of I/O register block
DDRD     EQU    $09             ; offset of DDRD from REGBAS
SPCR     EQU    $28
SPSR     EQU    $29
SPDR     EQU    $2A
SPCR_INI EQU    $54             ; value to initialize the SPCR register
DD_INI   EQU    $3A             ; value to set the DDRD register


         ORG    $00
digits   FCB    1,2,3,4,5,6     ; digits to be displayed
lp_cnt   RMB    1               ; loop count


         ORG    $C000           ; starting address of the program
         LDS    #$00FF          ; setup the stack pointer (set to $DFFF for EVB)
         LDX    #REGBAS
         LDAA   #DD_INI
         STAA   DDRD,X          ; set SPI pin directions
         LDAA   #SPCR_INI
         STAA   SPCR,X          ; initialize the SPI system
```

* The following seven instructions send 48 zeros to clear the LCD. This action allows the
* user to tell if the LCD circuit is connected properly.

```
         LDAB   #06             ; send out six zero bytes to
loop     CLRA                   ; initialize LCD driver
         STAA   SPDR,X          ; send out eight zero bits to clear LCD driver
*                               ; registers and latches
here     LDAA   SPSR,X          ; wait until SPI transfer is done
         BPL    here            ;         "
         DECB                   ; decrement the loop count
         BNE    loop
*******************************************************************
         LDY    #digits         ; point Y to the start of digits
         LDAB   #06
         STAB   lp_cnt          ; initialize loop count
loop1    LDAB   0,Y             ; get the digit to be displayed
         PSHY                   ; save the value of Y
```

```
              LDY    #lcdndp          ; put the base address of LCD table in Y
              ABY                     ; index into the LCD digit pattern table
              LDAA   0,Y              ; get the LCD pattern from table
              STAA   SPDR,X           ; send out the digit pattern
       lp2    LDAA   SPSR,X           ; wait until the digit has been shifted out
              BPL    lp2              ;        "
              PULY                    ; restore the digit pointer
              INY                     ; increment digit pointer
              DEC    lp_cnt           ; decrement loop count
              BNE    loop1
              ....                    ; do something else
```

* The following lines contain the table of LCD display patterns for the hex digits

```
lcdndp    FCB    $EB,$60,$C7,$E5,$6C,$AD,$2F,$E0,$EF,$EC
          FCB    $EE,$0A,$6A,$6B,$04,$00
          END
```

In C language,

```c
#include <hc11.h>
main ( )
{
    int i;
    char digits[] = {1,2,3,4,5,6};
    hex_pat [unsigned char] = {0xEB,0x60,0xC7,0xE5,0x6C,0xAD,0x2F,0xE0,0xEF,0xEC,0xEE,
                               0x0A,0x6A,0x6B,0x04,0x00};

    DDRD = 0x3A;                      /* configure SPI pin directions */
    SPCR = 0x54;                      /* configure SPI parameters */
    for (i = 0; i < 6; i++) {         /* blank the LCD displays */
        SPDR = 0;
        while (!(SPSR & 0x80));       /* wait until the SPI transfer is done */
    }
    for (i = 0; i < 6; i++) {
        SPDR = hex_pat [digits[i]];
        while (!(SPSR & 0x80));
    }
    return 0;
}
```

When there are more than six digits to be displayed, the user will need one master and several slave drivers. The master LCD driver supplies frontplane signals for six digits, and slave drivers supply the remaining frontplane signals. The master driver supplies backplane signals.

10.11   The MC68HC68T1 Real Time Clock with Serial Interface

The Motorola MC68HC68T1 is a dedicated clock chip that can keep track of seconds, minutes, hours, day-of-the-week, date, month, and year. It has a battery backup power supply input that can keep the device in operation during a power interruption, and its on-chip 32-byte RAM allows the user to save critical information when the power is interrupted. Its power control function can be used to sense the power transition, perform power-up or power-down, and reset the CPU when necessary. The 68HC68T1 is SPI-compatible and can be interfaced with the 68HC11 easily. The 68HC68T1 has a watchdog circuit similar to the COP system. When enabled, the watchdog circuit requires the microprocessor to toggle the slave-select (SS) pin of the 68HC68T1 periodically without performing a serial transfer. If this condition is not sensed, the CPUR line resets the CPU (or interrupts it, depending on how this signal is connected to the 68HC11).

10.11.1   MC68HC68T1 Signal Pins

The pin layout of the MC68HC68T1 is shown in Figure 10.34. The function of each signal is as follows:

$\overline{\text{CLK OUT}}$:   *Clock output.* This signal is the buffered clock output; it can provide one of the seven selectable frequencies.

$\overline{\text{CPUR}}$:   *CPU Reset (active low output).* This pin provides an open-drain output and requires an external pull-up resistor. This active low output can be used to drive the reset of the microprocessor to allow orderly power-up/power-down. The $\overline{\text{CPUR}}$ signal is low for 15 to -40 ms when the watchdog circuit detects a CPU failure.

$\overline{\text{INT}}$:   *Interrupt (active low output).* This active-low output is driven from a single N-channel transistor and must be tied to an external pull-up resistor. This interrupt is activated to a low level when any one of the following takes places:

1. Power sense operation is selected and a power failure occurs.
2. A previously set alarm time occurs. The alarm bit in the status register and the interrupt signal are delayed 30.5 ms when 32 KHz or 1 MHz operation is selected, 15.3 ms for 2 MHz operation, and 9.6 ms for 4 MHz operation.
3. A previously selected periodic interrupt signal activates.

The status register must be read to disable the interrupt output after the selected periodic interrupt occurs or when conditions 1 or 2 activates the interrupt.

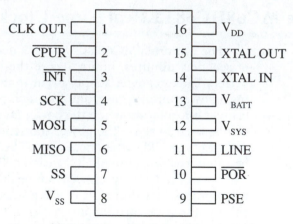

Figure 10.34 ■ The MC68HC68T1

SCK: *Serial clock.* This serial clock input is used to shift data into and out of the on-chip interface logic. SCK retains its previous state if the line driving it goes to a high-impedance state.

MOSI: *Master out slave in.* Data are shifted in from this pin, either on the rising or falling edge, with the most significant bit first. Data present at this pin are latched into the interface logic by SCK if the logic is enabled.

MISO: *Master in slave out.* The serial data present at this port are shifted out of the interface logic by SCK if the logic is enabled.

SS: *Slave select.* When high, the slave select input activates the interface logic, otherwise the logic is in a reset state and the MISO pin is in a high-impedance state. The watchdog circuit is toggled at this pin.

V_{ss}: *Negative power supply.*

PSE: *Power supply enable.* This pin is used to control system power supply and is enabled high under any one of the following conditions:

1. V_{sys} rises above the V_{batt} voltage after V_{sys} is reset low by a system failure.

2. An interrupt occurs (if the V_{sys} pin is powered up 0.7V above V_{batt})

3. A power-on reset occurs (if the V_{sys} pin is powered up 0.7 V above V_{batt}).

$\overline{POR}$: *Power-on reset.* This active low input generates an internal power-on reset using an external RC network.

LINE:	*Line sense.* The LINE sense input can be used to drive one of two functions. The first function utilizes the input signal as the frequency source for the timekeeping counters. The second function enables the LINE input to detect a power failure.
V_{sys}:	*System voltage.* This input is connected to system voltage.
V_{batt}:	Battery voltage.
XTAL IN, XTAL OUT:	*Crystal input/output.*
V_{DD}:	*Positive power supply.*

10.11.2 On-Chip RAM and Registers

The MC68HC68T1 has 32 bytes of on-chip RAM and 13 registers. Its address map is shown in Figure 10.35. When accessing these RAM locations and/or registers, the microprocessor first sends in an 8-bit address using the SPI or port pins and then performs the actual access. Some registers are read/writable. Each read/writable register and each RAM byte has two addresses: one for read access and the other for write access. The separation of read addresses and write addresses makes the read and write accesses possible without an R/W signal.

All time counters and alarm registers (see Table 10.5) are in BCD format. The time-of-day can be displayed in 12-hour or 24-hour format. The selection of the format is automatic: to use 24-hour format, the user enters the hour in the range 00-23; for 12-hour format, an hour in the range of 81 to 92 (for AM) or A1-B2 (for PM) is entered. The contents of the clock-control, interrupt-control, and status registers are described in the following subsections.

CLOCK CONTROL REGISTER

The Clock Control Register has two addresses. The read address is at $31, and the write address is at $B1. The function of each bit is as follows:

Start/stop:	A 1 written into this bit enables the counter stages of the clock circuitry. A 0 holds all bits reset in the divider chain from 32 Hz to 1 Hz. The clock out signal selected by bits D2, D1, and D0 is not affected by the stop function except for the 1 and 2 Hz outputs.
Line/$\overline{Xtal}$:	When this bit is high, clock operation uses a 50 or 60 cycle input present at the Line input pin. When this bit is low, the XTAL IN pin is the source of the time update.

XTAL select: Four possible crystal frequencies can be selected by the value in bits D5 and D4:

0 = 4.194304 MHz
1 = 2.097152 MHz
2 = 1.048576 MHz
3 = 32.768 KHz

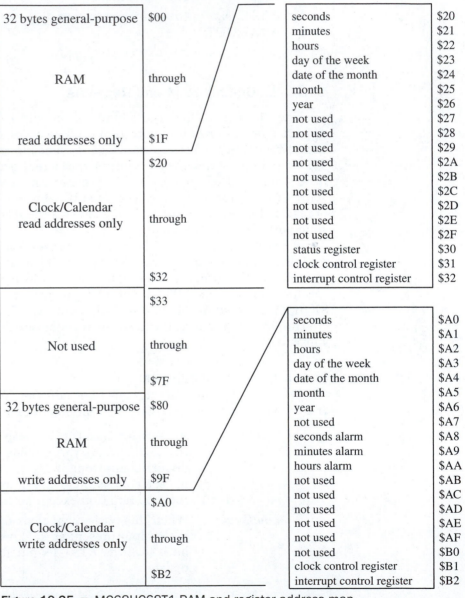

Figure 10.35 ■ MC68HC68T1 RAM and register address map
(Redrawn with permission of Motorola)

| Address Location | | | | | |
Read	Write	Function	Decimal range	BCD data range	BCD data example
$20	$A0	seconds	0-59	00-59	21
$21	$A1	minutes	0-59	00-59	40
$22	$A2	hours (12-hour mode)	1-12	81-92 (AM) A1-B2 (PM)	90
		hours (24-hour mode)	0-23	00-23	10
$23	$A3	day of week (Sunday = 1)	1-7	01-07	03
$24	$A4	date of month	1-31	01-31	16
$25	$A5	month (January = 1)	1-12	01-12	06
$26	$A6	year	0-99	00-99	87
N/A	$A8	seconds alarm	0-59	00-59	21
N/A	$A9	minutes alarm	0-59	00-59	40
N/A	$AA	hours alarm (12-hour mode)	1-12	01-12 (AM) 21-32 (PM)	10
		hours alarm (24-hour mode)	0-23	00-23	10

Table 10.5 ■ Clock/calendar and alarm data modes
(Redrawn with permission of Motorola)

At power up, the device sets up for a 4 MHz oscillator. The MC68HC68T1 has an on-chip 150K resistor that is switched in series with the internal inverter when 32 KHz is selected through the clock control register. The recommended crystal connection is shown in Figure 10.36. Resistor R1 should be 22 MΩ for 32 KHz operation. Resistor R2 is used for 32 KHz operation only. Use a 100 K to 300 K range as specified by the crystal manufacturer.

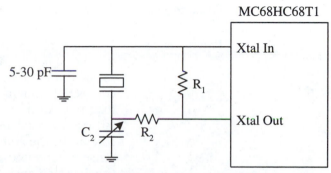

The values of C1 and C2 depend on the crystal frequency.

Figure 10.36 ■ Recommended oscillator circuit

Periodic select	Clock output frequency
0	Xtal
1	Xtal/2
2	Xtal/4
3	Xtal/8
4	disable
5	1 Hz
6	2 Hz
7	50/60 hz for line operation
	64 Hz for Xtal operation

Table 10.6 ■ Clock output frequencies

50/60 Hz.   When this bit is 1, 50 Hz is the frequency at the LINE input. Otherwise, 60 Hz is the LINE frequency.

CLK OUT 2-   These three bits specify one of the seven frequencies
CLK OUT 0.   to be used as the square-wave output at the CLK OUT pin (see Table 10.6).

INTERRUPT CONTROL REGISTER

This is a read/writable register. Its read address is $32, and its write address is $B2. The function of each bit is as follows:

D7	D6	D5	D4	D3	D2	D1	D0
Watch dog	Power down	Power sense	Alarm		Periodic select		

Watchdog.   When this bit is set to 1, the watchdog function is enabled.

Power down.   A power down operation is initiated when this bit is set to 1.

Power sense.   When set to 1, this bit enables the LINE input pin to sense a power failure. When power sense is selected, the input to the 50/60-Hz prescaler is disconnected, and crystal operation is therefore required. An interrupt is generated when a power failure is sensed and the power sense and interrupt true bit in the status register are set. When power sense is activated, a logic low must be written to this location and followed by a logic high to re-enable power sense.

D3-D0 value (hex)	Periodic interrupt output frequency	Frequency timebase	
		Xtal	Line
0	disable		
1	2048 Hz	x	
2	1024 Hz	x	
3	512 Hz	x	
4	256 Hz	x	
5	128 Hz	x	
6	64 Hz	x	
	50 or 60 Hz		x
7	32 Hz	x	
8	16 Hz	x	
9	8 Hz	x	
A	4 Hz	x	
B	2 Hz	x	x
C	1 Hz	x	x
D	1 cycle per minute	x	x
E	1 cycle per hour	x	x
F	1 cycle per day	x	x

Table 10.7 ■ Periodic interrupt output frequencies (at $\overline{INT}$ pin)

Alarm. The output of the alarm comparator is enabled when this bit is set to 1. When a comparison of the seconds, minutes, and hours time counters and the alarm latches finds that they are equal, the interrupt output is activated.

Periodic select. The values in these four bits (D3, D2, D1, and D0) select the frequency of the periodic output. Table 10.7 lists all available frequencies.

STATUS REGISTER

This read-only register is located at $30. The meaning of each bit in this register is as follows:

D7	D6	D5	D4	D3	D2	D1	D0
0	Watch dog	0	First time up	Inter-rupt true	Power sense INT	Alarm INT	Clock INT

Watchdog.	If this bit is high, the watchdog circuit has detected a CPU failure.
First time up.	This bit is set to high by a power-on reset to indicate that the data in the RAM and Clock registers are not valid and should be initialized. After the status register is read, this bit is cleared if the POR pin is high. Otherwise, it remains at high.
Interrupt true.	A high in this bit signifies that one of the three interrupts (power sense, alarm, or clock) is valid.
Power-sense interrupt.	This bit is set to high when the power sense circuit generates an interrupt. This bit is not cleared after a read of this register.
Alarm interrupt.	When the contents of the seconds, minutes, and hours time counters and alarm latches are equal, this bit is set to high. The status register must be read before loading the interrupt control register for a valid alarm indication when the alarm activates.
Clock interrupt.	A periodic interrupt sets this bit to high.

All bits except the first-time-up bit (which is set high) are reset low by a power-on reset. All bits except the power-sense bits are reset after a read of the status register.

10.11.3   The MC68HC68T1 Operation and Interfacing the 68HC68T1 to the 68HC11

The operations of the MC68HC68T1 chip are explained in the following subsections.

REGISTERS AND RAM ADDRESSING

The MC68HC68T1 has no address input. To select a particular RAM location or register, the microcontroller sends the corresponding address to the MC68HC68T1 via the SPI interface. The address and data bytes are shifted most significant bit first, into the serial data input (MOSI) and out of the serial data output (MISO). Any transfer of data requires the address of the byte, which specifies a write or read Clock/Calendar or RAM location; the address is followed by one or more bytes of data. Data transfer can occur one byte at a time or in multibyte burst mode. Before sending the address, the SS pin should be set to high. The high level at SS pin enables the MC68HC68T1. The first byte sent to the MC68HC68T1 is used to select either a location in RAM or a register. After that, the microcontroller can read or write single or multiple bytes from or into the MC68HC68T1. Each read- or write-cycle causes the Clock/Calendar register or RAM address to automatically increment by 1. *Incrementing continues after each byte transfer until the SS pin is set to low, which disables the serial interface logic.* If the RAM is selected, the address increments to $1F or $9F and then wraps to $00 and continues. When the Clock/Calendar is selected, the address wraps to $20 after incrementing to $32 or $B2.

Interfacing the 68HC68T1 to the 68HC11 is fairly straightforward. Figure 10.37 shows the circuit connections for using the clock function of the MC68HC68T1. A 32.760-KHz crystal oscillator supplies the clock signal to the 68HC68T1. An alarm transducer (a speaker) is added to inform the user when the preset alarm time is reached. The time-of-day and alarm time can be entered into the 68HC68T1 by using the keypad or switches. The alarm transducer is connected to the CLK OUT pin and will be driven by a square wave with a frequency of about 4KHz. A battery consisting of three rechargeable NiCd cells is connected to the V_{batt} pin so that the time of day information can be maintained even without the line power. The battery is charged through the resistor R_{charge} when line power is present.

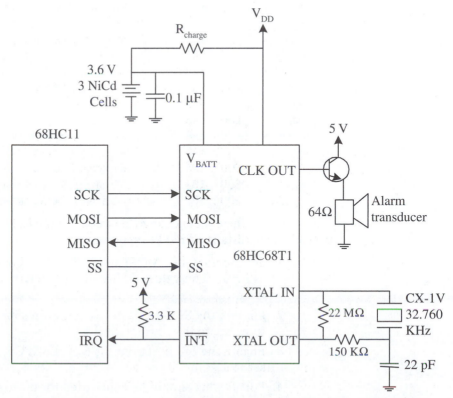

Figure 10.37 ■ Circuit connections between a 68HC11 and a 68HC68T1 for the clock function

The following constant definitions will be used in all the programming examples related to the 68HC68T1:

```
REGBAS     EQU    $1000        ; I/O register block base address
SPDR       EQU    $2A          ; offset of SPDR from REGBAS
SPCR       EQU    $28          ; offset of SPCR from REGBAS
SPSR       EQU    $29          ; offset of SPSR from REGBAS
DDRD       EQU    $09          ; offset of DDRD from REGBAS
PORTD      EQU    $08          ; offset of PORTD from REGBAS
ram_addr   EQU    $80          ; starting address to write the RAM
second_r   EQU    $20          ; read address for seconds
minute_r   EQU    $21          ; read address for minutes
hour_r     EQU    $22          ; read address for hours
day_wk_r   EQU    $23          ; read address for day-of-week
date_m_r   EQU    $24          ; read address for date-of-month
month_r    EQU    $25          ; read address for month
year_r     EQU    $26          ; read address for year
stat_reg   EQU    $30          ; address of status register
clk_ctlr   EQU    $31          ; read address for clock control register
int_ctlr   EQU    $32          ; read address for interrupt control register
second_w   EQU    $A0          ; write address for seconds
minute_w   EQU    $A1          ; write address for minutes
hour_w     EQU    $A2          ; write address for hours
day_wk_w   EQU    $A3          ; write address for day-of-week
date_m_w   EQU    $A4          ; write address for date-of-month
month_w    EQU    $A5          ; write address for month
year_w     EQU    $A6          ; write address for year
s_alarm    EQU    $A8          ; write address for seconds alarm
m_alarm    EQU    $A9          ; write address for minutes alarm
h_alarm    EQU    $AA          ; write address for hours alarm
clk_ctlw   EQU    $B1          ; write address for clock control register
int_ctlw   EQU    $B2          ; write address for clock interrupt control register
```

To enable the data exchange between the 68HC11 and the 68HC68T1, the user need to initialize the SPI system as follows:

1. Set the $\overline{SS}$, CLK, MOSI and TxD pins as output and the MISO pin as input. The value of $3A should be written into the DDRD register.

2. Disable the SPI interrupt—this requires bit 7 of the SPCR register to be set to 0.

3. Enable the SPI function. Bit 6 of the SPCR register must be set to 1.

4. Port D pins should be configured for normal CMOS pins—this requires bit 5 to be cleared to 0.

5. The clock (SCK) edge for shifting data in and out of the 68HC68T1 is unimportant. Let's choose the rising edge: set bits CPOL (bit 3) and CPHA (bit 2) of the SPCR register to 00.

6. Choose the highest usable data transfer rate because the 68HC68T1 can shift data in and out at a frequency higher than 2.1 MHz with a power supply of 5 V. To choose a 1 MHz frequency for the SCK clock, set bits SPR1 (bit 1) and SPR0 (bit 0) of the SPCR register to 00.

The following instruction sequence will set up these parameters:

```
LDX     #REGBAS      ; place register block base address into X
LDAA    #$3A         ; set up SPI pin directions
STAA    DDRD,X       ;      "
LDAA    #$50         ; initialize the SPI system
STAA    SPCR,X       ;      "
```

The time-of-day, day-of-the-week, date-of-month, year, seconds alarm, minutes alarm, and hours alarm can be written into the 68HC68T1 using the multibyte burst mode. To use the burst mode transfer, the SS input to the 68HC68T1 must be set to high during the whole transfer process. At the end of transfer, the SS signal should be returned to low.

Assume that the 24-hour mode is used in the 68HC68T1 and that *clk_mgt* contains the starting address of a memory block that stores seconds, minutes, hours, day-of-the-week, date-of-month, month, year, seconds alarm, minutes alarm, and hours alarm.

There are ten bytes in this block. Each piece of information occupies one byte and is stored in BCD format. The current time and the alarm can be entered interactively if the 68HC11 is connected to a PC using the SCI subsystem. The time-of-day and alarm time information can be defined by the following assembler directive:

```
clk_mgt    rmb    10        ; block of memory locations to store time and alarm info
```

This block of information can be written into the 68HC68T1 using the SPI function. Because the current times and alarm times are not in sequential addresses in the 68HC68T1, they must be sent to the 68HC68T1 in two bursts. The seconds write address and the current time are sent in the first burst, and the seconds alarm address and the alarm information are sent in the second burst, as shown in the following:

```
           ORG     $00
clk_mgt    rmb     10
           ORG     $C000
           BSET    PORTD,X $20    ; enable SPI transfer to the 68HC68T1
           LDAA    #second_w      ; load the second write address
           STAA    SPDR,X         ; send the second write address to the 68HC68T1
```

```
loop      LDAA    SPSR,X              ; wait until SPI transfer is done
          BPL     loop                ;         "
          LDAB    #7                  ; set the byte count to 7
          LDY     #clk_mgt            ; load the base address of the current time into Y
lp        LDAA    0,Y                 ; send one byte of time information to the 68HC68T1
          STAA    SPDR,X              ;         "
again     LDAA    SPSR,X              ; wait until SPI transfer complete
          BPL     again               ;         "
          INY                         ; move the time pointer
          DECB
          BNE     lp
          BCLR    PORTD,X $20         ; disable transfer to 68HC68T1
```

* The following instruction sequence sends out the alarm information into the 68HC68

```
          BSET    PORTD,X $20         ; set the SS pin to high to enable SPI transfer to
*                                     ; 68HC68T1
          LDAA    #s_alarm            ; load the write address of the seconds alarm
          STAA    SPDR,X              ; send the write address of the seconds alarm to the
*                                     ; 68HC68T1
loop1     LDAA    SPSR,X              ; wait until SPI transfer complete
          BPL     loop1               ;         "
          LDAB    #3                  ; set the byte count to 3
          LDY     #clk_mgt+7          ; load the base address of alarm into Y
lp1       LDAA    0,Y                 ; send one byte of alarm information to the 68HC68T1
          STAA    SPDR,X              ;         "
wait      LDAA    SPSR,X              ; wait until SPI transfer is complete
          BPL     wait                ;         "
          INY                         ; move the time pointer
          DECB
          BNE     lp1
          BCLR    PORTD,X $20         ; disable SPI transfer to the 68HC68T1
```

To check the status register, use the following instruction sequence:

```
          BSET    PORTD,X $20         ; enable SPI transfer to the 68HC68T1
          LDX     #REGBAS
          LDAA    #stat_reg
          STAA    SPDR,X              ; send status register address
lp2       LDAA    SPSR,X              ; wait until the address byte is shifted out
          BPL     lp2                 ;         "
          STAA    SPDR,X              ; start an SPI transfer
lp3       LDAA    SPSR,X              ; wait until the status register is shifted in
          BPL     lp3                 ;         "
          LDAA    SPDR,X              ; place the contents of the status register in A
          BCLR    PORTD,X $20         ; disable SPI transfer from the 68HC68T1
```

CLOCK/CALENDAR

The clock/calendar portion of this device consists of a long string of counters that are toggled by a 1 Hz input. The 1 Hz input can be derived from three different sources:

1. an external crystal oscillator applied between pins XTAL IN and XTAL OUT.
2. an external frequency source applied to XTAL IN.
3. a 50 or 60 Hz source that is connected to the LINE input.

The example in Figure 10.37 uses an external 32.760KHz crystal oscillator as the clock source. The time counters offer seconds, minutes, and hours data in 12- or 24-hour format. The values written into the time counters select the display format. An AM/PM indicator is available; once set, it toggles at 12:00 AM and 12:00 PM. The calendar counters contain day-of-the-week, date-of-month, month, and year information. The data in these counters are in BCD format. The hours counter uses BCD format for hours data plus bits for 12/24 hour and AM/PM modes. The seven time counters are readable at addresses $20 through $26 and can be written into at addresses $A0 through $A6.

ALARM

The alarm latches consist of seconds, minutes, and hours registers. When their outputs equal the values of the seconds, minutes, and hours time counters, an interrupt is generated. The interrupt output (INT) goes low if the alarm bit in the status register is set and the interrupt output is activated. To preclude a false interrupt when loading the time counters, the alarm interrupt bit in the control register should be reset. This procedure is not needed when loading the alarm time.

The alarm function can be used to drive an alarm transducer to remind the user when the preset alarm time is reached. One method for doing this is as follows:

1. Use the CLK OUT pin to drive an alarm transducer (by means of a transistor).
2. Disable the CLK OUT (set to low) if the alarm time has not been reached.
3. Enable the alarm interrupt.
4. When the alarm time is reached, the alarm interrupt-handling routine enables a pulse output from the CLK OUT pin for some specific amount of time (say, 3 minutes).

In order to turn the alarm on for a specific amount of time, the user needs to enable periodic interrupt of the 68HC68T1. The procedure is as follows:

1. Enable the 68HC68T1 to interrupt periodically (say, every second).

2. After setting up the alarm time, initialize an alarm counter. If the alarm is set to 3 minutes for periodic interrupt every second, then the alarm count is 180 (3×60).

3. Stay in a wait loop while checking whether the alarm count has decremented to 0. If yes, turn off the alarm by setting the CLK OUT signal to low.

4. The interrupt-handling routine performs the following operations:

    a. Checks the interrupt source by reading the status register.
    b. Returns if the interrupt is caused by the periodic interrupt and alarm is not turned on.
    c. Turns on the alarm if the interrupt is caused by the alarm function (i.e., alarm time has been reached).
    d. Decrements the alarm count by 1 and returns from interrupt if the interrupt is caused by the periodic interrupt and the alarm has been turned on.

Both the clock-control and interrupt-control registers must be set accordingly. The procedure for setting the interrupt-control register is as follows:

1. Disable the watchdog interrupt (set bit 7 to 0)

2. Disable the power-down (set bit 6 to 0)

3. Disable the power sense (set bit 5 to 0)

4. Enable the alarm interrupt (set bit 4 to 1)

5. Select a periodic interrupt frequency of 1 Hz (set bits 3-0 to 1100)

Thus the value %00011100 should be written into the interrupt-control register. The procedure for setting the clock-control register is as follows:

1. Set the start bit to start the clock (set bit 7 to 1)

2. Choose XTAL IN input to update the time (set bit 6 to 0)

3. Set the XTAL IN frequency to 32.760 KHz (set bit 5-4 to 11)

4. Bit 3 doesn't matter when XTAL IN is chosen to update the time (set it to 0 arbitrarily)

5. Disable CLK OUT output (set bits 2-0 to 100_2); to set the CLK OUT frequency to 32.760/8 ≈ 4KHz, set these three bits to 011.

Thus the value %10110100 should be written into the clock-control register to turn off the CLK OUT output. The value of %10110011 would be written into the clock-control register to set the alarm device with a frequency of 4 KHz. The following instruction sequence can set up the 68HC68T1 as desired:

```
LDX     #REGBAS
BSET    PORTD,X $20        ; enable SPI transfer to 68HC68T1
LDAA    #clk_ctlw
STAA    SPDR,X             ; send the write address of clock control register here
```

```
        BRCLR   SPSR,X $80 *          ; wait until the address has been shifted into the
*                                     ; 68HC68T1

* The interrupt control and the clock control registers are located at consecutive addresses
* and can be sent out in burst mode

        LDAA    #%10110100            ; send out control byte to clock control register
        STAA    SPDR,X
first_bt LDAA   SPSR,X                ; wait until the byte is shifted out
        BPL     first_bt              ;          "

        LDAA    #%00011100            ; send out the control byte to interrupt control register
        STAA    SPDR,X                ;          "
sec_bt  LDAA    SPSR,X                ; wait until the byte is shifted out
        BPL     sec_bt                ;          "
        BCLR    PORT,X $20            ; disable the SPI transfer
```

After initializing the 68HC68T1, the alarm count must be set up and updated to control the alarm interval. The 68HC68T1 will interrupt the 68HC11 every second once the alarm time is reached, so an interrupt-handling routine must be written to handle the interrupts. The following instruction sequence will set up and update the alarm count and clear the alarm flag when the alarm count is decremented to zero:

```
threemin EQU    180                  ; alarm count for three minutes
alarmcnt RMB    1                    ; memory location to store alarm count
alarmflg RMB    1                    ; alarm flag indicating if the alarm is on

set_alct LDAA   #threemin
         STAA   alarmcnt             ; initialize alarm count to 180
         CLR    alarmflg             ; initialize the alarm flag to 0
         .
         .
         .
```

With these settings, note that the periodic interrupt and the alarm interrupt may occur at the same time and that both are serviced by the same handling routine. The interrupt (IRQ) handling routine must therefore check the status register of the 68HC68T1 in order to identify the cause of the interrupt. The alarm interrupt should be checked before the periodic interrupt. The interrupt status bit will be cleared when the status register is read. The interrupt-handling routine will perform the following functions:

1. Read the clock status register to identify the cause of interrupt.
2. If the interrupt is caused by the alarm function, then turn on the alarm transducer and set the alarm flag to indicate that the alarm has been turned on.

3. If the interrupt is caused by periodic interrupt (every second) and the alarm is not turned on, then exit.

4. If the interrupt is caused by the periodic interrupt and the alarm flag is set, then decrement the alarm count by 1. If the alarm count is decremented to 0 at this step, then clear the alarm flag, reinitialize the alarm count, and disable the CLK OUT signal. Return to the interrupted program.

Since the CLK OUT pin drives the alarm device, a proper frequency for the CLK OUT must be selected. To be an alarm, the signal frequency must be slightly higher but still audible. We will choose 4 KHz as the frequency of the CLK OUT signal. The value %011 should be set to the lowest three bits of the clock-control register.

The clock interrupt handling routine is:

```
clk_ISR   LDX     #REGBAS
          BSET    PORTD,X $20      ; enable SPI transfer to the 68HC68T1
          LDAA    #stat_reg        ; send the status register address to 68HC68T1
          STAA    SPDR,X           ;           "
          BRCLR   SPSR,X $80 *     ; wait until the status register address is sent out
          STAA    SPDR,X           ; start SPI transfer to read the status register
          BRCLR   SPSR,X $80 *     ; wait until the status register is shifted in
          BCLR    PORTD,X $20      ; disable SPI transfer to the 68HC68T1
          BSET    PORTD,X $20      ; enable SPI transfer to the 68HC68T1 so that
     *                             ; new address can be sent to the 68HC68T1
          LDAA    SPDR,X           ; load the status register value into accumulator A
          ANDA    #$02             ; check the alarm interrupt flag
          BEQ     chkalarm         ; if interrupt is periodic, go and check the alarm flag

          LDAA    #clk_ctlw        ; sent out the clock-control register write address
          STAA    SPDR,X           ;           "
          BRCLR   SPSR,X $80 *     ; wait until the address is shifted out

          LDAA    #%10110011       ; send out a new control byte to 68HC68T1 to
          STAA    SPDR,X           ; enable the CLK OUT signal to turn on the alarm
     *                             ; and set the frequency of CLK OUT to 4 KHz
          BRCLR   SPSR,X $80 *     ; wait until the control byte is shifted out
          LDAA    #1
          STAA    alarmflg         ; set the alarm flag
          BRA     exit             ; return from interrupt

chkalarm  LDAA    alarmflg         ; check the alarm flag
          BEQ     exit             ; if alarm is not turned on then return
          DEC     alarmcnt         ; decrement the alarm count if the alarm has been
     *                             ; turned on
          BNE     exit             ; if the alarm count is not zero, return
```

* If the alarm count has been decremented to 0, then we need to disable CLK OUT output,
* reinitialize the alarm count, and clear the alarm flag

```
          LDAA      #threemin
          STAA      alarmcnt          ; initialize the alarm count to 180
          CLR       alarmflg          ; initialize the alarm flag to 0
```

* To disable CLK OUT, write the value %10110100 into the clock control register

```
          LDAA      #clk_ctlw         ; send write address of clock control to 68HC68T1
          STAA      SPDR,X            ; send out clock control register write address
          BRCLR     SPSR,X $80 *      ; wait until the address is shifted out
          LDAA      #%10110100        ; send a value to clock control register to disable
   *                                  ; CLK OUT
          STAA      SPDR,X            ; output from the 68HC68T1 to stop the alarm device
          BRCLR     SPSR,X $80 *      ; wait until the new control byte is shifted out

exit      BCLR      PORTD,X $20       ; disable SPI transfer
          RTI
```

The user can also add LCD drivers and display the time-of-day to make this clock function more useful.

Example 10.6

▼ ━━━

For the circuit in Figure 10.37, write a C function to send the current time-of-day and alarm time information to the 68HC68T1. This function must also enable the 68HC68T1 to generate periodic and alarm interrupt to the 68HC11.

Solution: Before we initialize the 68HC68T1 chip, we must initialize the SPI pin directions and SPI operation parameters. The caller of this function will pass the SPI pin directions, SPI parameters, current time-of-day, alarm times, and clock control and interrupt control information to this function. The function is as follows:

```c
#define     second_w    0xA0    /* seconds write address */
#define     s_alarm     0xA8    /* write address for seconds alarm */
#define     clk_ctlw    0xB1    /* clock-control register write address */
#include    <hc11.h>
init_68HC68T1 (char spi_dir, char spr_ctrl, char t_of_d [], char alarm [], char ccon [])
{
    int i;
    DDRD = spi_dir;                /* initialize SPI pin directions */
    SPCR = spr_ctrl;               /* initialize SPI parameters */
    PORTD |= 0x20;                 /* enable SPI transfer to the 68HC68T1 */
```

```
        SPDR = second_w;                /* send the write address for seconds */
        while (!(SPSR & 0x80));         /* wait until the SPI transfer is done */
        for (i = 0; i < 7; i++) {
            SPDR = t_of_d [i];          /* send the current time to 68HC68T1 */
            while (!(SPSR & 0x80));     /* wait until the SPI transfer is done */
        }
        PORTD &= 0xDF;                  /* disable SPI transfer so that new address can be sent to
                                           68HC68T1 */
        PORTD |= 0x20;                  /* enable SPI transfer to the 68HC68T1 */
        SPDR = s_alarm;                 /* send the second alarm address to the 68HC68T1 */
        while (!(SPSR & 0x80));         /* wait until the SPI transfer is done */
        for (i = 0; i < 3; i++) {
            SPDR = alarm [i];           /* send the new alarm time to the 68HC68T1 */
            while (!(SPSR & 0x80));     /* wait until the SPI transfer is done */
        }
        PORTD &= 0xDF;                  /* disable SPI transfer to the 68HC68T1 */
        PORTD |= 0x20;                  /* enable SPI transfer to the 68HC68T1 */
        SPDR = clk_ctlw;                /* send the write address of the clock-control register */
        while (!(SPSR & 0x80));         /* wait until the SPI transfer is done */
        for (i = 0; i < 2; i++) {
            SPDR = ccon[i];             /* send out the clock and interrupt control info */
            while (!(SPSR & 0x80));
        }
        PORTD &= 0xDF;                  /* disable SPI transfer to the 68HC68T1 */
}
```

▲

Example 10.7

▼

Write a main program and an interrupt service routine for the 68HC68T1. The main program will set up the $\overline{\text{IRQ}}$ interrupt vector, call the function in Example 10.6 to initialize the 68HC68T1, set up alarm count, clear alarm flag, and enable the interrupt. The $\overline{\text{IRQ}}$ interrupt service routine will perform the operations described earlier. Assume this program is to be run on the EVB, the EVBU, or the CMD-11A8 computer.

Solution: The main program and the interrupt service routine are as follows:

```
#include    <hc11.h>
#define     clk_ctlw        0xB1
#define     stat_reg        0x30    /* read address of the status register */
unsigned char alarm_cnt, alarm_flg;
unsigned char t_of_d [7], alarm [3], ccon [2];
void IRQ_ISR ( );
main ( )
```

```
    {
        alarm_cnt = 180;                    /* initialize the alarm count */
        alarm_flg = 0;                      /* indicate that the alarm device is not turned on yet */
        .
                                            /* read the current time-of-day and alarm time here */
        .
        ccon [0] = 0xB4;                    /* clock control byte */
        ccon [1] = 0x1C;                    /* interrupt control byte */
        init_68HC68T1 (0x3A, 0x50, t_of_d, alarm, ccon);

        *(unsigned char *)0xee = 0x7E;      /* set up IRQ interrupt jump vector */
        *(void (**)())0xef = IRQ_ISR;

        INTR_ON ( );                        /* enable interrupt */
        .
                                            /* do something else */
        .
        return 0;
    }

    #pragma interrupt_handler IRQ_ISR ( );
    void IRQ_ISR ( )
    {
    /* read the status register to identify the cause of interrupt */
        PORTD |= 0x20;                      /* enable SPI transfer to the 68HC68T1 */
        SPDR = stat_reg;                    /* send the status register read address to the
                                               68HC68T1 */

        while (!(SPSR & 0x80));
        SPDR = 0;                           /* start another SPI transfer to read the status register */
        while (!(SPSR & 0x80));             /* wait until the status register has been shifted in */
        PORTD &= 0xDF;                      /* disable SPI transfer */
        if (SPDR & 0x02) {                  /* interrupt is caused by alarm time */
            PORTD |= 0x20;
            SPDR = clk_ctlw;                /* send out the clock-control register write address */
            while (!(SPSR & 0x80));
            SPDR = 0xB3;                    /* turn on the alarm device */
            while (!(SPSR & 0x80));
            alarm_flg = 1;                  /* set this flag to indicate that alarm is turned on */
            PORTD &= 0xDF;                  /* disable SPI transfer to the 68HC68T1 */
            return;
        }
        if (!alarm_flg)
            return;                         /* interrupt is caused by periodic interrupt and alarm
                                               is not on */
    -- alarm_cnt;
```

```
        if (alarm_cnt)
            return;
        else {
            alarm_cnt = 180;
            alarm_flg = 0;
            PORTD |= 0x20;                    /* enable SPI transfer to the 68HC68T1 */
            SPDR = clk_ctlw;                  /* send clock-control register write address */
            while (!(SPSR & 0x80));
            SPDR = 0xB4;                      /* turn off alarm */
            while (!(SPSR & 0x80));
            PORTD &= 0xDF;                    /* disable SPI transfer to the 68HC68T1 */
            return;
        }
    }
```

▲

10.11.4   Power Control

Power control consists of two operations: power sense and power down/power up. Two pins are involved in power sensing, the LINE input pin and the INT output pin. Two additional pins, PSE and V_{sys}, are utilized during a power-down/power-up operation.

POWER SENSING

The power sense function is used mainly to detect power transitions. When power sensing is enabled, AC/DC transitions are sensed at the LINE input pin. Threshold detectors determine when transitions cease. After a delay of 2.68 to 4.64 ms plus the external input RC circuit time constant, an interrupt true bit in the status register is set high. This bit can be sampled to see whether system power has turned back on.

The power sense circuitry operates by sensing the level of the voltage present at the LINE input pin. This voltage is centered around V_{DD}, and as long as the voltage is either plus or minus a certain threshold (approximately 0.7 V) from V_{DD}, a power sense failure is not indicated. With an ac signal, remaining in this V_{DD} window longer than a maximum of 4.64 ms activates the power sense circuit. The larger the amplitude of the signal, the less likely a power failure is to be detected. A 50 or 60 Hz, 10 V peak-to-peak sine wave voltage is an acceptable signal to present at the LINE input pin to set up the power sense function.

POWER-DOWN

Power-down is an operation initiated by the processor. The microcontroller sets the power-down bit in the interrupt-control register to initiate the power-down operation. During power-down, the PSE, CLK OUT, and the CPUR output are placed low. In addition, the serial interface (MOSI and MISO) is disabled.

In general, a power-down procedure consists of the following steps:

1. Set power sense operation by setting bit 5 of the interrupt-control register to 1.
2. When an interrupt occurs, the CPU reads the status register to determine the interrupt source.
3. When a power failure is sensed, the CPU does the necessary housekeeping to prepare for shutdown.
4. The CPU reads the status register again after several milliseconds to determine the validity of the power failure.
5. The CPU sets power-down and disables all interrupts in the interrupt-control register when power-down is verified. This causes the CPU reset (CPUR pin) and Clock Out pins to be held low and disconnects the serial interface.
6. When power returns and V_{sys} rises above $V_{batt} + 0.7$ V, power-up is initiated. The CPU reset is released, and serial communication is established.

POWER-UP

At the end of a power-on reset, the POR signal will go high. If the V_{sys} input also goes high, then the 68HC68T1 initiates the power-up operation by placing PSE, CLK OUT, and CPUR to high, so that the CPU can start to boot.

10.11.5 Watchdog Function

When the *Watchdog* bit in the interrupt-control register is set to 1, the SS pin must be toggled at regular intervals without a serial transfer. If the SS pin is not toggled, the MC68HC68T1 supplies a CPU reset pulse at the CPUR pin and the watchdog bit in the status register is set to 1. The watchdog service time and the CPU reset duration are given in Table 10.8. *Service time* is the longest time interval within which the CPU must toggle the SS pin to prevent the CPUR signal from going low. *Reset time* is the amount of time that the CPUR signal will stay low once it goes low. This function is redundant for the 68HC11 because all the 68HC11 family members have a watchdog function. The 68HC11 COP timer may be better for some users because the time out period of the COP is programmable, whereas the later is fixed.

	50 Hz		60 Hz		XTAL	
	minimum	maximum	minimum	maximum	minimum	maximum
Service time (ms)	–	10	–	8.3	–	7.8
Reset time (ms)	20	40	16.7	33.3	15.6	31.3

Table 10.8 Watchdog service and reset time

10.12   Summary

Implementing the SPI function makes more of the limited number of signal pins in a microcontroller available for other functions. Two parties are involved in a SPI transfer: a master and one or more slaves. The SPI protocol uses a common clock supplied by the master to synchronize the shifting of data bits between the SPI master and slave units. Data bits are shifted in and out in the middle of a bit time. Both the master and the slave have a master-out-slave-in (MOSI) and master-in-slave-out (MISO) pins for data transfer. An SPI slave usually has a slave-select pin, when asserted, enables data shifting. The SPI function is mainly used in low-to-medium speed I/O transfers. Examples in this chapter illustrated the use of the SPI subsystem to interface shift registers, seven-segment displays, LCD drivers, and the timer chip with the 68HC11.

The HC589 has both a serial and parallel data inputs but only a serial data output. It can be used to expand the number of input ports to a microcontroller. The HC595 has a serial data input and parallel data outputs and thus can be used to expand the number of parallel output ports to a microcontroller. The MC14489 is an SPI-compatible seven-segment display driver chip. It can drive five seven-segment displays directly. Multiple MC14489s can be cascaded together to drive more than five seven-segment displays. The MC145000 and MC145001 are LCD displays driver chips with serial interface. The MC145000 is a master unit that provides both frontplane and backplane signals. The MC145001 is a slave unit and provides only frontplane signals. The MC145000 can drive up to six LCD digits directly. When more than six LCD digits need to be displayed, then we need to use one MC145000 and one or more MC145001s together. The MC68HC68T1 is an SPI-compatible real-time clock chip. It can keep track of seconds, minutes, hours, day-of-the-week, date, month, and year. It has battery backup power supply that can keep the device in operation during a power interruption, and its on-chip 32-byte RAM allows the user to save critical information when the power is interrupted. The 68HC68T1 can also provide alarm function. It allows the user to set the alarm time and provides signal output to drive the alarm device. Power control is another application of the 68HC68T1. It can detect the power transition and perform power-up and power-down operation. Finally, this device also incorporates an on-chip COP function similar to that of the 68HC11.

10.13   Exercises

E10.1. Is SPI a synchronous or an asynchronous serial data transfer protocol?
E10.2. Write an instruction sequence to set up the 68HC11 SPI with the following parameters: master mode, normal port D pins (no wired-or mode), disable the SPI interrupt, enable the SPI function, use the falling edge of the SCK clock to shift data in and out, set the data rate to 1 Mbits/sec at the 4 MHz E clock, and use the SS pin as a general output pin.

E10.3. Write an instruction sequence to output the value $2B to a peripheral chip using the following pins by simulating the SPI operation:

Din: serial data input

CLK: the falling edge of this clock signal is used to shift in the data.

EN: an active low signal to enable and disable the shifting of data.

Assume the chip has a 6-bit shift register.

E10.4. Write a program to read a byte from the HC589 #2 in Figure 10.7 into the accumulator A.

E10.5. Write a program to output the byte "$7F" to the HC595 #4 in Figure 10.10.

E10.6. Modify the program in Example 10.3 so that the circuit displays the value of 625.32.

E10.7. Modify the program in Example 10.3 so that the circuit will display one digit at a time. Each digit is displayed for 1 second and then turned off. The circuit displays a "1" in the first display for one second and blanks the other three displays, then displays a "2" in the second display for 1 second, etc. The operation is repeated forever.

E10.8. Use the same circuit as shown in Figure 10.14 but disable the SPI function. Use the simulation method as illustrated in section 10.5.3 to output the value 745.23 to five seven-segment displays. The data are shifted out bit-by-bit using the MOSI (PD3) pin. The SCK should be toggled to simulate the SCK clock whereas the SS pin is used to load data into the latch of the MC14489. Write a program to perform the simulation.

E10.9. Use two 1/4-multiplexing LCDs, one MC145000, two MC145001s, and a 68HC68T1 to keep track of and display the current time. The fields of the current time are separated by decimal points. Enable periodic and alarm interrupts. The periodic interrupt is to be generated every second, and the LCD display must be updated every second. Assume the current time and the alarm time are stored in a memory block that starts at $00 in the order that allows burst-mode transfer. Add an alarm transducer to the CLK OUT pin of the 68HC68T1. The alarm transducer should be turned on for 3 minutes when the preset alarm time is reached. Show the circuit connections and write a main program and an interrupt-handling routine to perform the specified functions.

E10.10. Use the SPI system to drive two MC14489s in cascade so that up to ten seven-segment displays can be driven. Draw the circuit connection and also write a program to display the value of 0123456789 continuously.

10.14 Lab Exercises and Assignments

L10.1. *Keyboard Scanning and Debouncing.* Connect the 16-key membrane keypad as shown in Figure 7.16 and use seven-segment displays and the MC14489 to display the entered BCD digits. The entered digits must be shifted from left-to-right when more digits are entered. Use one MC14489 so that five BCD digits can be displayed at a time. Write a main program to initialize the SPI system and start to scan the keyboard. The display should be blanked at the beginning. Whenever a key is pressed, the program should perform the debounce operation. If the debounce operation shows that the key was indeed pressed, then the main program should update the display. If the user has entered more than five digits, then only the latest five digits should be displayed.

L10.2. *Time-of-day display.* Use the circuit as in L10.1 to display the time-of-day. Use two MC14489 to control six seven-segment displays. The lab procedure is as follows:

1. Connect the circuit properly.

2. Write a program to
   a. Display six zeros as the prompt for the user to enter the time of day.
   b. Read two hours digits, then two minutes digits, and then two seconds digits. Display the entered digits immediately and also check the validity of the entered time. If invalid time is entered, the program will display six zeros again to prompt the user reenter the time.
   c. Update the time of day once every second using the output compare function.

11

68HC11 Analog to Digital Converter

11.1 Objectives

After completing this chapter you should be able to

- explain the A/D conversion process

- describe the precision, the various channels, and the operation modes of the 68HC11 A/D system

- interpret A/D conversion results

- describe the procedure for using the 68HC11 A/D system

- use the LM35 precision centigrade temperature sensors from National Semiconductor

- use the humidity sensor made by HyCal Engineering

- calculate the root-mean-square value of the A/D conversion results

- use external A/D converters

11.2   Introduction

An analog signal quantity has a continuous set of values over a given range, in contrast to the discrete values of digital signals. Almost any measurable quantity, for example, current, voltage, temperature, speed, and time, is analog in nature. To be processed by a digital computer, analog signals must be represented in the digital form; thus an analog-to-digital (A/D) converter is required.

An A/D converter can deal only with the electrical voltage. A nonelectric quantity must be converted into a voltage before A/D conversion can be performed. Conversion from a nonelectric quantity to a voltage requires the use of a *transducer*. For example, a temperature sensor is needed to convert a temperature into a voltage. To measure weight, a load cell is used to convert a weight into a voltage. However, the transducer output voltage may not be appropriate for processing by the A/D converter. A voltage level shifter and scaler are often needed to transform the voltage output into a range that can be handled by the A/D converter. The circuit that performs scaling and shifting of the transducer output is called *signal conditioning circuit*. The overall A/D process is illustrated in Figure 11.1.

The accuracy of an A/D converter is dictated by the number of bits used to represent the analog quantity. The greater the number of bits, the better the

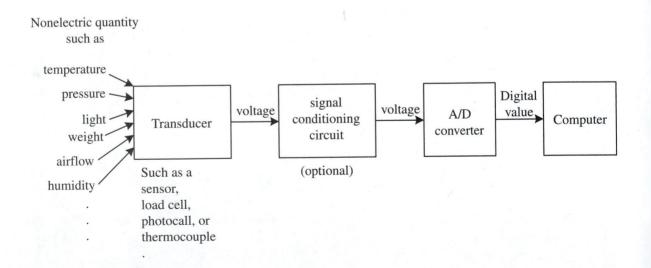

Figure 11.1 ■ The A/D conversion process

accuracy. Most of the 68HC11 members implement an 8-bit A/D converter. A few members, for example, 68HC11KW4, implement a 10-bit A/D converter.

Successive approximation method is the most widely used method for implementing A/D conversion in microcontrollers, and this method will be illustrated in the next section.

11.3   Successive Approximation Method

The block diagram of a successive approximation A/D converter is shown in Figure 11.2. A successive approximation A/D converter approximates the analog signal to n-bit code in n steps. It first initializes the successive approximation register (SAR) to zero and then performs a series of guesses, starting with the most significant bit and proceeding toward the least significant bit. The algorithm of the successive approximation method is illustrated in Figure 11.3. It assumes that the SAR register has n bits. For every bit of the SAR register, the algorithm

- guesses the bit to be 1
- converts the value of the SAR register to an analog voltage
- compares the D/A output with the analog input
- clears the bit to 0 if the D/A output is larger (which indicates that the guess is wrong)

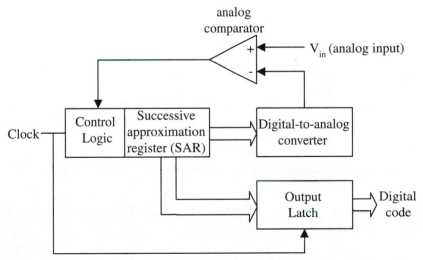

Figure 11.2 ■ Block diagram of a successive approximation A/D converter

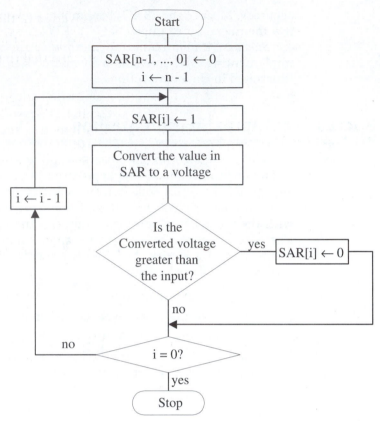

Figure 11.3 ■ Successive approximation A/D conversion method

11.4   Signal Conditioning Circuits

Not all the transducer outputs are appropriate for A/D converter. The user of an A/D converter often needs to use a circuit to scale and/or shift the transducer output to fit the input range of the A/D converter. This section discusses a few circuits that perform the voltage scaling and level shifting.

11.4.1   Optimal Voltage Range for the A/D Converter

An A/D converter needs a *low reference voltage* (V_{LREF}) and a *high reference voltage* (V_{HREF}) for performing the conversion. The low reference voltage is often set to 0 V, and the high reference voltage is often set to the power supply voltage V_{cc}. Most A/D converters are ratiometric, i.e.,

- a 0 V analog input is converted to the digital value 0
- a V_{cc} analog input is converted to the digital value $2^n - 1$
- an X volt input will be converted to the digital value $X \times (2^n - 1) \div V_{cc}$

where n is the number of bits the A/D converter uses to represent a conversion result. The value *n* is also called the *resolution* of the A/D converter.

Since the A/D converter is ratiometric, the optimal voltage range for the A/D converter is $0 \sim V_{cc}$ as long as the V_{LREF} and the V_{HREF} are set to 0 V and V_{cc}, respectively. The A/D conversion result x corresponds to an analog voltage given by the following equation:

$$V_x = V_{LREF} + (range \times x) \div (2^n - 1) \qquad\qquad 11\text{-}1$$

where,

V_x is the analog voltage corresponding to the converted result x.
$range = V_{HREF} - V_{LREF}$

Example 11.1

▼

Assume there is a 12-bit A/D converter with $V_{LREF} = 0$ V and $V_{HREF} = 5$ V. Find out the corresponding voltage values for A/D conversion results of 100, 400, 800, 1200, and 2400.

Solution: $range = V_{HREF} - V_{LREF} = 5\ V - 0\ V = 5\ V$

The voltages corresponding to the results of 100, 400, 800, 1200, and 2400 are:

$0\ V + (100 \times 5) \div (2^{12} - 1) = 0.12\ V$

$0\ V + (400 \times 5) \div (2^{12} - 1) = 0.49\ V$

$0\ V + (800 \times 5) \div (2^{12} - 1) = 0.98\ V$

$0\ V + (1200 \times 5) \div (2^{12} - 1) = 1.46\ V$

$0\ V + (2400 \times 5) \div (2^{12} - 1) = 2.93\ V$

▲

11.4.2   Voltage Scaling Circuit

There are situations in which the transducer output voltages are in the range of $0 \sim V_z$, where $V_z <$ power supply. The voltage scaling circuit can be used to improve the accuracy of the A/D conversion because it allows the A/D converter to utilize its full range. The diagram of this circuit is shown in Figure 11.4.

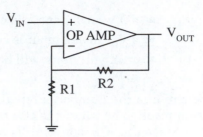

Figure 11.4 ■ A voltage scaler

The voltage gain of this circuit is given by the following equation:

$$A_v = \frac{R1 + R2}{R1} = 1 + \frac{R2}{R1}$$
11-2

Example 11.2

▼

Suppose the range of voltage output of an A/D sensor is 0 ~ 200 mV, choose appropriate standard resistor values to scale this output range to 0 to 5 V.

Solution:

5V ÷ 200 mV = 25
∴ R2/R1 = 24

By trial and error, choose 10 KΩ for R_1 and then R_2 is calculated to be 24 KΩ. Both R_1 and R_2 are standard resistors.

▲

11.4.3   Voltage Shifting and Scaling Circuit

There are transducers whose outputs are in the range of $V_1 \sim V_2$ (V_1 can be negative and $V_2 \neq V_{cc}$) instead of in the range of 0 V ~ V_{cc}. In order to improve the accuracy of A/D conversion, we can use a voltage shifting and scaling circuit to convert the transducer output so that it falls in the full range of 0 V ~ V_{cc}.

Voltage shifting and scaling can be achieved by using an OP AMP circuit. One such example is shown in Figure 11.5c. This circuit consists of a summing circuit (shown in Figure 11.5a) and an inverting voltage follower circuit (shown in Figure 11.5b). In Figure 11.5, the voltage V_{IN} comes from the transducer output and V_1 is an adjusting voltage. By choosing appropriate values for V_1 and resistors, the desired voltage scaling and shifting effect can be achieved. The power supply to the OP AMP 741 can be other values instead of 12 V. All resistors must be standard resistors.

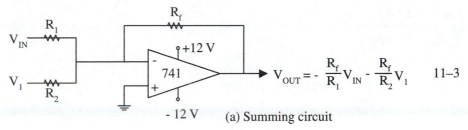

$$V_{OUT} = -\frac{R_f}{R_1}V_{IN} - \frac{R_f}{R_2}V_1 \qquad 11\text{--}3$$

(a) Summing circuit

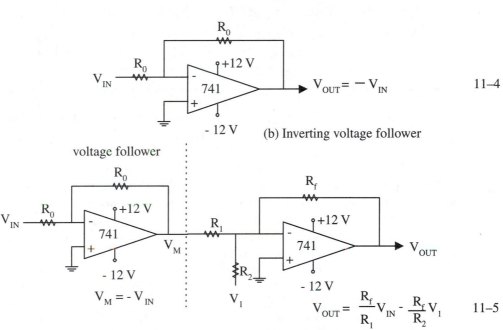

$$V_{OUT} = -V_{IN} \qquad 11\text{--}4$$

(b) Inverting voltage follower

$$V_M = -V_{IN}$$

$$V_{OUT} = \frac{R_f}{R_1}V_{IN} - \frac{R_f}{R_2}V_1 \qquad 11\text{--}5$$

(c) Level shifter and scaler

Figure 11.5 ■ Level shifting and scaling circuit

Example 11.3

Use the circuit shown in Figure 11.5 to shift the transducer output from $-2.5 \sim 2.5$ V to $0 \sim 5$ V.

Solution: Applying equation 11.5, we obtain

$$0 = R_f \div R_1 \times (-2.5) - R_f \div R_2 \times V_1$$
$$5 = R_f \div R_1 \times (2.5) - R_f \div R_2 \times V_1$$

By choosing $V_1 = -12$ V and $R_f = 5$ KΩ, R_1 and R_2 are solved to be 5 KΩ and 24 KΩ. R_0 is independent of these parameters, we arbitrarily set it to 5 KΩ.

Example 11.4

▼

Choose appropriate values for R_0, R_1, R_2, R_f, and V_1 in Figure 11.5 to shift and scale the voltage from -50 ~ 150 mV to 0 ~ 5 V. (All resistors must be standard resistors.)

Solution: Applying equation 11.5, we obtain

$$0 \text{ V} = R_f \div R_1 \times (-50 \text{ mV}) - (R_f \div R_2) \times V_1$$

$$5 \text{ V} = R_f \div R_1 \times (150 \text{ mV}) - (R_f \div R_2) \times V_1$$

By choosing $V_1 = -5\text{V}$ and $R_1 = 1.2$ KΩ, R_2 and R_f are solved to be 120 KΩ and 30 KΩ. R_0 is set to be 1.2 KΩ arbitrarily.

▲

It is not difficult to conclude from the above two examples that the selection of resistor values and the voltage V_1 is a trial-and-error process at best.

11.5   An Overview of the 68HC11 A/D Converter

The 68HC11 has an eight-channel, 8-bit, multiplexed input, successive-approximation, analog-to-digital converter with sample-and-hold circuitry to minimize conversion errors caused by rapidly changing input signals. Because it is equipped with a weighted array of capacitors, the 68HC11 uses an all-capacitive charge-redistribution technique to implement the successive-approximation A/D conversion method. Implementing the successive-approximation A/D conversion method in the hardware circuit requires the use of a clock signal. The charge in the capacitors will leak away and will not be able to produce correct results if the clock frequency is below 750 KHz.

Two dedicated lines (V_{RL}, V_{RH}) provide the reference voltages for the A/D conversion. The V_{RH} line is used as the high-reference voltage, and the V_{RL} line is used as the low-reference voltage. V_{RH} cannot be higher than V_{DD} by 0.1 V, and V_{RL} cannot be lower than V_{SS}. The difference between V_{RH} and V_{RL} cannot be smaller than 2.5 V, but accuracy is tested and guaranteed only for $\Delta V_R = 5$ V $\pm$ 10% ($\Delta V_R = V_{RH} - V_{RL}$).

The A/D converter is ratiometric. An input voltage equal to V_{RL} converts to $00, and an input voltage equal to V_{RH} converts to $FF (full scale), with no overflow indication.

The A/D converter must be enabled before it can be used. It is enabled (or disabled) by setting (or clearing) the A/D power-up (ADPU) control bit of the OPTION register. Setting the ADPU bit starts the charge pump circuit in the A/D converter system. It takes about 100 μs for the charge pump circuit to stabilize.

11.6   A/D System Options

The A/D converter has a multiplexor to select one of sixteen analog signals. Eight of these channels correspond to port E input lines to the 68HC11; four of the channels are for internal reference points or test functions, and four channels are reserved for future use. Therefore, only eight channels are available for normal use. The channels are selected by setting the channel select bits in the A/D control register (ADCTL).

The A/D system starts the conversion one clock cycle after a control byte is written into the ADCTL register. The A/D system can be configured to perform conversion sequences four times on a selected channel in the single-channel mode or one time on each of four channels (either channels 1 to 4 or channels 5 to 8) in the multiple-channel mode.

There are two variations of single-channel operation. In the first variation *(nonscan mode)*, four consecutive samples are taken from the single selected channel and are converted in turn, with the first result being stored in the A/D result register 1 (ADR1), the second result being stored in register ADR2, etc. After the fourth conversion is completed, all conversion activity is halted until a new conversion command is written into the ADCTL register. In the second variation *(scan mode)*, conversions are performed continually on the selected channel, with the fifth conversion being stored in register ADR1 (overwriting the first conversion result), the sixth conversion results overwriting ADR2, and so on.

There are also two variations of the multiple-channel operation. In the first variation (nonscan mode), the selected group of four channels are converted, one at a time, with the first result being stored in the register ADR1, the second result being stored in the register ADR2, and so on. After the fourth conversion is complete, all conversion activity is halted until a new conversion command is written into the ADCTL register. In the second variation (scan mode), conversions are performed continually on the selected group of four channels, with the fifth conversion being stored in register ADR1 (replacing the earlier conversion result for the first channel in the group), the sixth conversion overwriting the ADR2, and so on.

11.7   The Clock Frequency Issue

An A/D converter requires a clock signal to operate. The A/D system of the 68HC11 can select either the E clock signal or the internal RC circuit output as the clock source. The RC clock signal runs at about 1.5 MHz. The clock source is selected by setting or clearing the CSEL bit (bit 6) of the OPTION register. When the bit is 0, the E clock signal is selected. Otherwise, the RC clock source is selected. The RC clock requires 10 ms to start and settle. The

RC clock source should be selected only if the E clock frequency is lower than 750 KHz; otherwise, the capacitor array charge leakage will become too large to obtain correct A/D conversion results.

11.8   A/D System Registers

Six registers are involved in the A/D conversion process: the A/D control/status register (ADCTL), the OPTION register, and four A/D result registers (ADR1, ADR2, ADR3, & ADR4).

11.8.1   A/D Result Registers 1, 2, 3 and 4 (ADR1, ADR2, ADR3, and ADR4)

The A/D result registers are read-only registers that hold an 8-bit conversion result. Writes to these registers have no effect. Data in the A/D result registers are valid when the CCF flag bit in the ADCTL register is 1, indicating a conversion sequence is complete.

The A/D conversion process is started one clock cycle after the ADCTL register is written into. The conversion of one sample takes 32 clock cycles. The CCF bit in the ADCTL register will be set to 1 when all four A/D result registers contain valid conversion results, thus it will be set 129 clock cycles after the ADCTL register is written into.

The addresses of A/D result registers are as follows:

ADR1 is at $1031.
ADR2 is at $1032.
ADR3 is at $1033.
ADR4 is at $1034.

11.8.2   A/D Control/Status Register

The ADCTL register is located at $1030. All bits in this register may be read or written, except for bit 7, which is a read-only status indicator, and bit 6, which always reads as a zero. The contents of the ADCTL register are as follows:

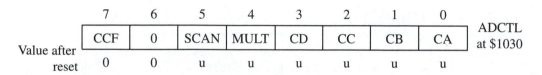

	7	6	5	4	3	2	1	0	
	CCF	0	SCAN	MULT	CD	CC	CB	CA	ADCTL at $1030
Value after reset	0	0	u	u	u	u	u	u	

CCF: *Conversion complete flag.* This read-only flag is set when all four A/D result registers contain valid conversion results. Each time the ADCTL register is written, this bit is automatically cleared to 0 and a conversion sequence is started.

SCAN: *Continuous scan control.* When this bit is 0, the four requested conversions are performed once to fill the four result registers. When this bit is 1, conversions continue in a round-robin fashion with the result registers being updated as data becomes available.

MULT: *Multiple-channel/single-channel control.* When this bit is 0, the A/D system performs four consecutive conversions on the single channel specified by the four channel select bits (CD through CA, bits 3-0 of the ADCTL register). When this bit is 1, the A/D system is configured to perform a conversion on each of four channels, with each result register corresponding to one channel.

CD: *Channel select D.*

CC: *Channel select C.*

CB: *Channel select B.*

CA: *Channel select A.*

These four bits are used to select one of the 16 A/D channels. When a multiple-channel mode is selected, the two least significant bits (CB and CA) have no meaning, and the CD and CC bits specify which group of four channels is to be converted. The channels selected by the four channel select bits are shown in Table 11.1.

CD	CC	CB	CA	Channel signal	Result in ADRx if MULT = 1
0	0	0	0	AN0	ADR1
0	0	0	1	AN1	ADR2
0	0	1	0	AN2	ADR3
0	0	1	1	AN3	ADR4
0	1	0	0	AN4*	ADR1
0	1	0	1	AN5*	ADR2
0	1	1	0	AN6*	ADR3
0	1	1	1	AN7*	ADR4
1	0	0	0	reserved	ADR1
1	0	0	1	reserved	ADR2
1	0	1	0	reserved	ADR3
1	0	1	1	reserved	ADR4
1	1	0	0	V_{RH}pin**	ADR1
1	1	0	1	V_{RL}pin**	ADR2
1	1	1	0	$(V_{RH})/2$**	ADR3
1	1	1	1	reserved**	ADR4

*Not available in 48-pin package.

**These channels are intended for factory testing.

Table 11.1 ■ Analog-to-digital channel assignments

11.8.3   OPTION Register

The OPTION register is located at $1039. Only bits 7 and 6 are related to the operation of the A/D converter. Bit 7 (ADPU) enables the A/D converter when it is set to 1. After setting this bit, the user must wait for at least 100 μs before using the A/D converter system to allow the charge pump and comparator circuit to stabilize. Bit 6 (CSEL) selects the clock source for the A/D converter. When CSEL = 1, the internal RC clock is selected. Otherwise, the E clock is selected. The contents of the OPTION register are as follows:

	ADPU	CSEL	IREQ	DLY	CME	0	CR1	CR2	OPTION at $1039
Value after reset	0	0	0	1	0	0	0	0	

11.9   Procedure for Using the A/D System

The procedure for using the 68HC11 A/D system is as follows:

Step 1
Connect the hardware properly. The high reference voltage V_{RH} should not be higher than V_{DD} (5 V). The low reference voltage V_{RL} should not be lower than V_{SS} (0 V). The analog signal to be converted should fall between V_{RL} and V_{RH}. If the value of the analog signal is not between V_{RL} and V_{RH}, then the user must scale and shift it to the desired range. Connect the analog signal (s) to the appropriate A/D input pin (s).

Step 2
Set the ADPU bit of the OPTION register to enable the A/D system. Set the CSEL bit to 0 or 1, depending on the E clock frequency. A 0 selects the E clock as the clock signal for the A/D conversion process.

Step 3
Wait for the charge pump to stabilize. The user can use a program loop to create a delay of at least 100 μs. If the internal RC oscillator output is selected as the clock source, then a 10 ms delay is required for the RC clock output to settle.

Step 4
Select the appropriate channel(s) and operation modes by programming the ADCTL register. The A/D conversion will be started one clock cycle after the ADCTL register is written into.

Step 5
Wait until the CCF bit in the ADCTL register is set, then collect the A/D conversion results and store them in memory. When the CCF bit is set, the value of the ADCTL register becomes negative and allows the user to decide if the A/D conversion is completed.

Example 11.5

▼

Write an instruction sequence to set up the following A/D conversion environment:

- nonscan mode
- single-channel mode
- select channel AN0
- choose the E clock as the clock source to the A/D converter
- enable the A/D converter

Solution: To set up the first three A/D parameters, we need to write a byte containing the following bits into the ADCTL register:

- to select nonscan mode, set bit 5 of the ADCTL register to 0
- to select single-channel mode, set bit 4 of the ADCTL register to 0
- to select channel AN0, set the bits 3-0 of the ADCTL register to 0000

Since the msb of the ADCTL register is a status bit, write a 0 into it. Therefore, we must write the byte %00000000 into the ADCTL register. To choose the E clock as the clock source for the A/D converter, set the CSEL bit (bit 6) of the OPTION register to 0. To enable the A/D converter, set the bit 7 of the OPTION register to 1 and wait for 100 μs. The following instruction sequence will set up the specified A/D conversion environment:

```
REGBAS  EQU   $1000            ; base address of I/O register block
ADCTL   EQU   $30              ; offset of ADCTL from REGBAS
OPTION  EQU   $39              ; offset of OPTION from REGBAS

        LDX   #REGBAS
        BCLR  OPTION,X $40      ; select E clock as the clock source for A/D converter
        BSET  OPTION,X $80      ; enable the A/D converter
        LDY   #30              ; delay 105 ms for charge pump to stabilize
wait    DEY                    ;      "
        BNE   wait             ;      "
        LDAB  #0               ; select non-scan, single channel mode and
*                             ; channel AN0
        STAB  ADCTL,X          ; and starts the A/D conversion
```

The following C statements will set up the same A/D conversion environment for the 68HC11:

```
OPTION &= 0xBF;         /* clear bit 6 */
OPTION |= 0x80;         /* set bit 7 to 1 */
TFLG1 = 0x40;           /* clear OC2F flag */
```

```
TOC2 = TCNT + 200;          /* start an OC2 operation with 200 E clock cycles as the delay */
while (!(TFLG1 & 0x40));
ADCTL = 0x00;               /* start an A/D conversion */
```

Example 11.6

▼

Write an instruction sequence to set up the following A/D conversion environment:

- nonscan mode
- multiple-channel mode
- select channels AN4-AN7
- choose the E clock as the clock source for the A/D converter
- enable the A/D converter

Solution: To set up the first three A/D parameters, we need to write a byte containing the following bits into the ADCTL register:

- to select non-scan mode, set bit 5 of the ADCTL register to 0
- to select multiple-channel mode, set bit 4 of the ADCTL register to 1
- to select channels AN4-AN7, set bits 3-0 of the ADCTL register to 0100 (bits 1 and 0 can be any value)

Because the msb of the ADCTL register is a status bit, write a 0 into it. Therefore, we must write the byte %00010100 into the ADCTL register.

To choose the E clock as the clock source for the A/D converter, set the CSEL bit (bit 6) of the OPTION register to 0. To enable the A/D converter, set bit 7 of the OPTION register to 1 and wait for 100 μs. The following instruction sequence will set up the specified A/D conversion environment:

```
REGBAS  EQU   $1000           ; base address of I/O register block
ADCTL   EQU   $30             ; offset of ADCTL from REGBAS
OPTION  EQU   $39             ; offset of OPTION from REGBAS

        LDX   #REGBAS
        BCLR  OPTION,X $40     ; select E clock as the clock source for A/D converter
        BSET  OPTION,X $80     ; enable the A/D converter
        LDY   #30              ; delay 105 μs for charge pump to stabilize
delay   DEY                    ;        "
        BNE   delay            ;        "
        LDAB  #%00010100       ; select nonscan, single channel mode and
*                              ; channel AN0
        STAB  ADCTL,X          ;        "
```

Example 11.7

Write an instruction sequence to convert the analog signal connected to channel AN0 into digital form. Perform four conversions and stop. Assume the frequency of the E clock is 2 MHz.

Solution: The circuit connections for this example are shown in Figure 11.6. The user should notice that the high reference voltage V_{RH} and the low reference voltage V_{RL} have been set to 5 V and 0 V, respectively.

Since only four A/D conversions are to be performed, the user needs only write one byte into the ADCTL register. The program is:

```
REGBAS   EQU    $1000              ; I/O registers base address
OPTION   EQU    $39                ; offset of OPTION from REGBAS
ADCTL    EQU    $30                ; offset of ADCTL from REGBAS
ADR1     EQU    $31                ; offset of ADR1 from REGBAS
ADR2     EQU    $32                ; offset of ADR2 from REGBAS
ADR3     EQU    $33                ; offset of ADR3 from REGBAS
ADR4     EQU    $34                ; offset of ADR4 from REGBAS
A2D_INI  EQU    $00                ; value to be written into ADCTL to select
*                                  ; single-channel, non-scan mode, & channel ANO
         ORG    $00
result   RMB    4                  ; reserved four bytes to store the results

         ORG    $C000              ; starting address of the program
         LDX    #REGBAS
         BSET   OPTION,X $80       ; enable the charge pump to start
*                                  ; A/D conversion
         BCLR   OPTION,X $40       ; select the E clock for A/D conversion
         LDY    #30                ; delay 105 µs to wait for charge pump to stabilize
delay    DEY                       ;        "
         BNE    delay              ;        "
         LDAA   #A2D_INI           ; initialize the ADCTL register and start
*                                  ; A/D conversion
         STAA   ADCTL,X            ;        "
again    LDAA   ADCTL,X            ; check CCF bit
         BPL    again              ; wait until the CCF bit is set
         LDAA   ADR1,X             ; save the first result
         STAA   result             ;        "
         LDAA   ADR2,X             ; save the second result
         STAA   result+1           ;        "
         LDAA   ADR3,X             ; save the third result
         STAA   result+2           ;        "
         LDAA   ADR4,X             ; save the fourth result
         STAA   result+3           ;        "
         END
```

68HC11

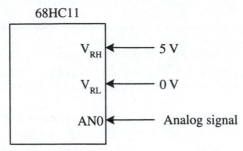

Figure 11.6 ■ Circuit connections for
Example 11.7

The C language version of this program is as follows:

```c
#include <hc11.h>
char result [4];
main ( )
{
    OPTION &= 0xBF;          /* select the E clock as the A/D clock source */
    OPTION | = 0x80;         /* enable A/D converter */
    TFLG1 = 0x40;            /* clear the OC2F flag */
    TOC2 = TCNT + 200;       /* start an OC2 operation with 200 E clock cycles as
                               the delay */
    while (!(TFLG1 & 0x40));
    ADCTL = 0x00;            /* start an A/D conversion */
    while (!(ADCTL & 0x80)); /* wait until A/D conversion is complete */
    result [0] = ADR1;
    result [1] = ADR2;
    result [2] = ADR3;
    result [3] = ADR4;
    return 0;
}
```

▲

Example 11.8

▼

Take 20 samples from each of the A/D channels AN0 to AN3, convert them
to digital form, and store them in memory locations from $D000 to $D04F.

Solution: The circuit connections are shown in Figure 11.7. The samples
must be taken at as regular intervals as possible. This goal can be achieved by
starting the next conversion (by writing a new byte into the ADCTL register)
immediately after the previous conversion has been completed before collect-
ing the result.

68HC11

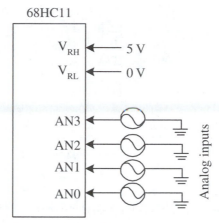

Figure 11.7 ■ Circuit connections for Example 11.8

The following program performs the conversion with this objective in mind:

```
N        EQU    20              ; number of samples
REGBAS   EQU    $1000           ; I/O registers base address
OPTION   EQU    $39             ; offset of OPTION from REGBAS
ADCTL    EQU    $30             ; offset of ADCTL from REGBAS
ADR1     EQU    $31             ; offset of ADR1 from REGBAS
ADR2     EQU    $32             ; offset of ADR2 from REGBAS
ADR3     EQU    $33             ; offset of ADR3 from REGBAS
ADR4     EQU    $34             ; offset of ADR4 from REGBAS
ADPU     EQU    $80             ; mask to set or clear the ADPU bit of the
*                               ; OPTION register
ADCLK    EQU    $40             ; mask to clear or set CSEL bit of the OPTION register
A2D_INI  EQU    $10             ; value to be written into the ADCTL register to select
*                               ; multi-channel, non-scan mode, and channels
*                               ; AN0-AN3

ORG      $D000
result   RMB    80              ; reserved 80 bytes to store the results
         ORG    $C000           ; starting address of the program
         LDX    #REGBAS
         BSET   OPTION,X ADPU   ; enable charge pump to start A/D conversion
         BCLR   OPTION,X ADCLK  ; select the E clock for A/D conversion
         LDY    #30             ; delay for 105 µs to wait for the charge pump
*                               ; to stabilize
delay    DEY                    ;       "
         BNE    delay           ;       "
         LDAA   #A2D_INI        ; initialize the ADCTL register
```

```
              STAA    ADCTL,X          ;       "
              LDAB    #N               ; number of samples to be collected in each channel
              LDY     #result          ; point Y to the result area
loop          LDAA    ADCTL,X          ; wait until the conversion of four samples
*                                      ; is completed
              BPL     loop             ;       "
```

* Start the next conversion immediately so that samples can be taken more uniformly in time

```
              LDAA    #A2D_INI         ; start the next conversion
              STAA    ADCTL,X          ;       "
```

* The following eight instructions collect the previous conversion results

```
              LDAA    ADR1,X           ; fetch the result from channel 1
              STAA    0,Y
              LDAA    ADR2,X           ; fetch the result from channel 2
              STAA    1,Y
              LDAA    ADR3,X           ; fetch the result from channel 3
              STAA    2,Y
              LDAA    ADR4,X           ; fetch the result from channel 4
              STAA    3,Y
              INY                      ; move the result pointer
              INY                      ;       "
              INY                      ;       "
              INY                      ;       "
              DECB                     ; decrement loop count
              BNE     loop
              END
```

In C language,

```c
#include <hc11.h>
unsigned char result [80];
main ( )
{
    int i;
    OPTION &= 0xBF;                 /* select E clock as the A/D clock source */
    OPTION |= 0x80;                 /* enable A/D converter */
    TFLG1 = 0x40;                   /* clear OC2F flag */
    TOC2 = TCNT + 200;              /* start an OC2 operation with 200 E clock cycles
                                       as the delay */
    while (!(TFLG1 & 0x40));
    ADCTL = 0x10;                   /* start an A/D conversion */
    for (i = 0; i < 20; i++) {
        while (!(ADCTL & 0x80));    /* wait until A/D conversion is complete */
```

```
        ADCTL = 0x10;              /* start the next A/D conversion */
        result[4*i ] = ADR1;
        result[4*i + 1] = ADR2;
        result[4*i + 2] = ADR3;
        result[4*i + 3] = ADR4;
    }
    return 0;
}
```

11.10   Measuring the Temperature

National Semiconductor produces an LM35 series of precision integrated-circuit temperature sensors whose output voltage is linearly proportional to the Celsius (centigrade) temperature. The LM35 does not require any external calibration or trimming to provide typical accuracies of $\pm^1/_4°$ C at room temperature and $\pm^3/_4°$ C over its complete $-55°$ to $+150°$ C temperature range. It can be used with single power supply, or with plus-and-minus supplies. Because it draws only 60 µA current from its supply, it has very low self-heating (less than 0.1° C in still air).

The LM35 is available in two packages: TO-46 metal can and TO-92 plastic package. It has three pins: V_{out} for voltage output, V_s for supply, and GND for ground. Two typical circuit connections for the LM35 are shown in Figure 11.8. These two circuits will sense the surface temperature of the LM35 and provide very accurate readings.

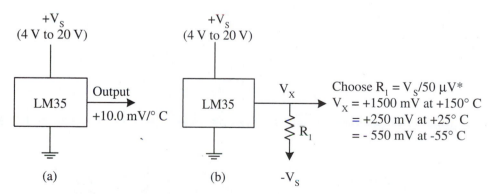

(a) Basic centigrade temperature sensor (12° C to 150° C)
(b) Full-range centigrade temperature sensor

Figure 11.8 ■ Circuit connections for the LM35 temperature sensor

Example 11.9

Digital Thermometer. Use the circuit shown in Figure 11.8b as a building block in a system to measure the room temperature. Display the result in two integral digits and one fractional digit using seven-segment displays. Assume that room temperature never goes below 0° C and never goes above 42.5° C so that the A/D converter of the 68HC11 can be used to perform the conversion and drive the seven-segment displays.

Solution: The voltage output from the circuit shown in Figure 11.8b will be 0 V at 0° C and 425 mV at 42.5° C. Higher precision can be obtained from the 68HC11 A/D converter if the voltage output corresponding to 42.5° C is scaled to 5 V. The circuit shown in Figure 11.9 will scaled V_x from the range of 0 to 425 mV to the range of 0 to 5 V. The CA3140 op amp is particularly suitable for microcontrollers and microprocessors because it has CMOS inputs, which use almost no current, and bipolar transistor outputs, which can supply plenty of current.

In this example, we will use the Motorola MC14489 seven-segment display driver along with three seven-segment displays to display the current temperature. The circuit connections between the 68HC11 and the MC14489 are shown in Figure 11.10.

Because room temperature does not change very fast, it should be adequate to measure the temperature once every second. We need to configure the MC14489 before using it. We will display the degree ° and C characters on banks 2 and 1, respectively. The control byte to be written into the configuration register is as follows.

bit 7: no decode, set to 0
bit 6: special decode, set to 1

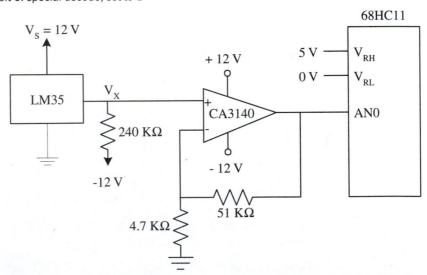

Figure 11.9 ■ Circuit connections between the LM35 and the 68HC11

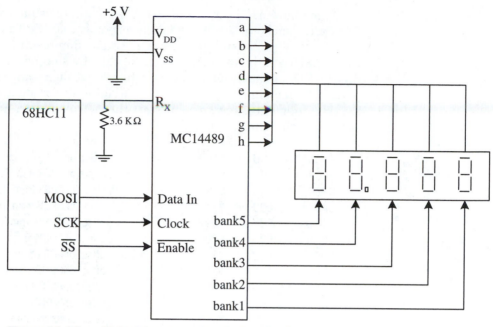

Figure 11.10 ■ Digital thermometer display circuit

bit 5: bank 5 hex decode, set to 0
bit 4: bank 4 hex decode, set to 0
bit 3: bank 3 hex decode, set to 0
bit 2: bank 2 special decode, set to 1
bit 1: bank 1 hex decode, set to 0
bit 0: normal mode, set to 1

Therefore, we must write the value $45 into the configuration register of the MC14489.

We need to reserve three bytes to hold the temperature data to be displayed. The format of the display data is shown in Figure 11.11.

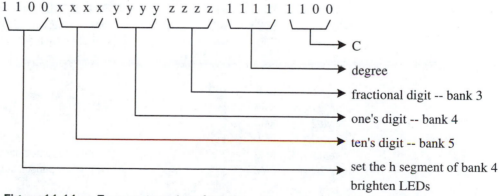

Figure 11.11 ■ Temperature data format

Since the 68HC11 A/D converter has 8-bit precision, the conversion result 255 corresponds to 42.5° C and 0 corresponds to 0° C. To obtain the temperature reading, we must divide the A/D conversion result by 6 (255 ÷ 42.5 = 6). In the 68HC11, the integer quotient of this division becomes the two integral digits of the temperature. The first fractional digit can be derived by multiplying the remainder by 10 and then dividing the product by 6.

The following program will start the A/D converter and display the temperature using three digits:

```
REGBAS    EQU    $1000           ; I/O register block base address
SPCR      EQU    $28             ; offset of SPCR from REGBAS
SPSR      EQU    $29             ; offset of SPSR from REGBAS
SPDR      EQU    $2A             ; offset of SPDR from REGBAS
DDRD      EQU    $09             ; offset of DDRD from REGBAS
ADCTL     EQU    $30             ; offset of ADCTL from REGBAS
ADR1      EQU    $31             ; offset of ADR1 from REGBAS
OPTION    EQU    $39             ; offset of OPTION from REGBAS
PORTB     EQU    $04             ; offset of PORTB from REGBAS
PORTD     EQU    $08             ; offset of PORTD from REGBAS
SP_DIR    EQU    %00111010       ; value to set SPI pins directions (set SS,
*                                ; MOSI, SCK, and TxD pins to be output, and
*                                ; MISO as input
SPCR_INI  EQU    %01010100       ; value to be written into SPCR to enable SPI,
*                                ; choose falling edge to shift data, master
*                                ; mode, normal port D output, and 1 Mbits/s
*                                ; data rate
ADPU      EQU    $80             ; mask to set the ADPU bit of the OPTION register
ADCLK     EQU    $40             ; mask to clear the CSEL bit of OPTION register
A2D_INI   EQU    $00             ; value to be written into ADCTL to set nonscan
*                                ; mode, select single-channel mode, and select
*                                ; channel AN0
TCNT      EQU    $0E             ; offset of TCNT from REGBAS
TOC2      EQU    $18             ; offset of TOC2 from REGBAS
TFLG1     EQU    $23             ; offset of TFLG1 from REGBAS
OC2       EQU    $40             ; mask to check the OC2F of TFLG1
OC2M      EQU    $BF             ; mask to clear the OC2F of TFLG1 for BCLR
*                                ; instruction

          ORG    $00
byte1     RMB    1               ; storage for results
byte2     RMB    1
byte3     RMB    1
remain    RMB    2               ; to hold the remainder of division
oc2cnt    RMB    1               ; output comparison count of OC2

          ORG    $C000
```

```
            LDX      #REGBAS
            LDAA     #$C0
            STAA     byte1
            LDAA     #$FC              ; store the value $FC in byte3
            STAA     byte3            ;       "
* The following four instructions initialize the SPI system to prepare shifting data out to
* MC14489
            LDAA     #SPDIR            ; set SPI pins directions
            STAA     DDRD,X           ;       "
            LDAA     #SPCR_INI         ; initialize the SPI system
            STAA     SPCR,X           ;       "
            BCLR     PORTD,X $20       ; enable SPI transfer to MC14489
            LDAA     #$45              ; send the configuration data to configure
*                                     ; MC14489
            STAA     SPDR,X           ;
            BRCLR    SPSR,X $80 *      ; wait until eight bits have been shifted out
            BSET     PORTD,X $20       ; load configuration data into the configuration
*                                     ; register

* The following five instructions enable the A/D converter and select E clock to control A/D
* conversion process and wait for the charge pump to stabilize

            BSET     OPTION,X ADPU     ; enable charge pump to start A/D conversion
            BCLR     OPTION,X ADCLK    ; select E clock for A/D conversion
            LDY      #30               ; delay 105 µs so that the charge pump
*                                     ; is stabilized
delay       DEY                       ;       "
            BNE      delay            ;       "
forever     LDAA     #A2D_INI          ; initialize the ADCTL and start an A/D conversion
            STAA     ADCTL,X          ;       "
here        LDAA     ADCTL,X           ; check the CCF bit
            BPL      here              ; wait until the A/D conversion is complete
            LDAB     ADR1,X            ; read the A/D conversion result

* The following three instructions convert the A/D result into temperature reading

            CLRA                      ; the upper byte of the dividend is 0
            LDX      #6                ; 1° C is equivalent to 6
            IDIV                      ; convert to temperature in Celsius
            STD      remain            ; save the remainder
            XGDX                      ; leave the integer part of the temperature in D
            LDX      #10               ; separate the upper and lower digits
            IDIV                      ; lower integer digit in D (actually in B
*                                     ; accumulator) upper integer digit in X register
            LSLB                      ; shift the lower integer digit to the upper four
*                                     ; bits of B
```

```
                LSLB                    ;      "
                LSLB                    ;      "
                LSLB                    ;      "
                STAB     byte2          ; the lower four bits in byte2 are 0000
                XGDX                    ; swap X and D so that the upper integer digit
        *                               ; is in B
                ADDB     byte1          ; combine the msd with decimal point specifier
                STAB     byte1          ; save the upper integer digit in memory
                LDD      remain         ; get back the remainder in the division.
        *                               ; The remainder only stay in accumulator B
        *                               ; because it is less than 6
                LDAA     #10            ; multiply the remainder by 10
                MUL                     ;      "
                LDX      #6             ; set the divisor to 6
                IDIV                    ; compute remainder x 10 ÷ 6 to derive the
        *                               ; first fractional digit
* The quotient will be less than 10 and will stay in the lowest four bits of accumulator B after
* the execution of the next instruction
                XGDX                    ; swap the quotient into A and B
                ADDB     byte2          ; combine the lower integer and the fractional
        *                               ; digits
                STAB     byte2          ;      "
                BCLR     PORTD,X $20    ; pull SS pin to low to enable SPI transfer
        *                               ; to MC14489
                LDAA     byte1          ; get byte1 and send it out
                STAA     SPDR,X         ;      "
                BRCLR    SPSR,X $80 *   ; wait until 8 bits have been shifted out
                LDAA     byte2          ; get byte2 and send it out
                STAA     SPDR,X         ;      "
                BRCLR    SPSR,X $80 *   ;      "
                LDAA     byte3          ; get byte3 and send it out
                STAA     SPDR,X         ;      "
                BRCLR    SPSR,X $80 *   ;      "
                BSET     PORTD,X $20    ; load data into the display register of the
        *                               ; MC14489


* The following instruction sequence uses the OC2 function 100 times to create a delay
* of 1 second. Each output compare operation on OC2 creates a delay of 10 ms.
                LDAB     #100
                STAB     oc2cnt         ; initialize oc2 count to 100 to create a
        *                               ; 1 second delay
                LDX      #REGBAS
                BCLR     TFLG1,X OC2M   ; clear the OC2F in TFLG1 before using OC2
        *                               ; function
                LDD      TCNT,X
```

```
repeat      ADDD    #20000          ; create a delay of 10 ms
            STD     TOC2,X          ;        "
wait        BRCLR   TFLG1,X OC2 *   ; check if the OC2F flag of the TFLG1 register is set
            BCLR    TFLG1,X OC2M    ; clear the OC2F flag
            LDD     TOC2,X
            DEC     oc2cnt
            BNE     repeat
            JMP     forever         ; repeat forever
            END
```

The C language version of the program is as follows:

```c
#include <hc11.h>
main ( )
{
    unsigned char byte1, byte2, byte3, temp;
    byte1 = 0xC0;
    byte3 = 0xFC;                       /* store characters "degree" and "C" in byte3 */
    DDRD = 0x3A;
    SPCR = 0x56;                        /* initialize the SPI function */
    OPTION |= 0x80;                     /* start the A/D charge pump */
    OPTION &= 0xBF;                     /* select E clock as the clock source for A/D
                                           converter */
    TFLG1 = 0x40;                       /* clear OC2F flag */
    TOC2 = TCNT + 200;                  /* create 100-µs delay so that the A/D charge pump
                                           is stabilized */

    while (!(TFLG1 & 0x40));
    PORTD &= 0xDF;                      /* enable SPI transfer */
    SPDR = 0x45;                        /* send configuration data to the MC14489 */
    while (!(SPSR & 0x80));             /* wait until SPI transfer is complete */
    PORTD |= 0x20;                      /* pull SS to high to transfer data to the configuration
                                           register */

    while (1) {
        ADCTL = 0x00;                   /* start an A/D conversion */
        while (!(ADCTL & 0x80));        /* wait until A/D conversion is complete */
        temp = ADR1/6;                  /* compute the tens digit of temperature */
        byte1 += temp / 10;             /* place the tens digit in the lower four bits of byte1 */
        byte2 = 4 << (temp % 10);       /* place the ones digit in the upper four bits of byte2 */
        temp = (ADR1 % 6) * 10 / 6;     /* compute the first fractional digit */
        byte2 += temp;                  /* combine the ones and fractional digits */
        PORTD &= 0xDF;                  /* pull SS pin to low to enable SPI transfer */
        SPDR = byte1;                   /* send out the first byte using SPI transfer */
        while (!(SPSR & 0x80));         /*          "        */
        SPDR = byte2;                   /* send out the second byte using SPI transfer */
        while (!(SPSR & 0x80));         /*          "        */
        SPRD = byte3;                   /* send out the third byte using SPI transfer */
```

```
        while (!(SPSR & 0x80));        /*       "       */
        PORTD |= 0x20;                 /* pull SS to high to transfer data to the display
                                          register */
        TOC2 = TCNT + 20000;           /* start an OC2 operation with 10 ms delay */
        TFLG1 = 0x40;                  /* clear OC2 flag */


   /* the following for loop creates a 1 second delay */


        for (temp = 0; temp < 100; temp −) {
            while (!(TFLG1 & 0x40));
            TOC2 += 20000;             /* start the next OC2 operation */
            TFLG1 = 0x40;              /* clear the OC2 flag */
        }
    }
    return 0;
}
```

11.11 Measuring the Humidity

The IH-3605 is a humidity sensor made by HyCal Engineering, which is a division of Honeywell. The IH-3605 humidity sensor provides a linear output from 0.8 to 3.9 V in the full range of 0 to 100% *relative humidity* (RH) when excited by a 5 V power supply. It can operate in a range of 0 to 100% RH, −40° to 185° F.

The pins of the IH-3605 are shown in Figure 11.12. The specifications of the IH-3605 are listed in Table 11.2. The IH-3605 is light sensitive and should be shielded from bright light for best results.

The IH-3605 can resist contaminant vapors such as organic solvents, chlorine, and ammonia. It is unaffected by water condensate as well. Because of this capability, the IH-3605 has been used in refrigeration, drying, instrumentation, meteorology and many other applications.

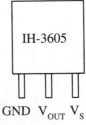

GND V_{OUT} V_S

Figure 11.12 ■ HyCal IH-3605
humidity sensor

Specification	Description
Total accuracy	±2% RH, 0-100% TH @ 25° C
Interchangeability	±5% RH up to 60% RH, ±8% RH at 90% RH
Operating temperature	–40 to 85° C (–40 to 185°F)
Storage temperature	–51 to 110° C (–60 to 230°F)
Linearity	±0.5% RH typical
Repeatability	±0.5% RH
Humidity stability	±1% RH typical at 50% RH in 5 years
Temp. effect on 0% RH voltage	±0.007% RH/° C (negligible)
Temp. effect on 100% span voltage	–0.22% RH/° C
Output voltage	$V_{OUT} = (V_s)$ (0.16 to 0.78) nominal relative to supply voltage for 0-100% RH; i.e., 1-4.9 V for 6.3 V supply; 0.8-3.9 V for 5 V supply; Sink capability 50 μA; drive capability 5 μA typical; low pass 1KH filter required. Turn on time <0.1 sec to full output.
V_s Supply requirement	4 to 9 V, regulated or use output/supply ratio; calibrated at 5 V
Current requirement	200 μA typical @ 5 V, increased to 2 mA at 9 V

Table 11.2 ■ Specifications of IH-3605

Example 11.10

▼

Construct a humidity measurement system that consists of the 68HC11, an IH-3605 humidity sensor, and four seven-segment displays.

Solution: The schematic of this system is shown in Figure 11.13. Since the humidity does not change very quickly, it is adequate to measure it once every second. A signal conditioning circuit is added to shift and scale the humidity sensor output voltage to the range of 0-5 V. The calculation of R_0, R_1, R_2, R_f, and V_1 will be left as an exercise problem. After adding this signal conditioning circuit, the A/D conversion results 0 and 255 correspond to the relative humidity 0 and 100%, respectively. A lookup table is used to simplify the conversion from an A/D result to its corresponding relative humidity. The conversion is based on the following equation:

RH = X × 100% ÷ 255   11-6

where, X is the A/D conversion result.

The procedure for measuring and displaying the humidity is as follows:

Step 1
Configure the A/D converter parameters.

Step 2
Take one humidity sample and convert it to digital value.

Step 3
Look up the humidity table and display it for one second.

Step 4
Repeat steps 2 and 3 forever.

In Figure 11.13, a 1 KΩ resistor and a 0.16 μF capacitor are used as the required 1 KHz low pass filter.

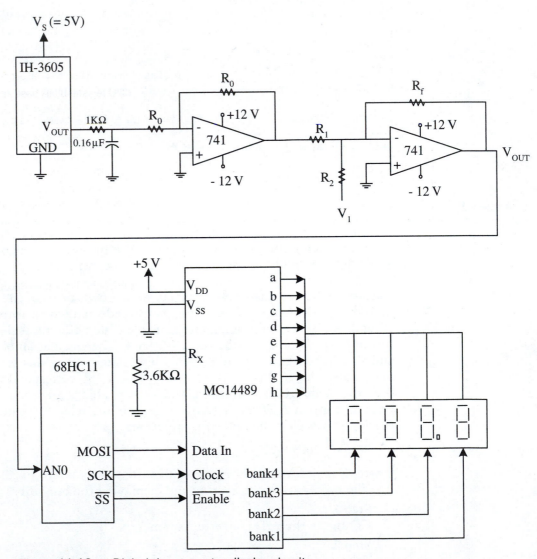

Figure 11.13 ■ Digital thermometer display circuit

In Figure 11.13, we use four seven-segment displays to display the current humidity. The most significant digit will be blank if the humidity is not 100%. Three bytes are used to hold the display data. The format of the display data is shown in Figure 11.14. Bank 5 is not used. Bank 4 controls the most significant digit. When the most significant digit is 0, we need to use the special decode mode to blank it. This is done by setting the bit 7 and bit 4 of the MC14489's configuration register to 1. Therefore, we must

1. send $91 as the configuration data if the most significant digit of the relative humidity is 0.

2. send $01 as the configuration data if the most significant digit of the relative humidity is 1.

The assembly program that performs the A/D conversion and display the humidity is as follows:

```
REGBAS    EQU    $1000        ; I/O register block base address
SPCR      EQU    $28          ; offset of SPCR from REGBAS
SPSR      EQU    $29          ; offset of SPSR from REGBAS
SPDR      EQU    $2A          ; offset of SPDR from REGBAS
DDRD      EQU    $09          ; offset of DDRD from REGBAS
ADR1      EQU    $31          ; offset of ADR1 from REGBAS
OPTION    EQU    $39          ; offset of OPTION from REGBAS
ADCTL     EQU    $30          ; offset of ADCTL from REGBAS
PORTB     EQU    $04          ; offset of PORTB from REGBAS
PORTD     EQU    $08          ; offset of PORTD from REGBAS
TCNT      EQU    $0E          ; offset of TCNT from REGBAS
TOC2      EQU    $18          ; offset of TOC2 from REGBAS
TFLG1     EQU    $23          ; offset of TFLG1 from REGBAS
```

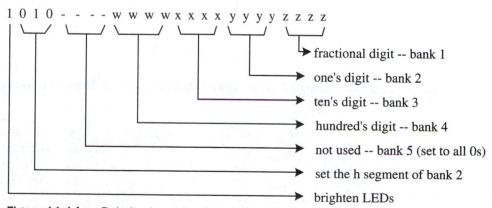

Figure 11.14 ■ Relative humidity data format

```
            SP_DIR      EQU     $3A                 ; value to set SPI pins directions (set SS,
            *                                       ; MOSI, SCK, and TxD pins to be output, and
            *                                       ; MISO as input)
            SPCR_INI    EQU     $54                 ; value to be written into SPCR to enable SPI,
            *                                       ; choose falling edge to shift data, master
            *                                       ; mode, normal port D output, and 1Mbits/s
            *                                       ; data rate
            ADPU        EQU     $80                 ; mask to set the ADPU bit of the OPTION
            *                                       ; register
            ADCLK       EQU     $40                 ; mask to clear the CSEL bit of OPTION register
            A2D_INI     EQU     $00                 ; value to be written into ADCTL to set nonscan
            *                                       ; mode, select single-channel mode, and
            *                                       ; select channel AN0
            OC2         EQU     $40                 ; mask to check the OC2F of TFLG1
            OC2M        EQU     $BF                 ; mask to clear the OC2F of TFLG1 for BCLR
            *                                       ; instruction
            CONF1       EQU     $91                 ; configuration data that will blank the most
            *                                       ; significant digit
            CONF2       EQU     $01                 ; configuration data that does not bland
            *                                       ; any digit

                        ORG     $00
            byte1       RMB     1                   ; storage for display data
            byte2       RMB     1
            byte3       RMB     1
            oc2cnt      RMB     1                   ; output comparison count of OC2

                        ORG     $C000
                        LDX     #REGBAS
                        LDAA    #$A0
                        STAA    byte1               ; first byte of the display data is a constant
                        LDAA    #SP_DIR             ; set SPI pins directions
                        STAA    DDRD,X              ;      "
                        LDAA    #SPCR_INI           ; initialize the SPI system
                        STAA    SPCR,X              ;      "
```

* The following five instructions enable the A/D converter and select E clock to control A/D
* conversion process and wait for the charge pump to stabilize

```
                        BSET    OPTION,X ADPU       ; enable charge pump to start A/D conversion
                        BCLR    OPTION,X ADCLK      ; select E clock for A/D conversion
                        LDY     #30                 ; delay 105 µs so that the charge pump
            *                                       ; is stabilized
            delay       DEY                         ;      "
                        BNE     delay               ;      "
```

```
forever     LDAA    #A2D_INI        ; initialize the ADCTL and start an A/D
*                                   ; conversion
            STAA    ADCTL,X         ;           "
here        LDAA    ADCTL,X         ; check the CCF bit
            BPL     here            ; wait until the A/D conversion is complete
            LDAB    ADR1,X          ; read the humidity data
            LDY     #humid_up
            ABY
            LDAA    0,Y             ; look up the humidity's hundreds and
*                                   ; tens digits
            STAA    byte2
            LDY     #humid_lo
            ABY
            LDAA    0,Y             ; look up the humidity's ones and tens digits
            STAA    byte3
            JSR     disp_humid      ; display the current humidity info
            JSR     wait_1s         ; wait for one second
            jmp     forever

disp_humid  PSHA
            PSHB
            PSHY
            PSHX
            LDX     #REGBAS
            BCLR    PORTD,X $20     ; enable SPI transfer to the MC14489
            LDAA    byte2
            ANDA    #$F0            ; check the upper four bits (most significant
*                                   ; digit)
            BEQ     blank
            LDAA    #01             ; choose normal hex decode
            BRA     send
blank       LDAA    #$91            ; choose special decode
send        STAA    SPDR,X
            BRCLR   SPSR,X $80 *    ; wait until the SPI transfer is complete
            BSET    PORTD,X $20     ; load data into the configuration register
            BCLR    PORTD,X $20     ; enable SPI transfer to the MC14489
            LDY     #byte1
            LDAB    #3
loop_disp   LDAA    0,Y
            STAA    SPDR,X
            BRCLR   SPSR,X $80 *    ; wait until SPI transfer is done
            INY
            DECB
            BNE     loop_disp
            BSET    PORTD,X $20     ; load data into display register
```

```
                    PULX
                    PULY
                    PULB
                    PULA
                    RTS


     wait_1s        PSHA
                    PSHB
                    PSHY
                    PSHX
                    LDX     #REGBAS
                    DES                       ; allocate one byte as OC2 count
                    TSY                       ; Y points to the top of the stack
                    LDAA    #50
                    STAA    0,Y               ; initialize OC2 count to 50
                    LDD     TCNT,X
     loop_50        ADDD    #40000
                    STD     TOC2,X
                    BCLR    TFLG1,X $BF       ; clear OC2F
                    BRCLR   TFLG1,X $40 *     ; wait until OC2F is set
                    LDD     TOC2,X
                    DEC     0,Y               ; decrement OC2 count
                    BNE     loop_50
                    INS
                    PULX
                    PULY
                    PULB
                    PULA
                    RTS


     humid_up  FCB  $00,$00,$00,$00,$00,$00,$00,$00,$00,$00,$00,$00,$00,$00,$00,$00
               FCB  $00,$00,$00,$00,$00,$00,$00,$00,$00,$00,$01,$01,$01,$01,$01,$01
               FCB  $01,$01,$01,$01,$01,$01,$01,$01,$01,$01,$01,$01,$01,$01,$01,$01
               FCB  $01,$01,$01,$02,$02,$02,$02,$02,$02,$02,$02,$02,$02,$02,$02,$02
               FCB  $02,$02,$02,$02,$02,$02,$02,$02,$02,$02,$02,$02,$02,$03,$03,$03
               FCB  $03,$03,$03,$03,$03,$03,$03,$03,$03,$03,$03,$03,$03,$03,$03,$03
               FCB  $03,$03,$03,$03,$03,$03,$40,$40,$40,$40,$40,$40,$40,$40,$40,$40
               FCB  $04,$40,$40,$40,$40,$40,$40,$40,$40,$40,$40,$40,$40,$40,$40,$40
               FCB  $05,$05,$05,$05,$05,$05,$05,$05,$05,$05,$05,$05,$05,$05,$05,$05
               FCB  $05,$05,$05,$05,$05,$05,$05,$05,$05,$06,$06,$06,$06,$06,$06,$06
               FCB  $06,$06,$06,$06,$06,$06,$06,$06,$06,$06,$06,$06,$06,$06,$06,$06
               FCB  $06,$06,$06,$07,$07,$07,$07,$07,$07,$07,$07,$07,$07,$07,$07,$07
               FCB  $07,$07,$07,$07,$07,$07,$07,$07,$07,$07,$07,$07,$08,$08,$08,$08
               FCB  $08,$08,$08,$08,$08,$08,$08,$08,$08,$08,$08,$08,$08,$08,$08,$08
               FCB  $08,$08,$08,$08,$08,$08,$09,$09,$09,$09,$09,$09,$09,$09,$09,$09
               FCB  $09,$09,$09,$09,$09,$09,$09,$09,$09,$09,$09,$09,$09,$09,$09,$10
```

```
humid_lo    FCB    $00,$04,$08,$12,$16,$20,$24,$27,$31,$35,$39,$43,$47,$51,$55,$59
            FCB    $63,$67,$71,$75,$78,$82,$86,$90,$94,$98,$02,$06,$10,$14,$18,$22
            FCB    $25,$29,$33,$37,$41,$45,$49,$53,$57,$61,$65,$69,$73,$76,$80,$84
            FCB    $88,$92,$96,$00,$04,$08,$12,$16,$20,$24,$27,$31,$35,$40,$43,$47
            FCB    $51,$55,$59,$63,$67,$71,$75,$78,$82,$86,$90,$94,$98,$02,$06,$10
            FCB    $14,$18,$22,$25,$29,$33,$37,$41,$45,$49,$53,$57,$61,$65,$69,$73
            FCB    $77,$80,$84,$88,$92,$96,$00,$04,$08,$12,$16,$20,$24,$27,$31,$35
            FCB    $39,$43,$47,$51,$55,$59,$63,$67,$71,$75,$78,$82,$86,$90,$94,$98
            FCB    $02,$06,$10,$14,$18,$22,$25,$29,$33,$37,$41,$45,$49,$53,$57,$61
            FCB    $65,$69,$73,$76,$80,$84,$88,$92,$96,$00,$04,$08,$12,$16,$20,$23
            FCB    $27,$31,$35,$39,$43,$47,$51,$55,$59,$63,$67,$71,$75,$78,$82,$86
            FCB    $90,$94,$98,$02,$06,$10,$14,$18,$22,$26,$29,$33,$37,$41,$45,$49
            FCB    $53,$57,$61,$65,$69,$73,$76,$80,$84,$88,$92,$96,$00,$04,$08,$12
            FCB    $16,$20,$24,$27,$31,$35,$39,$43,$47,$51,$55,$59,$63,$67,$71,$75
            FCB    $78,$82,$86,$90,$94,$98,$02,$06,$10,$14,$18,$22,$25,$29,$33,$37
            FCB    $41,$45,$49,$53,$57,$61,$65,$69,$73,$76,$80,$84,$88,$92,$96,$00
            END
```

In C language,

```c
#include <hc11.h>
unsigned char humid_up [] = {...};      /* upper byte of the humidity table */
unsigned char humid_lo [] = {...};      /* lower byte of the humidity table */
main ( )
{
    unsigned bytes[3];                  /* display data */
    unsigned char i, conf_dat;          /* loop index and configuration data */
    DDRD    = 0x3A;
    SPCR    = 0x54;
    OPTION &= 0xBF;                     /* select E clock to control A/D converter */
    OPTION |= 0x80;                     /* start the A/D charge pump */
    bytes[0] = 0xA0;                    /* most significant byte of display data is a constant */
    TOC2 = TCNT + 200;                 /* start an OC2 operation to create 100 µs delay */
    TFLG1 = 0x40;                      /* clear OC2F */
    while (!(TFLG1 & 0x40));            /* wait for 100 µs */
    while (1) {
        ADCTL = 0x00;                  /* start an A/D conversion */
        while (!(ADCTL & 0x80));       /* wait until A/D conversion is complete */
        if (!(humid_up [ADR1] & 0xF0))
            conf_dat = 0x91;           /* choose special decode mode to blank msd */
        else   conf_dat = 0x01;        /* choose normal decode mode */
        PORTD &= 0xDF;                 /* enable SPI transfer to the MC14489 */
        SPDR = conf_dat;               /* send out configuration data */
        while (!(SPSR & 0x80));         /* wait until SPI transfer is done */
        PORTD |= 0x20;                 /* load data into configuration register */
        bytes[1] = humid_up[ADR1];
        bytes[2] = humid_lo[ADR1];
```

```
        PORTD &= 0xDF;                /* enable SPI transfer to the MC14489 */
        for (i = 0; i < 3; i++) {     /* send out humidity data for display */
            SPDR = bytes[i];
            while(!(SPSR & 0x80));
        }
        PORTD |= 0x20;                /* load data into display register */
        TFLG1    = 0x40;              /* clear OC2F flag */
        TOC2     = TCNT + 20000;      /* start an OC2 operation */
            for (i = 0; i < 100; i++) {   /* create a delay of 1 second */
                while (!(TFLG1 & 0x40));
                TFLG1 = 0x40;
                TOC2 += 20000;
            }
    }
    return 0;
}
```

▲

11.12   Processing the A/D Results

The results of the A/D conversion often need to be processed. One of the most useful measurements of an AC signal is the root-mean-square (RMS) value. The root-mean-square value of an AC signal is defined as:

$$V_{RMS} = \sqrt{\frac{1}{T} \int_0^T V^2(t)dt} \qquad\qquad 11\text{-}7$$

Because the 68HC11 cannot calculate the integration, the following equation will be used to approximate the RMS value:

$$V_{RMS} = \sqrt{\frac{1}{N} \sum_{i=0}^{N-1} V^2(t)} \qquad\qquad 11\text{-}8$$

To make this equation a good approximation of the real RMS value, samples must be as equally spaced in time as possible. It is obvious that the more samples collected over the time period, the better the result.

Example 11.11

▼

Write a short program to compute the average of the squared values of 64 samples that are stored at the memory locations starting with the label *sample* and save the result at the memory locations starting with the label *sq_ave*.

Solution: The square of an 8-bit value requires two bytes to hold it, and three bytes (at locations sq_sum..sq_sum+2) are needed to store the sum of sixty-four 16-bit values. The average of these 64 values will need to be stored in two bytes. Division-by-64 can be performed by shifting the sum to the right by six positions. The program is:

```
            ORG     $D000
samples     RMB     64
sq_sum      RMB     3

sq_ave      RMB     2
            ORG     $C000       ; starting address of the program
            LDX     #samples    ; load the starting address of samples
            CLR     sq_sum      ; initialize the sum to 0
            CLR     sq_sum+1    ;       "
            CLR     sq_sum+2    ;       "
            LDY     #64         ; initialize the loop count
loop        LDAA    0,X         ; get a sample
            TAB
            MUL                 ; compute the square of the sample
            ADDD    sq_sum+1    ; add to the temporary sum
            STD     sq_sum+1    ; save the temporary sum
            LDAB    sq_sum      ; add the carry to the upper byte
            ADCB    #0          ;       "
            STAB    sq_sum      ;       "
            INX                 ; move to the next sample
            DEY                 ; decrement the loop count
            BNE     loop        ; repeat if loop count is not 0 yet
            LDAB    #6          ; prepare to shift right six times

* The next five instructions divide the sum by 64

loop1       LSR     sq_sum
            ROR     sq_sum+1
            ROR     sq_sum+2
            DECB
            BNE     loop1

* Transfer the result to sq_ave and sq_ave+1
            LDAA    sq_sum+1
            STAA    sq_ave
            LDAA    sq_sum+2
            STAA    sq_ave+1
            END
```

The next step in computing an RMS value is to compute the square root of the sum. Integer arithmetic can be used to perform this computation. The technique is based on the following equation:

$$\sum_{i=0}^{n-1} i = \frac{n(n-1)}{2}$$

11-9

This equation can be transformed into:

$$n^2 = \sum_{i=0}^{n-1} (2i + 1)$$

11-10

Equation 11.10 gives us a clue about how to compute the square root of a value. Suppose we want to compute the square root of p, and n is the integer value that is closest to the true square root. One of the following three relationships is satisfied:

$n^2 < p$
$n^2 = p$
$n^2 > p$

The logic shown in Figure 11.15 can be used to find a value for n, and this algorithm is translated into the program in Example 11.12. The flowchart in Figure 11.15 stops the iteration when $sum \geq p$. However, it may not find the closest square root when $sum > p$. This problem can be corrected by making a minor modification to the program, as will be illustrated in Example 11.13.

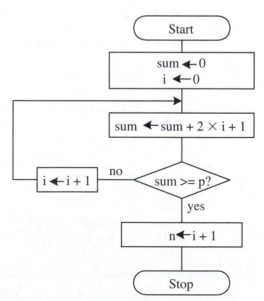

Figure 11.15 ■ The logic flow for finding the square root

Example 11.12

Write a program to compute the square root of the average value computed in Example 11.11 (stored at *sq_ave*) and save the square root at *sq_root*.

Solution: The program is as follows:

```
i           RMB     1                   ; index i
sq_root     RMB     2
            ...
            CLR     sq_root             ; initialize ∑(2i+1) to 0
            CLR     sq_root+1           ;      "
            LDAB    #-1
            STAB    i                   ; place a -1 into i
repeat      INC     i                   ; i starts from 0 because of the previous instruction
            LDAB    i
            CLRA
            LSLD                        ; compute 2i
            ADDD    #1                  ; compute 2i+1
            ADDD    sq_root             ; add 2i + 1 to accumulating sum
            STD     sq_root             ; update accumulating sum
            CPD     sq_ave              ; compare to n²
            BLO     repeat              ; repeat when ∑(2i+1) < n²
            INC     i                   ; add 1 to i to obtain the square root
            LDAA    i                   ; place the square root in sq_root
            STAA    sq_root             ;      "
            END                         ; square root is in memory location i
```

Example 11.13

Add a sequence of instructions to the program in Example 11.12 to find the closest square root of the square sum.

Solution: If n is the true square root of p (i.e., $n^2 = p$), then there is no problem. If n is not the true square root of p, then either n or $n - 1$ is the integer closest to the true square root of p. The choice between n and $n - 1$ can be made by comparing the following two expressions:

$$n^2 - p \qquad (1)$$
$$p - (n - 1)^2 \quad (2)$$

If the value of expression (1) is smaller, choose n; otherwise, choose $n - 1$. Note that both values are nonnegative.

This idea can be translated into the following instruction sequence:

```
sq_diff1    RMB     2
temp        RMB     2

            LDAA    i                   ; place n in A
            TAB                         ; also place n in B
            MUL                         ; compute n²
            CPD     sq_ave              ; compare to p
            BEQ     exit                ; n is the true square root of p
```

* We need to compare expression (1) and (2) in the following

```
            SUBD    sq_ave              ; compute n² - p
            STD     sq_diff1            ; save it in the memory
            LDAA    i
            DECA                        ; compute n - 1
            TAB                         ; place n - 1 in B
            MUL                         ; compute (n - 1)²
            STD     temp                ; save (n - 1)² in the memory
            LDD     sq_ave              ; place p in D
            SUBD    temp                ; compute p - (n - 1)²
            CPD     sq_diff1            ; compare p - (n - 1)² to n² - p
            BHI     exit                ; n is closer to the true square root
            DEC     i                   ; n - 1 is closer to the true square root
exit        .                           ; do something else
            .
            .
```

This instruction sequence should be added after the last instruction (INC i) of the program in Example 11.12.

▲

Example 11.14

▼

Write a C routine to compute the root-mean-square value of an array of 8-bit unsigned integers.

Solution: The starting address of the array and the array count are passed to this routine.

```
unsigned char root_mean_sq (unsigned char *samples, unsigned int n)
{
    int i;
    unsigned int sq_ave;
    unsigned long int sq_sum;
    unsigned int temp;
```

```
        sq_sum = 0;
        for (i = 0; i < n; i++)
            sq_sum += samples[i] * samples[i];
        sq_ave = sq_sum / n;
        temp = 0;
        i = 0;
        while (temp < sq_ave) {
            temp += 2 * i + 1;
            i ++;
        }
        if (temp == sq_ave) return i;
        if ((i * i - sq_ave) < (sq_ave - (i - 1) * (i - 1))
            return i;
        else return i - 1;
    }
```

11.13   Using the External A/D Converter MAX1241

The 68HC11 members have either an 8-bit or a 10-bit A/D converter.
Some applications require a resolution higher than 10 bits. Using an external
A/D converter becomes necessary for these applications. We will examine a
12-bit A/D converter from MAXIM.

The MAX1241 is a 12-bit A/D converter with serial interface that is compatible to the SPI function. The block diagram of this device is shown in
Figure 11.16. The MAX1241 has the following features:

- single analog input channel with track-and-hold circuit
- uses successive-approximation method to perform conversion and
  completes one conversion in 7.5 µs
- direct interface to the SPI interface
- single-supply operation: +2.7 V to +5.25 V
- analog input ranges from 0 to 5 V with 5 V supply

11.13.1   MAX1241 Signal Pins

The MAX1241 has eight signal pins. The function of each signal is as follows:

- $\overline{CS}$: *chip select.* The signal $\overline{CS}$ is active-low and it initiates conversions on the falling edge. When $\overline{CS}$ is high, the DOUT pin is
  in a high-impedance state
- DOUT: *serial data output.* Data changes state at SCLK's falling
  edge. DOUT is in a high impedance state when $\overline{CS}$ is high.
- AIN: *analog input.* 0 V to REF range.

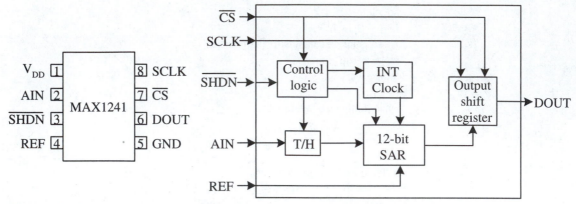

Figure 11.16 ■ MAX1241 12-bit A/D converter block diagram

- SCLK: *serial clock input.* SCLK clocks data out at rates up to 2.1 MHz.
- GND and V_{DD}: *device supply pins.* V_{DD} is connected to positive supply voltage.
- REF: *analog reference voltage.* Reference voltage for analog-to-digital conversion.
- $\overline{SHDN}$: *Shut-down input.* Pulling $\overline{SHDN}$ low shuts the MAX1241 down to 15 µA (max) supply current.

11.13.2 Chip Functioning

The MAX1241 uses an input track/hold (T/H) and successive-approximation register (SAR) circuitry to convert an analog input signal to a digital 12-bit output. The analog input must be in the 0 V to V_{REF} range. The conversion, including T/H acquisition time, takes 9 µs. The MAX1241 requires an external reference from 1.0 V to V_{DD}. The easiest way is to use V_{DD} as the reference voltage. The serial interface requires only three digital lines (SCLK, $\overline{CS}$, and DOUT) and provides an easy interface to microcontrollers.

The MAX1241 has two modes: *normal* and *shutdown.* Pulling $\overline{SHDN}$ low shuts the device down and reduces supply current below 10 µA, while pulling $\overline{SHDN}$ high or leaving it open puts the device into operational mode. Pulling $\overline{CS}$ low initiates a conversion. The conversion result is available at DOUT in unipolar serial interface. The serial data stream consists of a high bit, signaling the end of conversion (EOC), followed by the data bits (MSB first).

When power is first applied, and if $\overline{SHDN}$ is not pulled low, the internal reset time is 10 µs after the power supplies have stabilized. No conversions should be performed during these times.

To start a conversion, pull $\overline{CS}$ low. At $\overline{CS}$'s falling edge, the T/H enters its hold mode and a conversion is initiated. After an internally timed conversion period, the end of conversion is signaled by DOUT pulling high. Data can be shifted out serially with the external clock.

The actual conversion does not require the external clock. This allows the conversion result to be read back at the microprocessor's convenience at any clock rate up to 2.1 MHz. The clock duty cycle is unrestricted if each clock phase is at least 200 ns. Do not run the clock while a conversion is in progress.

11.13.3 Timing and Control

The $\overline{CS}$ and SCLK digital inputs control conversion-start and data-read operations. The timing diagrams of Figures 11.17 and 11.18 outline the serial-interface operation. The timing parameters are listed in Table 11.4.

A $\overline{CS}$ falling edge initiates a conversion sequence: the T/H stage holds the input voltage, the ADC begins to convert, and DOUT changes from high impedance to logic low. SCLK must be kept low during the conversion. An internal register stores the data when the conversion is in progress.

End of conversion (EOC) is signaled by DOUT going high. DOUT's rising edge can be used as a framing signal. We can use the SCLK signal to shift the data out of the shift register any time after the conversion is complete. After DOUT goes to high, it transitions on the first falling edge of the SCLK signal. The next falling clock edge shifts out the most significant bit of the conversion result at DOUT, followed by the remaining bits. Since there are 12 data bits and one leading high bit, at least 13 clock periods are needed to shift out

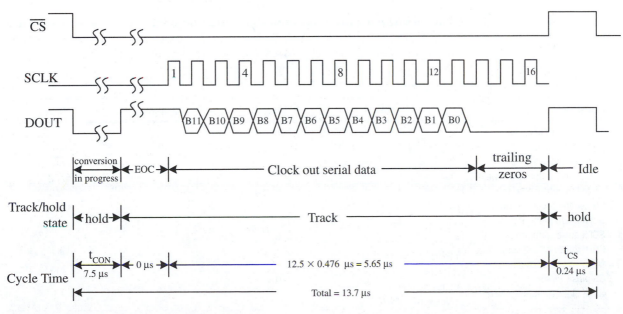

Figure 11.17 ■ Interface timing sequence

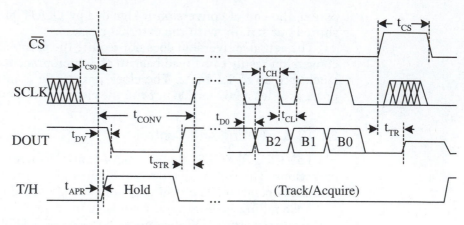

Figure 11.18 ■ Detailed serial-interface timing

a conversion result. Extra clock pulses occurring after the conversion result has been clocked out, and prior to a rising edge of the $\overline{CS}$ signal, produce trailing zeros at DOUT and have no effect on conversion operation.

Minimum cycle time is accomplished by using DOUT's rising edge as the EOC signal. Clock out the data with 12.5 clock cycles at full speed. Pull $\overline{CS}$ high after reading the conversion's least significant bit. After the specified minimum time (t_{cs}), $\overline{CS}$ can be pulled low again to initiate the next conversion.

11.13.4   Interfacing the MAX1241 with the 68HC11

The MAX1241 is fully compatible with the 68HC11 SPI system. To use the MAX1241, set the 68HC11's SPI in master mode. The circuit connections between the 68HC11 and the MAX1241 are shown in Figure 11.19.

Parameters	Symbol	Min (ns)	Typ (ns)	Max (ns)
Acquisition time	t_{ACQ}	1.5		
Conversion time	t_{CONV}	5.5		7.5
Aperture delay	t_{APR}		30	
SCLK fall to output data valid	t_{DO}	20		200
CS fall to output enable	t_{DV}			240
CS rise to output disable	t_{TR}			240
SCLK pulse width high	t_{CH}	200		
SCLK pulse width low	t_{CL}	200		
SCLK low to CS fall setup time	t_{CSO}	50		
DOUT rise to SCLK rise	t_{STR}	0		
CS pulse width	t_{CS}	240		
SCLK clock (frequency MHz)	f_{SCLK}	0		2.1

Table 11.4 ■ MAX1241 timing parameters

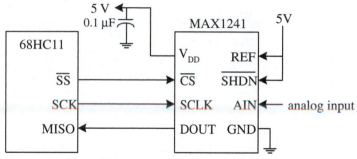

Figure 11.19 ■ SPI connection to the MAX1241

The procedure for A/D conversion is:

Step 1
Use a general-purpose I/O line ($\overline{\text{SS}}$ pin) to pull $\overline{\text{CS}}$ low. Keep SCLK low.

Step 2
Wait for the maximum conversion time specified before activating SCLK. Alternatively, look for a DOUT rising edge to determine the end of conversion.

Step 3
Activate SCLK for a minimum of 13 clock cycles. The first falling clock edge produces the MSB of the DOUT conversion. DOUT data is available MSB first and can be clocked into the 68HC11 on SCLK's rising edge. Since the SPI activates eight clock pulses with each byte written into the SPCR register, we need to write two bytes into the SPCR register. Sixteen clock pulses are sent to the MAX1241 on the SCLK pin. Two bytes will be shifted into the SPI with one leading 1 and 3 trailing 0s. This transaction is illustrated in Figure 11.20. Because the SCLK clock must be idle low, we need to set CPOL = CPHA = 0.

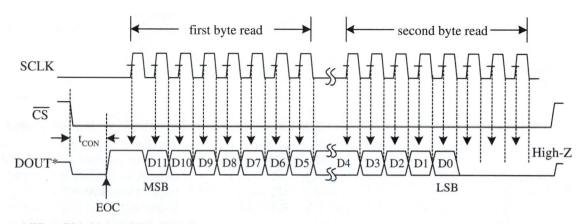

* When CS is high, DOUT = high-Z

Figure 11.20 ■ SPI serial interface timing (CPOL = CPHA = 0)

Step 4
Pull $\overline{CS}$ high at or after the 13th falling clock edge (actually the 16th clock edge).

Step 5
With $\overline{CS}$ = high, wait for the minimum specified time, t_{cs}, before initiating a new conversion by pulling $\overline{CS}$ low.

Example 11.15

Write a subroutine to perform an A/D conversion operation for the circuit shown in Figure 11.19 and return the result in double accumulator D.

Solution:

```
REGBAS      EQU     $1000           ; I/O register block base address
PORTD       EQU     $08
DDRD        EQU     $09             ; offset of DDRD from REGBAS
SPCR        EQU     $28             ; offset of SPCR from REGBAS
SPSR        EQU     $29             ; offset of SPSR from REGBAS
SPDR        EQU     $2A             ; offset of SPDR from REGBAS
CS          EQU     $20             ; value to select the CS pin
DOUT        EQU     $04             ; value to select the DOUT pin

get_sample  PSHX
            PSHY
            LDX     #REGBAS
            LDAA    #$3A
            STAA    DDRD,X          ; configure SPI pin directions
            LDAA    #$50            ; enable SPI, disable SPI interrupt, set
*                                   ; master mode,
            STAA    SPCR,X          ; set transfer rate to 1-Mbits/s, use rising edge
*                                   : to shift data
            BCLR    PORTD,X CS      ; start an A/D conversion
            BRCLR   PORTD,X DOUT *  ; wait until DOUT goes high
            STAA    SPDR,X          ; shift in the upper byte of the result using
*                                   ; SPI transfer
            BRCLR   SPSR,X $80 *    ; wait until the SPI transfer is complete
            LDAA    SPDR,X          ; place the upper byte of the A/D result in A
            STAB    SPDR,X          ; shift in the lower byte of the conversion result
            BRCLR   SPSR,X $80 *    ; wait until the SPI transfer is complete
            LDAB    SPDR,X          ; place the lower byte in B
            BSET    PORTD,X CS      ; prepare for the next A/D conversion
            LSLA                    ; clear the bit 7 of accumulator A to 0
            LSRA                    ;        "
```

```
            LSRD                    ; place the 12-bit A/D conversion result in
  *                                 ; the lower
            LSRD                    ; 12 bits of the double accumulator D
            LSRD                    ;        "
            PULY
            PULX
            RTS
```

In C language,

```c
unsigned int get_sample ( )
{
    unsigned char x1;
    unsigned a2d_result;

    DDRD = 0x3A;
    SPCR = 0x50;
    PORTD &= 0xDF;              /* start the A/D conversion */
    while (!(PORTD & 0x04));    /* wait until A/D conversion is done by
                                   checking DOUT */
    SPDR = 0x00;               /* shift in the upper byte of the A/D conversion
                                   result */

    while (!(SPSR & 0x80));     /*       "       */
    x1 = SPDR;                  /* place the upper byte of the A/D result in x1 */
    SPDR = 0x00;               /* shift in the lower byte of the A/D conversion
                                   result */

    while (!(SPSR & 0x80));     /*       "       */
    PORTD |= 0x20;             /* disable A/D converter */
    a2d_result = x1 * 256 + SPDR;  /* combine the upper and lower bytes of A/D
                                   result */

    a2d_result = a2d_result << 1;  /* place the 12-bit A/D result in the lower
                                   12 bits */

    a2d_result = a2d_result >> 4;  /*       "       */
    return a2d_result;
}
```

11.14 Summary

A data acquisition system consists of four major components: a transducer, a signal conditioning circuit, an A/D converter, and a computer. The transducer converts a nonelectric quantity into a voltage. The output of the transducer may not be appropriate for processing by the A/D converter. The signal conditioning circuit shifts and scales the output from a transducer to a range that can take advantage of the full capability of the A/D converter. The

A/D converter converts an electrical voltage into a digital value. The computer does the final processing.

The accuracy of an A/D converter is dictated by the number of bits used to represent the analog quantity. The greater the number of bits, the better the accuracy. Most microcontrollers use the successive-approximation method to implement their A/D converters.

The 68HC11 A series microcontrollers implement an 8-bit A/D converter. The highest and lowest analog input voltages are limited to +5 V and 0 V, respectively. The A/D converter needs a clock signal to perform the conversion. The user can choose either the E clock or the on-chip RC circuit output as the clock source. The RC-circuit output should be chosen when the E clock frequency is lower than 750 KHz. The conversion of one sample takes 32 clock cycles to complete. Four registers are provided to hold the conversion results.

This chapter uses the LM35 temperature sensor from National Semiconductor and the IH-3605 humidity sensor from HyCal Engineering as examples to demonstrate the A/D conversion process.

Some applications require an A/D resolution higher than ten bits, which cannot be provided by the on-chip A/D converter of most microcontrollers. We must use an external A/D conversion chip to satisfy this requirement. A/D converter chips may have a parallel or a serial interface. An A/D converter with a serial interface has fewer pins and smaller package. The MAX1241 is a 12-bit A/D converter that has an SPI-compatible serial interface. An example explains how to start an A/D conversion and shift the result into the 68HC11 using the SPI system.

11.15   Exercises

E11.1. Explain the difference between digital and analog signals.

E11.2. Survey the available D/A and A/D methods.

E11.3. Survey methods and devices for converting temperature, pressure, mass flow, humidity, height, and weight into electric voltage.

E11.4. How long does it take to complete the A/D conversion of four samples if the E clock frequency is 3 MHz and the E clock is used as the clock signal to the 68HC11 A/D converter?

E11.5. Design a signal conditioning circuit that can scale and shift the voltage range of 0 mV ~ +10 mV to 0 V ~ 5 V.

E11.6. Design a signal conditioning circuit to shift and scale the voltage range of −40~240 mV to 0 −5 V.

E11.7. Design a signaling conditioning circuit to shift the voltage range of 2~4 V to 0~5 V.

E11.8. What value should be written into the ADCTL register so that the 68HC11 A/D converter will convert signals from channel 3 in the scan mode continually?

E11.9. Assume that the V_{RH} and V_{RL} inputs to the 68HC11 A/D converter are 5 V and 0 V, respectively. What are the corresponding voltages for the conversion results of 25, 64, 100,150, and 200?

E11.10. For the circuit in Figure 11.19, what are the corresponding voltages for the conversion results of 128, 512, 768, 1024, 2400, and 3072?

E11.11. Use the 68HC11 A/D converter to sample an unknown signal applied at channel AN0 (pin PE0). Measure the signal every 10 ms and compute the average value at the end of 1 second. Display the average value using two seven-segment displays. Use port B to display the average voltage. Repeat the same measurement forever. The following components are used in this experiment:

- Two common cathode (or anode) seven-segment displays
- Two seven-segment display drivers. Use a 74LS47 for common-anode displays or a 74LS48 for common-cathode displays.
- Fourteen 100-Ω resistors.
- V_{RL} and V_{RH} are set to 0 V and 5 V, respectively.

E11.12. Design a digital thermometer so that it can measure a temperature range from $-10°$ C to $+50°$ C. The requirements are:

1. Display the temperature in a three-digit format: two integer digits and one fractional digit. The sign and the decimal point must be displayed.
2. Use four seven-segment displays to display the temperature. Segment h of the left-most display is used as the sign. When the temperature is negative, light the segment **h**; otherwise, turn off the left-most display.
3. Use a 12-bit A/D external converter such as the MAX1241 to achieve accuracy.
4. Build a signal conditioning circuit that can scale and shift the voltage output to the 0 V to 5 V range so that it can be properly handled by the external converter.
5. Measure the temperature ten times every second and display the average. Update the display once every second.
6. Write a program to perform the required control functions.

E11.13. *Measure the barometric pressure.* The Sensortechnics Inc. manufactures a Barometric Pressure Transducer (BPT) kit that has three external connectors: $+V_s$, GND, and V_{OUT}. The operating parameters are listed in Table 11E.1. This product can be found in an electronic components catalog prepared by Farnell, Inc.

The BPT is a calibrated and signal-conditioned transducer that provides a true 4.5 to 5.5 V output in the barometric pressure ranging from 800 to 1100 mbar. Internal voltage regulation allows the device to operate on a power

Parameter	Value
Reference conditions	V_s = 8.0 V, T (ambient) = 25° C, RL = 100 KΩ
Supply voltage	7 to 24 V DC
Operating pressure	800 to 1100 mBar
Breakdown pressure	2 Bar
Voltage output (span)	5.0 V ÷ 500 mV
Operating temperature	−40° C to 85° C
Compensated range	−10° C to 60° C
Nonlinearity and hysteresis	0.1% FSO (max.)
Repeatability	0.2% FSO (typ.)
Temperature shift (−10° C to 60° C)	0.3% FSO/10° C (max)
Response time	1 msec (typ.)
Long-term stability	0.1% FSO (typ.)

Table 11.E1 ■ BPT parameters

supply between 7 V and 24 V. Applications include barometry, weather stations, and absolute pressure compensation in sensitive equipment. A potentiometer is provided to adjust for changes in altitude. The transducer is designed for use in noncorrosive, nonionic media, e.g., dry air and gasses.

Design a circuit that can scale and shift the sensor output to 0~5 V and add four seven-segment displays to show the barometric pressure. Write a program to measure and update the display once every second.

11.16   Lab Exercises and Assignments

L11.1. *Simple A/D conversion experiment.* Use either the EVB, EVBU, or CMD-11A8 demo board and perform the following steps:

Step 1
Adjust the function generator output to between 0 and 5 V. Set the frequency to about 1 KHz.

Step 2
Connect the AN0 (PE0) pin to the function generator output.

Step 3
Set V_{RH} and V_{RL} inputs to the single board 68HC11 computer to 5 V and 0 V, respectively.

Step 4
Write a program that does the following operations:

1. starts the A/D converter
2. takes 32 samples, convert them, and stores the A/D conversion results at $00 ~ $1F
3. computes the average value and saves the result at $20

Step 5

Assemble the program and download it to the single-board computer for execution.

Step 6

Run the program for sine, triangular, and square waves. Compare the result with manual computation result.

L11.2. *Digital thermometer.* Use the National Semiconductor LM34 temperature sensor, the CA3140 op amp, and an MAX1241 12-bit external A/D converter to construct a digital thermometer. The temperature should be displayed in three integral and one fractional digits. Use the 14489 to drive seven-segment displays. The pin assignment and the circuit connections are shown in Figure 11.L1. The requirements are:

1. The temperature range is from 0° F to 212° F.

2. Use four seven-segment displays to display the temperature reading, which must include one fractional digit.

3. Use a water bath to change the ambient temperature to the LM34. You must insulate the LM34 so that it does not get wet.

4. Measure the temperature ten times in a second and display the average of the measurements.

5. Update the temperature once every second.

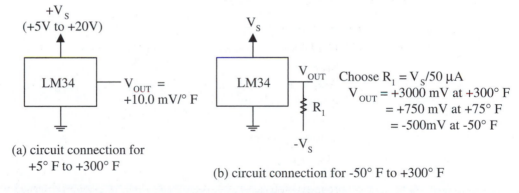

(a) circuit connection for $+5^\circ$ F to $+300^\circ$ F

(b) circuit connection for -50° F to $+300^\circ$ F

Figure 11.L1 ■ Circuit connection for the LM34

Appendix A

Number Systems

A.1 Objectives

After completing this chapter you should be able to

- convert a decimal number to a binary, octal, or hexadecimal number

- convert a binary, octal, or hexadecimal number to a decimal number

- convert numbers to their two's complement representation

- add, subtract, and negate binary, octal, and hexadecimal numbers

- describe the BCD, excess-3, 2-out-of-5, and Gray codes

A.2   An Introduction to Number Systems

For most of our lives, we have been exposed to the base 10 number system. Our preference for a 10-digit number system is probably based on our ten fingers. However, base 10 is not a natural system for digital hardware, where arithmetic is based on the binary digits 0 and 1. Binary numbers are used in digital electronics and computer and data communication applications to represent the logic high and low states of digital circuits. All information in a digital system is encoded and transmitted as binary numbers. This appendix will cover the main positional number systems related to digital hardware: the decimal, binary, octal, and hexadecimal systems. The concepts covered in this section are applied in many of the logic circuit examples throughout this text. Conversion between number systems and arithmetic in other systems will be discussed in the following sections.

In this appendix, the base of a number is written as a subscript on its lower right. For example, the binary number 1101 is written as 1101_2.

A.3   Binary, Octal, and Hexadecimal Numbers

In any system, a number is written from its most significant digit on the left to its least significant digit on the right. In general, an $(n + m)$-digit number in any number system can be represented by the mathematical equation

$$N = A_{n-1}r^{n-1} + A_{n-2}r^{n-2} + ... + A_0r^0 + A_{-1}r^{-1} + A_{-2}r^{-2} + ... + A_{-m}r^{-m} \qquad \text{A-1}$$

where the **A**s are any of the digits allowed in the number system and **r** is the *radix* or *base* of the system. The digits allowed in a number system run from 0 to $r - 1$. The allowable digits in decimal, binary, octal, and hexadecimal number systems are listed in Table A.1. Equation A-1 also provides a method for converting a number in any number system to decimal number representation.

Number system	Allowable digits
decimal	0, 1, 2, 3, 4, 5, 6, 7, 8, 9
binary	0, 1
octal	0, 1, 2, 3, 4, 5, 6, 7
hexadecimal	0, 1, 2, 3, 4, 5, 6, 7, 8, 9, A, B, C, D, E, F*

*A, B, C, D, E, and F correspond to the decimal numbers 10,11,12,13,14, and 15

Table A.1 ■ Allowable digits in the decimal, binary, octal, and hexadecimal systems

The primary reason for introducing hexadecimal and octal numbers is their natural correspondence with binary numbers. They permit compact representation of machine code instructions, data, and memory addresses. Hexadecimal numbers are used more often than octal numbers.

A.3.1 Binary-to-Decimal Conversion

A binary number can be easily converted to its decimal equivalent using equation A–1. For example, the decimal equivalent of the binary number 10010111_2 can be computed as follows:

$$10010111_2 = 1 \times 2^7 + 0 \times 2^6 + 0 \times 2^5 + 1 \times 2^4 + 0 \times 2^3 + 1 \times 2^2 + 1 \times 2^1 + 1 \times 2^0 = 151_{10}$$

Example A.1

Convert the binary number 100100110.101_2 to its decimal equivalent.

Solution: Simply apply equation A–1,

$$100100110.101_2 = 1 \times 2^8 + 0 \times 2^7 + 0 \times 2^6 + 1 \times 2^5 + 0 \times 2^4 + 0 \times 2^3 + 1 \times 2^2 + 1 \times 2^1$$
$$+ 0 \times 2^0 + 1 \times 2^{-1} + 0 \times 2^{-2} + 1 \times 2^{-3} = 294.625_{10}$$

A.3.2 Decimal-to-Binary Conversion

Converting a number from the decimal number system to the binary number system requires two separate methods: one to convert integers and one to convert fractional numbers. Decimal numbers containing both integer and fractional parts are converted to binary by applying those techniques separately to the integer and fractional parts of the number. Any value that is an integer in one number system will remain so in another number system, and the same is true for fractions.

Decimal integers are converted to binary integers by a method of repeated division by 2. The integer is divided by 2 and the remainder, either a 0 or a 1, is the binary value for the bit position of the least significant bit (LSB). The quotient resulting from the first division is the next value to be divided by 2. The remainder from the second division process becomes the second-least significant bit of the binary number, and so on. Successive divisions produce the bit values occupying positions farther and farther away from the binary point until the most significant bit (MSB) is reached. Division is continued until the quotient is 0 and no further division can take place.

Example A.2

▼

Convert the decimal integer 53_{10} to its binary equivalent.

Solution: Using the repeated division method and the result is shown in Table A.2.

Division	Quotient	Remainder	Binary equivalent	
53 ÷ 2	26	1	1	LSB
26 ÷ 2	13	0	0	
13 ÷ 2	6	1	1	
6 ÷ 2	3	0	0	
3 ÷ 2	1	1	1	
1 ÷ 2	0	1	1	MSB

Table A.2 ■ Division process for example A.2

We can multiply the resultant binary number by bit weights to verify our answer:

$$110101_2 = 1 \times 2^5 + 1 \times 2^4 + 0 \times 2^3 + 1 \times 2^2 + 0 \times 2^1 + 1 \times 2^0 = 53_{10}$$

The result is verified as correct.

▲

Decimal fractions are converted to binary fractions by a method of repeated multiplication by 2. The fraction is multiplied by 2, and the carry-out into the first integer position, either a 0 or 1, is the bit value for the MSB of the binary fraction. The fractional result from the first multiplication is then multiplied by 2, and its carry-out into the integer portion of the number is the next bit value. Each successive multiplication produces binary bit value to occupy the next position, moving from the binary point toward the LSB. The multiplication process continues until the fractional part of the number is 0 or until the desired level of precision has been met if exact conversion is impossible.

It is often necessary to decide how many multiplication steps must be performed in order to obtain the same degree of precision that the original decimal number had. Note that in the scientific representation, the right-most fractional digit is often inaccurate due to rounding. The number of multiplication steps to be performed can be determined as follows:

Suppose the decimal number has m fractional digits. Then the number has an inaccuracy of 10^{-m}. The number of multiplication steps to be performed (k) should make the resulting binary as precise as 10^{-m}. k can be computed by the following equation:

$$2^k \times 10^{-m} \geq 1$$

For example, if a decimal number has a fractional part with three digits, then the number of multiplication steps is computed as follows:

$2^k \times 10^{-3} \geq 1 \Rightarrow 2^k \geq 1000$
$\therefore k \geq 10$

Therefore, at least ten multiplication steps must be performed in order to obtain the binary equivalent with the same degree of precision as the original decimal number.

Example A.3

▼

Convert the decimal fraction $.24_{10}$ to its binary equivalent with the same degree of precision.

Solution: Multiply the fraction by 2. The carry-out is the bit value. Since there are 2 fractional digits in the decimal representation, the number of multiplication steps (k) to be performed should be such that $2^k \times 10^{-2} \geq 1$ if exact conversion is impossible. Therefore, $k \geq 7$.

Multiplication	Product	Carry-out	
0.24 x 2	0.48	0	MSB
0.48 x 2	0.96	0	
0.96 x 2	1.92	1	
0.92 x 2	1.84	1	
0.84 x 2	1.68	1	
0.68 x 2	1.36	1	
0.36 x 2	0.72	0	LSB

Table A.3 ■ Conversion table for Example A.3

▲

Example A.4

▼

Convert the decimal value 53.375_{10} to binary.

Solution: The integer and fraction must be converted separately.

$53.375_{10} = 53_{10} + .375_{10}$

Integer part:

$53_{10} = 110101_2$

Division	Quotient	Remainder	
53 ÷ 2	26	1	LSB
26 ÷ 2	13	0	
13 ÷ 2	6	1	
6 ÷ 2	3	0	
3 ÷ 2	1	1	
1 ÷ 2	0	1	MSB

Table A.4 ■ Integer part conversion result for Example A.4

Fraction part:

$.375_{10} = .011_2$

Multiplication	Product	Carry-out	
0.375 x 2	0.75	0	MSB
0.75 x 2	1.50	1	↓
0.50 x 2	1.00	1	LSB

Table A.5 ■ Fractional part conversion of Example A.4

Result: $53.375_{10} = 110101.011_2$

A.3.3   Octal to Decimal Conversion

An octal number can be converted to its decimal equivalent using equation A–1.

Example A.5

Convert the octal number 127.75_8 to its decimal equivalent.

Solution: Apply equation A–1,

$$127.75_8 = 1 \times 8^2 + 2 \times 8^1 + 7 \times 8^0 + 7 \times 8^{-1} + 5 \times 8^{-2} = 87.953125_{10}$$

Example A.6

Convert the octal number 376.444_8 to its decimal equivalent.

Solution: Apply equation A–1,

$$376.444_8 = 3 \times 8^2 + 7 \times 8^1 + 6 \times 8^0 + 4 \times 8^{-1} + 4 \times 8^{-2} + 4 \times 8^{-3} = 254.5703125_{10}$$

A.3.4 Decimal to Octal Conversion

As in decimal to binary conversion, a decimal integer is converted to its octal equivalent by a method of repeated division by 8, while a decimal fraction is converted to its octal equivalent by a method of repeated multiplication by 8. The decimal fraction to be converted is multiplied by 8, and the carry-out into the first integer position is the octal value for the MSB of the octal fraction. The fractional result from the first multiplication is then multiplied by 8, and its carry-out into the integer portion of the number is the next octal value. Each successive multiplication produces octal digit value to occupy the next position, moving from the fractional point toward the LSB. The multiplication process continues until the fractional part of the number is 0 or until the desired degree of precision has been met if the exact conversion is impossible.

Also as in decimal to binary conversion, it is often necessary to decide how many multiplication steps should be performed to obtain the same degree of precision as were found in the original decimal number, because the rightmost fractional digit is often inaccurate due to a rounding operation. The number of multiplication steps to carry out can be determined as follows:

If the decimal number has m fractional digits, it has an inaccuracy of 10^{-m}. The number of multiplication steps to be performed (k) should give the resultant octal the same degree of precision as 10^{-m}. k can be computed by the following equation:

$$8^k \times 10^{-m} \geq 1 \tag{A-3}$$

Example A.7

Convert the decimal number 100.625_{10} to its octal equivalent.

Solution: The integer and fractional parts must be converted separately.

$$100.625_{10} = 100_{10} + .625_{10}$$

Integer part: $100_{10} = 144_8$

Division	Quotient	Remainder	
100 ÷ 8	12	4	LSB
12 ÷ 8	1	4	↓
1 ÷ 8	0	1	MSB

Table A.6 ■ Integer part conversion for Example A.7

If an exact conversion is impossible for the fractional part, the minimal number of multiplication steps to be performed in order to maintain the same degree of precision as in the original decimal number is computed by using equation A–3. In this example, it is found to be four.

Fractional part: $.625_{10} = .5_8$

Multiplication	Product	Carry-out
0.625	5.00	5

Result: $100.625_{10} = 144.5_8$

Convert the decimal number 246.245_{10} to its octal equivalent. If an exact conversion for the fractional part cannot be done, perform as many multiplication steps as necessary to maintain the same degree of precision as in the decimal representation.

Solution: The integer and fractional parts must be converted separately.

$246.245_{10} = 246_{10} + .245_{10}$

Integer part (shown in Table A.1): $246_{10} = 366_8$

Division	Quotient	Remainder	
246 ÷ 8	30	6	LSB
30 ÷ 8	3	6	↓
3 ÷ 8	0	3	MSB

Table A.7 ■ Integer part conversion for Example A.8

There are three fractional digits in the decimal number, so the minimal number of multiplication steps needed to maintain the same accuracy as the decimal representation if the exact conversion is impossible is found to be four (using the equation A–3).

Fractional part (shown in Table A.8): $.245_{10} = .1753_8$

Multiplication	Product	Carry-out	
0.245 x 8	1.96	1	MSB
0.96 x 8	7.68	7	↓
0.68 x 8	5.44	5	↓
0.44 x 8	3.52	3	LSB

Table A.8 ■ Fractional part conversion for Example A.8

Result: $246.245_{10} = 366.1753_8$

A.3.5   Hexadecimal to Decimal Conversion

A hexadecimal number can be converted to its decimal equivalent by using equation A–1.

Example A.9

Convert the hexadecimal number $100.4A_{16}$ to its decimal equivalent.

Solution: Apply equation A–1,

$$100.4A_{16} = 1 \times 16^2 + 0 \times 16^1 + 0 \times 16^0 + 4 \times 16^{-1} + 10 \times 16^{-2} = 256.2890625_{10}$$

Example A.10

Convert the hexadecimal number $2FF.88_{16}$ to its decimal equivalent.

Solution: Apply equation A–1,

$$2FF.88_{16} = 2 \times 16^2 + 15 \times 16^1 + 15 \times 16^0 + 8 \times 16^{-1} + 8 \times 16^{-2} = 767.53125_{10}$$

A.3.6   Decimal to Hexadecimal Conversion

A decimal integer can be converted to its hexadecimal equivalent by a method of repeated division by 16, while the decimal fraction can be converted to its hexadecimal equivalent by a method of repeated multiplication by 16. The conversion from the decimal fraction to the hexadecimal fraction may not be exact, and again the user must decide how many digits are necessary. The number of multiplication steps to perform can be determined as follows:

If the decimal number has m fractional digits, it has an inaccuracy of 10^{-m}. The number of multiplication steps to be performed (k) should give the same degree of precision as 10^{-m}. k can be computed using the following equation:

$$16^k \times 10^{-m} \geq 1 \hspace{4em} \text{A–4}$$

Example A.11

Convert the decimal number 420.625_{10} to its hexadecimal equivalent and maintain the same degree of precision as in the decimal representation.

Solution: $420.625_{10} = 420_{10} + .625_{10}$

Integer part (shown in Table A.9): $420_{10} = 1A4_{16}$

Division	Quotient	Remainder	
420 ÷ 16	26	4	LSB
26 ÷ 16	1	10 (or A)	↓
1 ÷ 16	0	1	MSB

Table A.9 ■ Integer part conversion for Example A.11

Fractional part (shown in Table A.10): $.625_{10} = .A_{16}$

Multiplication	Product	Carry-out
0.625 x 16	10.00	10 (or A)

Table A.10 ■ Fractional part conversion for
Example A.11

Result: $420.625_{10} \times 1A4.A_{16}$

Example A.12

Convert the decimal number 980.475_{10} to its hexadecimal equivalent. Maintain the same degree of precision as the decimal representation if exact conversion is impossible.

Solution: $980.475_{10} = 980_{10} + .475_{10}$

Integer part (shown in Table A.11): $980_{10} = 3D4_{16}$

Division	Quotient	Remainder	
980 ÷ 16	61	4	LSB
61 ÷ 16	3	13 (or D)	↓
3 ÷ 16	0	3	MSB

Table A.11 ■ Integer part conversion for Example A.12

Fractional part (shown in Table A.12): $.475_{10} = .799_{16}$

The minimal number of multiplication steps to be performed to obtain the same degree of precision is computed using equation A–4 and is found to be three.

Multiplication	Product	Carry-out	
0.475 x 16	7.6	7	MSB
0.6 x 16	9.6	9	↓
0.6 x 16	9.6	9	LSB

Table A.12 ■ Fractional part conversion for
Example A.12

Result: $980.475_{10} \times 3D4.799_{16}$

A.3.7 Binary to Octal Conversion

Because an octal digit is in the range from 0 to 7, three bits are needed to represent an octal digit. The binary code of each octal digit is shown in Table A.13. The following procedure converts a binary number to its octal equivalent:

1. Start from the fractional point.
2. Partition the integer part from right to left into groups of three bits.
3. Partition the fractional part from left to right into groups of three bits
4. Write each group of three bits as an octal digit.
5. If the leftmost group in the integer part does not have three bits, then add one or two leading zeroes to make it a three-bit group.
6. If the rightmost group in the fractional part does not have three bits, then append one or two zeroes to its right to make it a three-bit group.

Octal digit	Binary code
0	000
1	001
2	010
3	011
4	100
5	101
6	110
7	111

Table A.13 ■ Binary code for
octal digit

Example A.13

▼

Convert the binary number 101011001.110011_2 to its octal equivalent.

Solution: The integer part of this binary number is 101011001. Partition it from right to left into groups of three bits, and write each group of three bits as an octal digit as follows:

$101,011,001_2 = 531_8$

Partition the fractional part from left to right into groups of three bits, and write each group of three bits as an octal digit as follows:

$.110011_2 = .110, 011_2 = .63_8$

Result: $101011001.110011_2 = 531.63_8$

▲

Example A.14

▼

Convert the binary number 1010101010.0010011_2 to its octal equivalent.

Solution: The integer part of this number is 1010101010. Partition it from right to left into groups of three bits, and write each group as an octal digit as follows:

$1010101010_2 = 1,010,101,010_2 = 001,010,101,010_2 = 1252_8$

Partition the fractional part from left to right into groups of three bits, and write each group of three bits as an octal digit as follows:

$.0010011_2 = .001,001,1_2 = .001,001,100_2 = .114_8$

Result: $1010101010.0010011_2 = 1252.114_8$

▲

A.3.8   Octal to Binary Conversion

To convert an octal number to its binary equivalent, simply convert each octal digit to its corresponding binary combination and then delete the leading and trailing zeroes.

Example A.15

▼

Convert the octal number 1375.426_8 to its binary equivalent.

Solution: Convert each octal digit to its corresponding binary code:

$1375.426_8 = 001011111101.100010110_2 = 1011111101.10001011_2$

Example A.16

Convert the octal number 364.231_8 to its binary equivalent.

Solution: Convert each octal digit to its corresponding binary code:

$364.231_8 = 011110100.010011001_2 = 11110100.010011001_2$

A.3.9   Binary to Hexadecimal Conversion

Because a hexadecimal digit is in the range from 0 to 15, four bits are needed to encode it. The binary code of each hexadecimal digit is shown in Table A.14. The following procedure can be used to convert a binary number to its hexadecimal equivalent:

1. Start from the fraction point.
2. Partition the integer part from right to left into groups of four bits.

Hexadecimal digit	Binary code
0	0000
1	0001
2	0010
3	0011
4	0100
5	0101
6	0110
7	0111
8	1000
9	1001
A	1010
B	1011
C	1100
D	1101
E	1110
F	1111

Table A.14 ■ Binary code for hexadecimal digit

3. Partition the fractional part from left to right into groups of four bits.

4. Write each group of four bits as a hexadecimal digit.

5. If the leftmost group in the integer part does not have four bits, then add zeroes to its left to make it a four-bit group.

6. If the rightmost group in the fractional part does not have four bits, then append zeroes to its right to make it a four-bit group.

Example A.17

▼

Convert the binary number $1000111000100100.101000010011_2$ to its hexadecimal equivalent.

Solution: The integer part of the given binary number is 1000111000100100_2. Partition it from right to left into four-bit groups, and write each group of four bits as its corresponding hexadecimal digit:

$1000111000100100_2 = 1000,1110,0010,0100_2 = 8E24_{16}$

The fraction part is $.101000010011_2$. Partition the fraction from left to right into four-bit groups and write each group of four bits into its corresponding hexadecimal digit:

$.101000010011_2 = .1010,0001,0011 = .A13_{16}$

Therefore, the equivalent hexadecimal number is $8E24.A13_{16}$.

▲

Example A.18

▼

Convert the binary number $100011100010.00101001001_2$ to its hexadecimal equivalent.

Solution: The integer part of the given number is 100011100010_2. Partition this number from right to left into four-bit groups, and write each group of four bits as its corresponding hexadecimal digit:

$100011100010_2 = 1000,1110,0010 = 8E2_{16}$

The fractional part is $.00101001001_2$. Partition the fraction from left to right into four-bit groups and write each group of four bits into its corresponding hexadecimal digit as follows:

$.00101001001_2 = .0010,1001,001_2 = .0010,1001,0010_2 = .292_{16}$

Therefore, the hexadecimal equivalent of the given number is $8E2.292_{16}$.

▲

A.3.10   Hexadecimal to Binary Conversion

To convert from hexadecimal to binary, simply convert each hexadecimal digit to its binary code and then delete the leading and trailing zeroes.

Example A.19

Convert the hexadecimal number $6AB2.04_{16}$ to its binary equivalent.

Solution: Convert each hexadecimal digit to its binary code.

$6AB2.04_{16} = 0110,1010,1011,0010.0000,0100_2 = 110101010110010.000001_2$

A.4   Binary Addition and Subtraction

Now that you are familiar with the basics of binary numbers, we will consider addition and subtraction. Addition in binary follows the familiar rules of decimal addition. When adding two numbers, add the successive bits and any carry. You will need only a few addition facts.

$$0 + 0 = 0$$
$$0 + 1 = 1$$
$$1 + 1 = 0 \quad \text{carry} = 1$$
$$1 + 1 + 1 = 1 \quad \text{carry} = 1$$

In this section, we will deal with positive binary numbers only, using as many bits as required to present a number. The carry generated in any bit position is added to the next higher bit.

Binary numbers are subtracted using the following rules:

$$0 - 0 = 0$$
$$0 - 1 = 1 \quad \text{with a borrow}$$
$$1 - 0 = 1$$
$$1 - 1 = 0$$

Example A.20

Add the following pairs of binary positive numbers:

(a) 1110110_2 and 1011101_2

(b) 01101_2 and 10001_2

Solution:

a. carry   1 1 1 1   b. carry   1

$$1110110_2$$
$$+ \ 1011101_2$$
$$\overline{110100011_2}$$

$$01101_2$$
$$+ \ 10001_2$$
$$\overline{11110_2}$$

Example A.21

▼

Perform the following binary subtractions.

(a) $11001_2 - 110_2$

(b) $110011_2 - 11101_2$

Solution: The subtraction is analyzed step by step as described below (the format may look a little bit confusing). Subtraction starts in column 0 and works toward the higher columns. A column with a 1 in the borrow row indicates that the column to its right has borrowed a 1 from it.

a.   4 3 2 1 0   column
    0 1 1 0 0   borrow
    1 1 0 0 1_2 minuend
       1 1 0_2 subtrahend
    1 0 0 1 1_2 difference

Column 0
$1 - 0 = 1$

Column 1
$0 - 1 = 1$. A borrow is generated in this step. Since the number in column 2 is also a 0, we need to borrow from column 3, which is not a 0. The 0 in column 2 should be changed to 1.

Column 2
Because this column has a borrow from column 3, this column becomes $1 - 1 = 0$.

Column 3
After the borrow, this column becomes $0 - 0 = 0$.

Column 4
$1 - 0 = 1$.

b.   5 4 3 2 1 0   column
    1 1 1 0 0 0   borrow
    1 1 0 0 1 1_2 minuend
  −   1 1 1 0 1_2 subtrahend
    0 1 0 1 1 0   difference

Column 0
$1 - 1 = 0$

Column 1
$1 - 0 = 1$

Column 2
$0 - 1 = 1$. A borrow is generated. Since the number in column 3 is also a 0, we need to borrow from column 4, which is not a 0. The 0 in column 3 is changed to 1.

Column 3
Because this column has a borrow from column 4, this column becomes $1 - 1 = 0$.

Column 4
After the borrow, this column becomes $0 - 1 = 1$. A borrow is generated.

Column 5
Because of the borrow, this column becomes $0 - 0 = 0$.

▲

A.5   Two's Complement Numbers

In a computer, the number of bits that can be used to represent a number if fixed, so computer arithmetic is performed on data stored in fixed-length memory locations. The size of the location is determined by the number of bits. The memory locations of most 8-bit computers, including the 68HC11, are eight bits in length.

Another restriction on the number representation in a computer is that both positive and negative numbers must be expressed. However, a computer does not include the plus and minus signs in a number. Instead, all modern computers use the two's complement number system to represent positive and negative numbers. In the two's complement system, all numbers that have a most significant bit (MSB) of 0 are positive, and all numbers with an MSB of 1 are negative. Positive two's complement numbers are identical to binary numbers, except that the MSB must be a 0. If N is a positive number, then its two's complement N_c is given by the following expression:

$$N_c = 2^n - N \qquad\qquad \text{A-5}$$

The two's complement of N is used to represent $-N$. Machines that use the two's complement number system can represent integers in the range

$$-2^{n-1} \le N \le 2^{n-1} - 1 \qquad\qquad \text{A-6}$$

where n is the number of bits available for representing N.

Example A.22

▼

Find the range of integers that can be represented by an 8-bit two's complement number system.

Solution: The range of integers that can be represented by the 8-bit two's complement number system is:

$-2^7 \leq N \leq 2^7 - 1$
$-128_{10} \leq N \leq 127_{10}$

▲

Example A.23

▼

Represent the negative binary number -11001_2 in 8-bit two's complement format.

Solution: Use equation A–5 and the subtraction method in section A.4.

$N_C = 2^8 - 11001 = 11100111_2$

▲

Example A.24

▼

Represent the negative binary number -1100101_2 in two's complement format.

Solution: Use equation A–5 and also the subtraction method in section A.4.

$N_C = 2^8 - 1100101_2 = 10011011_2$

It is obvious that a negative two's complement binary number does not represent the magnitude of the given number. However, the magnitude of the negative number can be found using the following equation:

$N = 2^8 - N_C$ A–7

where, N is the magnitude of the negative number and N_C is the negative two's complement number.

▲

Example A.25

▼

Find the magnitude of the two's complement numbers 11001100_2 and 10010010_2.

Solution: Use equation A–7, the magnitude of 11001100_2 is

$2^8 - 11001100_2 = 110100_2 = 52_{10}$.

and the magnitude of 10010010_2 is

$2^8 - 10010010_2 = 1101110_2 = 110_{10}$.

A.6   Two's Complement Negation, Addition, and Subtraction

In this section, we will discuss negation, addition, and subtraction of two's complement numbers.

A.6.1   Negating Two's Complement Numbers

To negate a number in the two's complement number system,

1. Invert all bits of the number.
2. Add 1.

Example A.26

Negate the numbers 00110101_2 and 10100011_2.

Solution:

To negate 00110101_2,

1. Invert all bits of $00110101_2 \rightarrow 11001010_2$
2. Add $1 \rightarrow 11001010_2 + 1_2 = 11001011_2$

To negate 10100011_2,

1. Invert all bits of $10100011_2 \rightarrow 01011100_2$
2. Add $1 \rightarrow 01011100_2 + 1_2 = 01011101_2$

A.6.2   Two's Complement Addition

In decimal arithmetic, we must be concerned about the signs of two operands. If two numbers are of the same sign, then we simply perform the addition. If they are of different signs, however, they must be subtracted. Two's complement addition is much easier because the signs of the numbers do not have to be considered. As long as overflow does not occur, the sign of the result will always be correct. The carry-out, if it occurs, can simply be discarded. Overflow will be discussed shortly.

Example A.27

▼

Add the decimal numbers 63_{10} and 27_{10} using two's complement arithmetic.

Solution: The 8-bit binary equivalent of 63_{10} is 00111111_2, and the 8-bit binary equivalent of 27_{10} is 00011011_2. Adding these two numbers together,

$$
\begin{array}{r}
0\,0\,1\,1\,1\,1\,1\,1_2 \\
+\ 0\,0\,0\,1\,1\,0\,1\,1_2 \\
\hline
0\,1\,0\,1\,1\,0\,1\,0_2 = 90_{10} = 63_{10} + 27_{10}
\end{array}
$$

▲

Example A.28

▼

Add the decimal numbers 97_{10} and -13_{10} using two's complement arithmetic.

Solution: The 8-bit binary equivalent of the decimal value 97_{10} is 01100001_2. The 8-bit two's complement equivalent of 13_{10} is 00001101_2. The 8-bit two's complement equivalent of decimal value -13 is obtained by negating 00001101_2 and is 11110011_2. Adding these two numbers together,

$$
\begin{array}{r}
0\,1\,1\,0\,0\,0\,0\,1_2 \\
+\ 1\,1\,1\,1\,0\,0\,1\,1_2 \\
\hline
1\,0\,1\,0\,1\,0\,1\,0\,0_2
\end{array}
$$
↑
carry-out to be discarded

The result is decimal value 84 (01010100_2), which is correct ($97_{10} - 13_{10} = 84_{10}$).

▲

A.6.3   Two's Complement Subtraction

Subtraction in the two's complement number system is performed by negating the subtrahend and then adding it to the minuend. The sign will be correct. The carry, if it occurs, is simply discarded.

Example A.29

▼

Subtract the decimal number 8 from the decimal number 15.

Solution: The two's complement equivalent of 15_{10} is represented in 8-bit format as 00001111_2. The two's complement equivalent of 8_{10} is represented in

8-bit format as 00001000_2; negation of this value yields 11111000_2. Adding these two numbers together,

$$
\begin{array}{r}
0\,0\,0\,0\,1\,1\,1\,1_2 \\
+\ 1\,1\,1\,1\,1\,0\,0\,0_2 \\
\hline
1\ 0\,0\,0\,0\,0\,1\,1\,1_2 \\
\end{array}
$$
↑
carry-out to be discarded

After discarding the carry out, the result is the decimal value 7_{10}, which is correct.

A.6.4 Overflow

Overflow can occur with either addition or subtraction in two's complement representation. During addition, overflow occurs when the sum of two numbers with like signs have a result with the opposite sign. Overflow never occurs when adding two numbers with unlike signs. In subtraction, overflow can occur when subtracting two numbers with unlike sign. If

negative – positive = positive

or

positive – negative = negative

then overflow has occurred.

Overflow never occurs when subtracting two numbers with like signs.

Example A.30

Does overflow occur in the following 8-bit operations?

(a) $01111111_2 - 00000111_2$
(b) $01100101_2 + 01100000_2$
(c) $10010001_2 - 01110000_2$

Solution:

a. Negation of 00000111_2 is 11111001_2. $01111111_2 - 00000111_2$ is performed as follows:

$$
\begin{array}{r}
0\,1\,1\,1\,1\,1\,1\,1_2 \\
+\ 1\,1\,1\,1\,1\,0\,0\,1_2 \\
\hline
1\ 0\,1\,1\,1\,1\,0\,0\,0_2 \\
\end{array}
$$
↑
carry-out to be discarded

The difference is $01111000_2 = 120_{10}$. There is no overflow.

b. The sum of these two numbers is as follows:

$$01100101_2$$
$$+\ 01100000_2$$
$$11000101_2$$

The sign has changed from positive to negative

Overflow has occurred.

c. The negation of 01110000_2 is 10010000_2. $10010001_2 - 01110000_2$ is performed as follows:

$$10010001_2$$
$$+\ 10010000_2$$
$$100100001_2$$

carry-out to   The sign bit is 0, which indicates that the difference is positive. This is impossible because we
be discarded   are subtracting a positive number from a negative number.

Overflow has occurred.

▲

A.7 Octal and Hexadecimal Arithmetic

Although we don't do octal and hexadecimal arithmetic in our daily life, we occasionally need to add, subtract, or negate hexadecimal numbers when studying microprocessors and microcontrollers. The arithmetic in these two number systems is not very different from arithmetic in decimal number system.

Octal numbers are added and subtracted in the same way as in the decimal number system, except when carries and borrows are generated. During addition, any sum that is equal to or larger than 8 is expressed by subtracting 8 from it and sending a carry to the next higher place. A *borrow* occurs when the subtrahend is larger than the minuend in any place of subtraction. A borrow has a value of 8, and the minuend in the next more significant place is decremented by 1. If this place is a 0, then it is changed to a 7 and the next higher place is decremented.

Hexadecimal numbers are added and subtracted by mentally replacing any letters (A, B, and so on) with their numeric equivalents. During addition, any sum that is equal to or larger than the decimal value 16 is expressed by subtracting 16 from it and sending a carry to the next more significant digit. During subtraction, a borrow has a decimal value of 16, and the minuend in the next more significant digit is decremented by 1. If this digit is a zero, then it is changed to F and the next higher digit is decremented by 1.

Example A.31

▼

Add the octal numbers 237_8 and 425_8.

Solution: Octal addition is similar to decimal addition.

$$
\begin{array}{rl}
2\ 1\ 0 & \text{column} \\
0\ 1\ 0 & \text{carry} \\
2\ 3\ 7_8 & \\
+\ 4\ 2\ 5_8 & \\
\hline
6\ 6\ 4_8 & \text{sum}
\end{array}
$$

Column 0
$5 + 7 = ?$ In decimal, the sum is 12. Because this number is larger than 8, there is a carry into the next higher digit. The sum is 8 less than 12 or 4 in octal.

Column 1
$3 + 2 = ?$ Here $3 + 2 = 5$, but the carry-in from the previous column increments this number to 6.

Column 2
$2 + 4 = ?$ Here $2 + 4 = 6$.
The sum is the octal value 664_8.

▲

Example A.32

▼

Subtract the octal value 236_8 from 432_8.

Solution: The difference is computed as follows:

$$
\begin{array}{rl}
2\ 1\ 0 & \text{column} \\
1\ 1\ 0 & \text{borrow} \\
4\ 3\ 2_8 & \text{minuend} \\
-\ 2\ 3\ 6_8 & \text{subtrahend} \\
\hline
1\ 7\ 4_8 & \text{difference}
\end{array}
$$

Column 0
$2 - 6 = ?$ Here the subtrahend is larger than the minuend, so a borrow occurs. The borrow allows us to add 8 to the minuend, so column 0 becomes $2 + 8 - 6 = 4$.

Column 1
The borrow from column 0 has changed the minuend in column 1 from 3 to 2. Hence the subtrahend is larger and a borrow occurs. The borrow allows us to add 8 to the minuend, so column 1 becomes $2 + 8 - 3 = 7$.

Column 2
The borrow from column 1 has changed the minuend in column 2 from 4 to 3. Hence the subtraction is simply $3 - 2 = 1$.
The difference is an octal value 174_8.

▲

Example A.33

Add the hexadecimal numbers 3215_{16} and $A07B_{16}$.

Solution: The sum is computed as follows:

```
3 2 1 0      column
0 0 1 0      carry
3 2 1 5₁₆
A 0 7 B₁₆
D 2 9 0₁₆    sum
```

Column 0
$5 + B = ?$ In decimal, the answer is 16, so a carry is generated. The sum is 16 less than 16 and is thus 0.

Column 1
$1 + 7 = ?$ $1 + 7 = 8$, and the carry-in from the previous column increments this number to 9.

Column 2
$2 + 0 = ?$ The sum is 2.

Column 3
$3 + A = ?$ $3 + 10 = 13_{10} = D_{16}$.
The sum is the hexadecimal value $D290_{16}$.

Example A.34

Subtract $3CF3_{16}$ from $663F_{16}$.

Solution: The difference is computed as follows:

```
3 2 1 0       column
1 1 0 0       carry
6 6 3 F₁₆     minuend
-3 C F 3₁₆    subtrahend
2 9 4 C₁₆     difference
```

Column 0
$F - 3 = ?$ The hexadecimal number F is 15 in decimal. Therefore, $F - 3 = 15 - 3 = 12_{10} = C_{16}$.

Column 1
$3 - F = ?$ Here the subtrahend is larger than the minuend, so a borrow occurs. The borrow allows us to add 16 to the minuend, so column 1 becomes $3 + 16 - 15 = 4$.

Column 2

The borrow from column 1 has changed the minuend in column 2 from 6 to 5. Hence the subtrahend is larger and a borrow occurs. The borrow allows us to add 16 to the minuend, so column 2 becomes $5 + 16 - 12 = 9$.

Column 3

The borrow in column 2 has changed the minuend in column 3 from 6 to 5. Hence the difference becomes $5 - 3 = 2$.

The difference is a hexadecimal value $294C_{16}$.

A.8   Negating Octal and Hexadecimal Numbers

Negating octal numbers or hexadecimal numbers means finding the negative equivalent of a given octal or hexadecimal number. Therefore, if an octal or hexadecimal number is positive (having an MSB of 0 in its binary equivalent), the result of the negation will be a negative number, and vice versa. In this section, we will consider only fixed-length numbers (either 8 bits or 16 bits). The 68HC11 can work only on 8-bit data, and a memory address is represented in 16 bits.

Because every octal digit is encoded into three binary bits, and neither 8 nor 16 is a multiple of 3, we will use the following steps to negate an octal number.

Step 1

Convert the octal number to its binary equivalent with an appropriate number of bits (either 8 or 16).

Step 2

Negate the binary equivalent.

Step 3

Convert the two's complement number to its octal equivalent.

Example A.35

Negate the following octal numbers.

   a. 144_8

   b. 06624_8

   c. 120006_8

Solution: Convert these octal numbers to binary and then negate them:

   a. Convert the octal number 144_8 to an 8-bit binary number: 01100100_2.
   Negate the binary equivalent: 10011100.
   Convert to octal: 234_8

b. Convert the octal number 06624_8 to the 16-bit binary number: 0000110110010100_2.
Negate the binary equivalent: 1111001001101100_2.
Convert to octal: 171154_8

c. Convert the octal number 120006_8 to the 16-bit binary number
Negate the binary equivalent: 0101111111111010_2.
1010000000000110_2.
Convert to octal: 057772_8.

We can use the same algorithm to negate hexadecimal numbers. Because each hexadecimal digit can be encoded in 4r bits, each 8-bit and 16-bit binary number can be converted to two and four hexadecimal digits. The following procedure will speed up the negation of the hexadecimal numbers:

Step 1
Start from the least significant nonzero digit. Subtract the digit from the hexadecimal number 10 and write down the difference.

Step 2
Subtract each higher digit from the hexadecimal value F and write down the difference.

▲

Example A.36

▼

Negate the following hexadecimal numbers:

a. $A107_{16}$

b. 2349_{16}

c. $44A0_{16}$

Solution:

a. The least significant nonzero digit is 7. Subtract it from the hexadecimal value 10. The difference is 9. Subtract each higher digit from F as follows:

F - 0 = F
F - 1 = E
F - A = 5

Therefore

$-A107_{16} = 5EF9_{16}$

b. The least significant nonzero digit is 9. Subtract it from the hexadecimal value 10.

$10_{16} - 9 = 7$

Subtract each higher digit from F as follows:

F - 4 = B
F - 3 = C
F - 2 = D

Therefore, $-2349_{16} = DCB7_{16}$

c. The least significant nonzero digit is A. Subtract it from the hexadecimal value 10.

$10_{16} - A = 6$

Subtract each higher digit from F as follows:

F - 4 = B
F - 4 = B

Therefore

$-44A0_{16} = BB60_{16}$

A.9   Coding of Decimal Numbers

Although virtually all digital systems are binary in the sense that all internal signals can take on only two values, some nevertheless perform arithmetic in the decimal system. In some circumstances, the identity of decimal numbers is retained to the extent that each decimal digit is individually represented by a binary code. There are ten decimal digits, so four binary bits are required in each code element. The most obvious choice is to let each decimal digit be represented by the corresponding four-bit binary number, as shown in Table A.15. This form of representation is known as the *binary-coded decimal* (BCD) representation or code.

Decimal digit	Binary representation
0	0000
1	0001
2	0010
3	0011
4	0100
5	0101
6	0110
7	0111
8	1000
9	1001

Table A.15 ■ Binary-coded decimal digits

Most input to and output from computer systems is in the decimal system because this system is most convenient for human users. On input, decimal numbers are converted into some binary form for processing, and this conversion is reversed on output. In a "straight binary" computer, a decimal number such as 46 is converted into its binary equivalent, that is, 101110. In a computer using the BCD system, 46 would be converted into 0100 0110. If the BCD format is used, it must be preserved in arithmetic processing. For example, addition in a BCD machine would be carried out as shown below.

```
decimal          BCD

    46         0100 0110
  + 38       + 0011 1000
  ----         ---------
    84         1000 0100
```

The 68HC11 handles BCD addition with the DAA instruction. DAA stands for *decimal adjust accumulator*; this instruction must be executed immediately following a BCD addition. The interested reader should refer to Section 2.8.4.

The principal advantage of the BCD system is the simplicity of input/output conversion; its principal disadvantage is the complexity of arithmetic processing. The choice (between binary and BCD) depends on the type of problems the system will be handling.

The BCD code is not the only possible coding method for decimal digits. The excess-3 code is larger than its corresponding decimal digit by 3. The 2-out-of-5 code represents each decimal digit by one of the ten possible combinations of two 1s and three 0s. The distinguishing feature of the Gray code is that successive coded characters never differ in more than one bit. The excess-3, 2-out-of-5, and Gray codes are shown in Table A.16.

Decimal digit	Excess-3				2-out-of-5					Gray code			
	d3	d2	d1	d0	d4	d3	d2	d1	d0	d3	d2	d1	d0
0	0	0	1	1	0	0	0	1	1	0	0	1	0
1	0	1	0	0	0	0	1	0	1	0	1	1	0
2	0	1	0	1	0	0	1	1	0	0	1	1	1
3	0	1	1	0	0	1	0	0	1	0	1	0	1
4	0	1	1	1	0	1	0	1	0	0	1	0	0
5	1	0	0	0	0	1	1	0	0	1	1	0	0
6	1	0	0	1	1	0	0	0	1	1	1	0	1
7	1	0	1	0	1	0	0	1	0	1	1	1	1
8	1	0	1	1	1	0	1	0	0	1	1	1	0
9	1	1	0	0	1	1	0	0	0	1	0	1	0

Table A.16 ■ Other coding methods for decimal digits

A.10   Representing Character Data

Computers use numbers to represent characters, but there is no natural correspondence between the characters and the numbers, so the assignment of number codes to characters must be defined. Two character code sets are in widespread use today: ASCII (American Standard Code for Information Interchange) and EBCDIC (Extended Binary Coded Decimal Interchange Code). Several types of Mainframe Computers use EBCDIC for internal storage and processing of characters. In EBCDIC, each character is represented by a unique 8-bit number, and a total of 256 different characters can be represented. Most microcomputer and minicomputer systems use the ASCII character set. A computer does not depend on any particular set, but most I/O devices that display characters require the use of ASCII codes. Only ASCII codes will be discussed in this text.

A.10.1   ASCII Codes

ASCII characters have a 7-bit code. They are usually stored in a fixed-length 8-bit number. The $2^7 = 128$ different codes are partitioned into 95 print characters and 33 control characters. The control characters define communication protocols and special operations on peripheral devices. The printable characters consists of the following:

> 26 uppercase letters (A-Z)
>
> 26 lowercase letters (a-z)
>
> 10 digits (0-9)
>
> 1 blank space
>
> 32 special-character symbols,
> > including ! @ # $ % ^ & * ( ) - _ = + ' [ ] ; : ''' , < . > / ? { }

The complete ASCII character set is shown in Table A.17. The binary number corresponding to a character is given in hexadecimal.

A.10.2   Control Characters

For efficient code conversion, the alphanumeric characters occupy contiguous binary ranges. This property is important for I/O code conversion and the design of I/O devices. Table A.18 shows the ranges for the control characters, the decimal digits, and the alphabetic characters.

BEL is one of the control characters in the ASCII set. It has a binary value of 00000111 (7), and its transmission to the terminal causes the bell to ring. Used infrequently, this character can remind a typist of a "near-end-of-line" condition in a word processing program or indicate an invalid editing stroke. Like other control characters, it has special terminal or data transmission applications. Table A.19 lists some noteworthy control characters, along with descriptions of their actions on typical peripheral devices.

Seven-bit hexadecimal code	character	Seven-bit hexadecimal code	character	Seven-bit hexadecimal code	character	Seven-bit hexadecimal code	character
00	NUL	20	SP	40	@	60	`
01	SOH	21	!	41	A	61	a
02	STX	22	"	42	B	62	b
03	ETX	23	#	43	C	63	c
04	EOT	24	$	44	D	64	d
05	ENQ	25	%	45	E	65	e
06	ACK	26	&	46	F	66	f
07	BEL	27	'	47	G	67	g
08	BS	28	(	48	H	68	h
09	HT	29	)	49	I	69	i
0A	LF	2A	*	4A	J	6A	j
0B	VT	2B	+	4B	K	6B	k
0C	FF	2C	,	4C	L	6C	l
0D	CR	2D	-	4D	M	6D	m
0E	SO	2E	.	4E	N	6E	n
0F	SI	2F	/	4F	O	6F	o
10	DLE	30	0	50	P	70	p
11	DC1	31	1	51	Q	71	q
12	DC2	32	2	52	R	72	r
13	DC3	33	3	53	S	73	s
14	DC4	34	4	54	T	74	t
15	NAK	35	5	55	U	75	u
16	SYN	36	6	56	V	76	v
17	ETB	37	7	57	W	77	w
18	CAN	38	8	58	X	78	x
19	EM	39	9	59	Y	79	y
1A	SUB	3A	:	5A	Z	7A	z
1B	ESC	3B	;	5B	[	7B	{
1C	FS	3C	<	5C	\	7C		
1D	GS	3D	=	5D	]	7D	}
1E	RS	3E	>	5E	^	7E	~
1F	US	3F	?	5F	_	7F	DEL

Table A.17 ■ ASCII code chart

Many of the other 27 control characters have special applications in conjunction with intelligent terminals and printers (for example, clearing the screen, inserting a line, or specifying cursor motion or special character fonts). Some of these characters are used as control codes in data communications for message switching and telegrams.

ASCII characters	Eight-bit code range		
	Decimal	Hexadecimal	Binary
Control characters	0-31	00-1F	00000000-00011111
Blank space	32	20	00100000
Decimal digits (0-9)	48-57	30-39	00110000-00111001
Uppercase letters (A-Z)	65-90	41-5A	01000001-01011010
Lowercase letters (a-z)	97-122	61-7A	01100001-01111010
Special characters		unused codes in the range 33-126	
Delete character	127		01111111

Table A.18 ■ ASCII character range

Symbol	Value		Meaning
	Hexadecimal	Decimal	
CR	0D	13	Carriage return—move the screen cursor to the beginning of the current line
LF	0A	10	Line feed—advance the screen cursor down one line
BS	08	8	Backspace—backspace the screen cursor
FF	0C	12	Form feed—advance the hard copy unit to the top of the next page
BEL	07	7	Ring the bell on the terminal
HT	09	9	Horizontal tab—advance the screen cursor to the next tab stop (as on a typewriter)

Table A.19 ■ Selected ASCII control characters

A.11   Summary

Binary numbers are used to represent data in digital systems. The binary values of 1 and 0 correspond to the two allowed states in digital electronics. The binary number system is a weighted base-2 system in which n bits can represent 2^n numbers.

The octal and hexadecimal codes are mainly used as shorthands for the binary code. Three binary bits can be represented by one octal digit, and four binary bits can be represented by one hexadecimal digit. Numbers can be converted among the binary, octal, decimal, and hexadecimal formats.

The binary number system is used in all arithmetic operations (addition, subtraction, multiplication, and division). Signed-number arithmetic operations can be accomplished entirely through addition by representing numbers in their two's complement form.

A.12 Exercises

EA.1. Convert the following binary numbers to their decimal equivalents:

    a. 1001011000_2

    b. 100011110_2

    c. 100000001_2

    d. 111111111_2

EA.2. Convert the following octal numbers to their decimal equivalents:

    a. 2335_8

    b. 167_8

    c. 4005.246_8

    d. 625.575_8

EA.3. Convert the following hexadecimal numbers to their decimal equivalents:

    a. 250_{16}

    b. 1200_{16}

    c. $4AE0_{16}$

    d. $FFFE.428_{16}$

    e. $245.62B_{16}$

EA.4. Convert the following decimal numbers to their binary, octal, and hexadecimal equivalents:

    a. 4135_{10}

    b. 625.625_{10}

    c. 910.125_{10}

    d. 1360.275_{10}

    e. 415_{10}

EA.5. Convert the following binary numbers to their octal and hexadecimal equivalents:

    a. 100100111110_2

    b. 101010111100_2

    c. 1000001110110_2

    d. 1001101100001_2

EA.6. Convert the following octal numbers to their binary equivalents:

    a. 423_8

    b. 1246_8

    c. 3314_8

    d. 5425_8

    e. 2400_8

EA.7. Convert the following hexadecimal numbers to their binary equivalents:

    a. 38_{16}

    b. $4AC0_{16}$

    c. $FF02_{16}$

    d. 1240_{16}

    e. $8B0_{16}$

EA.8. Compute the sums of the following pairs of binary numbers:

    a. 100100010_2 and 101101100_2
    b. 110010_2 and 100011_2

EA.9. Negate the following 8-bit binary numbers:

    a. 00100001_2
    b. 10101010_2
    c. 01100001_2
    d. 00010011_2
    e. 10011100_2

EA.10. Use 8-bit two's complement addition to perform the indicated operations:

    a. $01001011_2 - 00010111_2$
    b. $01000001_2 - 00010001_2$
    c. $10000001_2 - 00100000_2$

Does overflow occur in any of the operations?

EA.11. What range of integers can be represented by the 16-bit two's complement number system?

EA.12. What range of integers can be represented by the 32-bit two's complement number system?

EA.13. Perform the following octal operations:

    a. $75_8 - 12_8$
    b. $420_8 + 365_8$
    c. $330_8 - 225_8$
    d. $471_8 - 335_8$

EA.14. Perform the following hexadecimal operations:

    a. $A7_{16} - 66_{16}$
    b. $58_{16} - 3F_{16}$
    c. $459_{16} + 1FA_{16}$
    d. $330_{16} + 8FB_{16}$
    e. $12F4_{16} - 25F9_{16}$

EA.15. Negate the following hexadecimal numbers:

    a. $357A_{16}$
    b. $A03B_{16}$
    c. 2488_{16}
    d. 1239_{16}
    e. $F0A7_{16}$

Appendix B

Design Projects

B.1   Introduction

In chapters 1 to 11 we explained in detail the functions and applications of each subsystem of the 68HC11. Many lab exercises and assignments have also been given for familiarizing you with the functions of each subsystem. In this appendix, we will provide several design projects that will require the designer to utilize multiple subsystems of the 68HC11. These projects will provide an opportunity to the reader to learn and practice the skills of system analysis and design process.

When working on a design project, there are two possibilities about the project:

- *The requirements are well defined.* You know exactly what is required. What is left are the design and implementation. On the hardware side, you need to decide the implementation approach and figure out all the components that are needed in the final product. On the software side you need to break down the problem into modules, choose the language to implement the software, and write the software.

- *The requirements are not well defined.* This is the situation of most design projects. You have the freedom to make some decisions about the exact and detailed requirements. After the detailed requirements are agreed upon between you, your design team, and your customer, the rest of the work is identical to the first approach.

In the rest of this appendix, we will discuss hardware implementation approaches, software design issues, an example project, and design projects.

B.2   Hardware Implementation Approaches

You can use a prototyping board (also called demo board) to implement your project or start from scratch and build your own printed-circuit board. Each approach has its own advantages and drawbacks. You also need to know that, during the debugging stage, we need to download our programs into the SRAM for testing. However, after we have made sure that the software works correctly, we want to transfer the software into EPROM, EEPROM, or flash memory so that it will run automatically after the power is turned on.

B.2.1   Using a Demo Board to Implement Your Product

Using a demo board can dramatically reduce hardware design time because you don't need to go through the process of designing and fabricating a printed circuit board (PCB). On the downside, the resultant product may be bulky because the demo board was not designed for your purpose. Three different evaluation boards have been discussed in this book, and we will look at how to use them in a design project.

68HC11 EVB

EVB is one of the earliest demo boards designed for 68HC11. EVB is configured to operate in expanded mode from the factory. It is very useful during the software debugging process. A Buffalo monitor is located at the space from $E000 to $FFFF. An 8 KB SRAM is provided for user program execution. There is an empty socket at location U4 that allows you to add an additional SRAM (or ROM) in case 8 KB is not enough for running your program. This extra 8 KB memory can be configured to occupy the space at $6000-$7FFF (install jumper J3) or $A000-$BFFF (install jumper J7).

The 68HC11A1 CPU has a 512-byte internal EEPROM. If your application program can fit into 512 bytes, then it can be programmed into this EEPROM. To execute the program in EEPROM instead of the Buffalo monitor after power up, install the jumper J4 at the location between pins 2 and 3 (has a 10 K pullup resistor). Otherwise, install the jumper J4 between pins 1 and 2 (has a 10 K pulldown resistor). There are at least two methods to program the 68HC11A1 on-chip EEPROM. The first method is to use the Motorola 68HC11 EVM board. The EVM board has a PLCC and a DIP socket that can be used to program the on-chip EEPROM. You need to pop out the MCU on the EVB board and program it. After the MCU is programmed, put it back in the PLCC socket on the EVB. The 68HC11 EVM demo board is another earlier demo board for the 68HC11 and may not be available now. The second method is to use a commercial universal programmer. When using this approach, don't use the PE0 pin for analog-to-digital conversion input.

If your application needs more than 512 bytes but less than 8 KB, then you can program it in an EPROM (or an EEPROM) chip and replace the EPROM chip that holds the Buffalo monitor with your new EPROM chip. Another possibility is to program the on-chip 512-byte EEPROM with a jump instruction that jumps to the EPROM (or EEPROM) that is installed in socket U4. By changing the jumper position of J4, you can switch the program execution between your application and the Buffalo monitor after reset.

If your application program is larger than 8 KB but less than 16 KB, then you will need to store your application in two 8 KB EPROM (or EEPROM) chips and put them in socket U3 (the EPROM that holds the Buffalo monitor is in this socket) and U4 (or U5, which holds 8 KB SRAM).

The EVB allows the user to install up to 24 KB of user program memory without using extra breadboard or PCB to hold the memory.

You can also replace the 68HC11A1 MCU on the EVB with the 68HC711E9, 68HC811E2, or 68HC711E20. External ROM (EPROM or EEPROM) is not needed using this approach. These MCUs can be programmed by using the EVB or the 68HC711E9PGMR programmer. Please refer to the appropriate engineering bulletin reports from Motorola.

Remember that the reset vector must be stored at $FFFE and $FFFF. Your reset handling routine can be placed anywhere.

Although the EVB is configured to operate in the expanded mode, port B and port C are restored by the port replacement unit (PRU) of the EVB and are available for I/O purposes. All I/O signals are available on the P1 connector.

68HC11 EVBU

The 68HC11 EVBU is designed as a prototyping kit for implementing a 68HC11-based product. It is configured to operate in single-chip mode because it does not have external memory. However, users can reconfigure it to expanded or bootstrap mode. Small programs can be downloaded into the on-chip SRAM for execution. However, an external SRAM chip must be added in order to evaluate any significant application program.

When using the EVBU as a prototyping kit, we often use the following approach if the application program is no larger than 12 KB:

- Use the EVBU board to program the 68HC711E9 on-chip EPROM. The programming procedure is described in chapter 5. You need to remove the 68HC11E9 chip on the EVBU board and put in a 68HC711E chip. The user program will be executed when power is turned on.

- Connect the rest of the circuit.

- Power on the EVBU board.

The resultant product will be very compact because the EVBU is small and no external program memory is used.

If the application software is larger than 12 KB but smaller than 20 KB, then the 68HC711E20 can be used. This MCU can be programmed using the 68HC711R9PGMR programmer. External memory will be needed if your software needs are more than 20 KB.

CMD-11A8 AND OTHER DEMO BOARDS MADE BY AXIOM

The CMD-11A8 has three sockets for memory chips:

- U5: This socket can accommodate either 8 KB or 32 KB devices. An SRAM is installed in this socket by the factory.

- U6: This socket can accommodate either 8 KB or 32 KB devices. An EEPROM chip is installed in this socket by the factory.

- U7: This socket can accommodate either 8 KB or 32 KB EPROM or EEPROM. An EEPROM chip is installed in this socket by the factory. The Buffalo monitor is resident in this EEPROM chip.

Jumpers must be set properly for the size of the memory chip placed in each socket. The AX11 terminal program bundled with the CMD-11A8 demo board can be used to program the EEPROM chip on the U7 socket. When you are satisfied with the software operation, you can relocate your program to start at $E000 (for 8 KB EEPROM), then select "Program Code Memory" in the AX11 utility. When programming is complete, your program will run automatically when the CMD-11A8 is powered on or RESET is applied.

The CMD-711EX demo board uses the 68HC11E1 as the MCU. It provides three IC sockets for memory chips that allow the user to have either 44 KB or 52 KB usable SRAM and 12 KB or 20 KB usable EEPROM (the sum of these two types of memory is 64 KB). The CMD-711EX is configured to operate in expanded mode. A port replacement unit is included that restores both ports B and C and makes them available to the user. An optional programming socket is available for programming the 68HC711E9 or 68HC711E20 MCU. Like the CMD-11A8, the CMD-711EX also provides ports for connecting LCD and keypad. This demo board is more of a development board than a prototyping board for its capability and price ($199 at the time of this writing).

The CMD-11E9-EVBU demo board can emulate the capability of the Motorola EVBU board. It includes 32 KB of external SRAM and 8 KB of EEPROM. It provides the on-board programming capability for 68HC711E9/20 parts. This demo board also includes the following features:

- power supply adaptor
- an open memory chip socket
- mode selection jumpers
- LCD connector
- keypad connector
- prototyping area
- 60-pin connector that makes all 68HC11E1 signals available

The CMD-11E9-EVBU is inexpensive and is ideal for implementing the designs that use the 68HC11E series as the CPU.

B.2.2   Design and Construct Your Own PCB

This approach requires much more work than the previous approach. However, you can include only the components that are needed for your application. If a large quantity of end products is to be fabricated, this approach will require much lower cost.

To make the circuit work, the following few points are most critical:

- *Crystal oscillator circuit.* Without the clock signal, the MCU cannot operate. The crystal oscillator generates the sinusoidal waveform that will be converted into a square waveform inside the microcontroller. Motorola provides suggestions about the design of the crystal oscillator circuit. You can find these suggestions in section 2.2.3 and 2.5 of the M68HC11 reference manual.

- *Reset circuit.* The importance of this circuit has been explained in chapter 6. Figure 6.6 of this book is a usable reset circuit that will detect the condition of low power supply and also allow for manual reset.

- *Power supply circuit.* A 5 V power supply is required for the normal operation of the MCU. Usually, capacitors are needed to filter out the noise from the power supply that may be generated during normal operation. Please refer to Figures 2.22 and 2.23 of the M68HC11 reference manual.

- *Mode-select input.* MODA and MODB pins must be configured to appropriate voltage levels for the desired operation mode. Using jumpers is advisable if you want the operation mode to be selectable.

- *Unused input pins.* These pins should not be left unconnected because they may trigger unwanted conditions if they are not connected properly. Pull them to a voltage level that will either disable their functions or cause no harm.

B.3 Project 1: Multifunction Digital Meter

With the A/D converter and the timer function, the 68HC11 is ideal for implementing digital multifunction meters. Suppose you are assigned to design a digital meter with the following function:

1. voltage measurement
2. frequency measurement
3. period measurement
4. pulse width measurement
5. duty cycle measurement

B.3.1 Project Requirements

Suppose you have the freedom to decide the detailed requirements, what do you want to achieve in each of the above functions?

- **Voltage range**
  Most multimeters can measure at least ±1000 V. Let us use this as a goal. The root-mean-square value of the voltage must be displayed.

- **Frequency range**
  Using the pulse accumulation function along with the 2 MHz E clock, we can measure a frequency up to 630 KHz. Set this as the goal of the frequency measurement.

- **Period range**
  What are the longest and shortest periods that can be measured using a 2 MHz E clock? There is no limit on the longest period that can be measured. However, there is a limit on the shortest period that can be measured due to the instruction execution time. Try to write an instruction sequence that does the measurement (use the input capture function) and find out the shortest period that can be measured.

- **Pulse width range**
  This should be identical to the period range.

- **Duty cycle measurement**
  This meter should be able to measure a duty cycle larger than 0% and smaller than 100%. Set the range to 1% to 99%. How do you want to deal with 0 V and constant high voltage? We can treat a digital waveform as 0 if it is 0 V for longer than 1 second and as constant high if it stays at the same level for longer than 1 second.

B.3.2   Hardware Required to Implement the Project

VOLTAGE MEASUREMENT

What hardware is required to measure voltage? The voltage must be attenuated to between 0 and 5 V before an A/D converter can measure it. This can be done by a resistor circuit or a transformer along with an OP Amp circuit. A transformer is much heavier and expensive. A resistor circuit is a better choice. An Op Amp circuit is needed to scale and shift the resistor circuit output to between 0 and 5 V. We can also add a clamp circuit to limit the voltage output of the Op Amp circuit to protect the microcontroller from being damaged by a voltage surge.

The next consideration is the required resolution of the A/D converter. Suppose we want to limit the error to no greater than ± 0.1 V. How many bits are needed to represent the A/D conversion result? The dynamic range of the voltage is 2000 V. The total steps of the A/D converter will be $2000 \div 0.1 = 20000 > 2^{14}$. It is obvious that if we insist on achieving this accuracy, we will need an A/D converter that is higher than 14 bits. The closest commercial A/D converter is 16-bit. An A/D converter with serial interface has a smaller footprint and hence should be chosen.

To display the whole voltage range will require four integral digits and one fractional digit. A sign is also needed. The voltage output can be displayed by using seven-segment displays (six seven-segment displays are needed) or LCD display.

FREQUENCY MEASUREMENT

The most common periodic waveforms are sinusoidal, triangular, and square waveform. The rising and falling edges of the sinusoidal and triangular waveforms cannot be detected easily if they are connected directly to the pulse accumulator input. We need to "square them up" before we can measure them. A 74HC14 Schmidt-Trigger circuit is perfect for this job. The 74HC14 converts the sinusoidal and triangular waveform to a square wave between 0 and 5 V. Frequency can be displayed in six seven-segment displays or a single LCD display. The display should display six digits.

PERIOD MEASUREMENT

Same as frequency measurement. We can use the input capture function to measure the period of an unknown signal. Sinusoidal and triangular waveforms must be "squared up" before they can be measured. The period should be displayed in six digits and in the unit of microseconds or milliseconds.

PULSE WIDTH MEASUREMENT

Same as period measurement.

DUTY CYCLE MEASUREMENT

Hardware requirement is identical to period measurement.

B.3.3 System Design

FUNCTION SELECTION

We need to provide the capability that allows the user to select the measurement that he/she wants to perform. This can be done by several methods: DIP switches, push buttons, rotating knobs, and so on. We need to decide whether to allow more than one measurement to be done at a time. This will affect the amount of display hardware needed in the project. There should be a position to deselect every measurement.

SYSTEM OPERATION

The system will initialize the hardware after power is turned on and wait for the selection of measurement to be performed. All of the I/O subsystems that are used should be initialized. The software continuously checks to see if any function is selected. If the user selects a function, the system will perform the measurement and display the result.

After the measurement is done and the user does not change the selection, the system should continue to perform the same measurement and update the display. If all of the measurements are deselected, the system will stay in a polling loop to wait for a new selection. The system operation flow is shown in Figure B.1.

B.4 Other Projects

You are encouraged to complete all or part of the following design projects. You have the complete freedom to establish the final design goals and decide the hardware and software required to achieve these goals.

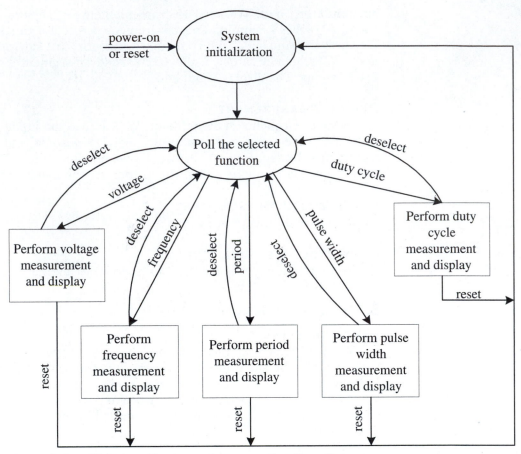

Figure B.1 ■ Digital multimeter system operation flow diagram

Project 2: Time-of-Day, Humidity, Pressure, and Temperature Display

Design and construct a system that will

1. allow the user to enter the current time-of-day.
2. update and display the current time-of-day in the format of hh:mm:ss:day:month:year.
3. measure and display the current ambient relative humidity (in percent).
4. measure and display the current barometric pressure (in mbar).
5. measure and display the current temperature (in Fahrenheit).

Project 3:   Digital Speedometer and Display

Design a microcontroller-controlled speedometer that can measure the speed of a bicycle or other vehicles. A sensor based on Hall-effect (for example, the 30137 Hall-effect transistor) along with magnets may be needed in this project. You can choose to measure the speed of a vehicle every half a second or every second and display the current speed in miles per hour, revolutions per minute, or both.

Project 4:   Digital Scale

Design a digital scale that can measure and display the weight of an object. This project is an application of the A/D converter. Depending on the range of weight to be measured, an appropriate load cell must be chosen. Several companies are making load cells. Hottinger Baldwin Measurements, Inc. in Massachusetts is one of them. Other companies can be found by searching the Internet.

Project 5:   3-D Cube Display

The 68HC11 is perfect for controlling a 3-D cube display. In this project, you are required to design a 3-dimensional LED cube that can display different words and patterns and has the capability to flash, rotate, and so on. The cube should have 512 LEDs.

Appendix C

68HC11A8 Register and Control Bit Summary

	7	6	5	4	3	2	1	0	
$i000	Bit 7	—	—	—	—	—	—	Bit 0	PORTA
$i001									Reserved
$i002	STAF	STAI	CWOM	HNDS	OIN	PLS	EGA	INVB	PIOC
$i003	Bit 7	—	—	—	—	—	—	Bit 0	PORTC
$i004	Bit 7	—	—	—	—	—	—	Bit 0	PORTB
$i005	Bit 7	—	—	—	—	—	—	Bit 0	PORTCL
$i006									Reserved
$i007	Bit 7	—	—	—	—	—	—	Bit 0	DDRC
$i008			Bit 5	—	—	—	—	Bit 0	PORTD
$i009			Bit 5	—	—	—	—	Bit 0	DDRD
$i00A	Bit 7	—	—	—	—	—	—	Bit 0	PORTE
$i00B	FOC1	FOC2	FOC3	FOC4	FOC5				COFRC
$i00C	OC1M7	OC1M6	OC1M5	OC1M4	OC1M3				OC1M
$i00D	OC1D7	OC1D6	OC1D5	OC1D4	OC1D3				OC1D
$i00E	Bit 15	—	—	—	—	—	—	Bit 8	TCNT
$i00F	Bit 7	—	—	—	—	—	—	Bit 0	
$i010	Bit 15	—	—	—	—	—	—	Bit 8	TIC1
$i011	Bit 7	—	—	—	—	—	—	Bit 0	
$i012	Bit 15	—	—	—	—	—	—	Bit 8	TIC2
$i013	Bit 7	—	—	—	—	—	—	Bit 0	
$i014	Bit 15	—	—	—	—	—	—	Bit 8	TIC3
$i015	Bit 7	—	—	—	—	—	—	Bit 0	
$i016	Bit 15	—	—	—	—	—	—	Bit 8	TOC1
$i017	Bit 7	—	—	—	—	—	—	Bit 0	
$i018	Bit 15	—	—	—	—	—	—	Bit 8	TOC2
$i019	Bit 7	—	—	—	—	—	—	Bit 0	
$i01A	Bit 15	—	—	—	—	—	—	Bit 8	TOC3
$i01B	Bit 7	—	—	—	—	—	—	Bit 0	
$i01C	Bit 15	—	—	—	—	—	—	Bit 8	TOC4
$i01D	Bit 7	—	—	—	—	—	—	Bit 0	
$i01E	Bit 15	—	—	—	—	—	—	Bit 8	TOC5
$i01F	Bit 7	—	—	—	—	—	—	Bit 0	

	7	6	5	4	3	2	1	0	
$i020	OM2	OL2	OM3	OL3	OM4	OL4	OM5	OL5	TCTL1
$i021			EDG1B	EDG1A	EDG2B	EDG2A	EDG3B	EDG3A	TCTL2
$i022	OC1I	OC2I	OC3I	OC4I	OC5I	IC1I	IC2I	IC3I	TMSK1
$i023	OC1F	OC2F	OC3F	OC4F	OC5F	IC1F	IC2F	IC3F	TFLG1
$i024	TOI	RTII	PAOVI	PAII			PR1	PR0	TMSK2
$i025	TOF	RTIF	PAOVF	PAIF					TFLG2
$i026	DDRA7	PAEN	PAMOD	PEDGE			RTR1	RTR0	PACTL
$i027	Bit 7	—	—	—	—	—	—	Bit 0	PACNT
$i028	SPIE	SPE	DWOM	MSTR	CPOL	CPHA	SPR1	SPR0	SPCR
$i029	SPIF	WCOL		MODF					SPSR
$i02A	Bit 7	—	—	—	—	—	—	Bit 0	SPDR
$i02B	TCLR		SCP1	SCP0	RCKB	SCR2	SCR1	SCR0	BAUD
$i02C	R8	T8		M	WAKE				SCCR1
$i02D	TIE	TCIE	RIE	ILIE	TE	RE	RWU	SBK	SCCR2
$i02E	TDRE	TC	RDRF	IDLE	OR	NF	FE		SCSR
$i02F	Bit 7	—	—	—	—	—	—	Bit 0	SCDR
$i030	CCF		SCAN	MULT	CD	CC	CB	CA	ADCTL
$i031	Bit 7	—	—	—	—	—	—	Bit 0	ADR1
$i032	Bit 7	—	—	—	—	—	—	Bit 0	ADR2
$i033	Bit 7	—	—	—	—	—	—	Bit 0	ADR3
$i034	Bit 7	—	—	—	—	—	—	Bit 0	ADR4
$i035									Reserved
$i036									Reserved
$i037									Reserved
$i038									Reserved
$i039	ADPU	CSEL	IRQE	DLY	CME		CR1	CR0	OPTION
$i03A	Bit 7	—	—	—	—	—	—	Bit 0	COPRST
$i03B	ODD	EVEN		BYTE	ROW	ERASE	EELAT	EEPGM	PPROG
$i03C	RBOOT	SMOD	MDA	IRV	PSEL3	PSEL2	PSEL1	PSEL0	HPRIO
$i03D	RAM3	RAM2	RAM1	RAM0	REG3	REG2	REG1	REG0	INIT
$i03E	TILOP		OCCR	CBYP	DISR	FCM	FCOP	TCON	TEST1
$i03F					NOSEC	NOCOP	ROMON	EEON	CONFIG

Appendix D

Instruction Listing by Alphabetical Order

Source forms	Operation	Boolean expression	Addressing mode for <opr>	Bytes	Cycles	S	X	H	I	N	Z	V	C
ABA	Add accumulators	A ← A + B	INH	1	2	–	–	↕	–	↕	↕	↕	↕
ABX	Add B to X	IX ← IX + 00:B	INH	1	3	–	–	–	–	–	–	–	–
ABY	Add B to Y	IY ← IY + 00:B	INH	2	4	–	–	–	–	–	–	–	–
ADCA	Add with carry to A	A ← A + M + C	IMM	2	2	–	–	↕	–	↕	↕	↕	↕
			DIR	2	3								
			EXT	3	4								
			IND, X	2	4								
			IND, Y	3	5								
ADCB	Add with carry to B	B ← B + M + C	IMM	2	2	–	–	↕	–	↕	↕	↕	↕
			DIR	2	3								
			EXT	3	4								
			IND, X	2	4								
			IND, Y	3	5								
ADDA <opr>	Add memory to A	A ← A + M	IMM	2	2	–	–	↕	–	↕	↕	↕	↕
			DIR	2	3								
			EXT	3	4								
			IND, X	2	5								
			IND, Y	3	5								
ADDB <opr>	Add memory to B	B ← B + M	IMM	2	2	–	–	↕	–	↕	↕	↕	↕
			DIR	2	3								
			EXT	3	4								
			IND, X	2	5								
			IND, Y	3	5								
ADDD <opr>	Add 16-bit to D	D ← D + M:M + 1	IMM	3	4	–	–	–	–	↕	↕	↕	↕
			DIR	2	5								
			EXT	3	6								
			IND, X	2	6								
			IND, Y	3	7								
ANDA <opr>	AND A with memory	A ← A · M	IMM	2	2	–	–	–	–	↕	↕	0	–
			DIR	2	3								
			EXT	3	4								
			IND, X	2	4								
			IND, Y	3	5								
ANDB <opr>	AND B with memory	B ← B · M	IMM	2	2	–	–	–	–	↕	↕	0	–
			DIR	2	3								
			EXT	3	4								
			IND, X	2	4								
			IND, Y	3	5								
ASL <opr>	Arithmetic Shift Left		EXT	3	6	–	–	–	–	↕	↕	↕	↕
			IND, X	2	6								
			IND, Y	3	7								
ASLA			INH	1	2	–	–	–	–	↕	↕	↕	↕
ASLB			INH	1	2	–	–	–	–	↕	↕	↕	↕
ASLD	Arithmetic shift left double		INH	1	3	–	–	–	–	↕	↕	↕	↕
ASR <opr>	Arithmetic shift right		EXT	3	6	–	–	–	–	↕	↕	↕	↕
			IND, X	2	6								
			IND, Y	3	7								

Source forms	Operation	Boolean expression	Addressing mode for \<opr\>	Bytes	Cycles	S	X	H	I	N	Z	V	C
ASRA	Arithmetic shift right A		INH	1	2	—	—	—	—	↕	↕	↕	↕
ASRB	Arithmetic shift right B		INH	1	2	—	—	—	—	↕	↕	↕	↕
BCC \<rel\>	Branch if carry clear	? C = 0	REL	2	3	—	—	—	—	—	—	—	—
BCLR \<opr\> \<msk\>	Clear bit (s)	$M \leftarrow M \cdot (\overline{mm})$	DIR	3	6	—	—	—	—	↕	↕	0	—
			IIND, X	3	7								
			IND, Y	4	8								
BCS \<rel\>	Branch if carry set	? C = 1	REL	2	3	—	—	—	—	—	—	—	—
BEQ \<rel\>	Branch if = Zero	? Z = 1	REL	2	3	—	—	—	—	—	—	—	—
BGE \<rel\>	Branch if ≥ Zero	? N ⊕ V = 0	REL	2	3	—	—	—	—	—	—	—	—
BGT \<rel\>	Branch if > Zero	? Z + (N ⊕ V) = 0	REL	2	3	—	—	—	—	—	—	—	—
BHI \<rel\>	Branch if higher	? C + Z = 0	REL	2	3	—	—	—	—	—	—	—	—
BHS \<rel\>	Branch if higher or same	? C = 0	REL	2	3	—	—	—	—	—	—	—	—
BITA \<opr\>	Bit(s) test A with memory	A · M	IMM	2	2	—	—	—	—	↕	↕	0	—
			DIR	2	3								
			EXT	3	4								
			IND, X	2	4								
			IND, Y	3	5								
BITB \<opr\>	Bit(s) test B with memory	B · M	IMM	2	2	—	—	—	—	↕	↕	0	—
			DIR	2	3								
			EXT	3	4								
			IND, X	2	4								
			IND, Y	3	5								
BLE \<rel\>	Branch if ≤ 0	? Z + (N ⊕ V) = 1	REL	2	3	—	—	—	—	—	—	—	—
BLO \<rel\>	Branch if lower	? C = 1	REL	2	3	—	—	—	—	—	—	—	—
BLS \<rel\>	Branch if lower or same	? C + Z = 1	REL	2	3	—	—	—	—	—	—	—	—
BLT \<rel\>	Branch if < zero	? N ⊕ V = 1	REL	2	3	—	—	—	—	—	—	—	—
BMI \<rel\>	Branch if minus	? N = 1	REL	2	3	—	—	—	—	—	—	—	—
BNE \<rel\>	Branch if not = zero	? Z = 0	REL	2	3	—	—	—	—	—	—	—	—
BPL \<rel\>	Branch if plus	? Z = 0	REL	2	3	—	—	—	—	—	—	—	—
BRA \<rel\>	Branch always	? 1 = 1	REL	2	3	—	—	—	—	—	—	—	—
BRCLR \<opr\> \<msk\> \<rel\>	Branch if bit(s) clear	? M · mm = 0	DIR	4	6	—	—	—	—	—	—	—	—
			IND, X	4	7								
			IND, Y	5	8								
BRN \<rel\>	Branch never	? 1 = 0	REL	2	3	—	—	—	—	—	—	—	—
BRSET \<opr\> \<msk\>\<rel\>	Branch if bit(s) set	? (M) · mm = 0	DIR	4	6	—	—	—	—	—	—	—	—
			IND, X	4	7								
			IND, Y	5	8								
BSET \<opr\> \<msk\>	Set bit(s)	M ← M + mm	DIR	3	6	—	—	—	—	↕	↕	0	—
			IND, X	3	7								
			IND, Y	4	8								
BSR \<rel\>	Branch to subroutine		REL	2	6	—	—	—	—	—	—	—	—
BVC \<rel\>	Branch if overflow clear	? V = 0	REL	2	3	—	—	—	—	—	—	—	—

Source forms	Operation	Boolean expression	Addressing mode for <opr>	Bytes	Cycles	S	X	H	I	N	Z	V	C
BVS <rel>	Branch if overflow set	? V = 1	REL	2	3	–	–	–	–	–	–	–	–
CBA	Compare B to A	A – B	INH	1	2	–	–	–	–	↕	↕	↕	↕
CLC	Compare A to B	C ← 0	INH	1	2	–	–	–	–	–	–	–	0
CLI	Clear interrupt mask	I ← 0	INH	1	2	–	–	–	0	–	–	–	–
CLR <opr>	Clear memory byte	M ← 0	EXT	3	6	–	–	–	–	0	1	0	0
			IND, X	2	6								
			IND, Y	3	7								
CLRA	Clear accumulator A	A ← 0	INH	1	2	–	–	–	–	0	1	0	0
CLRB	Clear accumulator B	B ← 0	INH	1	2	–	–	–	–	0	1	0	0
CLV	Clear overflow flag	V ← 0	INH	1	2	–	–	–	–	–	–	0	–
CMPA <opr>	Compare A to memory	A – M	IMM	2	2	–	–	–	–	↕	↕	↕	↕
			DIR	2	3								
			EXT	3	4								
			IND, X	2	4								
			IND, Y	3	5								
CMPB <opr>	Compare B to memory	B – M	IMM	2	2	–	–	–	–	↕	↕	↕	↕
			DIR	2	3								
			EXT	3	4								
			IND, X	2	4								
			IND, Y	3	5								
COM <opr>	1's complement memory byte	M ← $FF – M	EXT	3	6	–	–	–	–	↕	↕	0	1
			IND, X	2	6								
			IND, Y	3	7								
COMA	1's complement A	A ← $FF – A	INH	1	2	–	–	–	–	↕	↕	0	1
COMB	1's complement B	B ← $FF – B	INH	1	2	–	–	–	–	↕	↕	0	1
CPD <opr>	Compare D to memory 16-bit	D – M:M+1	IMM	4	5	–	–	–	–	↕	↕	↕	↕
			DIR	3	6								
			EXT	4	7								
			IND, X	3	7								
			IND, Y	3	7								
CPX <opr>	Compare X to memory 16-bit	1X – M:M+1	IMM	3	4	–	–	–	–	↕	↕	↕	↕
			DIR	2	5								
			EXT	3	6								
			IND, X	2	7								
			IND, Y	3	7								
CPY <opr>	Compare Y to memory 16-bit	IY – M:M+1	IMM	4	5	–	–	–	–	↕	↕	↕	↕
			DIR	3	6								
			EXT	4	7								
			IND, X	3	7								
			IND, Y	3	7								
DAA	Decimal adjust A	Adjust sum to BCD	INH	1	2	–	–	–	–	↕	↕	↕	↕
DEC <opr>	Decrement memory byte	M ← M – 1	EXT	3	6	–	–	–	–	↕	↕	↕	–
			IND, X	2	6								
			IND, Y	3	7								
DECA	Decrement accumulator A	A ← A – 1	INH	1	2	–	–	–	–	↕	↕	↕	–

Source forms	Operation	Boolean expression	Addressing mode for \<opr\>	Bytes	Cycles	S	X	H	I	N	Z	V	C
DECB	Decrement accumulator B	B ← B - 1	INH	1	2	–	–	–	–	↕	↕	↕	–
DES	Decrement stack pointer	SP ← SP - 1	INH	1	3	–	–	–	–	–	–	–	–
DEX	Decrement register IX	IX ← IX - 1	INH	1	3	–	–	–	–	–	↕	–	–
DEY	Decrement register IY	IY ← IY - 1	INH	2	4	–	–	–	–	–	↕	–	–
EORA \<opr\>	Exclusive OR A with memory	A ← A ⊕ M	IMM	2	2	–	–	–	–	↕	↕	0	–
			DIR	2	3								
			EXT	3	4								
			IND, X	2	4								
			IND, Y	3	5								
EORB	Exclusive OR B with memory	B ← B ⊕ M	IMM	2	2	–	–	–	–	↕	↕	0	–
			DIR	2	3								
			EXT	3	4								
			IND, X	2	4								
			IND, Y	3	5								
FDIV	Fractional divide 16-bit	IX ← D/IX, D ← r	INH	1	41	–	–	–	–	–	↕	↕	↕
IDIV	Integer divide 16-bit	IX ← D/IX, D ← r	INH	1	41	–	–	–	–	–	↕	0	↕
INC \<opr\>	Increment memory byte	M ← M + 1	EXT	3	6	–	–	–	–	↕	↕	↕	–
			IND, X	2	6								
			IND, Y	3	7								
INCA	Increment accumulator A	A ← A + 1	INH	1	2	–	–	–	–	↕	↕	↕	–
INCB	Increment accumulator B	B ← B + 1	INH	1	2	–	–	–	–	↕	↕	↕	–
INS	Increment stack pointer	SP ← SP + 1	INH	1	3	–	–	–	–	–	–	–	–
INX	Increment register IX	IX ← IX + 1	INH	1	3	–	–	–	–	–	↕	–	–
INY	Increment register IY	IY ← IY + 1	INH	1	4	–	–	–	–	–	↕	–	–
JMP \<opr\>	Jump		EXT	3	3	–	–	–	–	–	–	–	–
			IND, X	2	3								
			IND, Y	3	4								
JSR \<opr\>	Jump to subroutine		DIR	3	5	–	–	–	–	–	–	–	–
			EXT	2	6								
			IND, X	2	6								
			IND, Y	3	7								
LDAA \<opr\>	Load accumulator A	A ← M	IMM	2	2	–	–	–	–	↕	↕	0	–
			DIR	2	3								
			EXT	3	4								
			IND, X	2	4								
			IND, Y	3	5								
LDAB \<opr\>	Load accumulator B	B ← M	IMM	2	2	–	–	–	–	↕	↕	0	–
			DIR	2	3								
			EXT	3	4								
			IND, X	2	4								
			IND, Y	3	5								
LDD \<opr\>	Load double accumulator D	A ← M, B ← M + 1	IMM	3	3	–	–	–	–	↕	↕	0	–
			DIR	2	4								
			EXT	3	5								
			IND, X	2	5								
			IND, Y	3	6								

Source forms	Operation	Boolean expression	Addressing mode for \<opr\>	Bytes	Cycles	S	X	H	I	N	Z	V	C
LDS \<opr\>	Load stack pointer	Sp ← M:M + 1	IMM	3	3	–	–	–	–	↕	↕	0	–
			DIR	2	4								
			EXT	3	5								
			IND, X	2	5								
			IND, Y	3	6								
LDX \<opr\>	Load index register IX	IX ← M:M + 1	IMM	3	3	–	–	–	–	↕	↕	0	–
			DIR	2	4								
			EXT	3	5								
			IND, X	2	5								
			IND, Y	3	6								
LDY \<opr\>	Load index register IY	IY ← M:M + 1	IMM	4	4	–	–	–	–	↕	↕	0	–
			DIR	3	5								
			EXT	4	6								
			IND, X	3	6								
			IIND, Y	3	6								
LSL \<opr\>	Logical shift left		EXT	3	6	–	–	–	–	↕	↕	↕↕	
			IND, X	2	6								
			IND, Y	3	7								
LSLA	Logical shift left A		INH	1	2								
LSLB	Logical shift left B		INH	1	2								
LSLD	Logical shift left double		INH	1	3	–	–	–	–	↕	↕	↕	↕
LSR \<opr\>	Logical shift right		EXT	3	6	–	–	–	–	0	↕	↕	↕
			IND, X	2	6								
			IND, Y	3	7								
LSRA	Logical shift right A		INH	1	2								
LSRB	Logical shift right B		INH	1	2								
LSRD	Logical shift right double		INH	1	3	–	–	–	–	0	↕	↕	↕
MUL	Multiply 8 by 8	D ← A × B	INH	1	10	–	–	–	–	–	–	–	↕
NEG \<opr\>	2's complement memory byte	M ← 0 - M	EXT	3	6	–	–	–	–	↕	↕	↕	↕
			IND, X	2	6								
			IND, Y	3	7								
NEGA	2's complement A	A ← 0 - A	INH	1	2	–	–	–	–	↕	↕	↕	↕
NEGB	2's complement B	B ← 0 - B	INH	1	2	–	–	–	–	↕	↕	↕	↕
NOP	No operation	no operation	INH	1	2	–	–	–	–	–	–	–	–
ORAA \<opr\>	OR accumulator A (inclusive)	A ← A + M	IMM	2	2	–	–	–	–	↕	↕	0	–
			DIR	2	3								
			EXT	3	4								
			IND, X	2	4								
			IND, Y	3	5								
ORAB \<opr\>	OR accumulator B (inclusive)	B ← B + M	IMM	2	2	–	–	–	–	↕	↕	0	–
			DIR	2	3								
			EXT	3	4								
			IND, X	2	4								
			IND, Y	3	5								
PSHA	Push A onto stack	A ↑Stk, SP = SP - 1	INH	1	3	–	–	–	–	–	–	–	–

Source forms	Operation	Boolean expression	Addressing mode for <opr>	Bytes	Cycles	S	X	H	I	N	Z	V	C
PSHB	Push B onto stack	B ↑Stk, SP = SP – 1	INH	1	3	–	–	–	–	–	–	–	–
PSHX	Push X onto stack	IX ↑Stk, SP = SP – 2	INH	1	4	–	–	–	–	–	–	–	–
PSHY	Push Y onto stack	IY ↑Stk, SP = SP – 2	INH	2	5	–	–	–	–	–	–	–	–
PULA	Pull A from stack	SP = SP+1, A ↓Stk	INH	1	4	–	–	–	–	–	–	–	–
PULB	Pull B from stack	SP = SP+1, B ↓Stk	INH	1	4	–	–	–	–	–	–	–	–
PULX	Pull X from stack	SP=SP+2, IX ↓Stk	INH	1	5	–	–	–	–	–	–	–	–
PULY	Pull Y from stack	SP=SP+2, IY ↓Stk	INH	2	6	–	–	–	–	–	–	–	–
ROL <opr>	Rotate left	[C b7 ← b0 diagram]	EXT	3	6	–	–	–	–	↕	↕	↕	↕
			IND, X	2	6								
			IND, Y	3	7								
ROLA	Rotate left A		INH	1	2	–	–	–	–	↕	↕	↕	↕
ROLB	Rotate left B		INH	1	2	–	–	–	–	↕	↕	↕	↕
ROR <opr>	Rotate right	[b7 → b0 C diagram]	EXT	3	6	–	–	–	–	↕	↕	↕	↕
			IND, X	2	6								
			IND, Y	3	7								
RORA	Rotate right A		INH	1	2	–	–	–	–	↕	↕	↕	↕
RORB	Rotate right B		INH	1	2	–	–	–	–	↕	↕	↕	↕
RTI	Return from interrupt		INH	1	12	↕	↓	↕	↕	↕	↕	↕	↕
RTS	Return from subroutine		INH	1	5	–	–	–	–	–	–	–	–
SBA	Subtract B from A	A ← A – B	INH	1	2	–	–	–	↕	↕	↕	↕	↕
SBCA <opr>	Subtract with carry from A	A ← A – M – B	IMM	2	2	–	–	–	↕	↕	↕	↕	↕
			DIR	2	3								
			EXT	3	4								
			IND, X	2	4								
			IND, Y	3	5								
SBCB <opr>	Subtract with carry from B	B ← B – M – C	IMM	2	2	–	–	–	↕	↕	↕	↕	↕
			DIR	2	3								
			EXT	3	4								
			IND, X	2	4								
			IND, Y	3	5								
SEC	Set carry	C ← 1	INH	1	2	–	–	–	–	–	–	–	1
SEI	Set interrupt mask	I ← 1	INH	1	2	–	–	–	1	–	–	–	–
SEV	Set overflow flag	V ← 1	INH	1	2	–	–	–	–	–	–	1	–
STAA <opr>	Store accumulator A	M ← A	DIR	2	3	–	–	–	–	↕	↕	0	–
			EXT	3	4								
			IND, X	2	4								
			IND, Y	3	5								
STAB <opr>	Store accumulator B	M ← B	DIR	2	3	–	–	–	–	↕	↕	0	–
			EXT	3	4								
			IND, X	2	4								
			IND, Y	3	5								
STD <opr>	Store accumulator D	M ← A, M+1 ← B	DIR	2	4	–	–	–	–	↕	↕	0	–
			EXT	3	5								
			IND, X	2	5								
			IND, Y	3	6								

Source forms	Operation	Boolean expression	Addressing mode for \<opr\>	Bytes	Cycles	S	X	H	I	N	Z	V	C
STOP	Stop internal clock		INH	1	2	—	—	—	—	—	—	—	—
STS \<opr\>	Store stack pointer	M:M+1 ← SP	DIR	2	4	—	—	—	—	↕	↕	0	—
			EXT	3	5								
			IND, X	2	5								
			IND, Y	3	6								
STX \<opr\>	Store index register X	M:M+1 ← IX	DIR	2	4	—	—	—	—	↕	↕	0	—
			EXT	3	5								
			IND, X	2	5								
			IND, Y	3	6								
STY \<opr\>	Store index register Y	M:M+1 ← IY	DIR	2	5	—	—	—	—	↕	↕	0	—
			EXT	3	6								
			IND, X	2	6								
			IND, Y	3	6								
SUBA \<opr\>	Subtract memory from A	A ← A - M	IMM	2	2	—	—	—	—	↕	↕	↕	↕
			DIR	2	3								
			EXT	3	4								
			IND, X	2	4								
			IND, Y	3	5								
SUBB \<opr\>	Subtract memory from B	B ← B - M	IMM	2	2	—	—	—	—	↕	↕	↕	↕
			DIR	2	3								
			EXT	3	4								
			IND, X	2	4								
			IND, Y	3	5								
SUBD \<opr\>	Subtract memory from D	D ← D - M:M+1	IMM	3	4	—	—	—	—	↕	↕	↕	↕
			DIR	2	5								
			EXT	3	6								
			IND, X	2	6								
			IND, Y	3	7								
SWI	Software interrupt		INH	1	14	—	—	—	1	—	—	—	—
TAB	Transfer A to B	B ← A	INH	1	2	—	—	—	—	↕	↕	0	—
TAP	Transfer A to CCR	CCR ← A	INH	1	2	↕	↕	↕	↕	↕	↕	↕	↕
TBA	Transfer B to A	A ← B	INH	1	2	—	—	—	—	↕	↕	0	—

Source forms	Operation	Boolean expression	Addressing mode for <opr>	Bytes	Cycles	Condition codes S	X	H	I	N	Z	V	C
TEST	TEST (only in test mode)	AddressBusCount	INH	1	*	–	–	–	–	–	–	–	–
TPA	Transfer CCR to A	CCR ← A	INH	1	2	–	–	–	–	–	–	–	–
TST <opr>	Test for zero or negative	0 – M	EXT	3	6	–	–	–	–	↕	↕	0	–
			IND, X	2	6								
			IND, Y	3	7								
TSTA	Test A for zero or negative	A – 0	INH	1	2	–	–	–	–	↕	↕	0	–
TSTB	Test B for zero or negative	B – 0	INH	1	2	–	–	–	–	↕	↕	0	–
TSX	Transfer stack pointer to X	IX ← SP + 1	INH	1	3	–	–	↕	–	↕	↕	↕	↕
TSY	Transfer stack pointer to Y	IY ← SP + 1	INH	2	4	–	–	–	–	–	–	–	–
TXS	Transfer X to stack pointer	SP ← IX - 1	INH	1	3	–	–	–	–	–	–	–	–
TYS	Transfer Y to stack pointer	SP ← IY - 1	INH	2	4	–	–	–	–	–	–	–	–
WAI	Wait for interrupt	Stack regs and wait	INH	1	**	–	–	–	–	–	–	–	–
XGDX	Exchange D with X	D ← IX, IX ← D	INH	1	3	–	–	–	–	–	–	–	–
XGDY	Exchange D with Y	D ← IY, IY ← D	INH	2	4	–	–	–	–	–	–	–	–

Cycle:

* = Infinity or until occurs.

** = Twelve cycles are used beginning with the opcode fetch. A wait state is entered which remains in effect for an integer number of E clock cycle (n) until an interrupt is recognized. Finally, two additional cycles are used to fetch the appropriate interrupt vector (total = 14 + n).

– = Bit not changed.

0 = Always cleared (logic 0).

1 = Always set (logic 1).

↕ = Bit cleared or set depending on operation.

↓ = Bit may be cleared, cannot become set.

Appendix E

68HC11 Interrupt Vectors and Demo Board Vector Jump Table

Interrupt source	Vector address	Vector entry in EVB/ EVBU/CMD-11A8	Priority
SCI serial system	FFD6, D7	$00C4-$00C6	lowest
SPI serial transfer complete	FFD8, D9	$00C7-$00C9	
Pulse accumulator input edge	FFDA, DB	$00CA-$00CC	
Pulse accumulator overflow	FFDC, DD	$00CD-$00CF	
Timer overflow	FFDE, DF	$00D0-$00D2	
Timer output compare 5	FFE0, E1	$00D3-$00D5	
Timer output compare 4	FFE2, E3	$00D6-$00D8	
Timer output compare 3	FFE4, E5	$00D9-$00DB	
Timer output compare 2	FFE6, E7	$00DC-$00DE	
Timer output compare 1	FFE8, E9	$00DF-$00E1	
Timer input capture 3	FFEA, EB	$00E2-$00E4	
Timer input capture 2	FFEC, ED	$00E5-$00E7	
Timer input capture 1	FFEE, EF	$00E8-$00EA	
Real timer interrupt	FFF0, F1	$00EB-$00ED	
IRQ pin interrupt	FFF2, F3	$00EE-$00F0	
XIRQ pin interrupt	FFF4, F5	$00F1-$00F3	
SWI	FFF6, F7	$00F4-$00F6	
Illegal opcode trap	FFF8, F9	$00F7-$00F9	
COP failure	FFFA, FB	$00FA-$00FC	
COP clock monitor fail	FFFC, FD	$00FD-$00FF	
RESET	FFFE, FF	N/A	highest

Appendix F

SPI-Compatible Chips

The SPI (acronym of serial peripheral interface) is a protocol defined by Motorola that allows the microprocessor to exchange data with a peripheral chip using the serial format of data transfer. This protocol classes devices into two types: master and slave. There may be multiple slaves in a system. However, only one master device is allowed in a system. The master device initiates the data transfer and also provides the clock signal to synchronize the data transfer. When there are multiple slaves, an enable signal is required to select a slave device for data transfer. The master device is usually the MPU. However, an MPU can also be configured to be a slave device. The following signals are defined in the SPI protocol:

MISO: *master-in-slave-out.* This signal allows a slave to shift data out to the master device.

MOSI: *master-out-slave-in.* This signal allows the master device to shift data out to a slave device.

SCK: *shift clock.* This clock is generated by the master device to synchronize data shifting.

Microcontrollers from vendors other than Motorola may also implement the SPI protocol. For example, some of the PIC microcontrollers from Microchip and some 8051 variants from Atmel have incorporated the SPI interface.

Peripheral chips that may incorporate the SPI interface include: display (LED and LCD) driver chips, A/D and D/A converters, phase-lock loop devices, EEPROM, SRAM, timer/clock chips, etc. The peripheral chip that implements the SPI protocol is called *SPI-compatible.* The list of SPI-compatible chips is getting longer quickly. A partial list of these devices are in the following:

Vendor	Part number	Description
Motorola	MC28HC14	256-byte EEPROM in 8-pin DIP
	MC14021	8-bit input port
	MC74HC165	8-bit input port
	MC74LS165	8-bit input port
	MC74HC589	8-bit parallel-in/serial-out shift register
	MC74HC595	8-bit serial-in/parallel-out shift register
	MC74LS673	16-bit output port
	MC144115	16-segment non-muxed LCD driver
	MC144117	4-digit LCD driver
	MC14549	A/D converter successive approximation register
	MC14559	A/D converter successive approximation register
	MC14489	5-digit 7-segment LED display decoder/driver
	MC14499	4-digit 7-segment LED display decoder/driver
	MC145000	serial input multiplexed LCD driver (master)
	MC145001	serial input multiplexed LCD driver (slave)
	MC145453	LCD driver with serial interface
	MC144110	6-six-bit D/A converter
	MC144111	4-six-bit D/A converter
	MC144040	8-bit A/D converter
	MC144051	8-bit A/D converter
	MC145050	10-bit A/D converter
	MC145051	10-bit A/D converter
	MC145053	10-bit A/D converter
	MC68HC68T1	real-time clock
	MCCS1850	serial real-time clock
	MC145155	serial input PLL FS
	MC145156	serial input PLL FS
	MC145157	serial input PLL FS
	MC145158	serial input PLL FS
National Semiconductor	ADC0811	8-bit 11-channel A/D converter
	ADC0819	8-bit 19-channel A/D converter
	ADC0831	8-bit serial I/O A/D converter with mux option
	ADC0832	"
	ADC0834	"
	ADC0838	"
	ADC0833	8-bit 4-channel serial I/O A/D converter
	ADC08031	8-bit high-speed serial I/O A/D converter
	ADC08032	"
	ADC08034	"
	ADC08038	"
	ADC08131	"
	ADC08132	"
	ADC08134	"
	ADC08138	"
	ADC08231	"
	ADC08234	"
	ADC08238	"

Vendor	Part number	Description
National Semiconductor	ADC1031	10-bit A/D converter with analog mux
	ADC1034	"
	ADC1038	"
	ADC10731	10-bit plus sign A/D converter with analog mux
	ADC10732	"
	ADC10734	"
	ADC10738	"
	ADC10831	"
	ADC10832	"
	ADC10834	"
	ADC10838	"
	ADC12030	12-bit self-calibrating plus sign A/D converter
	ADC12032	"
	ADC12034	"
	ADC12038	"
	ADC12L030	3.3 V 12-bit self-calibrating plus sign A/D converter
	ADC12L032	"
	ADC12L034	"
	ADC12L038	"
	ADC12130	12-bit self-calibrating plus sign A/D converter
	ADC12132	"
	ADC12138	"
	DAC0854	quad 8-bit D/A converter with voltage output
	DAC1054	quad 10-bit D/A converter with voltage output
RCA	CDP68H68A1	10-bit A/D converter
	CDP68HC68R1	$128 \times$ 8-bit SRAM
	CDP68HC68R2	$256 \times$ 8-bit SRAM
	CDP68HC68T1	real-time clock
Signetics	PCx2100	40-segment LCD duplex driver
	PCx2110	60-segment LCD duplex driver
	PCx2111	64-segment LCD duplex driver
	PCx2112	32-segment LCD duplex driver
	PCx3311	DTMF generator with parallel inputs
	PCx3312	DTMF generator with I^{2}C bus inputs
	PCx8570A	256×8 SRAM
	PCx8571	128×8 SRAM
	PCx8573	clock/timer
	PCx8474	8-bit remote I/O expander
	PCx8476	1:4 mux LCD driver
	PCx8477	32/64 segment LCD driver
	PCx8491	8-bit, 4-channel A/D converter and D/A converter
	*SAA1057	PLL tuning circuit: 512 KHz to 120 MHz
	*SAA1060	32-segment LED driver
	*SAA1061	16-segment LED driver
	*SAA1062A	20-segment LED driver
	*SAA1063	fluorescent display driver
	SAA1300	switching circuit
	*SAA3019	clock/timer

Vendor	Part number	Description
Signetics	SAA3028	I/R transcoder
	*SAB3013	hex 6-bit D/A converter
	SAB3035	PLL digital tuning circuit with 8 D/A converters
	SAB3036	PLL digital tuning circuit
	SAB3037	PLL digital tuning circuit with 4 D/A converters
	SAA5240	teletext controller chip—625 line system
	TDA3820	digital stereo sound control IC
	TEA6000	MUSTI: FM/RF system
	TDA1534A	14-bit A/D converter—serial output
	TDA1540P, D	14-bit A/D converter—serial output
	NE5036	6-bit A/D converter—serial output
Sprague	UCN4810/5810	Power driving output port
	UAA2022/2023	16-bit power output port
Xicor	X2444	16×16 nonvolatile SRAM with serial I/O
	X2444I	"
	X2444M	" (military)
	X2402	256×8 serial E²PROM
	X2402I	"
	X2404	512×8 E²PROM with serial I/O
	X2404I	"
	X2404M	"
	X24C04	"
	X24C04I	"
	X24C16	2048×8 E²PROM with serial I/O
	X24C16I	"
	X24C16M	"
MAXIM	MAX5351	13-bit D/A converter voltage output using 3.3 V supply
	MAX5353	12-bit D/A converter voltage output using 3.3 V supply
	MAX551	12-bit D/A converter current output using 2.7 V supply
	MAX7543	12-bit D/A converter current output using 3.3V supply
	MAX5043	12-bit D/A converter current output using 3.3V supply
	MAX5151	13-bit D/A converter voltage output using 2.7 ~ 3.6V supply
	MAX5153	13-bit D/A converter voltage output using 2.7 ~ 3.6V supply
	MAX5155	12-bit D/A converter voltage output using 2.7 ~ 3.6V supply
	MAX5157	12-bit D/A converter voltage output using 2.7 ~ 3.6V supply
	MAX1110	8-channel 8-bit A/D converter using 2.7 ~ 5.5V supply
	MAX1111	4-channel 8-bit A/D converter using 2.7 ~ 5.5B supply
	MAX1241	1-channel 12-bit A/D converter using 3V ~ 5.5V supply
	MAX1247	4-channel 12-bit A/D converter using 2.7V ~ 5.25V supply
	MAX147	8-channel 12-bit A/D converter using 2.7V ~ 5.25 V supply
	MAX1240	1-channel 12-bit A/D converter using 3V ~ 5.5V supply
	MAX1246	4-channel 12-bit A/D converter using 2.7V ~ 5.25 V supply
	MAX146	8-channel 12-bit A/D converter using 2.7V ~ 5.25 V supply
	MAX1243	1-channel 10-bit A/D converter using 2.7V ~ 5.25V supply
	MAX1249	4-channel 10-bit A/D converter using 2.7V ~ 5.25V supply
	MAX148	8-channel 10-bit A/D converter using 2.7V ~ 5.25V supply
	MAX1242	1-channel 10-bit A/D converter using 2.7V ~ 5.25V supply
	MAX1248	4-channel 10-bit A/D converter using 2.7V ~ 5.25V supply
	MAX149	8-channel 10-bit A/D converter using 2.7V ~ 5.25V supply

Appendix G

68HC11 Development Tool Vendors

Company	Products	Contact address
Software tools		
Avocet	C compiler, assembler, simulator, integrated development environment	(800) 448-8500 http://www.avocetsystems.com/

COSMIC Software Inc.	C compiler, assembler, simulator, debugger, and integrated development environment	400 West Cummings Park STE 6000
		    Woburn, MA 01801-6512
		    Phone (781) 932-2556
		    http://www.cosmic-software.com

GAIO U.S. Office	C compiler	Phone: (408) 752-0268
		    Web: http://www.gaio.com
		    E-MAIL: gaio@gol.com

HIWARE Inc.	C compiler, assembler, debugger, simulator	8608 Barasinga Trail, Austin, TX 78749
		    phone: (512) 282-4435
		    http://www.hiware.com/

IAR System	C compiler, assembler, debugger, simulator	One Maritime Plaza, San Francisco, CA 94111
		    Phone: (415) 765-5500
		    http://www.iar.se/

ImageCraft	C compiler, IDE	706 Colorado Ave, Palo Alto, CA 94303
		    Phone (650) 493-9326
		    http://www.imagecraft.com/

Hardware tools		
Avocet	in-circuit emulator	Phone: (800) 448-8500 http://www.avocetsystems.com/

Ashling	in-circuit emulator	National Technological Park, Limerick, Ireland
		    Tel: +353-61-334466
		    http://www.ashling.com/

Axiom Manufacturing	68HC11 evaluation boards	717 Lingco Dr., Suite 209, Richardson, TX 75081
		    Phone: (972) 994-9676
		    http://www.axman.com

Hitex Development Tools	in-circuit emulator	http://www.hitex.de/

Company	Products	Contact address
Huntsville Microsystem	in-circuit emulator, debugger	Phone: (800) 847-1998 http://www.hmi.com/
TECI Inc.	68HC11 evaluation boards in-circuit emulator, programmer	RR#3, Box 8C, Barton, VT 05822 Phone: (802) 525-3458 http://www.tec-I.com

Hardware tools

Company	Products	Contact address
iSystem	in-circuit emulator	16776 Bernardo Center Drive, Suite 203 San Diego, CA 92128, USA Phone: (888) 543-5300 http://www.isystem.com
Lauterbach	in-circuit emulator	5 Mount Royal Ave., Marlborough, MA 01752 Phone: (508) 303-6812 http://www.lauterbach.com/
MetaLink	in-circuit emulator	325 East Elliot Rd. Suite 23, Chandler, AZ 85225. Phone (480) 926-0797 http://www.metaice.com
Nohau	in-circuit emulator	Corporation, 51 E. Campbell Ave., Campbell, CA 95008 Phone: (1-888-886-6428) http://www.nohau.com
Orion	in-circuit emulator	1376 Borregas Avenue, Sunnyvale, CA 94089-1004 Phone: (800) 729-7700 http://www.oritools.com
BP System	Programmer	1000 North Post Oak Road, Suite 225, Houston, Texas 77055-7237. Phone: (800) 225-2102 or (713) 688-2620 www.bpmicro.com.
Data I/O	Programmer	Phone: (800) 247-5700 http://www.dataio.com

Appendix H

Standard Values of Commercially Available Resistors

Ohms (Ω)					Kilohms (KΩ)		Megaohms (MΩ)	
0.10	1.0	10	100	1000	10	100	1.0	10.0
0.11	1.1	11	110	1100	11	110	1.1	11.0
0.12	1.2	12	120	1200	12	120	1.2	12.0
0.13	1.3	13	130	1300	13	130	1.3	13.0
0.15	1.5	15	150	1500	15	150	1.5	15.0
0.16	1.6	16	160	1600	16	160	1.6	16.0
0.18	1.8	18	180	1800	18	180	1.8	18.0
0.20	2.0	20	200	2000	20	200	2.0	20.0
0.22	2.2	22	220	2200	22	220	2.2	22.0
0.24	2.4	24	240	2400	24	240	2.4	
0.27	2.7	27	270	2700	27	270	2.7	
0.30	3.0	30	300	3000	30	300	3.0	
0.33	3.3	33	330	3300	33	330	3.3	
0.36	3.6	36	360	3600	36	360	3.6	
0.39	3.9	39	390	3900	39	390	3.9	
0.43	4.3	43	430	4300	43	430	4.3	
0.47	4.7	47	470	4700	47	470	4.7	
0.51	5.1	51	510	5100	51	510	5.1	
0.56	5.6	56	560	5600	56	560	5.6	
0.62	6.2	62	620	6200	62	620	6.2	
0.68	6.8	68	680	6800	68	680	6.8	
0.75	7.5	75	750	7500	75	750	7.5	
0.82	8.2	82	820	8200	82	820	8.2	
0.91	9.1	91	910	9100	91	910	9.1	

Appendix I

Data Sheets

Appendix I.1

68HC11 Instruction Set

Appendix I.1   68HC11 Instruction Set Details

I.1 Introduction

This appendix contains complete detailed information for all MC 68HC11 instructions. The instructions are arranged in alphabetical order with the instruction mnemonic set in larger type for easy reference.

I.2 Nomenclature

The following nomenclature is used in the subsequent defintions.

(a) Operators

( )	= Contents of Register Shown Inside Parentheses
◀	= Is Transferred to
▲	= Is Pulled from Stack
▼	= Is Pushed onto Stack
•	= Boolean AND
+	= Arithmetic Addition Symbol Except Where Used as Inclusive-OR Symbol in Boolean Formula
⊕	= Exclusive-OR
×	= Multiply
:	= Concatenation
–	= Arithmetic Subtraction Symbol or Negation Symbol (Twos Complement)

(b) Registers in the MPU

ACCA	= Accumulator A
ACCB	= Accumulator B
ACCX	= Accumulator ACCA or ACCB
ACCD	= Double Accumulator — Accumulator A Concatenated with Accumulator B Where A is the Most Significant Byte
CCR	= Condition Code Register
IX	= Index Register X, 16 Bits
IXH	= Index Register X, Higher Order 8 Bits
IXL	= Index Register X, Lower Order 8 Bits
IY	= Index Register Y, 16 Bits
IYH	= Index Register Y, Higher Order 8 Bits
IYL	= Index Register Y, Lower Order 8 Bits
PC	= Program Counter, 16 Bits
PCH	= Program Counter, Higher Order (Most Significant) 8 Bits
PCL	= Program Counter, Lower Order (Least Significant) 8 Bits
SP	= Stack Pointer, 16 Bits
SPH	= Stack Pointer, Higher Order 8 Bits
SPL	= Stack Pointer, Lower Order 8 Bits

(c) Memory and Addressing
    M   = A Memory Location (One Byte)
    M + 1   = The Byte of Memory at $0001 Plus the Address of the Memory Location
               Indicated by "M"
    Rel   = Relative Offset (i.e., the Twos Complement Number Stored in the Last Byte
               of Machine Code Corresponding to a Branch Instruction)
    (opr)   = Operand
    (msk)   = Mask Used in Bit Manipulation Instructions
    (rel)   = Relative Offset Used in Branch Instructions

(d) Bits 7–0 of the Condition Code Register
    S   = Stop Disable, Bit 7
    X   = X Interrupt Mask, Bit 6
    H   = Half Carry, Bit 5
    I   = I Interrupt Mask, Bit 4
    N   = Negative Indicator, Bit 3
    Z   = Zero Indicator, Bit 2
    V   = Twos Complement Overflow Indicator, Bit 1
    C   = Carry/Borrow, Bit 0

(e) Status of Individual Bits BEFORE Execution of an Instruction
    An   = Bit n of ACCA (n = 7, 6, 5 . . . 0)
    Bn   = Bit n of ACCB (n = 7, 6, 5 . . . 0)
    Dn   = Bit n of ACCD (n = 15, 14, 13 . . . 0)
               Where Bits 15–8 Refer to ACCA and Bit 7–0 Refer to ACCB
    IXn   = Bit n of IX (n = 15, 14, 13 . . . 0)
    IXHn   = Bit n of IXH (n = 7, 6, 5 . . . 0)
    IXLn   = Bit n of IXL (n = 7, 6, 5 . . . 0)
    IYn   = Bit n of IY (n = 15, 14, 13 . . . 0)
    IYHn   = Bit n of IYH (n = 7, 6, 5 . . . 0)
    IYLn   = Bit n of IYL (n = 7, 6, 5 . . . 0)
    Mn   = Bit n of M (n = 7, 6, 5 . . . 0)
    SPHn   = Bit n of SPH (n = 7, 6, 5 . . . 0)
    SPLn   = Bit n of SPL (n = 7, 6, 5 . . . 0)
    Xn   = Bit n of ACCX (n = 7, 6, 5 . . . 0)

(f) Status of Individual Bits of RESULT of Execution of an Instruction
    (i) For 8-Bit Results
        Rn   = Bit n of the Result (n = 7, 6, 5 . . . 0)
               This applies to instructions which provide a result contained in a single
               byte of memory or in an 8-bit register.
    (ii) For 16-Bit Results
        RHn = Bit n of the Most Significant Byte of the Result (n = 7, 6, 5 . . . 0)
        RLn = Bit n of the Least Significant Byte of the Result (n = 7, 6, 5 . . . 0)
               This applies to instructions which provide a result contained in two
               consecutive bytes of memory or in a 16-bit register.
        Rn   = Bit n of the Result (n = 15, 14, 13 . . . 0)

(g) Notation Used in CCR Activity Summary Figures

—	= Bit Not Affected
0	= Bit Forced to Zero
1	= Bit Forced to One
♦	= Bit Set or Cleared According to Results of Operation
♦	= Bit may change from one to zero, remain zero, or remain one as a result of this operation, but cannot change from zero to one.

(h) Notation Used in Cycle-by-Cycle Execution Tables

—	= Irrelevant Data
ii	= One Byte of Immediate Data
jj	= High-Order Byte of 16-Bit Immediate Data
kk	= Low-Order Byte of 16-Bit Immediate Data
hh	= High-Order Byte of 16-Bit Extended Address
ll	= Low-Order Byte of 16-Bit Extended Address
dd	= Low-Order 8 Bits of Direct Address $0000–$00FF (High Byte Assumed to be $00)
mm	= 8-Bit Mask (Set Bits Correspond to Operand Bits Which Will Be Affected)
ff	= 8-Bit Forward Offset $00 (0) to $FF (255) (Is Added to Index)
rr	= Signed Relative Offset $80 ($-128$) to $7F ($+127$) (Offset Relative to Address Following Machine Code Offset Byte)
OP	= Address of Opcode Byte
OP + n	= Address of nth Location after Opcode Byte
SP	= Address Pointed to by Stack Pointer Value (at the Start of an Instruction)
SP + n	= Address of nth Higher Address Past That Pointed to by Stack Pointer
SP − n	= Address of nth Lower Address Before That Pointed to by Stack Pointer
Sub	= Address of Called Subroutine
Nxt op	= Opcode of Next Instruction
Rtn hi	= High-Order Byte of Return Address
Rtn lo	= Low-Order Byte of Return Address
Svc hi	= High-Order Byte of Address for Service Routine
Svc lo	= Low-Order Byte of Address for Service Routine
Vec hi	= High-Order Byte of Interrupt Vector
Vec lo	= Low-Order Byte of Interrupt Vector

ABA   Add Accumulator B to Accumulator A   **ABA**

Operation: ACCA ◆ (ACCA) + (ACCB)

Description: Adds the contents of accumulator B to the contents of accumulator A and places the result in accumulator A. Accumulator B is not changed. This instruction affects the H condition code bit so it is suitable for use in BCD arithmetic operations (see DAA instruction for additional information).

Condition Codes and Boolean Formulae:

S	X	H	I	N	Z	V	C
—	—	↕	—	↕	↕	↕	↕

H $A3 \cdot B3 + B3 \cdot \overline{R3} + \overline{R3} \cdot A3$
    Set if there was a carry from bit 3; cleared otherwise.

N   R7
    Set if MSB of result is set; cleared otherwise.

Z $\overline{R7} \cdot \overline{R6} \cdot \overline{R5} \cdot \overline{R4} \cdot \overline{R3} \cdot \overline{R2} \cdot \overline{R1} \cdot \overline{R0}$
    Set if result is $00; cleared otherwise.

V $A7 \cdot B7 \cdot \overline{R7} + \overline{A7} \cdot \overline{B7} \cdot R7$
    Set if a twos complement overflow resulted from the operation; cleared otherwise.

C $A7 \cdot B7 + B7 \cdot \overline{R7} + \overline{R7} \cdot A7$
    Set if there was a carry from the MSB of the result; cleared otherwise.

Source Forms: ABA

Addressing Modes, Machine Code, and Cycle-by-Cycle Execution:

Cycle	ABA (INH)		
	Addr	Data	R/W̄
1	OP	1B	1
2	OP + 1	—	1

ABX   Add Accumulator B to Index Register X   ABX

Operation: IX ◀ (IX) + (ACCB)

Description: Adds the 8-bit unsigned contents of accumulator B to the contents of index register X (IX) considering the possible carry out of the low-order byte of the index register X; places the result in index register X (IX). Accumulator B is not changed. There is no equivalent instruction to add accumulator A to an index register.

Condition Codes and Boolean Formulae:

S	X	H	I	N	Z	V	C
—	—	—	—	—	—	—	—

None affected

Source Forms: ABX

Addressing Modes, Machine Code, and Cycle-by-Cycle Execution:

Cycle	ABX (INH)		
	Addr	Data	R/W̄
1	OP	3A	1
2	OP + 1	—	1
3	FFFF	—	1

ABY

Add Accumulator B to Index Register Y

ABY

Operation: IY ◆ (IY) + (ACCB)

Description: Adds the 8-bit unsigned contents of accumulator B to the contents of index register Y (IY) considering the possible carry out of the low-order byte of index register Y; places the result in index register Y (IY). Accumulator B is not changed. There is no equivalent instruction to add accumulator A to an index register.

Condition Codes and Boolean Formulae:

S	X	H	I	N	Z	V	C
—	—	—	—	—	—	—	—

None affected

Source Forms: ABY

Addressing Modes, Machine Code, and Cycle-by-Cycle Execution:

Cycle	ABY (INH)		
	Addr	Data	R/W̄
1	OP	18	1
2	OP+1	3A	1
3	OP+2	—	1
4	FFFF	—	1

Reprinted with permission of Motorola

ADC   Add with Carry   ADC

Operation: ACCX ⬅ (ACCX) + (M) + (C)

Description: Adds the contents of the C bit to the sum of the contents of ACCX and M and places the result in ACCX. This instruction affects the H condition code bit so it is suitable for use in BCD arithmetic operations (see DAA instruction for additional information).

Condition Codes and Boolean Formulae:

S	X	H	I	N	Z	V	C
—	—	⬍	—	⬍	⬍	⬍	⬍

H $X3 \cdot M3 + M3 \cdot \overline{R3} + \overline{R3} \cdot X3$
   Set if there was a carry from bit 3; cleared otherwise.

N   R7
   Set if MSB of result is set; cleared otherwise.

Z $\overline{R7} \cdot \overline{R6} \cdot \overline{R5} \cdot \overline{R4} \cdot \overline{R3} \cdot \overline{R2} \cdot \overline{R1} \cdot \overline{R0}$
   Set if result is $00; cleared otherwise.

V $X7 \cdot M7 \cdot \overline{R7} + \overline{X7} \cdot \overline{M7} \cdot R7$
   Set if a twos complement overflow resulted from the operation; cleared otherwise.

C $X7 \cdot M7 + M7 \cdot \overline{R7} + \overline{R7} \cdot X7$
   Set if there was a carry from the MSB of the result; cleared otherwise.

Source Forms: ADCA (opr); ADCB (opr)

Addressing Modes, Machine Code, and Cycle-by-Cycle Execution:

Cycle	ADCA (IMM) Addr	Data	R/$\overline{W}$	ADCA (DIR) Addr	Data	R/$\overline{W}$	ADCA (EXT) Addr	Data	R/$\overline{W}$	ADCA (IND, X) Addr	Data	R/$\overline{W}$	ADCA (IND, Y) Addr	Data	R/$\overline{W}$
1	OP	89	1	OP	99	1	OP	B9	1	OP	A9	1	OP	18	1
2	OP+1	ii	1	OP+1	dd	1	OP+1	hh	1	OP+1	ff	1	OP+1	A9	1
3				00dd	(00dd)	1	OP+2	ll	1	FFFF	—	1	OP+2	ff	1
4							hhll	(hhll)	1	X+ff	(X+ff)	1	FFFF	—	1
5													Y+ff	(Y+ff)	1

Cycle	ADCB (IMM) Addr	Data	R/$\overline{W}$	ADCB (DIR) Addr	Data	R/$\overline{W}$	ADCB (EXT) Addr	Data	R/$\overline{W}$	ADCB (IND, X) Addr	Data	R/$\overline{W}$	ADCB (IND, Y) Addr	Data	R/$\overline{W}$
1	OP	C9	1	OP	D9	1	OP	F9	1	OP	E9	1	OP	18	1
2	OP+1	ii	1	OP+1	dd	1	OP+1	hh	1	OP+1	ff	1	OP+1	E9	1
3				00dd	(00dd)	1	OP+2	ll	1	FFFF	—	1	OP+2	ff	1
4							hhll	(hhll)	1	X+ff	(X+ff)	1	FFFF	—	1
5													Y+ff	(Y+ff)	1

ADD

Add without Carry

ADD

Operation: ACCX ◀ (ACCX) + (M)

Description: Adds the contents of M to the contents of ACCX and places the result in ACCX. This instruction affects the H condition code bit so it is suitable for use in the BCD arithmetic operations (see DAA instruction for additional information).

Condition Codes and Boolean Formulae:

S	X	H	I	N	Z	V	C
—	—	↕	—	↕	↕	↕	↕

H $X3 \cdot M3 + M3 \cdot \overline{R3} + \overline{R3} \cdot X3$
   Set if there was a carry from bit 3; cleared otherwise.

N   R7
   Set if MSB of result is set; cleared otherwise.

Z $\overline{R7} \cdot \overline{R6} \cdot \overline{R5} \cdot \overline{R4} \cdot \overline{R3} \cdot \overline{R2} \cdot \overline{R1} \cdot \overline{R0}$
   Set if result is $00; cleared otherwise.

V $X7 \cdot M7 \cdot \overline{R7} + \overline{X7} \cdot \overline{M7} \cdot R7$
   Set if a twos complement overflow resulted from the operation; cleared otherwise.

C $X7 \cdot M7 + M7 \cdot \overline{R7} + \overline{R7} \cdot X7$
   Set if there was a carry from the MSB of the result; cleared otherwise.

Source Forms: ADDA (opr); ADDB (opr)

Addressing Modes, Machine Code, and Cycle-by-Cycle Execution:

Cycle	ADDA (IMM)			ADDA (DIR)			ADDA (EXT)			ADDA (IND, X)			ADDA (IND, Y)		
	Addr	Data	R/W̄	Addr	Data	R/W̄	Addr	Data	R/W̄	Addr	Data	R/W̄	Addr	Data	R/W̄
1	OP	8B	1	OP	9B	1	OP	BB	1	OP	AB	1	OP	18	1
2	OP+1	ii	1	OP+1	dd	1	OP+1	hh	1	OP+1	ff	1	OP+1	AB	1
3				00dd	(00dd)	1	OP+2	ll	1	FFFF	—	1	OP+2	ff	1
4							hhll	(hhll)	1	X+ff	(X+ff)	1	FFFF	—	1
5													Y+ff	(Y+ff)	1

Cycle	ADDB (IMM)			ADDB (DIR)			ADDB (EXT)			ADDB (IND, X)			ADDB (IND, Y)		
	Addr	Data	R/W̄	Addr	Data	R/W̄	Addr	Data	R/W̄	Addr	Data	R/W̄	Addr	Data	R/W̄
1	OP	CB	1	OP	DB	1	OP	FB	1	OP	EB	1	OP	18	1
2	OP+1	ii	1	OP+1	dd	1	OP+1	hh	1	OP+1	ff	1	OP+1	EB	1
3				00dd	(00dd)	1	OP+2	ll	1	FFFF	—	1	OP+2	ff	1
4							hhll	(hhll)	1	X+ff	(X+ff)	1	FFFF	—	1
5													Y+ff	(Y+ff)	1

ADDD   Add Double Accumulator   ADDD

Operation: ACCD ◆ (ACCD) + (M:M + 1)

Description: Adds the contents of M concatenated with M + 1 to the contents of ACCD and places the results in ACCD. Accumulator A corresponds to the high-order half of the 16-bit double accumulator D.

Condition Codes and Boolean Formulae:

S	X	H	I	N	Z	V	C
—	—	—	—	↕	↕	↕	↕

N   R15
    Set if MSB of result is set; cleared otherwise.

Z $\overline{R15} \cdot \overline{R14} \cdot \overline{R13} \cdot \overline{R12} \cdot \overline{R11} \cdot \overline{R10} \cdot \overline{R9} \cdot \overline{R8} \cdot \overline{R7} \cdot \overline{R6} \cdot \overline{R5} \cdot \overline{R4} \cdot \overline{R3} \cdot \overline{R2} \cdot \overline{R1} \cdot \overline{R0}$
    Set if result is $0000; cleared otherwise.

V $D15 \cdot M15 \cdot \overline{R15} + \overline{D15} \cdot \overline{M15} \cdot R15$
    Set if a twos complement overflow resulted from the operation; cleared otherwise.

C $D15 \cdot M15 + M15 \cdot \overline{R15} + \overline{R15} \cdot D15$
    Set if there was a carry from the MSB of the result; cleared otherwise.

Source Form: ADDD (opr)

Addressing Modes, Machine Code, and Cycle-by-Cycle Execution:

Cycle	ADDD (IMM) Addr	Data	R/W̄	ADDD (DIR) Addr	Data	R/W̄	ADDD (EXT) Addr	Data	R/W̄	ADDD (IND, X) Addr	Data	R/W̄	ADDD (IND, Y) Addr	Data	R/W̄
1	OP	C3	1	OP	D3	1	OP	F3	1	OP	E3	1	OP	18	1
2	OP + 1	jj	1	OP + 1	dd	1	OP + 1	hh	1	OP + 1	ff	1	OP + 1	E3	1
3	OP + 2	kk	1	00dd	(00dd)	1	OP + 2	ll	1	FFFF	—	1	OP + 2	ff	1
4	FFFF	—	1	00dd + 1	(00dd + 1)	1	hhll	(hhll)	1	X + ff	(X + ff)	1	FFFF	—	1
5				FFFF	—	1	hhll + 1	(hhll + 1)	1	X + ff + 1	(X + ff + 1)	1	Y + ff	(Y + ff)	1
6							FFFF	—	1	FFFF	—	1	Y · ff + 1	(Y · ff + 1)	1
7													FFFF	—	1

AND

Logical AND

AND

Operation: ACCX ◀ (ACCX) · (M)

Description: Performs the logical AND between the contents of ACCX and the contents of M and places the result in ACCX. (Each bit of ACCX after the operation will be the logical AND of the corresponding bits of M and of ACCX before the operation.)

Condition Codes and Boolean Formulae:

S	X	H	I	N	Z	V	C
—	—	—	—	⬍	⬍	0	—

N R7
   Set if MSB of result is set; cleared otherwise.

Z $\overline{R7} \cdot \overline{R6} \cdot \overline{R5} \cdot \overline{R4} \cdot \overline{R3} \cdot \overline{R2} \cdot \overline{R1} \cdot \overline{R0}$
   Set if result is $00; cleared otherwise.

V 0
   Cleared

Source Forms: ANDA (opr); ANDB (opr)

Addressing Modes, Machine Code, and Cycle-by-Cycle Execution:

Cycle	ANDA (IMM)			ANDA (DIR)			ANDA (EXT)			ANDA (IND, X)			ANDA (IND, Y)		
	Addr	Data	R/W̄	Addr	Data	R/W̄	Addr	Data	R/W̄	Addr	Data	R/W̄	Addr	Data	R/W̄
1	OP	84	1	OP	94	1	OP	B4	1	OP	A4	1	OP	18	1
2	OP+1	ii	1	OP+1	dd	1	OP+1	hh	1	OP+1	ff	1	OP+1	A4	1
3				00dd	(00dd)	1	OP+2	ll	1	FFFF	—	1	OP+2	ff	1
4							hhll	(hhll)	1	X+ff	(X+ff)	1	FFFF	—	1
5													Y+ff	(Y+ff)	1

Cycle	ANDB (IMM)			ANDB (DIR)			ANDB (EXT)			ANDB (IND, X)			ANDB (IND, Y)		
	Addr	Data	R/W̄	Addr	Data	R/W̄	Addr	Data	R/W̄	Addr	Data	R/W̄	Addr	Data	R/W̄
1	OP	C4	1	OP	D4	1	OP	F4	1	OP	E4	1	OP	18	1
2	OP+1	ii	1	OP+1	dd	1	OP+1	hh	1	OP+1	ff	1	OP+1	E4	1
3				00dd	(00dd)	1	OP+2	ll	1	FFFF	—	1	OP+2	ff	1
4							hhll	(hhll)	1	X+ff	(X+ff)	1	FFFF	—	1
5													Y+ff	(Y+ff)	1

ASL

Arithmetic Shift Left
(Same as LSL)

ASL

Operation:

$$C \leftarrow \boxed{b7 - - - - - b0} \leftarrow 0$$

Description: Shifts all bits of the ACCX or M one place to the left. Bit 0 is loaded with a zero. The C bit in the CCR is loaded from the most significant bit of ACCX or M.

Condition Codes and Boolean Formulae:

S	X	H	I	N	Z	V	C
—	—	—	—	↕	↕	↕	↕

N   R7
    Set if MSB of result is set; cleared otherwise.

Z $\overline{R7} \cdot \overline{R6} \cdot \overline{R5} \cdot \overline{R4} \cdot \overline{R3} \cdot \overline{R2} \cdot \overline{R1} \cdot \overline{R0}$
    Set if result is $00; cleared otherwise.

V $N \oplus C = [N \cdot \overline{C}] + [\overline{N} \cdot C]$ (for N and C after the shift)
    Set if (N is set and C is clear) or (N is clear and C is set); cleared otherwise (for values of N and C after the shift).

C   M7
    Set if, before the shift, the MSB of ACCX or M was set; cleared otherwise.

Source Forms: ASLA; ASLB; ASL (opr)

Addressing Modes, Machine Code, and Cycle-by-Cycle Execution:

Cycle	ASLA (INH)			ASLB (INH)			ASL (EXT)			ASL (IND, X)			ASL (IND, Y)		
	Addr	Data	R/W̄	Addr	Data	R/W̄	Addr	Data	R/W̄	Addr	Data	R/W̄	Addr	Data	R/W̄
1	OP	48	1	OP	58	1	OP	78	1	OP	68	1	OP	18	1
2	OP+1	—	1	OP+1	—	1	OP+1	hh	1	OP+1	ff	1	OP+1	68	1
3							OP+2	ll	1	FFFF	—	1	OP+2	ff	1
4							hhll	(hhll)	1	X+ff	(X+ff)	1	FFFF	—	1
5							FFFF	—	1	FFFF	—	1	Y+ff	(Y+ff)	1
6							hhll	result	0	X+ff	result	0	FFFF	—	1
7													Y+ff	result	0

ASLD

Arithmetic Shift Left Double Accumulator
(Same as LSLD)

ASLD

Operation:

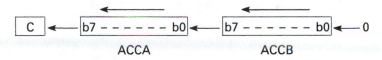

$$C \longleftarrow \boxed{b7 ------ b0} \longleftarrow \boxed{b7 ------ b0} \longleftarrow 0$$

ACCA   ACCB

Description: Shifts all bits of ACCD one place to the left. Bit 0 is loaded with a zero. The C bit in the CCR is loaded from the most significant bit of ACCD.

Condition Codes and Boolean Formulae:

S	X	H	I	N	Z	V	C
—	—	—	—	‡	‡	‡	‡

N R15
   Set if MSB of result is set; cleared otherwise.

Z $\overline{R15} \cdot \overline{R14} \cdot \overline{R13} \cdot \overline{R12} \cdot \overline{R11} \cdot \overline{R10} \cdot \overline{R9} \cdot \overline{R8} \cdot \overline{R7} \cdot \overline{R6} \cdot \overline{R5} \cdot \overline{R4} \cdot \overline{R3} \cdot \overline{R2} \cdot \overline{R1} \cdot \overline{R0}$
   Set if result is $0000; cleared otherwise.

V $N \oplus C = [N \cdot \overline{C}] + [\overline{N} \cdot C]$ (for N and C after the shift)
   Set if (N is set and C is clear) or (N is clear and C is set); cleared otherwise (for values of N and C after the shift).

C D15
   Set if, before the shift, the MSB of ACCD was set; cleared otherwise.

Source Form: ASLD

Addressing Modes, Machine Code, and Cycle-by-Cycle Execution:

Cycle	ASLD (INH)		
	Addr	Data	R/$\overline{W}$
1	OP	05	1
2	OP + 1	—	1
3	FFFF	—	1

ASR
Arithmetic Shift Right
ASR

Operation:

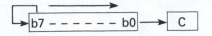

Description: Shifts all of ACCX or M one place to the right. Bit 7 is held constant. Bit 0 is loaded into the C bit of the CCR. This operation effectively divides a twos complement value by two without changing its sign. The carry bit can be used to round the result.

Condition Codes and Boolean Formulae:

S	X	H	I	N	Z	V	C
—	—	—	—	↕	↕	↕	↕

N R7
Set if MSB of result is set; cleared otherwise.

Z $\overline{R7} \cdot \overline{R6} \cdot \overline{R5} \cdot \overline{R4} \cdot \overline{R3} \cdot \overline{R2} \cdot \overline{R1} \cdot \overline{R0}$
Set if result is $00; cleared otherwise.

V $N \oplus C = [N \cdot \overline{C}] + [\overline{N} \cdot C]$ (for N and C after the shift)
Set if (N is set and C is clear) or (N is clear and C is set); cleared otherwise (for values of N and C after the shift).

C M0
Set if, before the shift, the LSB of ACCX or M was set; cleared otherwise.

Source Forms: ASRA; ASRB; ASR (opr)

Addressing Modes, Machine Code, and Cycle-by-Cycle Execution:

Cycle	ASRA (INH)			ASRB (INH)			ASR (EXT)			ASR (IND, X)			ASR (IND, Y)		
	Addr	Data	R/W̄	Addr	Data	R/W̄	Addr	Data	R/W̄	Addr	Data	R/W̄	Addr	Data	R/W̄
1	OP	47	1	OP	57	1	OP	77	1	OP	67	1	OP	18	1
2	OP+1	—	1	OP+1	—	1	OP+1	hh	1	OP+1	ff	1	OP+1	67	1
3							OP+2	ll	1	FFFF	—	1	OP+2	ff	1
4							hhll	(hhll)	1	X+ff	(X+ff)	1	FFFF	—	1
5							FFFF	—	1	FFFF	—	1	Y+ff	(Y+ff)	1
6							hhll	result	0	X+ff	result	0	FFFF	—	1
7													Y+ff	result	0

BCC

**Branch if Carry Clear
(Same as BHS)**

BCC

Operation: PC ◀ (PC) + $0002 + Rel   if (C) = 0

Description: Tests the state of the C bit in the CCR and causes a branch if C is clear.

See BRA instruction for further details of the execution of the branch.

Condition Codes and Boolean Formulae:

S	X	H	I	N	Z	V	C
—	—	—	—	—	—	—	—

None affected

Source Form: BCC (rel)

Addressing Modes, Machine Code, and Cycle-by-Cycle Execution:

Cycle	BCC (REL)		
	Addr	Data	R/$\overline{\text{W}}$
1	OP	24	1
2	OP + 1	rr	1
3	FFFF	—	1

The following table is a summary of all branch instructions.

Test	Boolean	Mnemonic	Opcode	Complementary		Branch	Comment
r>m	Z+(N ⊕ V)=0	BGT	2E	r≤m	BLE	2F	Signed
r≥m	N ⊕ V=0	BGE	2C	r<m	BLT	2D	Signed
r=m	Z=1	BEQ	27	r≠m	BNE	26	Signed
r≤m	Z+(N ⊕ V)=1	BLE	2F	r>m	BGT	2E	Signed
r<m	N ⊕ V=1	BLT	2D	r≥m	BGE	2C	Signed
r>m	C+Z=0	BHI	22	r≤m	BLS	23	Unsigned
r≥m	C=0	BHS/BCC	24	r<m	BLO/BCS	25	Unsigned
r=m	Z=1	BEQ	27	r≠m	BNE	26	Unsigned
r≤m	C+Z=1	BLS	23	r>m	BHI	22	Unsigned
r<m	C=1	BLO/BCS	25	r≥m	BHS/BCC	24	Unsigned
Carry	C=1	BCS	25	No Carry	BCC	24	Simple
Negative	N=1	BMI	2B	Plus	BPL	2A	Simple
Overflow	V=1	BVS	29	No Overflow	BVC	28	Simple
r=0	Z=1	BEQ	27	r≠0	BNE	26	Simple
Always	—	BRA	20	Never	BRN	21	Unconditional

BCLR   Clear Bit(s) in Memory   BCLR

Operation: M ← (M)•($\overline{PC+2}$)

M ← (M)•($\overline{PC+3}$)   (for IND, Y address mode only)

Description: Clear multiple bits in location M. The bit(s) to be cleared are specified by ones in the mask byte. All other bits in M are rewritten to their current state.

Condition Codes and Boolean Formulae:

S	X	H	I	N	Z	V	C
—	—	—	—	↕	↕	0	—

N R7
   Set if MSB of result is set; cleared otherwise.

Z $\overline{R7} \cdot \overline{R6} \cdot \overline{R5} \cdot \overline{R4} \cdot \overline{R3} \cdot \overline{R2} \cdot \overline{R1} \cdot \overline{R0}$
   Set if result is $00; cleared otherwise.

V 0
   Cleared

Source Forms: BCLR (opr) (msk)

Addressing Modes, Machine Code, and Cycle-by-Cycle Execution:

Cycle	BCLR (DIR)			BCLR (IND, X)			BCLR (IND, Y)		
	Addr	Data	R/$\overline{W}$	Addr	Data	R/$\overline{W}$	Addr	Data	R/$\overline{W}$
1	OP	15	1	OP	1D	1	OP	18	1
2	OP+1	dd	1	OP+1	ff	1	OP+1	1D	1
3	00dd	(00dd)	1	FFFF	—	1	OP+2	ff	1
4	OP+2	mm	1	X+ff	(X+ff)	1	FFFF	—	1
5	FFFF	—	1	OP+2	mm	1	(IY)+ff	(Y+ff)	1
6	00dd	result	0	FFFF	—	1	OP+3	mm	1
7				X+ff	result	0	FFFF	—	1
8							Y+ff	result	0

BCS

**Branch if Carry Set
(Same as BLO)**

BCS

Operation: PC ◀ (PC) + $0002 + Rel   if (C) = 1

Description: Tests the state of the C bit in the CCR and causes a branch if C is set.

See BRA instruction for further details of the execution of the branch.

Condition Codes and Boolean Formulae:

S	X	H	I	N	Z	V	C
—	—	—	—	—	—	—	—

None affected

Source Form: BCS (rel)

Addressing Modes, Machine Code, and Cycle-by-Cycle Execution:

Cycle	BCS (REL) Addr	Data	R/$\overline{\text{W}}$
1	OP	25	1
2	OP + 1	rr	1
3	FFFF	—	1

The following table is a summary of all branch instructions.

Test	Boolean	Mnemonic	Opcode	Complementary		Branch	Comment
r>m	$Z + (N \oplus V) = 0$	BGT	2E	r≤m	BLE	2F	Signed
r≥m	$N \oplus V = 0$	BGE	2C	r<m	BLT	2D	Signed
r=m	$Z = 1$	BEQ	27	r≠m	BNE	26	Signed
r≤m	$Z + (N \oplus V) = 1$	BLE	2F	r>m	BGT	2E	Signed
r<m	$N \oplus V = 1$	BLT	2D	r≥m	BGE	2C	Signed
r>m	$C + Z = 0$	BHI	22	r≤m	BLS	23	Unsigned
r≥m	$C = 0$	BHS/BCC	24	r<m	BLO/BCS	25	Unsigned
r=m	$Z = 1$	BEQ	27	r≠m	BNE	26	Unsigned
r≤m	$C + Z = 1$	BLS	23	r>m	BHI	22	Unsigned
r<m	$C = 1$	BLO/BCS	25	r≥m	BHS/BCC	24	Unsigned
Carry	$C = 1$	BCS	25	No Carry	BCC	24	Simple
Negative	$N = 1$	BMI	2B	Plus	BPL	2A	Simple
Overflow	$V = 1$	BVS	29	No Overflow	BVC	28	Simple
r=0	$Z = 1$	BEQ	27	r≠0	BNE	26	Simple
Always	—	BRA	20	Never	BRN	21	Unconditional

BEQ   Branch if Equal   BEQ

Operation: PC ◆ (PC) + $0002 + Rel   if (Z) = 1

Description: Tests the state of the Z bit in the CCR and causes a branch if Z is set.

See BRA instruction for further details of the execution of the branch.

Condition Codes and Boolean Formulae:

S	X	H	I	N	Z	V	C
—	—	—	—	—	—	—	—

None affected

Source Form: BEQ (rel)

Addressing Modes, Machine Code, and Cycle-by-Cycle Execution:

Cycle	BEQ (REL)		
	Addr	Data	R/W̄
1	OP	27	1
2	OP + 1	rr	1
3	FFFF	—	1

The following table is a summary of all branch instructions.

Test	Boolean	Mnemonic	Opcode	Complementary		Branch	Comment
r>m	Z + (N ⊕ V) = 0	BGT	2E	r≤m	BLE	2F	Signed
r≥m	N ⊕ V = 0	BGE	2C	r<m	BLT	2D	Signed
r = m	Z = 1	BEQ	27	r ≠ m	BNE	26	Signed
r≤m	Z + (N ⊕ V) = 1	BLE	2F	r>m	BGT	2E	Signed
r<m	N ⊕ V = 1	BLT	2D	r≥m	BGE	2C	Signed
r>m	C + Z = 0	BHI	22	r≤m	BLS	23	Unsigned
r≥m	C = 0	BHS/BCC	24	r<m	BLO/BCS	25	Unsigned
r = m	Z = 1	BEQ	27	r ≠ m	BNE	26	Unsigned
r≤m	C + Z = 1	BLS	23	r>m	BHI	22	Unsigned
r<m	C = 1	BLO/BCS	25	r≥m	BHS/BCC	24	Unsigned
Carry	C = 1	BCS	25	No Carry	BCC	24	Simple
Negative	N = 1	BMI	2B	Plus	BPL	2A	Simple
Overflow	V = 1	BVS	29	No Overflow	BVC	28	Simple
r = 0	Z = 1	BEQ	27	r ≠ 0	BNE	26	Simple
Always	—	BRA	20	Never	BRN	21	Unconditional

BGE   Branch if Greater than or Equal to Zero   BGE

Operation: PC ◀ (PC) + $0002 + Rel   if $(N)\oplus(V)=0$
i.e., if (ACCX)≥(M)   (twos-complement "signed" numbers)

Description: If the BGE instruction is executed immediately after execution of any of the instructions, CBA, CMP(A, B, or D), CP(X or Y), SBA, SUB(A, B, or D), the branch will occur if and only if the twos-complement number represented by the ACCX was greater than or equal to the two-complement number represented by M.

See BRA instruction for further details of the execution of the branch.

Condition Codes and Boolean Formulae:

S	X	H	I	N	Z	V	C
—	—	—	—	—	—	—	—

None affected

Source Form: BGE (rel)

Addressing Modes, Machine Code, and Cycle-by-Cycle Execution:

Cycle	BGE (REL)		
	Addr	Data	R/W̄
1	OP	2C	1
2	OP + 1	rr	1
3	FFFF	—	1

The following table is a summary of all branch instructions.

Test	Boolean	Mnemonic	Opcode	Complementary		Branch	Comment
r>m	$Z+(N \oplus V)=0$	BGT	2E	r≤m	BLE	2F	Signed
r≥m	$N \oplus V=0$	BGE	2C	r<m	BLT	2D	Signed
r=m	Z=1	BEQ	27	r≠m	BNE	26	Signed
r≤m	$Z+(N \oplus V)=1$	BLE	2F	r>m	BGT	2E	Signed
r<m	$N \oplus V=1$	BLT	2D	r≥m	BGE	2C	Signed
r>m	C+Z=0	BHI	22	r≤m	BLS	23	Unsigned
r≥m	C=0	BHS/BCC	24	r<m	BLO/BCS	25	Unsigned
r=m	Z=1	BEQ	27	r≠m	BNE	26	Unsigned
r≤m	C+Z=1	BLS	23	r>m	BHI	22	Unsigned
r<m	C=1	BLO/BCS	25	r≥m	BHS/BCC	24	Unsigned
Carry	C=1	BCS	25	No Carry	BCC	24	Simple
Negative	N=1	BMI	2B	Plus	BPL	2A	Simple
Overflow	V=1	BVS	29	No Overflow	BVC	28	Simple
r=0	Z=1	BEQ	27	r≠0	BNE	26	Simple
Always	—	BRA	20	Never	BRN	21	Unconditional

BGT   Branch if Greater than Zero   BGT

Operation: PC ◆ (PC) + $0002 + Rel   if (Z) + [(N) ⊕ (V)] = 0
               i.e., if (ACCX)>(M)   (twos-complement signed numbers)

Description: If the BGT instruction is executed immediately after execution of any of the instructions, CBA, CMP(A, B, or D), CP(X or Y), SBA, SUB(A, B, or D), the branch will occur if and only if the twos-complement number represented by ACCX was greater than the twos-complement number represented by M.

See BRA instruction for further details of the execution of the branch.

Condition Codes and Boolean Formulae:

S	X	H	I	N	Z	V	C
—	—	—	—	—	—	—	—

None affected

Source Form: BGT (rel)

Addressing Modes, Machine Code, and Cycle-by-Cycle Execution:

Cycle	BGT (REL)		
	Addr	**Data**	**R/W̄**
1	OP	2E	1
2	OP + 1	rr	1
3	FFFF	—	1

The following table is a summary of all branch instructions.

Test	Boolean	Mnemonic	Opcode	Complementary		Branch	Comment
r>m	Z + (N ⊕ V) = 0	BGT	2E	r≤m	BLE	2F	Signed
r≥m	N ⊕ V = 0	BGE	2C	r<m	BLT	2D	Signed
r = m	Z = 1	BEQ	27	r ≠ m	BNE	26	Signed
r≤m	Z + (N ⊕ V) = 1	BLE	2F	r>m	BGT	2E	Signed
r<m	N ⊕ V = 1	BLT	2D	r≥m	BGE	2C	Signed
r>m	C + Z = 0	BHI	22	r≤m	BLS	23	Unsigned
r≥m	C = 0	BHS/BCC	24	r<m	BLO/BCS	25	Unsigned
r = m	Z = 1	BEQ	27	r ≠ m	BNE	26	Unsigned
r≤m	C + Z = 1	BLS	23	r>m	BHI	22	Unsigned
r<m	C = 1	BLO/BCS	25	r≥m	BHS/BCC	24	Unsigned
Carry	C = 1	BCS	25	No Carry	BCC	24	Simple
Negative	N = 1	BMI	2B	Plus	BPL	2A	Simple
Overflow	V = 1	BVS	29	No Overflow	BVC	28	Simple
r = 0	Z = 1	BEQ	27	r ≠ 0	BNE	26	Simple
Always	—	BRA	20	Never	BRN	21	Unconditional

BHI

Branch if Higher

BHI

Operation: PC ◀ (PC) + $0002 + Rel   if (C) + (Z) = 0
                   i.e., if (ACCX)>(M)   (unsigned binary numbers)

Description: If the BHI instruction is executed immediately after execution of any of the instructions, CBA, CMP(A, B, or D), CP(X or Y), SBA, SUB(A, B, or D), the branch will occur if and only if the unsigned binary number represented by ACCX was greater than the unsigned binary number represented by M. Generally not useful after INC/DEC, LD/ST, TST/CLR/COM because these instructions do not affect the C bit in the CCR.

See BRA instruction for further details of the execution of the branch.

Condition Codes and Boolean Formulae:

S	X	H	I	N	Z	V	C
—	—	—	—	—	—	—	—

None affected

Source Form: BHI (rel)

Addressing Modes, Machine Code, and Cycle-by-Cycle Execution:

Cycle	BHI (REL)		
	Addr	Data	R/$\overline{W}$
1	OP	22	1
2	OP + 1	rr	1
3	FFFF	—	1

The following table is a summary of all branch instructions.

Test	Boolean	Mnemonic	Opcode	Complementary		Branch	Comment
r>m	Z + (N ⊕ V) = 0	BGT	2E	r≤m	BLE	2F	Signed
r≥m	N ⊕ V = 0	BGE	2C	r<m	BLT	2D	Signed
r = m	Z = 1	BEQ	27	r ≠ m	BNE	26	Signed
r≤m	Z + (N ⊕ V) = 1	BLE	2F	r>m	BGT	2E	Signed
r<m	N ⊕ V = 1	BLT	2D	r≥m	BGE	2C	Signed
r>m	C + Z = 0	BHI	22	r≤m	BLS	23	Unsigned
r≥m	C = 0	BHS/BCC	24	r<m	BLO/BCS	25	Unsigned
r = m	Z = 1	BEQ	27	r ≠ m	BNE	26	Unsigned
r≤m	C + Z = 1	BLS	23	r>m	BHI	22	Unsigned
r<m	C = 1	BLO/BCS	25	r≥m	BHS/BCC	24	Unsigned
Carry	C = 1	BCS	25	No Carry	BCC	24	Simple
Negative	N = 1	BMI	2B	Plus	BPL	2A	Simple
Overflow	V = 1	BVS	29	No Overflow	BVC	28	Simple
r = 0	Z = 1	BEQ	27	r ≠ 0	BNE	26	Simple
Always	—	BRA	20	Never	BRN	21	Unconditional

BHS

Branch if Higher or Same
(Same as BCC)

BHS

Operation: PC ◆ (PC) + $0002 + Rel   if (C) = 0
i.e., if (ACCX) ≥ (M)   (unsigned binary numbers)

Description: If the BHS instruction is executed immediately after execution of any of the instructions, CBA, CMP(A, B, or D), CP(X or Y), SBA, SUB(A, B, or D), the branch will occur if and only if the unsigned binary number represented by ACCX was greater than or equal to the unsigned binary number represented by M. Generally not useful after INC/DEC, LD/ST, TST/CLR/COM because these instructions do not affect the C bit in the CCR.

See BRA instruction for further details of the execution of the branch.

Condition Codes and Boolean Formulae:

S	X	H	I	N	Z	V	C
—	—	—	—	—	—	—	—

None affected

Source Form: BHS (rel)

Addressing Modes, Machine Code, and Cycle-by-Cycle Execution:

Cycle	BHS (REL)		
	Addr	Data	R/W̄
1	OP	24	1
2	OP + 1	rr	1
3	FFFF	—	1

The following table is a summary of all branch instructons.

Test	Boolean	Mnemonic	Opcode	Complementary		Branch	Comment
r > m	Z + (N ⊕ V) = 0	BGT	2E	r ≤ m	BLE	2F	Signed
r ≥ m	N ⊕ V = 0	BGE	2C	r < m	BLT	2D	Signed
r = m	Z = 1	BEQ	27	r ≠ m	BNE	26	Signed
r ≤ m	Z + (N ⊕ V) = 1	BLE	2F	r > m	BGT	2E	Signed
r < m	N ⊕ V = 1	BLT	2D	r ≥ m	BGE	2C	Signed
r > m	C + Z = 0	BHI	22	r ≤ m	BLS	23	Unsigned
r ≥ m	C = 0	BHS/BCC	24	r < m	BLO/BCS	25	Unsigned
r = m	Z = 1	BEQ	27	r ≠ m	BNE	26	Unsigned
r ≤ m	C + Z = 1	BLS	23	r > m	BHI	22	Unsigned
r < m	C = 1	BLO/BCS	25	r ≥ m	BHS/BCC	24	Unsigned
Carry	C = 1	BCS	25	No Carry	BCC	24	Simple
Negative	N = 1	BMI	2B	Plus	BPL	2A	Simple
Overflow	V = 1	BVS	29	No Overflow	BVC	28	Simple
r = 0	Z = 1	BEQ	27	r ≠ 0	BNE	26	Simple
Always	—	BRA	20	Never	BRN	21	Unconditional

BIT   Bit Test   BIT

Operation: (ACCX)•(M)

Description: Performs the logical AND operation between the contents of ACCX and the contents of M and modifies the condition codes accordingly. Neither the contents of ACCX or M operands are affected. (Each bit of the result of the AND would be the logical AND of the corresponding bits of ACCX and M.)

Condition Codes and Boolean Formulae:

S	X	H	I	N	Z	V	C
—	—	—	—	↕	↕	0	—

N R7
   Set if MSB of result is set; cleared otherwise.

Z $\overline{R7} \cdot \overline{R6} \cdot \overline{R5} \cdot \overline{R4} \cdot \overline{R3} \cdot \overline{R2} \cdot \overline{R1} \cdot \overline{R0}$
   Set if result is $00; cleared otherwise.

V 0
   Cleared

Source Forms: BITA (opr); BITB (opr)

Addressing Modes, Machine Code, and Cycle-by-Cycle Execution:

Cycle	BITA (IMM)			BITA (DIR)			BITA (EXT)			BITA (IND, X)			BITA (IND, Y)		
	Addr	Data	R/$\overline{W}$	Addr	Data	R/$\overline{W}$	Addr	Data	R/$\overline{W}$	Addr	Data	R/$\overline{W}$	Addr	Data	R/$\overline{W}$
1	OP	85	1	OP	95	1	OP	B5	1	OP	A5	1	OP	18	1
2	OP+1	ii	1	OP+1	dd	1	OP+1	hh	1	OP+1	ff	1	OP+1	A5	1
3				00dd	(00dd)	1	OP+2	ll	1	FFFF	—	1	OP+2	ff	1
4							hhll	(hhll)	1	X+ff	(X+ff)	1	FFFF	—	1
5													Y+ff	(Y+ff)	1

Cycle	BITB (IMM)			BITB (DIR)			BITB (EXT)			BITB (IND, X)			BITB (IND, Y)		
	Addr	Data	R/$\overline{W}$	Addr	Data	R/$\overline{W}$	Addr	Data	R/$\overline{W}$	Addr	Data	R/$\overline{W}$	Addr	Data	R/$\overline{W}$
1	OP	C5	1	OP	D5	1	OP	F5	1	OP	E5	1	OP	18	1
2	OP+1	ii	1	OP+1	dd	1	OP+1	hh	1	OP+1	ff	1	OP+1	E5	1
3				00dd	(00dd)	1	OP+2	ll	1	FFFF	—	1	OP+2	ff	1
4							hhll	(hhll)	1	X+ff	(X+ff)	1	FFFF	—	1
5													Y+ff	(Y+ff)	1

BLE

Branch if Less than or Equal to Zero

BLE

Operation: PC ◀ (PC) + $0002 + Rel   if (Z) + [(N)⊕(V)] = 1
i.e., if (ACCX) ≤ (M)   (twos-complement signed numbers)

Description: If the BLE instruction is executed immediately after execution of any of the instructions, CBA, CMP(A, B, or D), CP(X, or Y), SBA, SUB(A, B, or D), the branch will occur if and only if the twos-complement number represented by ACCX was less than or equal to the twos-complement number represented by M.

See BRA instruction for further details of the execution of the branch.

Condition Codes and Boolean Formulae:

S	X	H	I	N	Z	V	C
—	—	—	—	—	—	—	—

None affected

Source Form: BLE (rel)

Addressing Modes, Machine Code, and Cycle-by-Cycle Execution:

Cycle	BLE (REL)		
	Addr	Data	R/W̄
1	OP	2F	1
2	OP + 1	rr	1
3	FFFF	—	1

The following table is a summary of all branch instructions.

Test	Boolean	Mnemonic	Opcode	Complementary		Branch	Comment
r>m	Z + (N ⊕ V) = 0	BGT	2E	r≤m	BLE	2F	Signed
r≥m	N ⊕ V = 0	BGE	2C	r<m	BLT	2D	Signed
r = m	Z = 1	BEQ	27	r ≠ m	BNE	26	Signed
r≤m	Z + (N ⊕ V) = 1	BLE	2F	r>m	BGT	2E	Signed
r<m	N ⊕ V = 1	BLT	2D	r≥m	BGE	2C	Signed
r>m	C + Z = 0	BHI	22	r≤m	BLS	23	Unsigned
r≥m	C = 0	BHS/BCC	24	r<m	BLO/BCS	25	Unsigned
r = m	Z = 1	BEQ	27	r ≠ m	BNE	26	Unsigned
r≤m	C + Z = 1	BLS	23	r>m	BHI	22	Unsigned
r<m	C = 1	BLO/BCS	25	r≥m	BHS/BCC	24	Unsigned
Carry	C = 1	BCS	25	No Carry	BCC	24	Simple
Negative	N = 1	BMI	2B	Plus	BPL	2A	Simple
Overflow	V = 1	BVS	29	No Overflow	BVC	28	Simple
r = 0	Z = 1	BEQ	27	r ≠ 0	BNE	26	Simple
Always	—	BRA	20	Never	BRN	21	Unconditional

BLO Branch if Lower (Same as BCS) BLO

Operation: $PC \leftarrow (PC) + \$0002 + Rel$ if $(C) = 1$
i.e., if $(ACCX) < (M)$ (unsigned binary numbers)

Description: If the BLO instruction is executed immediately after execution of any of the instructions, CBA, CMP(A, B, or D), CP(X or Y), SBA, SUB(A, B, or D), the branch will occur if and only if the unsigned binary number represented by ACCX was less than the unsigned binary number represented by M. Generally not useful after INC/DEC, LD/ST, TST/CLR/COM because these instructions do not affect the C bit in the CCR.

See BRA instruction for further details of the execution of the branch.

Condition Codes and Boolean Formulae:

S	X	H	I	N	Z	V	C
—	—	—	—	—	—	—	—

None affected

Source Form: BLO (rel)

Addressing Modes, Machine Code, and Cycle-by-Cycle Execution:

Cycle	BLO (REL)		
	Addr	Data	R/$\overline{W}$
1	OP	25	1
2	OP+1	rr	1
3	FFFF	—	1

The following table is a summary of all branch instructions.

Test	Boolean	Mnemonic	Opcode	Complementary Branch			Comment
r>m	Z+(N ⊕ V)=0	BGT	2E	r≤m	BLE	2F	Signed
r≥m	N ⊕ V=0	BGE	2C	r<m	BLT	2D	Signed
r=m	Z=1	BEQ	27	r≠m	BNE	26	Signed
r≤m	Z+(N ⊕ V)=1	BLE	2F	r>m	BGT	2E	Signed
r<m	N ⊕ V=1	BLT	2D	r≥m	BGE	2C	Signed
r>m	C+Z=0	BHI	22	r≤m	BLS	23	Unsigned
r≥m	C=0	BHS/BCC	24	r<m	BLO/BCS	25	Unsigned
r=m	Z=1	BEQ	27	r≠m	BNE	26	Unsigned
r≤m	C+Z=1	BLS	23	r>m	BHI	22	Unsigned
r<m	C=1	BLO/BCS	25	r≥m	BHS/BCC	24	Unsigned
Carry	C=1	BCS	25	No Carry	BCC	24	Simple
Negative	N=1	BMI	2B	Plus	BPL	2A	Simple
Overflow	V=1	BVS	29	No Overflow	BVC	28	Simple
r=0	Z=1	BEQ	27	r≠0	BNE	26	Simple
Always	—	BRA	20	Never	BRN	21	Unconditional

BLS   Branch if Lower or Same   BLS

Operation: PC ◀ (PC) + $0002 + Rel   if (C) + (Z) = 1
                   i.e., if (ACCX) ≤ (M)   (unsigned binary numbers)

Description: If the BLS instruction is executed immediately after execution of any of the instructions, CBA, CMP(A, B, or D), CP(X or Y), SBA, SUB(A, B, or D), the branch will occur if and only if the unsigned binary number represented by ACCX was less than or equal to the unsigned binary number represented by M. Generally not useful after INC/DEC, LD/ST, TST/CLR/COM because these instructions do not affect the C bit in the CCR.

See BRA instruction for further details of the execution of the branch.

Condition Codes and Boolean Formulae:

S	X	H	I	N	Z	V	C
—	—	—	—	—	—	—	—

None affected

Source Form: BLS (rel)

Addressing Modes, Machine Code, and Cycle-by-Cycle Execution:

Cycle	BLS (REL)		
	Addr	Data	R/W̅
1	OP	23	1
2	OP + 1	rr	1
3	FFFF	—	1

The following table is a summary of all branch instructions.

Test	Boolean	Mnemonic	Opcode	Complementary		Branch	Comment
r>m	Z + (N ⊕ V) = 0	BGT	2E	r≤m	BLE	2F	Signed
r≥m	N ⊕ V = 0	BGE	2C	r<m	BLT	2D	Signed
r = m	Z = 1	BEQ	27	r ≠ m	BNE	26	Signed
r≤m	Z + (N ⊕ V) = 1	BLE	2F	r>m	BGT	2E	Signed
r<m	N ⊕ V = 1	BLT	2D	r≥m	BGE	2C	Signed
r>m	C + Z = 0	BHI	22	r≤m	BLS	23	Unsigned
r≥m	C = 0	BHS/BCC	24	r<m	BLO/BCS	25	Unsigned
r = m	Z = 1	BEQ	27	r ≠ m	BNE	26	Unsigned
r≤m	C + Z = 1	BLS	23	r>m	BHI	22	Unsigned
r<m	C = 1	BLO/BCS	25	r≥m	BHS/BCC	24	Unsigned
Carry	C = 1	BCS	25	No Carry	BCC	24	Simple
Negative	N = 1	BMI	2B	Plus	BPL	2A	Simple
Overflow	V = 1	BVS	29	No Overflow	BVC	28	Simple
r = 0	Z = 1	BEQ	27	r ≠ 0	BNE	26	Simple
Always	—	BRA	20	Never	BRN	21	Unconditional

BLT

Branch if Less than Zero

BLT

Operation: PC ← (PC) + $0002 + Rel   if (N)⊕(V) = 1
                   i.e., if (ACCX)<(M)   (twos-complement signed numbers)

Description: If the BLT instruction is executed immediately after execution of any of the instructons, CBA, CMP(A, B, or D), CP(X or Y), SBA, SUB(A, B, or D), the branch will occur if and only if the twos-complement number represented by ACCX was less than the twos-complement number represented by M.

See BRA instruction for further details of the execution of the branch.

Condition Codes and Boolean Formulae:

S	X	H	I	N	Z	V	C
—	—	—	—	—	—	—	—

None affected

Source Form: BLT (rel)

Addressing Modes, Machine Code, and Cycle-by-Cycle Execution:

Cycle	BLT (REL)		
	Addr	Data	R/W̄
1	OP	2D	1
2	OP+1	rr	1
3	FFFF	—	1

The following table is a summary of all branch instructions.

Test	Boolean	Mnemonic	Opcode	Complementary		Branch	Comment
r>m	Z+(N ⊕ V) = 0	BGT	2E	r≤m	BLE	2F	Signed
r≥m	N ⊕ V = 0	BGE	2C	r<m	BLT	2D	Signed
r=m	Z = 1	BEQ	27	r≠m	BNE	26	Signed
r≤m	Z+(N ⊕ V) = 1	BLE	2F	r>m	BGT	2E	Signed
r<m	N ⊕ V = 1	BLT	2D	r≥m	BGE	2C	Signed
r>m	C+Z = 0	BHI	22	r≤m	BLS	23	Unsigned
r≥m	C = 0	BHS/BCC	24	r<m	BLO/BCS	25	Unsigned
r=m	Z = 1	BEQ	27	r≠m	BNE	26	Unsigned
r≤m	C+Z = 1	BLS	23	r>m	BHI	22	Unsigned
r<m	C = 1	BLO/BCS	25	r≥m	BHS/BCC	24	Unsigned
Carry	C = 1	BCS	25	No Carry	BCC	24	Simple
Negative	N = 1	BMI	2B	Plus	BPL	2A	Simple
Overflow	V = 1	BVS	29	No Overflow	BVC	28	Simple
r=0	Z = 1	BEQ	27	r≠0	BNE	26	Simple
Always	—	BRA	20	Never	BRN	21	Unconditional

BMI

Branch if Minus

BMI

Operation: PC ◀ (PC) + $0002 + Rel   if (N) = 1

Description: Tests the state of the N bit in the CCR and causes a branch if N is set.

See BRA instruction for further details of the execution of the branch.

Condition Codes and Boolean Formulae:

S	X	H	I	N	Z	V	C
—	—	—	—	—	—	—	—

None affected

Source Form: BMI (rel)

Addressing Modes, Machine Code, and Cycle-by-Cycle Execution:

Cycle	BMI (REL)		
	Addr	Data	R/W̄
1	OP	2B	1
2	OP + 1	rr	1
3	FFFF	—	1

The following table is a summary of all branch instructions.

Test	Boolean	Mnemonic	Opcode	Complementary		Branch	Comment
r>m	Z + (N ⊕ V) = 0	BGT	2E	r≤m	BLE	2F	Signed
r≥m	N ⊕ V = 0	BGE	2C	r<m	BLT	2D	Signed
r = m	Z = 1	BEQ	27	r ≠ m	BNE	26	Signed
r≤m	Z + (N ⊕ V) = 1	BLE	2F	r>m	BGT	2E	Signed
r<m	N ⊕ V = 1	BLT	2D	r≥m	BGE	2C	Signed
r>m	C + Z = 0	BHI	22	r≤m	BLS	23	Unsigned
r≥m	C = 0	BHS/BCC	24	r<m	BLO/BCS	25	Unsigned
r = m	Z = 1	BEQ	27	r ≠ m	BNE	26	Unsigned
r≤m	C + Z = 1	BLS	23	r>m	BHI	22	Unsigned
r<m	C = 1	BLO/BCS	25	r≥m	BHS/BCC	24	Unsigned
Carry	C = 1	BCS	25	No Carry	BCC	24	Simple
Negative	N = 1	BMI	2B	Plus	BPL	2A	Simple
Overflow	V = 1	BVS	29	No Overflow	BVC	28	Simple
r = 0	Z = 1	BEQ	27	r ≠ 0	BNE	26	Simple
Always	—	BRA	20	Never	BRN	21	Unconditional

BNE   **Branch if Not Equal to Zero**   # BNE

Operation: PC ⬥ (PC) + $0002 + Rel   if (Z) = 0

Description: Tests the state of the Z bit in the CCR and causes a branch if Z is clear.

See BRA instruction for further details of the execution of the branch.

Condition Codes and Boolean Formulae:

S	X	H	I	N	Z	V	C
—	—	—	—	—	—	—	—

None affected

Source Form: BNE (rel)

Addressing Modes, Machine Code, and Cycle-by-Cycle Execution:

Cycle	BNE (REL)		
	Addr	Data	R/W̄
1	OP	26	1
2	OP + 1	rr	1
3	FFFF	—	1

The following table is a summary of all branch instructions.

Test	Boolean	Mnemonic	Opcode	Complementary		Branch	Comment
r>m	Z + (N ⊕ V) = 0	BGT	2E	r≤m	BLE	2F	Signed
r≥m	N ⊕ V = 0	BGE	2C	r<m	BLT	2D	Signed
r = m	Z = 1	BEQ	27	r ≠ m	BNE	26	Signed
r≤m	Z + (N ⊕ V) = 1	BLE	2F	r>m	BGT	2E	Signed
r<m	N ⊕ V = 1	BLT	2D	r≥m	BGE	2C	Signed
r>m	C + Z = 0	BHI	22	r≤m	BLS	23	Unsigned
r≥m	C = 0	BHS/BCC	24	r<m	BLO/BCS	25	Unsigned
r = m	Z = 1	BEQ	27	r ≠ m	BNE	26	Unsigned
r≤m	C + Z = 1	BLS	23	r>m	BHI	22	Unsigned
r<m	C = 1	BLO/BCS	25	r≥m	BHS/BCC	24	Unsigned
Carry	C = 1	BCS	25	No Carry	BCC	24	Simple
Negative	N = 1	BMI	2B	Plus	BPL	2A	Simple
Overflow	V = 1	BVS	29	No Overflow	BVC	28	Simple
r = 0	Z = 1	BEQ	27	r ≠ 0	BNE	26	Simple
Always	—	BRA	20	Never	BRN	21	Unconditional

BPL

Branch if Plus

BPL

Operation: PC ◄ (PC) + $0002 + Rel   if (N) = 0

Description: Tests the state of the N bit in the CCR and causes a branch if N is clear.

See BRA instruction for details of the execution of the branch.

Condition Codes and Boolean Formulae:

S	X	H	I	N	Z	V	C
—	—	—	—	—	—	—	—

None affected

Source Form: BPL (rel)

Addressing Modes, Machine Code, and Cycle-by-Cycle Execution:

Cycle	BPL (REL)		
	Addr	Data	R/$\overline{W}$
1	OP	2A	1
2	OP + 1	rr	1
3	FFFF	—	1

The following table is a summary of all branch instructions.

Test	Boolean	Mnemonic	Opcode	Complementary		Branch	Comment
r>m	Z + (N ⊕ V) = 0	BGT	2E	r≤m	BLE	2F	Signed
r≥m	N ⊕ V = 0	BGE	2C	r<m	BLT	2D	Signed
r = m	Z = 1	BEQ	27	r≠m	BNE	26	Signed
r≤m	Z + (N ⊕ V) = 1	BLE	2F	r>m	BGT	2E	Signed
r<m	N ⊕ V = 1	BLT	2D	r≥m	BGE	2C	Signed
r>m	C + Z = 0	BHI	22	r≤m	BLS	23	Unsigned
r≥m	C = 0	BHS/BCC	24	r<m	BLO/BCS	25	Unsigned
r = m	Z = 1	BEQ	27	r≠m	BNE	26	Unsigned
r≤m	C + Z = 1	BLS	23	r>m	BHI	22	Unsigned
r<m	C = 1	BLO/BCS	25	r≥m	BHS/BCC	24	Unsigned
Carry	C = 1	BCS	25	No Carry	BCC	24	Simple
Negative	N = 1	BMI	2B	Plus	BPL	2A	Simple
Overflow	V = 1	BVS	29	No Overflow	BVC	28	Simple
r = 0	Z = 1	BEQ	27	r≠0	BNE	26	Simple
Always	—	BRA	20	Never	BRN	21	Unconditional

BRA   Branch Always   # BRA

Operation: PC ◀ (PC) + $0002 + Rel

Description: Unconditional branch to the address given by the foregoing formula, in which Rel is the relative offset stored as a twos complement number in the second byte of machine code corresponding to the branch instruction.

The source program specifies the destination of any branch instruction by its absolute address, either as a numerical value or as a symbol or expression, that can be numerically evaluated by the assembler. The assembler obtains the relative address, Rel, from the absolute address and the current value of the location counter.

Condition Codes and Boolean Formulae:

S	X	H	I	N	Z	V	C
—	—	—	—	—	—	—	—

None affected

Source Form: BRA (rel)

Addressing Modes, Machine Code, and Cycle-by-Cycle Execution:

Cycle	BRA (REL)		
	Addr	Data	R/W̄
1	OP	20	1
2	OP + 1	rr	1
3	FFFF	—	1

The following table is a summary of all branch instructions.

Test	Boolean	Mnemonic	Opcode	Complementary		Branch	Comment
r>m	Z+(N ⊕ V)=0	BGT	2E	r≤m	BLE	2F	Signed
r≥m	N ⊕ V=0	BGE	2C	r<m	BLT	2D	Signed
r=m	Z=1	BEQ	27	r≠m	BNE	26	Signed
r≤m	Z+(N ⊕ V)=1	BLE	2F	r>m	BGT	2E	Signed
r<m	N ⊕ V=1	BLT	2D	r≥m	BGE	2C	Signed
r>m	C+Z=0	BHI	22	r≤m	BLS	23	Unsigned
r≥m	C=0	BHS/BCC	24	r<m	BLO/BCS	25	Unsigned
r=m	Z=1	BEQ	27	r≠m	BNE	26	Unsigned
r≤m	C+Z=1	BLS	23	r>m	BHI	22	Unsigned
r<m	C=1	BLO/BCS	25	r≥m	BHS/BCC	24	Unsigned
Carry	C=1	BCS	25	No Carry	BCC	24	Simple
Negative	N=1	BMI	2B	Plus	BPL	2A	Simple
Overflow	V=1	BVS	29	No Overflow	BVC	28	Simple
r=0	Z=1	BEQ	27	r≠0	BNE	26	Simple
Always	—	BRA	20	Never	BRN	21	Unconditional

BRCLR   Branch if Bit(s) Clear   BRCLR

Operation: PC ◄ (PC) + $0004 + Rel   if (M)•(PC + 2) = 0
                   PC ◄ (PC) + $0005 + Rel   if (M)•(PC + 3) = 0 (for IND, Y address mode only)

Description: Performs the logical AND of location M and the mask supplied with the instruction, then branches if the result is zero (only if all bits corresponding to ones in the mask byte are zeros in the tested byte).

Condition Codes and Boolean Formulae:

S	X	H	I	N	Z	V	C
—	—	—	—	—	—	—	—

None affected

Source Form: BRCLR (opr) (msk) (rel)

Addressing Modes, Machine Code, and Cycle-by-Cycle Execution:

Cycle	BRCLR (DIR)			BRCLR (IND, X)			BRCLR (IND, Y)		
	Addr	Data	R/W̄	Addr	Data	R/W̄	Addr	Data	R/W̄
1	OP	13	1	OP	1F	1	OP	18	1
2	OP + 1	dd	1	OP + 1	ff	1	OP + 1	1F	1
3	00dd	(00dd)	1	FFFF	—	1	OP + 2	ff	1
4	OP + 2	mm	1	X + ff	(X + ff)	1	FFFF	—	1
5	OP + 3	rr	1	OP + 2	mm	1	(IY) + ff	(Y + ff)	1
6	FFFF	—	1	OP + 3	rr	1	OP + 3	mm	1
7				FFFF	—	1	OP + 4	rr	1
8							FFFF	—	1

BRN   Branch Never   BRN

Operation: PC ◀ (PC) + $0002

Description: Never branches. In effect, this instruction can be considered as a two-byte NOP (no operation) requiring three cycles for execution. Its inclusion in the instruction set is to provide a complement for the BRA instruction. The instruction is useful during program debug to negate the effect of another branch instruction without disturbing the offset byte. Having a complement for BRA is also useful in compiler implementations.

Condition Codes and Boolean Formulae:

S	X	H	I	N	Z	V	C
—	—	—	—	—	—	—	—

None affected

Source Form: BRN (rel)

Addressing Modes, Machine Code, and Cycle-by-Cycle Execution:

Cycle	BRN (REL)		
	Addr	Data	R/$\overline{\text{W}}$
1	OP	21	1
2	OP + 1	rr	1
3	FFFF	—	1

The following table is a summary of all branch instructions.

Test	Boolean	Mnemonic	Opcode	Complementary		Branch	Comment
r>m	Z + (N ⊕ V) = 0	BGT	2E	r≤m	BLE	2F	Signed
r≥m	N ⊕ V = 0	BGE	2C	r<m	BLT	2D	Signed
r = m	Z = 1	BEQ	27	r ≠ m	BNE	26	Signed
r≤m	Z + (N ⊕ V) = 1	BLE	2F	r>m	BGT	2E	Signed
r<m	N ⊕ V = 1	BLT	2D	r≥m	BGE	2C	Signed
r>m	C + Z = 0	BHI	22	r≤m	BLS	23	Unsigned
r≥m	C = 0	BHS/BCC	24	r<m	BLO/BCS	25	Unsigned
r = m	Z = 1	BEQ	27	r ≠ m	BNE	26	Unsigned
r≤m	C + Z = 1	BLS	23	r>m	BHI	22	Unsigned
r<m	C = 1	BLO/BCS	25	r≥m	BHS/BCC	24	Unsigned
Carry	C = 1	BCS	25	No Carry	BCC	24	Simple
Negative	N = 1	BMI	2B	Plus	BPL	2A	Simple
Overflow	V = 1	BVS	29	No Overflow	BVC	28	Simple
r = 0	Z = 1	BEQ	27	r ≠ 0	BNE	26	Simple
Always	—	BRA	20	Never	BRN	21	Unconditional

BRSET

Branch if Bit(s) Set

BRSET

Operation: $\text{PC} \leftarrow (\text{PC}) + \$0004 + \text{Rel} \quad \text{if } \overline{(M)} \cdot (\text{PC} + 2) = 0$
$\text{PC} \leftarrow (\text{PC}) + \$0005 + \text{Rel} \quad \text{if } \overline{(M)} \cdot (\text{PC} + 3) = 0 \text{ (for IND, Y address mode only)}$

Description: Performs the logical AND of location M inverted and the mask supplied with the instruction, then branches if the result is zero (only if all bits corresponding to ones in the mask byte are ones in the tested byte).

Condition Codes and Boolean Formulae:

S	X	H	I	N	Z	V	C
—	—	—	—	—	—	—	—

None affected

Source Form: BRSET (opr) (msk) (rel)

Addressing Modes, Machine Code, and Cycle-by-Cycle Execution:

Cycle	BRSET (DIR)			BRSET (IND, X)			BRSET (IND, Y)		
	Addr	Data	R/$\overline{\text{W}}$	Addr	Data	R/$\overline{\text{W}}$	Addr	Data	R/$\overline{\text{W}}$
1	OP	12	1	OP	1E	1	OP	18	1
2	OP+1	dd	1	OP+1	ff	1	OP+1	1E	1
3	00dd	(00dd)	1	FFFF	—	1	OP+2	ff	1
4	OP+2	mm	1	X+ff	(X+ff)	1	FFFF	—	1
5	OP+3	rr	1	OP+2	mm	1	(IY)+ff	(Y+ff)	1
6	FFFF	—	1	OP+3	rr	1	OP+3	mm	1
7				FFFF	—	1	OP+4	rr	1
8							FFFF	—	1

BSET

Set Bit(s) in Memory

BSET

Operation: M ◊ (M) + (PC + 2)

M ◊ (M) + (PC + 3)   (for IND, Y address mode only)

Description: Set multiple bits in location M. The bit(s) to be set are specified by ones in the mask byte (last machine code byte of the instruction). All other bits in M are unaffected.

Condition Codes and Boolean Formulae:

S	X	H	I	N	Z	V	C
—	—	—	—	↕	↕	0	—

N R7
Set if MSB of result is set; cleared otherwise.

Z $\overline{R7} \cdot \overline{R6} \cdot \overline{R5} \cdot \overline{R4} \cdot \overline{R3} \cdot \overline{R2} \cdot \overline{R1} \cdot \overline{R0}$
Set if result is $00; cleared otherwise.

V 0
Cleared

Source Form: BSET (opr) (msk)

Addressing Modes, Machine Code, and Cycle-by-Cycle Execution:

Cycle	BSET (DIR)			BSET (IND; X)			BSET (IND, Y)		
	Addr	Data	R/W̄	Addr	Data	R/W̄	Addr	Data	R/W̄
1	OP	14	1	OP	1C	1	OP	18	1
2	OP + 1	dd	1	OP + 1	ff	1	OP + 1	1C	1
3	00dd	(00dd)	1	FFFF	—	1	OP + 2	ff	1
4	OP + 2	mm	1	X + ff	(X + ff)	1	FFFF	—	1
5	FFFF	—	1	OP + 2	mm	1	(IY) + ff	(Y + ff)	1
6	00dd	result	0	FFFF	—	1	OP + 3	mm	1
7				X + ff	result	0	FFFF	—	1
8							Y + ff	result	0

BSR   Branch to Subroutine   BSR

Operation:

PC ⬅ (PC) + $0002	Advance PC to return address
⬅(PCL)	Push low-order return onto stack
SP ⬅ (SP) − 0001	
⬅(PCH)	Push high-order return onto stack
SP ⬅ (SP) − $0001	
PC ⬅ (PC) + Rel	Load start address of requested subroutine

Description: The program counter is incremented by two (this will be the return address). The least significant byte of the contents of the program counter (low-order return address) is pushed onto the stack. The stack pointer is then decremented by one. The most significant byte of the contents of the program counter (high-order return address) is pushed onto the stack. The stack pointer is then decremented by one. A branch then occurs to the location specified by the branch offset.

See BRA instruction for further details of the execution of the branch.

Condition Codes and Boolean Formulae:

S	X	H	I	N	Z	V	C
—	—	—	—	—	—	—	—

None affected

Source Form: BSR (rel)

Addressing Modes, Machine Code, and Cycle-by-Cycle Execution:

Cycle	BSR (REL)		
	Addr	Data	R/W̄
1	OP	8D	1
2	OP + 1	rr	1
3	FFFF	—	1
4	Sub	Nxt op	1
5	SP	Rtn lo	0
6	SP − 1	Rtn hi	0

BVC

Branch if Overflow Clear

BVC

Operation: PC ◄ (PC) + $0002 + Rel   if (V) = 0

Description: Tests the state of the V bit in the CCR and causes a branch if V is clear.

Used after an operation on twos-complement binary values, this instruction will cause a branch if there was NO overflow. That is, branch if the twos-complement result was valid.

See BRA instruction for further details of the execution of the branch.

Condition Codes and Boolean Formulae:

S	X	H	I	N	Z	V	C
—	—	—	—	—	—	—	—

None affected

Source Form: BVC (rel)

Addressing Modes, Machine Code, and Cycle-by-Cycle Execution:

Cycle	BVC (REL)		
	Addr	Data	R/W̄
1	OP	28	1
2	OP + 1	rr	1
3	FFFF	—	1

The following table is a summary of all branch instructions.

Test	Boolean	Mnemonic	Opcode	Complementary		Branch	Comment
r>m	Z + (N ⊕ V) = 0	BGT	2E	r≤m	BLE	2F	Signed
r≥m	N ⊕ V = 0	BGE	2C	r<m	BLT	2D	Signed
r = m	Z = 1	BEQ	27	r ≠ m	BNE	26	Signed
r≤m	Z + (N ⊕ V) = 1	BLE	2F	r>m	BGT	2E	Signed
r<m	N ⊕ V = 1	BLT	2D	r≥m	BGE	2C	Signed
r>m	C + Z = 0	BHI	22	r≤m	BLS	23	Unsigned
r≥m	C = 0	BHS/BCC	24	r<m	BLO/BCS	25	Unsigned
r = m	Z = 1	BEQ	27	r ≠ m	BNE	26	Unsigned
r≤m	C + Z = 1	BLS	23	r>m	BHI	22	Unsigned
r<m	C = 1	BLO/BCS	25	r≥m	BHS/BCC	24	Unsigned
Carry	C = 1	BCS	25	No Carry	BCC	24	Simple
Negative	N = 1	BMI	2B	Plus	BPL	2A	Simple
Overflow	V = 1	BVS	29	No Overflow	BVC	28	Simple
r = 0	Z = 1	BEQ	27	r ≠ 0	BNE	26	Simple
Always	—	BRA	20	Never	BRN	21	Unconditional

Reprinted with permission of Motorola

BVS   Branch if Overflow Set   BVS

Operation: PC ◄ (PC) + $0002 + Rel   if (V) = 1

Description: Tests the state of the V bit in the CCR and causes a branch if V is set.

Used after an operation on twos-complement binary values, this instruction will cause a branch if there was an overflow. That is, branch if the twos-complement result was invalid.

See BRA instruction for details of the execution of the branch.

Condition Codes and Boolean Formulae:

S	X	H	I	N	Z	V	C
—	—	—	—	—	—	—	—

None affected

Source Form: BVS (rel)

Addressing Modes, Machine Code, and Cycle-by-Cycle Execution:

Cycle	BVS (REL)		
	Addr	Data	R/$\overline{W}$
1	OP	29	1
2	OP + 1	rr	1
3	FFFF	—	1

The following table is a summary of all branch instructions.

Test	Boolean	Mnemonic	Opcode	Complementary		Branch	Comment
r>m	Z + (N ⊕ V) = 0	BGT	2E	r≤m	BLE	2F	Signed
r≥m	N ⊕ V = 0	BGE	2C	r<m	BLT	2D	Signed
r = m	Z = 1	BEQ	27	r ≠ m	BNE	26	Signed
r≤m	Z + (N ⊕ V) = 1	BLE	2F	r>m	BGT	2E	Signed
r<m	N ⊕ V = 1	BLT	2D	r≥m	BGE	2C	Signed
r>m	C + Z = 0	BHI	22	r≤m	BLS	23	Unsigned
r≥m	C = 0	BHS/BCC	24	r<m	BLO/BCS	25	Unsigned
r = m	Z = 1	BEQ	27	r ≠ m	BNE	26	Unsigned
r≤m	C + Z = 1	BLS	23	r>m	BHI	22	Unsigned
r<m	C = 1	BLO/BCS	25	r≥m	BHS/BCC	24	Unsigned
Carry	C = 1	BCS	25	No Carry	BCC	24	Simple
Negative	N = 1	BMI	2B	Plus	BPL	2A	Simple
Overflow	V = 1	BVS	29	No Overflow	BVC	28	Simple
r = 0	Z = 1	BEQ	27	r ≠ 0	BNE	26	Simple
Always	—	BRA	20	Never	BRN	21	Unconditional

CBA

Compare Accumulators

CBA

Operation: (ACCA) – (ACCB)

Description: Compares the contents of ACCA to the contents of ACCB and sets the condition codes, which may be used for arithmetic and logical conditional branches. Both operands are unaffected.

Condition Codes and Boolean Formulae:

S	X	H	I	N	Z	V	C
—	—	—	—	↕	↕	↕	↕

N R7
  Set if MSB of result is set; cleared otherwise.

Z $\overline{R7} \cdot \overline{R6} \cdot \overline{R5} \cdot \overline{R4} \cdot \overline{R3} \cdot \overline{R2} \cdot \overline{R1} \cdot \overline{R0}$
  Set if result is $00; cleared otherwise.

V $A7 \cdot \overline{B7} \cdot \overline{R7} + \overline{A7} \cdot B7 \cdot R7$
  Set if a twos complement overflow resulted from the operation; cleared otherwise.

C $\overline{A7} \cdot B7 + B7 \cdot R7 + R7 \cdot \overline{A7}$
  Set if there was a borrow from the MSB of the result; cleared otherwise.

Source Form: CBA

Addressing Modes, Machine Code, and Cycle-by-Cycle Execution:

Cycle	CBA (INH)		
	Addr	Data	R/W̄
1	OP	11	1
2	OP+1	—	1

CLC

Clear Carry

CLC

Operation: C bit ◀ 0

Description: Clears the C bit in the CCR.

CLC may be used to set up the C bit prior to a shift or rotate instruction involving the C bit.

Condition Codes and Boolean Formulae:

S	X	H	I	N	Z	V	C
—	—	—	—	—	—	—	0

C 0
   Cleared

Source Form: CLC

Addressing Modes, Machine Code, and Cycle-by-Cycle Execution:

Cycle	CLC (INH)		
	Addr	Data	R/W̄
1	OP	0C	1
2	OP+1	—	1

CLI   **Clear Interrupt Mask**   # CLI

Operation: I bit ⬥ 0

Description: Clears the interrupt mask bit in the CCR. When the I bit is clear, interrupts are enabled. There is a one E-clock cycle delay in the clearing mechanism for the I bit so that, if interrupts were previously disabled, the next instruction after a CLI will always be executed, even if there was an interrupt pending prior to execution of the CLI instruction.

Condition Codes and Boolean Formulae:

S	X	H	I	N	Z	V	C
—	—	—	0	—	—	—	—

I   0
   Cleared

Source Form: CLI

Addressing Modes, Machine Code, and Cycle-by-Cycle Execution:

Cycle	CLI (INH)		
	Addr	Data	R/W̄
1	OP	0E	1
2	OP+1	—	1

CLR

Clear

CLR

Operation: ACCX ◀ 0   **or:**   M ◀ 0

Description; The contents of ACCX or M are replaced with zeros.

Condition Codes and Boolean Formulae:

S	X	H	I	N	Z	V	C
—	—	—	—	0	1	0	0

N 0
Cleared

Z 1
Set

V 0
Cleared

C 0
Cleared

Source Forms: CLRA; CLRB; CLR (opr)

Addressing Modes, Machine Code, and Cycle-by-Cycle Execution:

Cycle	CLRA (INH)			CLRB (INH)			CLR (EXT)			CLR (IND, X)			CLR (IND, Y)		
	Addr	Data	R/W̄	Addr	Data	R/W̄	Addr	Data	R/W̄	Addr	Data	R/W̄	Addr	Data	R/W̄
1	OP	4F	1	OP	5F	1	OP	7F	1	OP	6F	1	OP	18	1
2	OP+1	—	1	OP+1	—	1	OP+1	hh	1	OP+1	ff	1	OP+1	6F	1
3							OP+2	ll	1	FFFF	—	1	OP+2	ff	1
4							hhll	(hhll)	1	X+ff	(X+ff)	1	FFFF	—	1
5							FFFF	—	1	FFFF	—	1	Y+ff	(Y+ff)	1
6							hhll	00	0	X+ff	00	0	FFFF	—	1
7													Y+ff	00	0

CLV
Clear Twos-Complement Overflow Bit
CLV

Operation: V bit ⬧ 0

Description: Clears the twos complement overflow bit in the CCR.

Condition Codes and Boolean Formulae:

S	X	H	I	N	Z	V	C
—	—	—	—	—	—	0	—

V 0
    Cleared

Source Form: CLV

Addressing Modes, Machine Code, and Cycle-by-Cycle Execution:

Cycle	CLV (INH)		
	Addr	Data	R/W̄
1	OP	0A	1
2	OP+1	—	1

CMP

Compare

CMP

Operation: (ACCX) – (M)

Description: Compares the contents of ACCX to the contents of M and sets the condition codes, which may be used for arithmetic and logical conditional branching. Both operands are unaffected.

Condition Codes and Boolean Formulae:

S	X	H	I	N	Z	V	C
—	—	—	—	↕	↕	↕	↕

N R7
   Set if MSB of result is set; cleared otherwise.

Z $\overline{R7} \cdot \overline{R6} \cdot \overline{R5} \cdot \overline{R4} \cdot \overline{R3} \cdot \overline{R2} \cdot \overline{R1} \cdot \overline{R0}$
   Set if result is $00; cleared otherwise.

V $X7 \cdot \overline{M7} \cdot \overline{R7} + \overline{X7} \cdot M7 \cdot R7$
   Set if a twos complement overflow resulted from the operation; cleared otherwise.

C $\overline{X7} \cdot M7 + M7 \cdot R7 + R7 \cdot \overline{X7}$
   Set if there was a borrow from the MSB of the result; cleared otherwise.

Source Forms: CMPA (opr); CMPB (opr)

Addressing Modes, Machine Code, and Cycle-by-Cycle Execution:

Cycle	CMPA (IMM)			CMPA (DIR)			CMPA (EXT)			CMPA (IND, X)			CMPA (IND, Y)		
	Addr	Data	R/W̄	Addr	Data	R/W̄	Addr	Data	R/W̄	Addr	Data	R/W̄	Addr	Data	R/W̄
1	OP	81	1	OP	91	1	OP	B1	1	OP	A1	1	OP	18	1
2	OP+1	ii	1	OP+1	dd	1	OP+1	hh	1	OP+1	ff	1	OP+1	A1	1
3				00dd	(00dd)	1	OP+2	ll	1	FFFF	—	1	OP+2	ff	1
4							hhll	(hhll)	1	X+ff	(X+ff)	1	FFFF	—	1
5													Y+ff	(Y+ff)	1

Cycle	CMPB (IMM)			CMPB (DIR)			CMPB (EXT)			CMPB (IND, X)			CMPB (IND, Y)		
	Addr	Data	R/W̄	Addr	Data	R/W̄	Addr	Data	R/W̄	Addr	Data	R/W̄	Addr	Data	R/W̄
1	OP	C1	1	OP	D1	1	OP	F1	1	OP	E1	1	OP	18	1
2	OP+1	ii	1	OP+1	dd	1	OP+1	hh	1	OP+1	ff	1	OP+1	E1	1
3				00dd	(00dd)	1	OP+2	ll	1	FFFF	—	1	OP+2	ff	1
4							hhll	(hhll)	1	X+ff	(X+ff)	1	FFFF	—	1
5													Y+ff	(Y+ff)	1

COM
Complement
COM

Operation: ACCX ◀ $(\overline{ACCX}) = \$FF - (ACCX)$ or: M ◀ $(\overline{M}) = \$FF - (M)$

Description: Replaces the contents of ACCX or M with its ones complement. (Each bit of the contents of ACCX or M is replaced with the complement of that bit.) To complement a value without affecting the C-bit, EXclusive-OR the value with $FF.

Condition Codes and Boolean Formulae:

S	X	H	I	N	Z	V	C
—	—	—	—	↕	↕	0	1

N R7
   Set if MSB of result is set; cleared otherwise.

Z $\overline{R7} \cdot \overline{R6} \cdot \overline{R5} \cdot \overline{R4} \cdot \overline{R3} \cdot \overline{R2} \cdot \overline{R1} \cdot \overline{R0}$
   Set if result is $00; cleared otherwise.

V 0
   Cleared

C 1
   Set (For compatibility with M6800)

Source Forms: COMA; COMB; COM (opr)

Addressing Modes, Machine Code, and Cycle-by-Cycle Execution:

Cycle	COMA (INH) Addr	Data	R/W̄	COMB (INH) Addr	Data	R/W̄	COM (EXT) Addr	Data	R/W̄	COM (IND, X) Addr	Data	R/W̄	COM (IND, Y) Addr	Data	R/W̄
1	OP	43	1	OP	53	1	OP	73	1	OP	63	1	OP	18	1
2	OP+1	—	1	OP+1	—	1	OP+1	hh	1	OP+1	ff	1	OP+1	63	1
3							OP+2	ll	1	FFFF	—	1	OP+2	ff	1
4							hhll	(hhll)	1	X+ff	(X+ff)	1	FFFF	—	1
5							FFFF	—	1	FFFF	—	1	Y+ff	(Y+ff)	1
6							hhll	result	0	X+ff	result	0	FFFF	—	1
7													Y+ff	result	0

CPD   **Compare Double Accumulator**   **CPD**

Operation: $(ACCD) - (M:M+1)$

Description: Compares the contents of accumulator D with a 16-bit value at the address specified and sets the condition codes accordingly. The compare is accomplished internally by doing a 16-bit subtract of $(M:M+1)$ from accumulator D without modifying either accumulator D or $(M:M+1)$.

Condition Codes and Boolean Formulae:

S	X	H	I	N	Z	V	C
—	—	—	—	↕	↕	↕	↕

N   R15
   Set if MSB of result is set; cleared otherwise.

Z $\overline{R15} \cdot \overline{R14} \cdot \overline{R13} \cdot \overline{R12} \cdot \overline{R11} \cdot \overline{R10} \cdot \overline{R9} \cdot \overline{R8} \cdot \overline{R7} \cdot \overline{R6} \cdot \overline{R5} \cdot \overline{R4} \cdot \overline{R3} \cdot \overline{R2} \cdot \overline{R1} \cdot \overline{R0}$
   Set if result is $0000; cleared otherwise.

V $D15 \cdot \overline{M15} \cdot \overline{R15} + \overline{D15} \cdot M15 \cdot R15$
   Set if a twos complement overflow resulted from the operation; cleared otherwise.

C $\overline{D15} \cdot M15 + M15 \cdot R15 + R15 \cdot \overline{D15}$
   Set if the absolute value of the contents of memory is larger than the absolute value of the accumulator; cleared otherwise.

Source Form: CPD (opr)

Addressing Modes, Machine Code, and Cycle-by-Cycle Execution:

Cycle	CPD (IMM)			CPD (DIR)			CPD (EXT)			CPD (IND, X)			CPD (IND, Y)		
	Addr	Data	R/W̄	Addr	Data	R/W̄	Addr	Data	R/W̄	Addr	Data	R/W̄	Addr	Data	R/W̄
1	OP	1A	1	OP	1A	1	OP	1A	1	OP	1A	1	OP	CD	1
2	OP+1	83	1	OP+1	93	1	OP+1	B3	1	OP+1	A3	1	OP+1	A3	1
3	OP+2	jj	1	OP+2	dd	1	OP+2	hh	1	OP+2	ff	1	OP+2	ff	1
4	OP+3	kk	1	00dd	(00dd)	1	OP+3	ll	1	FFFF	—	1	FFFF	—	1
5	FFFF	—	1	00dd+1	(00dd+1)	1	hhll	(hhll)	1	X+ff	(X+ff)	1	Y+ff	(Y+ff)	1
6				FFFF	—	1	hhll+1	(hhll+1)	1	X+ff+1	(X+ff+1)	1	Y+ff+1	(Y+ff+1)	1
7							FFFF	—	1	FFFF	—	1	FFFF	—	1

CPX

Compare Index Register X

CPX

Operation: $(IX) - (M:M+1)$

Description: Compares the contents of the index register X with a 16-bit value at the address specified and sets the condition codes accordingly. The compare is accomplished internally by doing a 16-bit subtract of $(M:M+1)$ from index register X without modifying either index register X or $(M:M+1)$.

Condition Codes and Boolean Formulae:

S	X	H	I	N	Z	V	C
—	—	—	—	↕	↕	↕	↕

N R15
  Set if MSB of result is set; cleared otherwise.

Z $\overline{R15} \cdot \overline{R14} \cdot \overline{R13} \cdot \overline{R12} \cdot \overline{R11} \cdot \overline{R10} \cdot \overline{R9} \cdot \overline{R8} \cdot \overline{R7} \cdot \overline{R6} \cdot \overline{R5} \cdot \overline{R4} \cdot \overline{R3} \cdot \overline{R2} \cdot \overline{R1} \cdot \overline{R0}$
  Set if result is $0000; cleared otherwise.

V $IX15 \cdot \overline{M15} \cdot \overline{R15} + \overline{IX15} \cdot M15 \cdot R15$
  Set if a twos complement overflow resulted from the operation; cleared otherwise.

C $\overline{IX15} \cdot M15 + M15 \cdot R15 + R15 \cdot \overline{IX15}$
  Set if the absolute value of the contents of memory is larger than the absolute value of the index register; cleared otherwise.

Source Form: CPX (opr)

Addressing Modes, Machine Code, and Cycle-by-Cycle Execution:

Cycle	CPX (IMM) Addr	Data	R/W̄	CPX (DIR) Addr	Data	R/W̄	CPX (EXT) Addr	Data	R/W̄	CPX (IND, X) Addr	Data	R/W̄	CPX (IND, Y) Addr	Data	R/W̄
1	OP	8C	1	OP	9C	1	OP	BC	1	OP	AC	1	OP	CD	1
2	OP+1	jj	1	OP+1	dd	1	OP+1	hh	1	OP+1	ff	1	OP+1	AC	1
3	OP+2	kk	1	00dd	(00dd)	1	OP+2	ll	1	FFFF	—	1	OP+2	ff	1
4	FFFF	—	1	00dd+1	(00dd+1)	1	hhll	(hhll)	1	X+ff	(X+ff)	1	FFFF	—	1
5				FFFF	—	1	hhll+1	(hhll+1)	1	X+ff+1	(X+ff+1)	1	Y+ff	(Y+ff)	1
6							FFFF	—	1	FFFF	—	1	Y+ff+1	(Y+ff+1)	1
7													FFFF	—	1

CPY   Compare Index Register Y   CPY

Operation: $(IY) - (M:M + 1)$

Description: Compares the contents of the index register Y with a 16-bit value at the
address specified and sets the condition codes accordingly. The compare is accomplished internally by doing a 16-bit subtract of $(M:M + 1)$ from index register Y without
modifying either index register Y or $(M:M + 1)$.

Condition Codes and Boolean Formulae:

S	X	H	I	N	Z	V	C
—	—	—	—	↕	↕	↕	↕

N R15
   Set if MSB of result is set; cleared otherwise.

Z $\overline{R15} \cdot \overline{R14} \cdot \overline{R13} \cdot \overline{R12} \cdot \overline{R11} \cdot \overline{R10} \cdot \overline{R9} \cdot \overline{R8} \cdot \overline{R7} \cdot \overline{R6} \cdot \overline{R5} \cdot \overline{R4} \cdot \overline{R3} \cdot \overline{R2} \cdot \overline{R1} \cdot \overline{R0}$
   Set if result is $0000; cleared otherwise.

V $IY15 \cdot \overline{M15} \cdot \overline{R15} + \overline{IY15} \cdot M15 \cdot R15$
   Set if a twos complement overflow resulted from the operation; cleared otherwise.

C $\overline{IY15} \cdot M15 + M15 \cdot R15 + R15 \cdot \overline{IY15}$
   Set if the absolute value of the contents of memory is larger than the absolute
   value of the index register; cleared otherwise.

Source Form: CPY (opr)

Addressing Modes, Machine Code, and Cycle-by-Cycle Execution:

Cycle	CPY (IMM)			CPY (DIR)			CPY (EXT)			CPY (IND, X)			CPY (IND, Y)		
	Addr	Data	R/W̄	Addr	Data	R/W̄	Addr	Data	R/W̄	Addr	Data	R/W̄	Addr	Data	R/W̄
1	OP	18	1	OP	18	1	OP	18	1	OP	1A	1	OP	18	1
2	OP+1	8C	1	OP+1	9C	1	OP+1	BC	1	OP+1	AC	1	OP+1	AC	1
3	OP+2	jj	1	OP+2	dd	1	OP+2	hh	1	OP+2	ff	1	OP+2	ff	1
4	OP+3	kk	1	00dd	(00dd)	1	OP+3	ll	1	FFFF	—	1	FFFF	—	1
5	FFFF	—	1	00dd+1	(00dd+1)	1	hhll	(hhll)	1	X+ff	(X+ff)	1	Y+ff	(Y+ff)	1
6				FFFF	—	1	hhll+1	(hhll+1)	1	X+ff+1	(X+ff+1)	1	Y+ff+1	(Y+ff+1)	1
7							FFFF	—	1	FFFF	—	1	FFFF	—	1

DAA   Decimal Adjust ACCA   # DAA

Operation: The following table summarizes the operation of the DAA instruction for all legal combinations of input operands. A correction factor (column 5 in the following table) is added to ACCA to restore the result of an addition of two BCD operands to a valid BCD value and set or clear the carry bit.

State of C Bit Before DAA (Column 1)	Upper Half-Byte of ACCA (Bits 7–4) (Column 2)	Initial Half-Carry H Bit from CCR (Column 3)	Lower Half-Byte of ACCA (Bits 3–0) (Column 4)	Number Added to ACCA by DAA (Column 5)	State of C Bit After DAA (Column 6)
0	0-9	0	0-9	00	0
0	0-8	0	A-F	06	0
0	0-9	1	0-3	06	0
0	A-F	0	0-9	60	1
0	9-F	0	A-F	66	1
0	A-F	1	0-3	66	1
1	0-2	0	0-9	60	1
1	0-2	0	A-F	66	1
1	0-3	1	0-3	66	1

NOTE

Columns (1) through (4) of the above table represent all possible cases which can result from any of the operations ABA, ADD, or ADC, with initial carry either set or clear, applied to two binary-coded-decimal operands. The table shows hexadecimal values.

Description: If the contents of ACCA and the state of the carry/borrow bit C and the state of the half-carry bit H are all the result of applying any of the operations ABA, ADD, or ADC to binary-coded-decimal operands, with or without an initial carry, the DAA operation will adjust the contents of ACCA and the carry bit C in the CCR to represent the correct binary-coded-decimal sum and the correct state of the C bit.

Condition Codes and Boolean Formulae:

S	X	H	I	N	Z	V	C
—	—	—	—	↕	↕	?	↕

N R7
   Set if MSB of result is set; cleared otherwise.

Z $\overline{R7} \cdot \overline{R6} \cdot \overline{R5} \cdot \overline{R4} \cdot \overline{R3} \cdot \overline{R2} \cdot \overline{R1} \cdot R0$
   Set if result is $00; cleared otherwise.

V ?
   Not defined

C See table above.

Reprinted with permission of Motorola

DAA

Decimal Adjust ACCA
(Continued)

DAA

Source Form: DAA

Addressing Modes, Machine Code, and Cycle-by-Cycle Execution:

Cycle	DAA (INH)		
	Addr	Data	R/W̄
1	OP	19	1
2	OP+1	—	1

For the purpose of illustration, consider the case where the BCD value $99 was just added to the BCD value $22. The add instruction is a binary operation, which yields the result $BB with no carry (C) or half carry (H). This corresponds to the fifth row of the table on the previous page. The DAA instruction will therefore add the correction factor $66 to the result of the addition, giving a result of $21 with the carry bit set. This result corresponds to the BCD value $121, which is the expected BCD result.

DEC

Decrement

DEC

Operation: ACCX ⬦ (ACCX) − $01   or: M ⬦ (M) − $01

Description: Subtract one from the contents of ACCX or M.

The N, Z, and V bits in the CCR are set or cleared according to the results of the operation. The C bit in the CCR is not affected by the operation, thus allowing the DEC instruction to be used as a loop counter in multiple-precision computations.

When operating on unsigned values, only BEQ and BNE branches can be expected to perform consistently. When operating on twos-complement values, all signed branches are available.

Condition Codes and Boolean Formulae:

S	X	H	I	N	Z	V	C
—	—	—	—	⬍	⬍	⬍	—

N R7
  Set if MSB of result is set; cleared otherwise.

Z $\overline{R7} \cdot \overline{R6} \cdot \overline{R5} \cdot \overline{R4} \cdot \overline{R3} \cdot \overline{R2} \cdot \overline{R1} \cdot \overline{R0}$
  Set if result is $00; cleared otherwise

V $X7 \cdot \overline{X6} \cdot \overline{X5} \cdot \overline{X4} \cdot X3 \cdot \overline{X2} \cdot \overline{X1} \cdot \overline{X0} = \overline{R7} \cdot R6 \cdot R5 \cdot R4 \cdot R3 \cdot R2 \cdot R1 \cdot R0$
  Set if there was a twos complement overflow as a result of the operation; cleared otherwise. Twos complement overflow occurs if and only if (ACCX) or (M) was $80 before the operation.

Source Forms: DECA; DECB; DEC (opr)

Addressing Modes, Machine Code, and Cycle-by-Cycle Execution:

Cycle	DECA (INH)			DECB (INH)			DEC (EXT)			DEC (IND, X)			DEC (IND, Y)		
	Addr	Data	R/$\overline{W}$	Addr	Data	R/$\overline{W}$	Addr	Data	R/$\overline{W}$	Addr	Data	R/$\overline{W}$	Addr	Data	R/$\overline{W}$
1	OP	4A	1	OP	5A	1	OP	7A	1	OP	6A	1	OP	18	1
2	OP+1	—	1	OP+1	—	1	OP+1	hh	1	OP+1	ff	1	OP+1	6A	1
3							OP+2	ll	1	FFFF	—	1	OP+2	ff	1
4							hhll	(hhll)	1	X+ff	(X+ff)	1	FFFF	—	1
5							FFFF	—	1	FFFF	—	1	Y+ff	(Y+ff)	1
6							hhll	result	0	X+ff	result	0	FFFF	—	1
7													Y+ff	result	0

DES

Decrement Stack Pointer

DES

Operation: SP ◆ (SP) − $0001

Description: Subtract one from the stack pointer.

Condition Codes and Boolean Formulae:

S	X	H	I	N	Z	V	C
—	—	—	—	—	—	—	—

None affected

Source Form: DES

Addressing Modes, Machine Code, and Cycle-by-Cycle Execution:

Cycle	DES (INH)		
	Addr	Data	R/W̄
1	OP	34	1
2	OP + 1	—	1
3	SP	—	1

DEX   Decrement Index Register X   **DEX**

Operation: IX ⬕ (IX) − $0001

Description: Subtract one from the index register X.

Only the Z bit is set or cleared according to the result of this operation.

Condition Codes and Boolean Formulae:

S	X	H	I	N	Z	V	C
—	—	—	—	—	⬕	—	—

Z $\overline{R15} \cdot \overline{R14} \cdot \overline{R13} \cdot \overline{R12} \cdot \overline{R11} \cdot \overline{R10} \cdot \overline{R9} \cdot \overline{R8} \cdot \overline{R7} \cdot \overline{R6} \cdot \overline{R5} \cdot \overline{R4} \cdot \overline{R3} \cdot \overline{R2} \cdot \overline{R1} \cdot \overline{R0}$
Set if result is $0000; cleared otherwise.

Source Form: DEX

Addressing Modes, Machine Code, and Cycle-by-Cycle Execution:

Cycle	DEX (INH)		
	Addr	Data	R/W̄
1	OP	09	1
2	OP + 1	—	1
3	FFFF	—	1

DEY

Decrement Index Register Y

DEY

Operation: IY ◀ (IY) − $0001

Description: Subtract one from the index register Y.

Only the Z bit is set or cleared according to the result of this operation.

Condition Codes and Boolean Formulae:

S	X	H	I	N	Z	V	C
—	—	—	—	—	✸	—	—

Z $\overline{R15} \cdot \overline{R14} \cdot \overline{R13} \cdot \overline{R12} \cdot \overline{R11} \cdot \overline{R10} \cdot \overline{R9} \cdot \overline{R8} \cdot \overline{R7} \cdot \overline{R6} \cdot \overline{R5} \cdot \overline{R4} \cdot \overline{R3} \cdot \overline{R2} \cdot \overline{R1} \cdot \overline{R0}$
Set if result is $0000; cleared otherwise.

Source Form: DEY

Addressing Modes, Machine Code, and Cycle-by-Cycle Execution:

Cycle	DEY (INH)		
	Addr	Data	R/W̄
1	OP	18	1
2	OP+1	09	1
3	OP+2	—	1
4	FFFF	—	1

Reprinted with permission of Motorola

EOR

Exclusive-OR

EOR

Operation: ACCX ◀ (ACCX) ⊕ (M)

Description: Performs the logical exclusive-OR between the contents of ACCX and the contents of M and places the result in ACCX. (Each bit of ACCX after the operation will be the logical exclusive-OR of the corresponding bits of M and ACCX before the operation.)

Condition Codes and Boolean Formulae:

S	X	H	I	N	Z	V	C
—	—	—	—	♦	♦	0	—

N R7
   Set if MSB of result is set; cleared otherwise.

Z $\overline{R7} \cdot \overline{R6} \cdot \overline{R5} \cdot \overline{R4} \cdot \overline{R3} \cdot \overline{R2} \cdot \overline{R1} \cdot \overline{R0}$
   Set if result is $00; cleared otherwise

V 0
   Cleared

Source Forms: EORA (opr); EORB (opr)

Addressing Modes, Machine Code, and Cycle-by-Cycle Execution:

Cycle	EORA (IMM)			EORA (DIR)			EORA (EXT)			EORA (IND, X)			EORA (IND, Y)		
	Addr	Data	R/W̄	Addr	Data	R/W̄	Addr	Data	R/W̄	Addr	Data	R/W̄	Addr	Data	R/W̄
1	OP	88	1	OP	98	1	OP	B8	1	OP	A8	1	OP	18	1
2	OP+1	ii	1	OP+1	dd	1	OP+1	hh	1	OP+1	ff	1	OP+1	A8	1
3				00dd	(00dd)	1	OP+2	ll	1	FFFF	—	1	OP+2	ff	1
4							hhll	(hhll)	1	X+ff	(X+ff)	1	FFFF	—	1
5													Y+ff	(Y+ff)	1

Cycle	EORB (IMM)			EORB (DIR)			EORB (EXT)			EORB (IND, X)			EORB (IND, Y)		
	Addr	Data	R/W̄	Addr	Data	R/W̄	Addr	Data	R/W̄	Addr	Data	R/W̄	Addr	Data	R/W̄
1	OP	C8	1	OP	D8	1	OP	F8	1	OP	E8	1	OP	18	1
2	OP+1	ii	1	OP+1	dd	1	OP+1	hh	1	OP+1	ff	1	OP+1	E8	1
3				00dd	(00dd)	1	OP+2	ll	1	FFFF	—	1	OP+2	ff	1
4							hhll	(hhll)	1	X+ff	(X+ff)	1	FFFF	—	1
5													Y+ff	(Y+ff)	1

FDIV   Fractional Divide   FDIV

Operation: (ACCD)/(IX); IX ⬍ Quotient, ACCD ⬍ Remainder

Description: Performs an usigned fractional divide of the 16-bit numerator in the D accumulator by the 16-bit denominator in the index register X and sets the condition codes accordingly. The quotient is placed in the index register X, and the remainder is placed in the D accumulator. The radix point is assumed to be in the same place for both the numerator and the denominator. The radix point is to the left of bit 15 for the quotient. The numerator is assumed to be less than the denominator. In the case of overflow (denominator is less than or equal to the numerator) or divide by zero, the quotient is set to $FFFF, and the remainder is indeterminate.

FDIV is equivalent to multiplying the numerator by 2^{16} and then performing a 32×16-bit integer divide. The result is interpreted as a binary-weighted fraction, which resulted from the division of a 16-bit integer by a larger 16-bit integer. A result of $0001 corresponds to 0.000015, and $FFFF corresponds to 0.99998. The remainder of an IDIV instruction can be resolved into a binary-weighted fraction by an FDIV instruction. The remainder of an FDIV instruction can be resolved into the next 16-bits of binary-weighted fraction by another FDIV instruction.

Condition Codes and Boolean Formulae:

S	X	H	I	N	Z	V	C
—	—	—	—	—	⬍	⬍	⬍

Z $\overline{R15} \cdot \overline{R14} \cdot \overline{R13} \cdot \overline{R12} \cdot \overline{R11} \cdot \overline{R10} \cdot \overline{R9} \cdot \overline{R8} \cdot \overline{R7} \cdot \overline{R6} \cdot \overline{R5} \cdot \overline{R4} \cdot \overline{R3} \cdot \overline{R2} \cdot \overline{R1} \cdot \overline{R0}$
Set if quotient is $0000; cleared otherwise.

V   1 if IX≤D
Set if denominator was less than or equal to the numerator; cleared otherwise.

C $\overline{IX15} \cdot \overline{IX14} \cdot \overline{IX13} \cdot \overline{IX12} \cdot \overline{IX11} \cdot \overline{IX10} \cdot \overline{IX9} \cdot \overline{IX8} \cdot$
$\overline{IX7} \cdot \overline{IX6} \cdot \overline{IX5} \cdot \overline{IX4} \cdot \overline{IX3} \cdot \overline{IX2} \cdot \overline{IX1} \cdot \overline{IX0}$
Set if denominator was $0000; cleared otherwise.

Source Form: FDIV

Addressing Modes, Machine Code, and Cycle-by-Cycle Execution:

Cycle	FDIV (INH)		
	Addr	Data	R/$\overline{W}$
1	OP	03	1
2	OP + 1	—	1
3–41	FFFF	—	1

IDIV

Integer Divide

IDIV

Operation: (ACCD)/(IX); IX ◀ Quotient, ACCD ◀ Remainder

Description: Performs an unsigned integer divide of the 16-bit numerator in D accumulator by the 16-bit denominator in index register X and sets the condition codes accordingly. The quotient is placed in index register X, and the remainder is placed in accumulator D. The radix point is assumed to be in the same place for both the numerator and the denominator. The radix point is to the right of bit zero for the quotient. In the case of divide by zero, the quotient is set to $FFFF, and the remainder is indeterminate.

Condition Codes and Boolean Formulae:

S	X	H	I	N	Z	V	C
—	—	—	—	—	↕	0	↕

Z $\overline{R15} \cdot \overline{R14} \cdot \overline{R13} \cdot \overline{R12} \cdot \overline{R11} \cdot \overline{R10} \cdot \overline{R9} \cdot \overline{R8} \cdot \overline{R7} \cdot \overline{R6} \cdot \overline{R5} \cdot \overline{R4} \cdot \overline{R3} \cdot \overline{R2} \cdot \overline{R1} \cdot \overline{R0}$
Set if result is $0000; cleared otherwise.

V 0
Cleared.

C $\overline{IX15} \cdot \overline{IX14} \cdot \overline{IX13} \cdot \overline{IX12} \cdot \overline{IX11} \cdot \overline{IX10} \cdot \overline{IX9} \cdot \overline{IX8} \cdot$
$\overline{IX7} \cdot \overline{IX6} \cdot \overline{IX5} \cdot \overline{IX4} \cdot \overline{IX3} \cdot \overline{IX2} \cdot \overline{IX1} \cdot \overline{IX0}$
Set if denominator was $0000; cleared otherwise.

Source Form: IDIV

Addressing Modes, Machine Code, and Cycle-by-Cycle Execution:

Cycle	IDIV (INH)		
	Addr	Data	R/W̅
1	OP	02	1
2	OP + 1	—	1
3–41	FFFF	—	1

INC

Increment

INC

Operation: ACCX ◀ (ACCX) + $01   **or:**   M ◀ (M) + $01

Description: Add one to the contents of ACCX or M.

The N, Z, and V bits in the CCR are set or cleared according to the results of the operation. The C bit in the CCR is not affected by the operation, thus allowing the INC instruction to be used as a loop counter in multiple-precision computations.

When operating on unsigned values, only BEQ and BNE branches can be expected to perform consistently. When operating on twos-complement values, all signed branches are available.

Condition Codes and Boolean Formulae:

S	X	H	I	N	Z	V	C
—	—	—	—	✦	✦	✦	—

N R7
   Set if MSB of result is set; cleared otherwise.

Z $\overline{R7} \cdot \overline{R6} \cdot \overline{R5} \cdot \overline{R4} \cdot \overline{R3} \cdot \overline{R2} \cdot \overline{R1} \cdot \overline{R0}$
   Set if result is $00; cleared otherwise.

V $\overline{X7} \cdot X6 \cdot X5 \cdot X4 \cdot X3 \cdot X2 \cdot X1 \cdot X0$
   Set if there is a twos complement overflow as a result of the operation; cleared otherwise. Twos complement overflow occurs if and only if (ACCX) or (M) was $7F before the operation.

Source Forms: INCA; INCB; INC (opr)

Addressing Modes, Machine Code, and Cycle-by-Cycle Execution:

Cycle	INCA (INH) Addr	Data	R/W̄	INCB (INH) Addr	Data	R/W̄	INC (EXT) Addr	Data	R/W̄	INC (IND, X) Addr	Data	R/W̄	INC (IND, Y) Addr	Data	R/W̄
1	OP	4C	1	OP	5C	1	OP	7C	1	OP	6C	1	OP	18	1
2	OP+1	—	1	OP+1	—	1	OP+1	hh	1	OP+1	ff	1	OP+1	6C	1
3							OP+2	ll	1	FFFF	—	1	OP+2	ff	1
4							hhll	(hhll)	1	X+ff	(X+ff)	1	FFFF	—	1
5							FFFF	—	1	FFFF	—	1	Y+ff	(Y+ff)	1
6							hhll	result	0	X+ff	result	0	FFFF	—	1
7													Y+ff	result	0

INS   Increment Stack Pointer   INS

Operation: SP ◀ (SP) + $0001

Description: Add one to the stack pointer.

Condition Codes and Boolean Formulae:

S	X	H	I	N	Z	V	C
—	—	—	—	—	—	—	—

None affected

Source Form: INS

Addressing Modes, Machine Code, and Cycle-by-Cycle Execution:

Cycle	INS (INH)		
	Addr	Data	R/W̅
1	OP	31	1
2	OP + 1	—	1
3	SP	—	1

INX
Increment Index Register X
INX

Operation: IX ◀ (IX) + $0001

Description: Add one to index register X.

Only the Z bit is set or cleared according to the result of this operation.

Condition Codes and Boolean Formulae:

S	X	H	I	N	Z	V	C
—	—	—	—	—	✦	—	—

Z $\overline{R15} \cdot \overline{R14} \cdot \overline{R13} \cdot \overline{R12} \cdot \overline{R11} \cdot \overline{R10} \cdot \overline{R9} \cdot \overline{R8} \cdot \overline{R7} \cdot \overline{R6} \cdot \overline{R5} \cdot \overline{R4} \cdot \overline{R3} \cdot \overline{R2} \cdot \overline{R1} \cdot \overline{R0}$
Set if result is $0000; cleared otherwise.

Source Form: INX

Addressing Modes, Machine Code, and Cycle-by-Cycle Execution:

Cycle	INX (INH)		
	Addr	Data	R/$\overline{W}$
1	OP	08	1
2	OP+1	—	1
3	FFFF	—	1

INY

Increment Index Register Y

INY

Operation: IY ◀ (IY) + $0001

Description: Add one to index register Y.

Only the Z bit is set or cleared according to the result of this operation.

Condition Codes and Boolean Formulae:

S	X	H	I	N	Z	V	C
—	—	—	—	—	⬦	—	—

Z $\overline{R15} \cdot \overline{R14} \cdot \overline{R13} \cdot \overline{R12} \cdot \overline{R11} \cdot \overline{R10} \cdot \overline{R9} \cdot \overline{R8} \cdot \overline{R7} \cdot \overline{R6} \cdot \overline{R5} \cdot \overline{R4} \cdot \overline{R3} \cdot \overline{R2} \cdot \overline{R1} \cdot \overline{R0}$
Set if result is $0000; cleared otherwise.

Source Form: INY

Addressing Modes, Machine Code, and Cycle-by-Cycle Execution:

Cycle	INY (INH)		
	Addr	Data	R/$\overline{W}$
1	OP	18	1
2	OP+1	08	1
3	OP+2	—	1
4	FFFF	—	1

JMP   Jump   JMP

Operation: PC ◄ Effective Address

Description: A jump occurs to the instruction stored at the effective address. The effective address is obtained according to the rules for EXTended or INDexed addressing.

Condition Codes and Boolean Formulae:

S	X	H	I	N	Z	V	C
—	—	—	—	—	—	—	—

None affected

Source Form: JMP (opr)

Addressing Modes, Machine Code, and Cycle-by-Cycle Execution:

Cycle	JMP (EXT)			JMP (IND, X)			JMP (IND, Y)		
	Addr	Data	R/W̄	Addr	Data	R/W̄	Addr	Data	R/W̄
1	OP	7E	1	OP	6E	1	OP	18	1
2	OP+1	hh	1	OP+1	ff	1	OP+1	6E	1
3	OP+2	ll	1	FFFF	—	1	OP+2	ff	1
4							FFFF	—	1

JSR

Jump to Subroutine

JSR

Operation:

PC ◀ (PC) + $0003   (for EXTended or INDexed, Y addressing) **or:**
PC ◀ (PC) + $0002   (for DIRect or INDexed, X addressing)
◀(PCL)   Push low-order return address onto stack
SP ◀ (SP) − $0001
◀(PCH)   Push high-order return address onto stack
SP ◀ (SP) − $0001
PC ◀ Effective Addr Load start address of requested subroutine

Description: The program counter is incremented by three or by two, depending on the addressing mode, and is then pushed onto the stack, eight bits at a time, least significant byte first. The stack pointer points to the next empty location in the stack. A jump occurs to the instruction stored at the effective address. The effective address is obtained according to the rules for EXTended, DIRect, or INDexed addressing.

Condition Codes and Boolean Formulae:

S	X	H	I	N	Z	V	C
—	—	—	—	—	—	—	—

None affected

Source Form: JSR (opr)

Addressing Modes, Machine Code, and Cycle-by-Cycle Execution:

Cycle	JSR (DIR)			JSR (EXT)			JSR (IND, X)			JSR (IND, Y)		
	Addr	Data	R/W̄	Addr	Data	R/W̄	Addr	Data	R/W̄	Addr	Data	R/W̄
1	OP	9D	1	OP	BD	1	OP	AD	1	OP	18	1
2	OP+1	dd	1	OP+1	hh	1	OP+1	ff	1	OP+1	AD	1
3	00dd	(00dd)	1	OP+2	ll	1	FFFF	—	1	OP+2	ff	1
4	SP	Rtn lo	0	hhll	(hhll)	1	X+ff	(X+ff)	1	FFFF	—	1
5	SP−1	Rtn hi	0	SP	Rtn lo	0	SP	Rtn lo	0	Y+ff	(Y+ff)	1
6				SP−1	Rtn hi	0	SP−1	Rtn hi	0	SP	Rtn lo	0
7										SP−1	Rtn hi	0

LDA

Load Accumulator

LDA

Operation: ACCX ◆ (M)

Description: Loads the contents of memory into the 8-bit accumulator. The condition codes are set according to the data.

Condition Codes and Boolean Formulae:

S	X	H	I	N	Z	V	C
—	—	—	—	↕	↕	0	—

N   R7
Set if MSB of result is set; cleared otherwise.

Z $\overline{R7} \cdot \overline{R6} \cdot \overline{R5} \cdot \overline{R4} \cdot \overline{R3} \cdot \overline{R2} \cdot \overline{R1} \cdot \overline{R0}$
Set if result is $00; cleared otherwise

V   0
Cleared

Source Form: LDAA (opr); LDAB (opr)

Addressing Modes, Machine Code, and Cycle-by-Cycle Execution:

Cycle	LDAA (IMM)			LDAA (DIR)			LDAA (EXT)			LDAA (IND, X)			LDAA (IND, Y)		
	Addr	Data	R/W̄	Addr	Data	R/W̄	Addr	Data	R/W̄	Addr	Data	R/W̄	Addr	Data	R/W̄
1	OP	86	1	OP	96	1	OP	B6	1	OP	A6	1	OP	18	1
2	OP+1	ii	1	OP+1	dd	1	OP+1	hh	1	OP+1	ff	1	OP+1	A6	1
3				00dd	(00dd)	1	OP+2	ll	1	FFFF	—	1	OP+2	ff	1
4							hhll	(hhll)	1	X+ff	(X+ff)	1	FFFF	—	1
5													Y+ff	(Y+ff)	1

Cycle	LDAB (IMM)			LDAB (DIR)			LDAB (EXT)			LDAB (IND, X)			LDAB (IND, Y)		
	Addr	Data	R/W̄	Addr	Data	R/W̄	Addr	Data	R/W̄	Addr	Data	R/W̄	Addr	Data	R/W̄
1	OP	C6	1	OP	D6	1	OP	F6	1	OP	E6	1	OP	18	1
2	OP+1	ii	1	OP+1	dd	1	OP+1	hh	1	OP+1	ff	1	OP+1	E6	1
3				00dd	(00dd)	1	OP+2	ll	1	FFFF	—	1	OP+2	ff	1
4							hhll	(hhll)	1	X+ff	(X+ff)	1	FFFF	—	1
5													Y+ff	(Y+ff)	1

LDD   Load Double Accumulator   LDD

Opeation: ACCD ⬧ (M:M + 1); ACCA ⬧ (M), ACCB ⬧ (M + 1)

Description: Loads the contents of memory locations M and M + 1 into the double accumulator D. The condition codes are set according to the data. The information from location M is loaded into accumulator A, and the information from location M + 1 is loaded into accumulator B.

Condition Codes and Boolean Formulae:

S	X	H	I	N	Z	V	C
—	—	—	—	⬧	⬧	0	

N R15
   Set if MSB of result is set; cleared otherwise.

Z $\overline{R15} \cdot \overline{R14} \cdot \overline{R13} \cdot \overline{R12} \cdot \overline{R11} \cdot \overline{R10} \cdot \overline{R9} \cdot \overline{R8} \cdot \overline{R7} \cdot \overline{R6} \cdot \overline{R5} \cdot \overline{R4} \cdot \overline{R3} \cdot \overline{R2} \cdot \overline{R1} \cdot \overline{R0}$
   Set if result is $0000; cleared otherwise.

V 0
   Cleared

Source Form: LDD (opr)

Addressing Modes, Machine Code, and Cycle-by-Cycle Execution:

Cycle	LDD (IMM)			LDD (DIR)			LDD (EXT)			LDD (IND, X)			LDD (IND, Y)		
	Addr	Data	R/$\overline{W}$	Addr	Data	R/$\overline{W}$	Addr	Data	R/$\overline{W}$	Addr	Data	R/$\overline{W}$	Addr	Data	R/$\overline{W}$
1	OP	CC	1	OP	DC	1	OP	FC	1	OP	EC	1	OP	18	1
2	OP + 1	jj	1	OP + 1	dd	1	OP + 1	hh	1	OP + 1	ff	1	OP + 1	EC	1
3	OP + 2	kk	1	00dd	(00dd)	1	OP + 2	ll	1	FFFF	—	1	OP + 2	ff	1
4				00dd + 1	(00dd + 1)	1	hhll	(hhll)	1	X + ff	(X + ff)	1	FFFF	—	1
5							hhll + 1	(hhll + 1)	1	X + ff + 1	(X + ff + 1)	1	Y + ff	(Y + ff)	1
6													Y + ff + 1	(Y + ff + 1)	1

LDS   Load Stack Pointer   LDS

Operation: SPH ◀ (M), SPL ◀ (M + 1)

Description: Loads the most significant byte of the stack pointer from the byte of memory at the address specified by the program, and loads the least significant byte of the stack pointer from the next byte of memory at one plus the address specified by the program.

Condition Codes and Boolean Formulae:

S	X	H	I	N	Z	V	C
—	—	—	—	✸	✸	0	—

N   R15
Set if MSB of result is set; cleared otherwise.

Z $\overline{R15} \cdot \overline{R14} \cdot \overline{R13} \cdot \overline{R12} \cdot \overline{R11} \cdot \overline{R10} \cdot \overline{R9} \cdot \overline{R8} \cdot \overline{R7} \cdot \overline{R6} \cdot \overline{R5} \cdot \overline{R4} \cdot \overline{R3} \cdot \overline{R2} \cdot \overline{R1} \cdot \overline{R0}$
Set if result is $0000; cleared otherwise.

V   0
Cleared

Source Form: LDS (opr)

Addressing Modes, Machine Code, and Cycle-by-Cycle Execution:

Cycle	LDS (IMM)			LDS (DIR)			LDS (EXT)			LDS (IND, X)			LDS (IND, Y)		
	Addr	Data	R/W̄	Addr	Data	R/W̄	Addr	Data	R/W̄	Addr	Data	R/W̄	Addr	Data	R/W̄
1	OP	8E	1	OP	9E	1	OP	BE	1	OP	AE	1	OP	18	1
2	OP + 1	jj	1	OP + 1	dd	1	OP + 1	hh	1	OP + 1	ff	1	OP + 1	AE	1
3	OP + 2	kk	1	00dd	(00dd)	1	OP + 2	ll	1	FFFF	—	1	OP + 2	ff	1
4				00dd + 1	(00dd + 1)	1	hhll	(hhll)	1	X + ff	(X + ff)	1	FFFF	—	1
5							hhll + 1	(hhll + 1)	1	X + ff + 1	(X + ff + 1)	1	Y + ff	(Y + ff)	1
6													Y + ff + 1	(Y + ff + 1)	1

LDX

Load Index Register X

LDX

Operation;   IXH ◀ (M), IXL ◀ (M + 1)

Description: Loads the most significant byte of index register X from the byte of memory at the address specified by the program, and loads the least significant byte of index register X from the next byte of memory at one plus the address specified by the program.

Condition Codes and Boolean Formulae:

S	X	H	I	N	Z	V	C
—	—	—	—	✸	✸	0	—

N R15
Set if MSB of result is set; cleared otherwise.

Z $\overline{R15} \cdot \overline{R14} \cdot \overline{R13} \cdot \overline{R12} \cdot \overline{R11} \cdot \overline{R10} \cdot \overline{R9} \cdot \overline{R8} \cdot \overline{R7} \cdot \overline{R6} \cdot \overline{R5} \cdot \overline{R4} \cdot \overline{R3} \cdot \overline{R2} \cdot \overline{R1} \cdot \overline{R0}$
Set if result is $0000; cleared otherwise.

V 0
Cleared

Source Form: LDX (opr)

Addressing Modes, Machine Code, and Cycle-by-Cycle Execution:

Cycle	LDX (IMM) Addr	Data	R/W̄	LDX (DIR) Addr	Data	R/W̄	LDX (EXT) Addr	Data	R/W̄	LDX (IND, X) Addr	Data	R/W̄	LDX (IND, Y) Addr	Data	R/W̄
1	OP	CE	1	OP	DE	1	OP	FE	1	OP	EE	1	OP	CD	1
2	OP+1	jj	1	OP+1	dd	1	OP+1	hh	1	OP+1	ff	1	OP+1	EE	1
3	OP+2	kk	1	00dd	(00dd)	1	OP+2	ll	1	FFFF	—	1	OP+2	ff	1
4				00dd+1	(00dd+1)	1	hhll	(hhll)	1	X+ff	(X+ff)	1	FFFF	—	1
5							hhll+1	(hhll+1)	1	X+ff+1	(X+ff+1)	1	Y+ff	(Y+ff)	1
6													Y+ff+1	(Y+ff+1)	1

LDY

Load Index Register Y

LDY

Operation: IYH ◀ (M), IYL ◀ (M + 1)

Description: Loads the most significant byte of index register Y from the byte of memory at the address specified by the program, and loads the least significant byte of index register Y from the next byte of memory at one plus the address specified by the program.

Condition Codes and Boolean Formulae:

S	X	H	I	N	Z	V	C
—	—	—	—	⭳	⭳	0	—

N R15
Set if MSB of result is set; cleared otherwise.

Z $\overline{R15} \cdot \overline{R14} \cdot \overline{R13} \cdot \overline{R12} \cdot \overline{R11} \cdot \overline{R10} \cdot \overline{R9} \cdot \overline{R8} \cdot \overline{R7} \cdot \overline{R6} \cdot \overline{R5} \cdot \overline{R4} \cdot \overline{R3} \cdot \overline{R2} \cdot \overline{R1} \cdot \overline{R0}$
Set if result is $0000; cleared otherwise.

V 0
Cleared

Source Form: LDY (opr)

Addressing Modes, Machine Code, and Cycle-by-Cycle Execution:

Cycle	LDY (IMM)			LDY (DIR)			LDY (EXT)			LDY (IND, X)			LDY (IND, Y)		
	Addr	Data	R/W̄	Addr	Data	R/W̄	Addr	Data	R/W̄	Addr	Data	R/W̄	Addr	Data	R/W̄
1	OP	18	1	OP	18	1	OP	18	1	OP	1A	1	OP	18	1
2	OP + 1	CE	1	OP + 1	DE	1	OP + 1	FE	1	OP + 1	EE	1	OP + 1	EE	1
3	OP + 2	jj	1	OP + 2	dd	1	OP + 2	hh	1	OP + 2	ff	1	OP + 2	ff	1
4	OP + 3	kk	1	00dd	(00dd)	1	OP + 3	ll	1	FFFF	—	1	FFFF	—	1
5				00dd + 1	(00dd + 1)	1	hhll	(hhll)	1	X + ff	(X + ff)	1	Y + ff	(Y + ff)	1
6							hhll + 1	(hhll + 1)	1	X + ff + 1	(X + ff + 1)	1	Y + ff + 1	(Y + ff + 1)	1

LSL

**Logical Shift Left
(Same as ASL)**

LSL

Operation:

Description: Shifts all bits of the ACCX or M one place to the left. Bit 0 is loaded with zero. The C bit is loaded from the most significant bit of ACCX or M.

Condition Codes and Boolean Formulae:

S	X	H	I	N	Z	V	C
—	—	—	—	⇕	⇕	⇕	⇕

N R7
Set if MSB of result is set; cleared otherwise.

Z $\overline{R7} \cdot \overline{R6} \cdot \overline{R5} \cdot \overline{R4} \cdot \overline{R3} \cdot \overline{R2} \cdot \overline{R1} \cdot \overline{R0}$
Set if result is $00; cleared otherwise.

V $N \oplus C = [N \cdot \overline{C}] + [\overline{N} \cdot C]$ (for N and C after the shift)
Set if (N is set and C is clear) or (N is clear and C is set); cleared otherwise (for values of N and C after the shift).

C M7
Set if, before the shift, the MSB of ACCX or M was set; cleared otherwise.

Source Forms: LSLA; LSLB; LSL (opr)

Addressing Modes, Machine Code, and Cycle-by-Cycle Execution:

Cycle	LSLA (INH)			LSLB (INH)			LSL (EXT)			LSL (IND, X)			LSL (IND, Y)		
	Addr	Data	R/$\overline{W}$	Addr	Data	R/$\overline{W}$	Addr	Data	R/$\overline{W}$	Addr	Data	R/$\overline{W}$	Addr	Data	R/$\overline{W}$
1	OP	48	1	OP	58	1	OP	78	1	OP	68	1	OP	18	1
2	OP+1	—	1	OP+1	—	1	OP+1	hh	1	OP+1	ff	1	OP+1	68	1
3							OP+2	ll	1	FFFF	—	1	OP+2	ff	1
4							hhll	(hhll)	1	X+ff	(X+ff)	1	FFFF	—	1
5							FFFF	—	1	FFFF	—	1	Y+ff	(Y+ff)	1
6							hhll	result	0	X+ff	result	0	FFFF	—	1
7													Y+ff	result	0

LSLD

**Logical Shift Left Double
(Same as ASLD)**

LSLD

Operation:

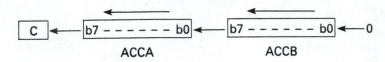

Description: Shifts all of ACCD one place to the left. Bit 0 is loaded with zero. The C bit is loaded from the most significant bit of ACCD.

Condition Codes and Boolean Formulae:

S	X	H	I	N	Z	V	C
—	—	—	—	↕	↕	↕	↕

N R15
   Set if MSB of result is set; cleared otherwise.

Z $\overline{R15} \cdot \overline{R14} \cdot \overline{R13} \cdot \overline{R12} \cdot \overline{R11} \cdot \overline{R10} \cdot \overline{R9} \cdot \overline{R8} \cdot \overline{R7} \cdot \overline{R6} \cdot \overline{R5} \cdot \overline{R4} \cdot \overline{R3} \cdot \overline{R2} \cdot \overline{R1} \cdot \overline{R0}$
   Set if result is $0000; cleared otherwise.

V $N \oplus C = [N \cdot \overline{C}] + [\overline{N} \cdot C]$ (for N and C after the shift)
   Set if (N is set and C is clear) or (N is clear and C is set); cleared otherwise (for values of N and C after the shift).

C D15
   Set if, before the shift, the MSB of ACCD was set; cleared otherwise.

Source Form: LSLD

Addressing Modes, Machine Code, and Cycle-by-Cycle Execution:

Cycle	LSLD (INH)		
	Addr	Data	R/$\overline{W}$
1	OP	05	1
2	OP+1	—	1
3	FFFF	—	1

LSR

Logical Shift Right

LSR

Operation:
$$0 \rightarrow \boxed{b7 - - - - - - b0} \rightarrow \boxed{C}$$

Description: Shifts all bits of ACCX or M one place to the right. Bit 7 is loaded with zero. The C bit is loaded from the least significant bit of ACCX or M.

Condition Codes and Boolean Formulae:

S	X	H	I	N	Z	V	C
—	—	—	—	0	⬍	⬍	⬍

N 0
Cleared.

Z $\overline{R7} \cdot \overline{R6} \cdot \overline{R5} \cdot \overline{R4} \cdot \overline{R3} \cdot \overline{R2} \cdot \overline{R1} \cdot \overline{R0}$
Set if result is $00; cleared otherwise.

V $N \oplus C = [N \cdot \overline{C}] + [\overline{N} \cdot C]$ (for N and C after the shift)
Since N = 0, this simplifies to C (after the shift).

C M0
Set if, before the shift, the LSB of ACCX or M was set; cleared otherwise.

Source Forms: LSRA; LSRB; LSR (opr)

Addressing Modes, Machine Code, and Cycle-by-Cycle Execution:

Cycle	LSRA (INH)			LSRB (INH)			LSR (EXT)			LSR (IND, X)			LSR (IND, Y)		
	Addr	Data	R/$\overline{W}$	Addr	Data	R/$\overline{W}$	Addr	Data	R/$\overline{W}$	Addr	Data	R/$\overline{W}$	Addr	Data	R/$\overline{W}$
1	OP	44	1	OP	54	1	OP	74	1	OP	64	1	OP	18	1
2	OP+1	—	1	OP+1	—	1	OP+1	hh	1	OP+1	ff	1	OP+1	64	1
3							OP+2	ll	1	FFFF	—	1	OP+2	ff	1
4							hhll	(hhll)	1	X+ff	(X+ff)	1	FFFF	—	1
5							FFFF	—	1	FFFF	—	1	Y+ff	(Y+ff)	1
6							hhll	result	0	X+ff	result	0	FFFF	—	1
7													Y+ff	result	0

LSRD   Logical Shift Right Double Accumulator   LSRD

Operation:

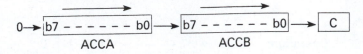

ACCA   ACCB

Description: Shifts all bits of ACCD one place to the right. Bit 15 (MSB of ACCA) is loaded with zero. The C bit is loaded from the least significant bit of ACCD (LSB of ACCB).

Condition Codes and Boolean Formulae:

S	X	H	I	N	Z	V	C
—	—	—	—	0	⬍	⬍	⬍

N   0
Cleared

Z $\overline{R15} \cdot \overline{R14} \cdot \overline{R13} \cdot \overline{R12} \cdot \overline{R11} \cdot \overline{R10} \cdot \overline{R9} \cdot \overline{R8} \cdot \overline{R7} \cdot \overline{R6} \cdot \overline{R5} \cdot \overline{R4} \cdot \overline{R3} \cdot \overline{R2} \cdot \overline{R1} \cdot \overline{R0}$
Set if result is $0000; cleared otherwise.

V   D0
Set if, after the shift operaton, C is set; cleared otherwise.

C   D0
Set if, before the shift, the least significant bit of ACCD was set; cleared otherwise.

Source Form: LSRD

Addressing Modes, Machine Code, and Cycle-by-Cycle Execution:

Cycle	LSRD (INH)		
	Addr	Data	R/$\overline{W}$
1	OP	04	1
2	OP + 1	—	1
3	FFFF	—	1

MUL

Multiply Unsigned

MUL

Operation: ACCD ◀ (ACCA) × (ACCB)

Description: Multiplies the 8-bit unsigned binary value in accumulator A by the 8-bit unsigned binary value in accumulator B to obtain a 16-bit unsigned result in the double accumulator D. Unsigned multiply allows multiple-precision operations. The carry flag allows rounding the most significant byte of the result through the sequence: MUL, ADCA #0.

Condition Codes and Boolean Formulae:

S	X	H	I	N	Z	V	C
—	—	—	—	—	—	—	⇕

C   R7
    Set if bit 7 of the result (ACCB bit 7) is set; cleared otherwise.

Source Form: MUL

Addressing Modes, Machine Code, and Cycle-by-Cycle Execution:

Cycle	MUL (INH)		
	Addr	Data	R/$\overline{W}$
1	OP	3D	1
2	OP + 1	—	1
3–10	FFFF	—	1

NEG   Negate   NEG

Operation: $(ACCX) \blacklozenge - (ACCX) = \$00 - (ACCX)$ **or:** $(M) \blacklozenge - (M) = \$00 - (M)$

Description: Replaces the contents of ACCX or M with its twos complement; the value $80 is left unchanged.

Condition Codes and Boolean Formulae:

S	X	H	I	N	Z	V	C
—	—	—	—	↕	↕	↕	↕

N R7
   Set if MSB of result is set; cleared otherwise.

Z $\overline{R7} \cdot \overline{R6} \cdot \overline{R5} \cdot \overline{R4} \cdot \overline{R3} \cdot \overline{R2} \cdot \overline{R1} \cdot \overline{R0}$
   Set if result is $00; cleared otherwise.

V $R7 \cdot \overline{R6} \cdot \overline{R5} \cdot \overline{R4} \cdot \overline{R3} \cdot \overline{R2} \cdot \overline{R1} \cdot \overline{R0}$
   Set if there is a twos complement overflow from the implied subtraction from zero; cleared otherwise. A twos complement overflow will occur if and only if the contents of ACCX or M is $80.

C $R7 + R6 + R5 + R4 + R3 + R2 + R1 + R0$
   Set if there is a borrow in the implied subtraction from zero; cleared otherwise. The C bit will be set in all cases except when the contents of ACCX or M is $00.

Source Forms: NEGA; NEGB; NEG (opr)

Addressing Modes, Machine Code, and Cycle-by-Cycle Execution:

Cycle	NEGA (INH)			NEGB (INH)			NEG (EXT)			NEG (IND, X)			NEG (IND, Y)		
	Addr	Data	R/W̄	Addr	Data	R/W̄	Addr	Data	R/W̄	Addr	Data	R/W̄	Addr	Data	R/W̄
1	OP	40	1	OP	50	1	OP	70	1	OP	60	1	OP	18	1
2	OP+1	—	1	OP+1	—	1	OP+1	hh	1	OP+1	ff	1	OP+1	60	1
3							OP+2	ll	1	FFFF	—	1	OP+2	ff	1
4							hhll	(hhll)	1	X+ff	(X+ff)	1	FFFF	—	1
5							FFFF	—	1	FFFF	—	1	Y+ff	(Y+ff)	1
6							hhll	result	0	X+ff	result	0	FFFF	—	1
7													Y+ff	result	0

NOP   No Operation   NOP

Description: This is a single-byte instruction that causes only the program counter to be incremented. No other registers are affected. This instruction is typically used to produce a time delay although some software disciplines discourage CPU frequency-based time delays. During debug, NOP instructions are sometimes used to temporarily replace other machine code instructions, thus disabling the replaced instruction(s).

Condition Codes and Boolean Formulae:

S	X	H	I	N	Z	V	C
—	—	—	—	—	—	—	—

None affected

Source Form: NOP

Addressing Modes, Machine Code, and Cycle-by-Cycle Execution:

Cycle	NOP (INH)		
	Addr	Data	R/W̄
1	OP	01	1
2	OP + 1	—	1

ORA   Inclusive-OR   **ORA**

Operation: ACCX ◖ (ACCX) + (M)

Description: Performs the logical inclusive-OR between the contents of ACCX and the contents of M and places the result in ACCX. (Each bit of ACCX after the operation will be the logical inclusive-OR of the corresponding bits of M and of ACCX before the operation.)

Condition Codes and Boolean Formulae:

S	X	H	I	N	Z	V	C
—	—	—	—	↕	↕	0	—

N R7
   Set if MSB of result is set; cleared otherwise.

Z $\overline{R7} \cdot \overline{R6} \cdot \overline{R5} \cdot \overline{R4} \cdot \overline{R3} \cdot \overline{R2} \cdot \overline{R1} \cdot \overline{R0}$
   Set if result is $00; cleared otherwise

V 0
   Cleared

Source Forms: ORAA (opr); ORAB (opr)

Addressing Modes, Machine Code, and Cycle-by-Cycle Execution:

Cycle	ORAA (IMM)			ORAA (DIR)			ORAA (EXT)			ORAA (IND, X)			ORAA (IND, Y)		
	Addr	Data	R/W̄	Addr	Data	R/W̄	Addr	Data	R/W̄	Addr	Data	R/W̄	Addr	Data	R/W̄
1	OP	8A	1	OP	9A	1	OP	BA	1	OP	AA	1	OP	18	1
2	OP+1	ii	1	OP+1	dd	1	OP+1	hh	1	OP+1	ff	1	OP+1	AA	1
3				00dd	(00dd)	1	OP+2	ll	1	FFFF	—	1	OP+2	ff	1
4							hhll	(hhll)	1	X+ff	(X+ff)	1	FFFF	—	1
5													Y+ff	(Y+ff)	1

Cycle	ORAB (IMM)			ORAB (DIR)			ORAB (EXT)			ORAB (IND, X)			ORAB (IND, Y)		
	Addr	Data	R/W̄	Addr	Data	R/W̄	Addr	Data	R/W̄	Addr	Data	R/W̄	Addr	Data	R/W̄
1	OP	CA	1	OP	DA	1	OP	FA	1	OP	EA	1	OP	18	1
2	OP+1	ii	1	OP+1	dd	1	OP+1	hh	1	OP+1	ff	1	OP+1	EA	1
3				00dd	(00dd)	1	OP+2	ll	1	FFFF	—	1	OP+2	ff	1
4							hhll	(hhll)	1	X+ff	(X+ff)	1	FFFF	—	1
5													Y+ff	(Y+ff)	1

PSH Push Data onto Stack PSH

Operation: ⇒ACCX, SP ⇐ (SP) − $0001

Description: The contents of ACCX are stored on the stack at the address contained in the stack pointer. The stack pointer is then decremented.

Push instructions are commonly used to save the contents of one or more CPU registers at the start of a subroutine. Just before returning from the subroutine, corresponding pull instructions are used to restore the saved CPU registers so the subroutine will appear not to have affected these registers.

Condition Codes and Boolean Formulae:

S	X	H	I	N	Z	V	C
—	—	—	—	—	—	—	—

None affected

Source Forms: PSHA; PSHB

Addressing Modes, Machine Code, and Cycle-by-Cycle Execution:

Cycle	PSHA (INH)			PSHB (INH)		
	Addr	Data	R/$\overline{W}$	Addr	Data	R/$\overline{W}$
1	OP	36	1	OP	37	1
2	OP+1	—	1	OP+1	—	1
3	SP	(A)	0	SP	(B)	0

PSHX   Push Index Register X onto Stack   PSHX

Operation: ➡(IXL), SP ⬍ (SP) – $0001
➡(IXH), SP ⬍ (SP) – $0001

Description: The contents of the index register X are pushed onto the stack (low-order byte first) at the address contained in the stack pointer. The stack pointer is then decremented by two.

Push instructions are commonly used to save the contents of one or more CPU registers at the start of a subroutine. Just before returning from the subroutine, corresponding pull instructions are used to restore the saved CPU registers so the subroutine will appear not to have affected these registers.

Condition Codes and Boolean Formulae:

S	X	H	I	N	Z	V	C
—	—	—	—	—	—	—	—

None affected

Source Form: PSHX

Addressing Modes, Machine Code, and Cycle-by-Cycle Execution:

Cycle	PSHX (INH)		
	Addr	Data	R/W̄
1	OP	3C	1
2	OP + 1	—	1
3	SP	(IXL)	0
4	SP – 1	(IXH)	0

PSHY

Push Index Register Y onto Stack

PSHY

Operation: →(IYL), SP ↓ (SP) − $0001
→(IYH), SP ↓ (SP) − $0001

Description: The contents of the index register Y are pushed onto the stack (low-order byte first) at the address contained in the stack pointer. The stack pointer is then decremented by two.

Push instructions are commonly used to save the contents of one or more CPU registers at the start of a subroutine. Just before returning from the subroutine, corresponding pull instructions are used to restore the saved CPU registers so the subroutine will appear not to have affected these registers.

Condition Codes and Boolean Formulae:

S	X	H	I	N	Z	V	C
—	—	—	—	—	—	—	—

None affected

Source Form: PSHY

Addressing Modes, Machine Code, and Cycle-by-Cycle Execution:

Cycle	PSHY (INH)		
	Addr	Data	R/W̅
1	OP	18	1
2	OP + 1	3C	1
3	OP + 2	—	1
4	SP	(IYL)	0
5	SP − 1	(IYH)	0

PUL   Pull Data from Stack   PUL

Operation: SP ⬆ (SP) + $0001, ⬅(ACCX)

Description: The stack pointer is incremented. The ACCX is then loaded from the stack at the address contained in the stack pointer.

Push instructions are commonly used to save the contents of one or more CPU registers at the start of a subroutine. Just before returning from the subroutine, corresponding pull instructions are used to restore the saved CPU registers so the subroutine will appear not to have affected these registers.

Condition Codes and Boolean Formulae:

S	X	H	I	N	Z	V	C
—	—	—	—	—	—	—	—

None affected

Source Forms: PULA; PULB

Addressing Modes, Machine Code, and Cycle-by-Cycle Execution:

Cycle	PULA (INH)			PULB (INH)		
	Addr	Data	R/W̄	Addr	Data	R/W̄
1	OP	32	1	OP	33	1
2	OP + 1	—	1	OP + 1	—	1
3	SP	—	1	SP	—	1
4	SP + 1	get A	1	SP + 1	get B	1

PULX

Pull Index Register X from Stack

PULX

Operation: SP ◆ (SP) + $0001; ▲(IXH)
SP ◆ (SP) + $0001; ▲(IXL)

Description: The index register X is pulled from the stack (high-order byte first), beginning at the address contained in the stack pointer plus one. The stack pointer is incremented by two in total.

Push instructions are commonly used to save the contents of one or more CPU registers at the start of a subroutine. Just before returning from the subroutine, corresponding pull instructions are used to restore the saved CPU registers so the subroutine will appear not to have affected these registers.

Condition Codes and Boolean Formulae:

S	X	H	I	N	Z	V	C
—	—	—	—	—	—	—	—

None affected

Source Form: PULX

Addressing Modes, Machine Code, and Cycle-by-Cycle Execution:

Cycle	PULX (INH)		
	Addr	Data	R/W̄
1	OP	38	1
2	OP + 1	—	1
3	SP	—	1
4	SP + 1	get IXH	1
5	SP + 2	get IXL	1

PULY
Pull Index Register Y from Stack
PULY

Operation: SP ◀ (SP) + $0001; ◀(IYH)
SP ◀ (SP) + $0001; ◀(IYL)

Description: The index register Y is pulled from the stack (high-order byte first) beginning at the address contained in the stack pointer plus one. The stack pointer is
incremented by two in total.

Push instructions are commonly used to save the contents of one or more CPU registers at the start of a subroutine. Just before returning from the subroutine, corresponding pull instructions are used to restore the saved CPU registers so the subroutine
will appear not to have affected these registers.

Condition Codes and Boolean Formulae:

S	X	H	I	N	Z	V	C
—	—	—	—	—	—	—	—

None affected

Source Form: PULY

Addressing Modes, Machine Code, and Cycle-by-Cycle Execution:

Cycle	PULY (INH)		
	Addr	Data	R/$\overline{W}$
1	OP	18	1
2	OP + 1	38	1
3	OP + 2	—	1
4	SP	—	1
5	SP + 1	get IYH	1
6	SP + 2	get IYL	1

ROL   **Rotate Left**   # ROL

Operation:

Description: Shifts all bits of ACCX or M one place to the left. Bit 0 is loaded from the C bit. The C bit is loaded from the most significant bit of ACCX or M. The rotate operations include the carry bit to allow extension of the shift and rotate operations to multiple bytes. For example, to shift a 24-bit vaue left one bit, the sequence ASL LOW, ROL MID, ROL HIGH could be used where LOW, MID, and HIGH refer to the low-order, middle, and high-order bytes of the 24-bit value, respectively.

Condition Codes and Boolean Formulae:

S	X	H	I	N	Z	V	C
—	—	—	—	↕	↕	↕	↕

N   R7
Set if MSB of result is set; cleared otherwise.

Z $\overline{R7} \cdot \overline{R6} \cdot \overline{R5} \cdot \overline{R4} \cdot \overline{R3} \cdot \overline{R2} \cdot \overline{R1} \cdot \overline{R0}$
Set if result is $00; cleared otherwise.

V $N \oplus C = [N \cdot \overline{C}] + [\overline{N} \cdot C]$ (for N and C after the rotate)
Set if (N is set and C is clear) or (N is clear and C is set); cleared otherwise (for values of N and C after the rotate).

C   M7
Set if, before the rotate, the MSB of ACCX or M was set; cleared otherwise.

Source Forms: ROLA; ROLB; ROL (opr)

Addressing Modes, Machine Code, and Cycle-by-Cycle Execution:

Cycle	ROLA (INH)			ROLB (INH)			ROL (EXT)			ROL (IND, X)			ROL (IND, Y)		
	Addr	Data	R/$\overline{W}$	Addr	Data	R/$\overline{W}$	Addr	Data	R/$\overline{W}$	Addr	Data	R/$\overline{W}$	Addr	Data	R/$\overline{W}$
1	OP	49	1	OP	59	1	OP	79	1	OP	69	1	OP	18	1
2	OP+1	—	1	OP+1	—	1	OP+1	hh	1	OP+1	ff	1	OP+1	69	1
3							OP+2	ll	1	FFFF	—	1	OP+2	ff	1
4							hhll	(hhll)	1	X+ff	(X+ff)	1	FFFF	—	1
5							FFFF	—	1	FFFF	—	1	Y+ff	(Y+ff)	1
6							hhll	result	0	X+ff	result	0	FFFF	—	1
7													Y+ff	result	0

ROR   Rotate Right   ROR

Operation:

$$C \rightarrow \boxed{b7 - - - - - - b0} \rightarrow \boxed{C}$$

Description: Shift all bits of ACCX or M one place to the right. Bit 7 is loaded from the C bit. The C bit is loaded from the least significant bit of ACCX or M. The rotate operations include the carry bit to allow extension of the shift and rotate operations to multiple bytes. For example, to shift a 24-bit value right one bit, the sequence LSR HIGH, ROR MID, ROR LOW could be used where LOW, MID, and HIGH refer to the low-order, middle, and high-order bytes of the 24-bit value, respectively. The first LSR could be replaced by ASR to maintain the original value of the sign bit (MSB of high-order byte) of the 24-bit value.

Condition Codes and Boolean Formulae:

S	X	H	I	N	Z	V	C
—	—	—	—	↕	↕	↕	↕

N   R7
   Set if MSB of result is set; cleared otherwise.

Z $\overline{R7} \cdot \overline{R6} \cdot \overline{R5} \cdot \overline{R4} \cdot \overline{R3} \cdot \overline{R2} \cdot \overline{R1} \cdot \overline{R0}$
   Set if result is $00; cleared otherwise.

V $N \oplus C = [N \cdot \overline{C}] + [\overline{N} \cdot C]$ (for N and C after the rotate)
   Set if (N is set and C is clear) or (N is clear and C is set); cleared otherwise (for values of N and C after the rotate).

C   M0
   Set if, before the rotate, the LSB of ACCX or M was set; cleared otherwise.

Source Forms: RORA; RORB; ROR (opr)

Addressing Modes, Machine Code, and Cycle-by-Cycle Execution:

Cycle	RORA (INH)			RORB (INH)			ROR (EXT)			ROR (IND, X)			ROR (IND, Y)		
	Addr	Data	R/$\overline{W}$	Addr	Data	R/$\overline{W}$	Addr	Data	R/$\overline{W}$	Addr	Data	R/$\overline{W}$	Addr	Data	R/$\overline{W}$
1	OP	46	1	OP	56	1	OP	76	1	OP	66	1	OP	18	1
2	OP+1	—	1	OP+1	—	1	OP+1	hh	1	OP+1	ff	1	OP+1	66	1
3							OP+2	ll	1	FFFF	—	1	OP+2	ff	1
4							hhll	(hhll)	1	X+ff	(X+ff)	1	FFFF	—	1
5							FFFF	—	1	FFFF	—	1	Y+ff	(Y+ff)	1
6							hhll	result	0	X+ff	result	0	FFFF	—	1
7													Y+ff	result	0

RTI   Return from Interrupt   RTI

Operation:

$$SP \blacktriangleleft (SP) + \$0001, \blacktriangleleft (CCR)$$
$$SP \blacktriangleleft (SP) + \$0001, \blacktriangleleft (ACCB)$$
$$SP \blacktriangleleft (SP) + \$0001, \blacktriangleleft (ACCA)$$
$$SP \blacktriangleleft (SP) + \$0001, \blacktriangleleft (IXH)$$
$$SP \blacktriangleleft (SP) + \$0001, \blacktriangleleft (IXL)$$
$$SP \blacktriangleleft (SP) + \$0001, \blacktriangleleft (IYH)$$
$$SP \blacktriangleleft (SP) + \$0001, \blacktriangleleft (IYL)$$
$$SP \blacktriangleleft (SP) + \$0001, \blacktriangleleft (PCH)$$
$$SP \blacktriangleleft (SP) + \$0001, \blacktriangleleft (PCL)$$

Description: The condition code, accumulators B and A, index registers X and Y, and the program counter will be restored to a state pulled from the stack. The X bit in the CCR may be cleared as a result of an RTI instruction but may not be set if it was cleared prior to execution of the RTI instruction.

Condition Codes and Boolean Formulae:

S	X	H	I	N	Z	V	C
↕	↝	↕	↕	↕	↕	↕	↕

Condition code bits take on the value of the corresponding bit of the unstacked CCR except that the X bit may not change from a zero to a one. Software can leave X set, leave X clear, or change X from one to zero. The XIRQ interrupt mask can only become set as a $\overline{RESET}$ of a reset or recognition of an XIRQ interrupt.

Source Form: RTI

Addressing Modes, Machine Code, and Cycle-by-Cycle Execution:

Cycle	RTI (INH)		
	Addr	Data	R/$\overline{W}$
1	OP	3B	1
2	OP+1	—	1
3	SP	—	1
4	SP+1	get CC	1
5	SP+2	get B	1
6	SP+3	get A	1
7	SP+4	get IXH	1
8	SP+5	get IXL	1
9	SP+6	get IXH	1
10	SP+7	get IXL	1
11	SP+8	Rtn hi	1
12	SP+9	Rtn lo	1

RTS   **Return from Subroutine**   RTS

Operation: SP ◆ (SP) + $0001, ◀(PCH)
SP ◆ (SP) + $0001, ◀(PCL)

Description: The stack pointer is incremented by one. The contents of the byte of memory, at the address now contained in the stack pointer, are loaded into the high-order
eight bits of the program counter. The stack pointer is again incremented by one. The
contents of the byte of memory, at the address now contained in the stack pointer,
are loaded into the low-order eight bits of the program counter.

Condition Codes and Boolean Formulae:

S	X	H	I	N	Z	V	C
—	—	—	—	—	—	—	—

None affected

Source Form: RTS

Addressing Modes, Machine Code, and Cycle-by-Cycle Execution:

Cycle	RTS (INH)		
	Addr	Data	R/$\overline{W}$
1	OP	39	1
2	OP+1	—	1
3	SP	—	1
4	SP+1	Rtn hi	1
5	SP+2	Rtn lo	1

SBA

Subtract Accumulators

SBA

Operation: ACCA ◀ (ACCA) − (ACCB)

Description: Subtracts the contents of ACCB from the contents of ACCA and places the result in ACCA. The contents of ACCB are not affected. For subtract instructions, the C bit in the CCR represents a borrow.

Condition Codes and Boolean Formulae:

S	X	H	I	N	Z	V	C
—	—	—	—	$\updownarrow$	$\updownarrow$	$\updownarrow$	$\updownarrow$

N R7
   Set if MSB of result is set; cleared otherwise.

Z $\overline{R7} \cdot \overline{R6} \cdot \overline{R5} \cdot \overline{R4} \cdot \overline{R3} \cdot \overline{R2} \cdot \overline{R1} \cdot \overline{R0}$
   Set if result is $00; cleared otherwise.

V $A7 \cdot \overline{B7} \cdot \overline{R7} + \overline{A7} \cdot B7 \cdot R7$
   Set if a twos complement overflow resulted from the operation; cleared otherwise.

C $\overline{A7} \cdot B7 + B7 \cdot R7 + R7 \cdot \overline{A7}$
   Set if the absolute value of ACCB is larger than the absolute value of ACCA; cleared otherwise.

Source Form: SBA

Addressing Modes, Machine Code, and Cycle-by-Cycle Execution:

Cycle	SBA (INH)		
	Addr	Data	R/$\overline{\text{W}}$
1	OP	10	1
2	OP+1	—	1

SBC

Subtract with Carry

SBC

Operation: ACCX ⬅ (ACCX) − (M) − (C)

Description: Subtracts the contents of M and the contents of C from the contents of ACCX and places the result in ACCX. For subtract instructions the C bit in the CCR represents a borrow.

Condition Codes and Boolean Formulae:

S	X	H	I	N	Z	V	C
—	—	—	—	⬍	⬍	⬍	⬍

N R7
   Set if MSB of result is set; cleared otherwise.

Z $\overline{R7} \cdot \overline{R6} \cdot \overline{R5} \cdot \overline{R4} \cdot \overline{R3} \cdot \overline{R2} \cdot \overline{R1} \cdot \overline{R0}$
   Set if result is $00; cleared otherwise.

V $X7 \cdot \overline{M7} \cdot \overline{R7} + \overline{X7} \cdot M7 \cdot R7$
   Set if a twos complement overflow resulted from the operation; cleared otherwise.

C $\overline{X7} \cdot M7 + M7 \cdot R7 + R7 \cdot \overline{X7}$
   Set if the absolute value of the contents of memory plus previous carry is larger than the absolute value of the accumulator; cleared otherwise.

Source Forms: SBCA (opr); SBCB (opr)

Addressing Modes, Machine Code, and Cycle-by-Cycle Execution:

Cycle	SBCA (IMM)			SBCA (DIR)			SBCA (EXT)			SBCA (IND, X)			SBCA (IND, Y)		
	Addr	Data	R/W̄	Addr	Data	R/W̄	Addr	Data	R/W̄	Addr	Data	R/W̄	Addr	Data	R/W̄
1	OP	82	1	OP	92	1	OP	B2	1	OP	A2	1	OP	18	1
2	OP+1	ii	1	OP+1	dd	1	OP+1	hh	1	OP+1	ff	1	OP+1	A2	1
3				00dd	(00dd)	1	OP+2	ll	1	FFFF	—	1	OP+2	ff	1
4							hhll	(hhll)	1	X+ff	(X+ff)	1	FFFF	—	1
5													Y+ff	(Y+ff)	1

Cycle	SBCB (IMM)			SBCB (DIR)			SBCB (EXT)			SBCB (IND, X)			SBCB (IND, Y)		
	Addr	Data	R/W̄	Addr	Data	R/W̄	Addr	Data	R/W̄	Addr	Data	R/W̄	Addr	Data	R/W̄
1	OP	C2	1	OP	D2	1	OP	F2	1	OP	E2	1	OP	18	1
2	OP+1	ii	1	OP+1	dd	1	OP+1	hh	1	OP+1	ff	1	OP+1	E2	1
3				00dd	(00dd)	1	OP+2	ll	1	FFFF	—	1	OP+2	ff	1
4							hhll	(hhll)	1	X+ff	(X+ff)	1	FFFF	—	1
5													Y+ff	(Y+ff)	1

SEC

Set Carry

SEC

Operation: C bit ◀ 1

Description: Sets the C bit in the CCR.

Condition Codes and Boolean Formulae:

S	X	H	I	N	Z	V	C
—	—	—	—	—	—	—	1

C 1
   Set

Source Form: SEC

Addressing Modes, Machine Code, and Cycle-by-Cycle Execution:

Cycle	SEC (INH)		
	Addr	Data	R/W̄
1	OP	0D	1
2	OP+1	—	1

SEI   Set Interrupt Mask   SEI

Operation: I bit ⬥ 1

Description: Sets the interrupt mask bit in the CCR. When the I bit is set, all maskable interrupts are inhibited, and the MPU will recognize only non-maskable interrupt sources or an SWI.

Condition Codes and Boolean Formulae:

S	X	H	I	N	Z	V	C
—	—	—	1	—	—	—	—

I   1
    Set

Source Form: SEI

Addressing Modes, Machine Code, and Cycle-by-Cycle Execution:

Cycle	SEI (INH)		
	Addr	Data	R/W̄
1	OP	0F	1
2	OP+1	—	1

SEV

Set Twos Complement Overflow Bit

SEV

Operation: V bit ◀ 1

Description: Sets the twos complement overflow bit in the CCR.

Condition Codes and Boolean Formulae:

S	X	H	I	N	Z	V	C
—	—	—	—	—	—	1	—

V 1
   Set

Source Form: SEV

Addressing Modes, Machine Code, and Cycle-by-Cycle Execution:

Cycle	SEV (INH)		
	Addr	Data	R/$\overline{W}$
1	OP	0B	1
2	OP+1	—	1

STA

Store Accumulator

STA

Operation: M ◆ (ACCX)

Description: Stores the contents of ACCX in memory. The contents of ACCX remains unchanged.

Condition Codes and Boolean Formulae:

S	X	H	I	N	Z	V	C
—	—	—	—	✿	✿	0	—

N   X7
   Set if MSB of result is set; cleared otherwise.

Z $\overline{X7} \cdot \overline{X6} \cdot \overline{X5} \cdot \overline{X4} \cdot \overline{X3} \cdot \overline{X2} \cdot \overline{X1} \cdot \overline{X0}$
   Set if result is $00; cleared otherwise.

V   0
   Cleared

Source Forms: STAA (opr); STAB (opr)

Addressing Modes, Machine Code, and Cycle-by-Cycle Execution:

Cycle	STAA (DIR)			STAA (EXT)			STAA (IND, X)			STAA (IND, Y)		
	Addr	Data	R/W̄	Addr	Data	R/W̄	Addr	Data	R/W̄	Addr	Data	R/W̄
1	OP	97	1	OP	B7	1	OP	A7	1	OP	18	1
2	OP+1	dd	1	OP+1	hh	1	OP+1	ff	1	OP+1	A7	1
3	00dd	(A)	0	OP+2	ll	1	FFFF	—	1	OP+2	ff	1
4				hhll	(A)	0	X+ff	(A)	0	FFFF	—	1
5										Y+ff	(A)	0

Cycle	STAB (DIR)			STAB (EXT)			STAB (IND, X)			STAB (IND, Y)		
	Addr	Data	R/W̄	Addr	Data	R/W̄	Addr	Data	R/W̄	Addr	Data	R/W̄
1	OP	D7	1	OP	F7	1	OP	E7	1	OP	18	1
2	OP+1	dd	1	OP+1	hh	1	OP+1	ff	1	OP+1	E7	1
3	00dd	(B)	0	OP+2	ll	1	FFFF	—	1	OP+2	ff	1
4				hhll	(B)	0	X+ff	(B)	0	FFFF	—	1
5										Y+ff	(B)	0

STD   Store Double Accumulator   STD

Operation: M:M + 1 ◀ (ACCD); M ◀ (ACCA), M + 1 ◀ (ACCB)

Description: Stores the contents of double accumulator ACCD in memory. The contents of ACCD remain unchanged.

Condition Codes and Boolean Formulae:

S	X	H	I	N	Z	V	C
—	—	—	—	✸	✸	0	—

N D15
   Set if MSB of result is set; cleared otherwise.

Z $\overline{D15} \cdot \overline{D14} \cdot \overline{D13} \cdot \overline{D12} \cdot \overline{D11} \cdot \overline{D10} \cdot \overline{D9} \cdot \overline{D8} \cdot \overline{D7} \cdot \overline{D6} \cdot \overline{D5} \cdot \overline{D4} \cdot \overline{D3} \cdot \overline{D2} \cdot \overline{D1} \cdot \overline{D0}$
   Set if result is $0000; cleared otherwise.

V 0
   Cleared

Source Form: STD (opr)

Addressing Modes, Machine Code, and Cycle-by-Cycle Execution:

Cycle	STD (DIR)			STD (EXT)			STD (IND, X)			STD (IND, Y)		
	Addr	Data	R/$\overline{W}$	Addr	Data	R/$\overline{W}$	Addr	Data	R/$\overline{W}$	Addr	Data	R/$\overline{W}$
1	OP	DD	1	OP	FD	1	OP	ED	1	OP	18	1
2	OP + 1	dd	1	OP + 1	hh	1	OP + 1	ff	1	OP + 1	ED	1
3	00dd	(A)	0	OP + 2	ll	1	FFFF	—	1	OP + 2	ff	1
4	00dd + 1	(B)	0	hhll	(A)	0	X + ff	(A)	0	FFFF	—	1
5				hhll + 1	(B)	0	X + ff + 1	(B)	0	Y + ff	(A)	0
6										Y + ff + 1	(B)	0

STOP

Stop Processing

STOP

Description: If the S bit in the CCR is set, then the STOP instruction is disabled and operates like the NOP instruction. If the S bit in the CCR is clear, the STOP instruction causes all system clocks to halt, and the system is placed in a minimum-power standby mode. All CPU registers remain unchanged. I/O pins also remain unaffected.

Recovery from STOP may be accomplished by $\overline{\text{RESET}}$, $\overline{\text{XIRQ}}$, or an unmasked $\overline{\text{IRQ}}$. When recovering from STOP with $\overline{\text{XIRQ}}$, if the X bit in the CCR is clear, execution will resume with the stacking operations for the $\overline{\text{XIRQ}}$ interrupt. If the X bit in the CCR is set, masking $\overline{\text{XIRQ}}$ interrupts, execution will resume with the opcode fetch for the instruction which follows the STOP instruction (continue).

An error in some mask sets of the M68HC11 caused incorrect recover from STOP under very specific unusual conditions. If the opcode of the instruction before the STOP instruction came from column 4 or 5 of the opcode map, the STOP instruction was incorrectly interpreted as a two-byte instruction. A simple way to avoid this potential problem is to put a NOP instruction (which is a column 0 opcode) immediately before any STOP instruction.

Condition Codes and Boolean Formulae:

S	X	H	I	N	Z	V	C
—	—	—	—	—	—	—	—

None affected

Source Form: STOP

Addressing Modes, Machine Code, and Cycle-by-Cycle Execution:

Cycle	STOP (INH)		
	Addr	Data	R/$\overline{\text{W}}$
1	OP	CF	1
2	OP + 1	—	1

STS

Store Stack Pointer

STS

Operation: M ◀ (SPH), M + 1 ◀ (SPL)

Description: Stores the most significant byte of the stack pointer in memory at the address specified by the program and stores the least significant byte of the stack pointer at the next location in memory, at one plus the address specified by the program.

Condition Codes and Boolean Formulae:

S	X	H	I	N	Z	V	C
—	—	—	—	↕	↕	0	—

N SP15
   Set if MSB of result is set; cleared otherwise.

Z $\overline{SP15} \cdot \overline{SP14} \cdot \overline{SP13} \cdot \overline{SP12} \cdot \overline{SP11} \cdot \overline{SP10} \cdot \overline{SP9} \cdot \overline{SP8} \cdot$
   $\overline{SP7} \cdot \overline{SP6} \cdot \overline{SP5} \cdot \overline{SP4} \cdot \overline{SP3} \cdot \overline{SP2} \cdot \overline{SP1} \cdot \overline{SP0}$
   Set if result is $0000; cleared otherwise.

V 0
   Cleared

Source Form: STS (opr)

Addressing Modes, Machine Code, and Cycle-by-Cycle Execution:

Cycle	STS (DIR)			STS (EXT)			STS (IND, X)			STS (IND, Y)		
	Addr	Data	R/W̄	Addr	Data	R/W̄	Addr	Data	R/W̄	Addr	Data	R/W̄
1	OP	9F	1	OP	BF	1	OP	AF	1	OP	18	1
2	OP + 1	dd	1	OP + 1	hh	1	OP + 1	ff	1	OP + 1	AF	1
3	00dd	(SPH)	0	OP + 2	ll	1	FFFF	—	1	OP + 2	ff	1
4	oodd + 1	(SPL)	0	hhll	(SPH)	0	X + ff	(SPH)	0	FFFF	—	1
5				hhll + 1	(SPL)	0	X + ff + 1	(SPL)	0	Y + ff	(SPH)	0
6										Y + ff + 1	(SPL)	0

STX Store Index Register X STX

Operation: M ⧫ (IXH), M+1 ⧫ (IXL)

Description: Stores the most significant byte of index register X in memory at the address specified by the program and stores the least significant byte of index register X at the next location in memory, at one plus the address specified by the program.

Condition Codes and Boolean Formulae:

S	X	H	I	N	Z	V	C
—	—	—	—	⧫	⧫	0	—

N IX15
Set if MSB of result is set; cleared otherwise.

Z $\overline{IX15} \cdot \overline{IX14} \cdot \overline{IX13} \cdot \overline{IX12} \cdot \overline{IX11} \cdot \overline{IX10} \cdot \overline{IX9} \cdot \overline{IX8} \cdot \overline{IX7} \cdot \overline{IX6} \cdot \overline{IX5} \cdot \overline{IX4} \cdot \overline{IX3} \cdot \overline{IX2} \cdot \overline{IX1} \cdot \overline{IX0}$
Set if result is $0000; cleared otherwise.

V 0
Cleared

Source Form: STX (opr)

Addressing Modes, Machine Code, and Cycle-by-Cycle Execution:

Cycle	STX (DIR)			STX (EXT)			STX (IND, X)			STX (IND, Y)		
	Addr	Data	R/$\overline{W}$	Addr	Data	R/$\overline{W}$	Addr	Data	R/$\overline{W}$	Addr	Data	R/$\overline{W}$
1	OP	DF	1	OP	FF	1	OP	EF	1	OP	CD	1
2	OP+1	dd	1	OP+1	hh	1	OP+1	ff	1	OP+1	EF	1
3	00dd	(IXH)	0	OP+2	ll	1	FFFF	—	1	OP+2	ff	1
4	oodd+1	(IXL)	0	hhll	(IXH)	0	X+ff	(IXH)	0	FFFF	—	1
5				hhll+1	(IXL)	0	X+ff+1	(IXL)	0	Y+ff	(IXH)	0
6										Y+ff+1	(IXL)	0

STY

Store Index Register Y

STY

Operation: M $\Leftarrow$ (IYH), M + 1 $\Leftarrow$ (IYL)

Description: Stores the most significant byte of index register Y in memory at the address specified by the program and stores the least significant byte of index register Y at the next location in memory, at one plus the address specified by the program.

Condition Codes and Boolean Formulae:

S	X	H	I	N	Z	V	C
—	—	—	—	$\updownarrow$	$\updownarrow$	0	—

N   IY15
   Set if MSB of result is set; cleared otherwise.

Z $\overline{IY15} \cdot \overline{IY14} \cdot \overline{IY13} \cdot \overline{IY12} \cdot \overline{IY11} \cdot \overline{IY10} \cdot \overline{IY9} \cdot \overline{IY8} \cdot$
   $\overline{IY7} \cdot \overline{IY6} \cdot \overline{IY5} \cdot \overline{IY4} \cdot \overline{IY3} \cdot \overline{IY2} \cdot \overline{IY1} \cdot \overline{IY0}$
   Set if result is \$0000; cleared otherwise.

V   0
   Cleared

Source Form: STY (opr)

Addressing Modes, Machine Code, and Cycle-by-Cycle Execution:

Cycle	STY (DIR)			STY (EXT)			STY (IND, X)			STY (IND, Y)		
	Addr	Data	R/$\overline{W}$	Addr	Data	R/$\overline{W}$	Addr	Data	R/$\overline{W}$	Addr	Data	R/$\overline{W}$
1	OP	18	1	OP	18	1	OP	1A	1	OP	18	1
2	OP + 1	DF	1	OP + 1	FF	1	OP + 1	EF	1	OP + 1	EF	1
3	OP + 2	dd	1	OP + 2	hh	1	OP + 2	ff	1	OP + 2	ff	1
4	00dd	(IYH)	0	OP + 3	ll	1	FFFF	—	1	FFFF	—	1
5	00dd + 1	(IYL)	0	hhll	(IYH)	0	X + ff	(IYH)	0	Y + ff	(IYH)	0
6				hhll + 1	(IYL)	0	X + ff + 1	(IYL)	0	Y + ff + 1	(IYL)	0

SUB

Subtract

SUB

Operation: ACCX ◄ (ACCX) – (M)

Description: Subtracts the contents of M from the contents of ACCX and places the result in ACCX. For subtract instructions, the C bit in the CCR represents a borrow.

Condition Codes and Boolean Formulae:

S	X	H	I	N	Z	V	C
—	—	—	—	↕	↕	↕	↕

N R7
Set if MSB of result is set; cleared otherwise.

Z $\overline{R7} \cdot \overline{R6} \cdot \overline{R5} \cdot \overline{R4} \cdot \overline{R3} \cdot \overline{R2} \cdot \overline{R1} \cdot \overline{R0}$
Set if result is $00; cleared otherwise.

V $X7 \cdot \overline{M7} \cdot \overline{R7} + \overline{X7} \cdot M7 \cdot R7$
Set if a twos complement overflow resulted from the operation; cleared otherwise.

C $\overline{X7} \cdot M7 + M7 \cdot R7 + R7 \cdot \overline{X7}$
Set if the absolute value of the contents of memory are larger than the absolute value of the contents of the accumulator; cleared otherwise.

Source Forms: SUBA (opr); SUBB (opr)

Addressing Modes, Machine Code, and Cycle-by-Cycle Execution:

Cycle	SUBA (IMM) Addr	Data	R/$\overline{W}$	SUBA (DIR) Addr	Data	R/$\overline{W}$	SUBA (EXT) Addr	Data	R/$\overline{W}$	SUBA (IND, X) Addr	Data	R/$\overline{W}$	SUBA (IND, Y) Addr	Data	R/$\overline{W}$
1	OP	80	1	OP	90	1	OP	B0	1	OP	A0	1	OP	18	1
2	OP+1	ii	1	OP+1	dd	1	OP+1	hh	1	OP+1	ff	1	OP+1	A0	1
3				00dd	(00dd)	1	OP+2	ll	1	FFFF	—	1	OP+2	ff	1
4							hhll	(hhll)	1	X+ff	(X+ff)	1	FFFF	—	1
5													Y+ff	(Y+ff)	1

Cycle	SUBB (IMM) Addr	Data	R/$\overline{W}$	SUBB (DIR) Addr	Data	R/$\overline{W}$	SUBB (EXT) Addr	Data	R/$\overline{W}$	SUBB (IND, X) Addr	Data	R/$\overline{W}$	SUBB (IND, Y) Addr	Data	R/$\overline{W}$
1	OP	C0	1	OP	D0	1	OP	F0	1	OP	E0	1	OP	18	1
2	OP+1	ii	1	OP+1	dd	1	OP+1	hh	1	OP+1	ff	1	OP+1	E0	1
3				00dd	(00dd)	1	OP+2	ll	1	FFFF	—	1	OP+2	ff	1
4							hhll	(hhll)	1	X+ff	(X+ff)	1	FFFF	—	1
5													Y+ff	(Y+ff)	1

SUBD   Subtract Double Accumulator   SUBD

Operation: ACCD ◀ (ACCD) − (M:M + 1)

Description: Subtracts the contents of M:M + 1 from the contents of double accumulator D and places the result in ACCD. For subtract instructions, the C bit in the CCR represents a borrow.

Condition Codes and Boolean Formulae:

S	X	H	I	N	Z	V	C
—	—	—	—	↕	↕	↕	↕

N   R15
Set if MSB of result is set; cleared otherwise.

Z $\overline{R15} \cdot \overline{R14} \cdot \overline{R13} \cdot \overline{R12} \cdot \overline{R11} \cdot \overline{R10} \cdot \overline{R9} \cdot \overline{R8} \cdot \overline{R7} \cdot \overline{R6} \cdot \overline{R5} \cdot \overline{R4} \cdot \overline{R3} \cdot \overline{R2} \cdot \overline{R1} \cdot \overline{R0}$
Set if result is $0000; cleared otherwise.

V $D15 \cdot \overline{M15} \cdot \overline{R15} + \overline{D15} \cdot M15 \cdot R15$
Set if a twos complement overflow resulted from the operation; cleared otherwise.

C $\overline{D15} \cdot M15 + M15 \cdot R15 + R15 \cdot \overline{D15}$
Set if the absolute value of the contents of memory is larger than the absolute value of the accumulator; cleared otherwise.

Source Form: SUBD (opr)

Addressing Modes, Machine Code, and Cycle-by-Cycle Execution:

Cycle	SUBD (IMM) Addr	Data	R/$\overline{W}$	SUBD (DIR) Addr	Data	R/$\overline{W}$	SUBD (EXT) Addr	Data	R/$\overline{W}$	SUBD (IND, X) Addr	Data	R/$\overline{W}$	SUBD (IND, Y) Addr	Data	R/$\overline{W}$
1	OP	83	1	OP	93	1	OP	B3	1	OP	A3	1	OP	18	1
2	OP + 1	jj	1	OP + 1	dd	1	OP + 1	hh	1	OP + 1	ff	1	OP + 1	A3	1
3	OP + 2	kk	1	00dd	(00dd)	1	OP + 2	ll	1	FFFF	—	1	OP + 2	ff	1
4	FFFF	—	1	00dd + 1	(00dd + 1)	1	hhll	(hhll)	1	X + ff	(X + ff)	1	FFFF	—	1
5				FFFF	—	1	hhll + 1	(hhll + 1)	1	X + ff + 1	(X + ff + 1)	1	Y + ff	(Y + ff)	1
6							FFFF	—	1	FFFF	—	1	Y + ff + 1	(Y + ff + 1)	1
7													FFFF	—	1

SWI   Software Interrupt   SWI

Operation: PC ◆ (PC) + $0001
◆ (PCL), SP ◆ (SP) − $0001
◆ (PCH), SP ◆ (SP) − $0001
◆ (IYL), SP ◆ (SP) − $0001
◆ (IYH), SP ◆ (SP) − $0001
◆ (IXL), SP ◆ (SP) − $0001
◆ (IXH), SP ◆ (SP) − $0001
◆ (ACCA), SP ◆ (SP) − $0001
◆ (ACCB), SP ◆ (SP) − $0001
◆ (CCR), SP ◆ (SP) − $0001
I ◆ 1, PC ◆ (SWI vector)

Description: The program counter is incremented by one. The program counter, index registers Y and X, and accumulators A and B are pushed onto the stack. The CCR is then pushed onto the stack. The stack pointer is decremented by one after each byte of data is stored on the stack. The I bit in the CCR is then set. The program counter is loaded with the address stored at the SWI vector, and instruction execution resumes at this location. This instruction is not maskable by the I bit.

Condition Codes and Boolean Formulae:

S	X	H	I	N	Z	V	C
—	—	—	1	—	—	—	—

I   1
   Set

Source Form: SWI

Addressing Modes, Machine Code, and Cycle-by-Cycle Execution:

Cycle	SWI (INH)		
	Addr	Data	R/W̄
1	OP	3F	1
2	OP + 1	—	1
3	SP	Rtn lo	0
4	SP − 1	Rtn hi	0
5	SP − 2	(IYL)	0
6	SP − 3	(IYH)	0
7	SP − 4	(IXL)	0
8	SP − 5	(IXH)	0
9	SP − 6	(A)	0
10	SP − 7	(B)	0
11	SP − 8	(CCR)	0
12	SP − 8	(CCR)	1
13	Vec hi	Svc hi	1
14	Vec lo	Svc lo	1

TAB Transfer from Accumulator A to Accumulator B TAB

Operation: ACCB ⬥ (ACCA)

Description: Moves the contents of ACCA to ACCB. The former contents of ACCB are lost; the contents of ACCA are not affected.

Condition Codes and Boolean Formulae:

S	X	H	I	N	Z	V	C
—	—	—	—	⬥	⬥	0	—

N R7
    Set if MSB of result is set; cleared otherwise.

Z $\overline{R7} \cdot \overline{R6} \cdot \overline{R5} \cdot \overline{R4} \cdot \overline{R3} \cdot \overline{R2} \cdot \overline{R1} \cdot \overline{R0}$
    Set if result is $00; cleared otherwise

V 0
    Cleared

Source Form: TAB

Addressing Modes, Machine Code, and Cycle-by-Cycle Execution:

Cycle	TAB (INH)		
	Addr	Data	R/$\overline{W}$
1	OP	16	1
2	OP+1	—	1

TAP

Transfer from Accumulator A to Condition Code Register

TAP

Operation: CCR ◀ (ACCA)

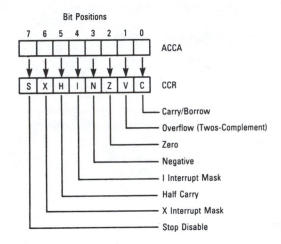

Description: Transfers the contents of bit positions 7–0 of accumulator A to the corresponding bit positions of the CCR. The contents of accumulator A remain unchanged. The X bit in the CCR may be cleared as a result of a TAP instruction but may not be set if it was clear prior to execution of the TAP instruction.

Condition Codes and Boolean Formulae:

S	X	H	I	N	Z	V	C
↕	↤	↕	↕	↕	↕	↕	↕

Condition code bits take on the value of the corresponding bit of accumulator A except that the X bit may not change from a zero to a one. Software can leave X set, leave X clear or change X from one to zero. The $\overline{XIRQ}$ interrupt mask can only become set as a result of a $\overline{RESET}$ or recognition of an $\overline{XIRQ}$ interrupt.

Source Form: TAP

Addressing Modes, Machine Code, and Cycle-by-Cycle Execution:

Cycle	TAP (INH)		
	Addr	Data	R/$\overline{W}$
1	OP	06	1
2	OP + 1	—	1

TBA   Transfer from Accumulator B to Accumulator A   TBA

Operation: ACCA ◀ (ACCB)

Description: Moves the contents of ACCB to ACCA. The former contents of ACCA are lost; the contents of ACCB are not affected.

Condition Codes and Boolean Formulae:

S	X	H	I	N	Z	V	C
—	—	—	—	↕	↕	0	—

N   R7
   Set if MSB of result is set; cleared otherwise.

Z $\overline{R7} \cdot \overline{R6} \cdot \overline{R5} \cdot \overline{R4} \cdot \overline{R3} \cdot \overline{R2} \cdot \overline{R1} \cdot \overline{R0}$
   Set if result is $00; cleared otherwise.

V   0
   Cleared

Source Form: TBA

Addressing Modes, Machine Code, and Cycle-by-Cycle Execution:

Cycle	TBA (INH)		
	Addr	Data	R/W̄
1	OP	17	1
2	OP+1	—	1

TEST

Test Operation
(Test Mode Only)

TEST

Description: This is a single-byte instruction that causes the program counter to be continuously incremented. It can only be executed while in the test mode. The MPU must be reset to exit this instruction. Code execution is suspended during this instruction. This is an illegal opcode when not in test mode.

Condition Codes and Boolean Formulae:

S	X	H	I	N	Z	V	C
—	—	—	—	—	—	—	—

None affected

Source Form: TEST

Addressing Modes, Machine Code, and Cycle-by-Cycle Execution:

Cycle	TEST (INH)		
	Addr	Data	R/W̄
1	OP	00	1
2	OP + 1	—	1
3	OP + 2	—	1
4	OP + 3	—	1
5 – n	PREV – 1	(PREV – 1)	1

TPA

Transfer from Condition Code Register to Accumulator A

TPA

Operation: (ACCA) ◀ (CCR)

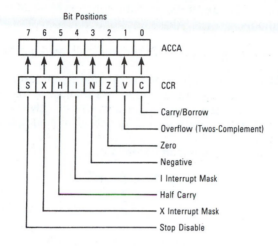

Description: Transfers the contents of the CCR to corresponding bit positions of accumulator A. The CCR remains unchanged.

Condition Codes and Boolean Formulae:

S	X	H	I	N	Z	V	C
—	—	—	—	—	—	—	—

None affected

Source Form: TPA

Addressing Modes, Machine Code, and Cycle-by-Cycle Execution:

Cycle	TPA (INH)		
	Addr	Data	R/$\overline{W}$
1	OP	07	1
2	OP+1	—	1

TST   Test   TST

Operation: (ACCX) – $00 **or:**   (M) – $00

Description: Subtracts $00 from the contents of ACCX or M and sets the condition codes accordingly.

The subtraction is accomplished internally without modifying either ACCX or M.

The TST instruction provides only minimum information when testing unsigned values. Since no unsigned value is less than zero, BLO and BLS have no utility. While BHI could be used after TST, it provides exactly the same control as BNE, which is preferred. After testing signed values, all signed branches are available.

Condition Codes and Boolean Formulae:

S	X	H	I	N	Z	V	C
—	—	—	—	↕	↕	0	0

N M7
   Set if MSB of result is set; cleared otherwise.

Z $\overline{M7} \cdot \overline{M6} \cdot \overline{M5} \cdot \overline{M4} \cdot \overline{M3} \cdot \overline{M2} \cdot \overline{M1} \cdot \overline{M0}$
   Set if result is $00; cleared otherwise

V 0
   Cleared

C 0
   Cleared

Source Forms: TSTA; TSTB; TST (opr).

Addressing Modes, Machine Code, and Cycle-by-Cycle Execution:

Cycle	TSTA (INH)			TSTB (INH)			TST (EXT)			TST (IND, X)			TST (IND, Y)		
	Addr	Data	R/W̄	Addr	Data	R/W̄	Addr	Data	R/W̄	Addr	Data	R/W̄	Addr	Data	R/W̄
1	OP	4D	1	OP	5D	1	OP	7D	1	OP	6D	1	OP	18	1
2	OP+1	—	1	OP+1	—	1	OP+1	hh	1	OP+1	ff	1	OP+1	6D	1
3							OP+2	ll	1	FFFF	—	1	OP+2	ff	1
4							hhll	(hhll)	1	X+ff	(X+ff)	1	FFFF	—	1
5							FFFF	—	1	FFFF	—	1	Y+ff	(Y+ff)	1
6							FFFF	—	1	FFFF	—	1	FFFF	—	1
7													FFFF	—	1

TSX

Transfer from Stack Pointer to Index Register X

TSX

Operation: IX ◀ (SP) + $0001

Description: Loads the index register X with one plus the contents of the stack pointer. The contents of the stack pointer remain unchanged. After a TSX instruction the index register X points at the last value that was stored on the stack.

Condition Codes and Boolean Formulae:

S	X	H	I	N	Z	V	C
—	—	—	—	—	—	—	—

None affected

Source Form: TSX

Addressing Modes, Machine Code, and Cycle-by-Cycle Execution:

Cycle	TSX (INH)		
	Addr	Data	R/W̄
1	OP	30	1
2	OP+1	—	1
3	SP	—	1

TSY   Transfer from Stack Pointer to Index Register Y   TSY

Operation: IY ◄ (SP) + $0001

Description: Loads the index register Y with one plus the contents of the stack pointer. The contents of the stack pointer remain unchanged. After a TSY instruction the index register Y points at the last value that was stored on the stack.

Condition Codes and Boolean Formulae:

S	X	H	I	N	Z	V	C
—	—	—	—	—	—	—	—

None affected

Source Form: TSY

Addressing Modes, Machine Code, and Cycle-by-Cycle Execution:

Cycle	TSY (INH)		
	Addr	Data	R/W̄
1	OP	18	1
2	OP+1	30	1
3	OP+2	—	1
4	SP	—	1

TXS

Transfer from Index Register X to Stack Pointer

TXS

Operation: SP ◂ (IX) − $0001

Description: Loads the stack pointer with the contents of the index register X minus one. The contents of the index register X remain unchanged.

Condition Codes and Boolean Formulae:

S	X	H	I	N	Z	V	C
—	—	—	—	—	—	—	—

None affected

Source Form: TXS

Addressing Modes, Machine Code, and Cycle-by-Cycle Execution:

Cycle	TXS (INH)		
	Addr	Data	R/W̄
1	OP	35	1
2	OP+1	—	1
3	FFFF	—	1

TYS   Transfer from Index Register Y to Stack Pointer   TYS

Operation; $SP \Leftarrow (IY) - \$0001$

Description: Loads the stack pointer with the contents of the index register Y minus one. The contents of the index register Y remain unchanged.

Condition Codes and Boolean Formulae:

S	X	H	I	N	Z	V	C
—	—	—	—	—	—	—	—

None affected

Source Form: TYS

Addressing Modes, Machine Code, and Cycle-by-Cycle Execution:

Cycle	TYS (INH)		
	Addr	Data	R/W̄
1	OP	18	1
2	OP + 1	35	1
3	OP + 2	—	1
4	FFFF	—	1

Reprinted with permission of Motorola

WAI   Wait for Interrupt   WAI

Operation;   PC ← (PC) + \$0001
⬇(PCL), SP ← (SP) − \$0001
⬇(PCH), SP ← (SP) − \$0001
⬇(IYL), SP ← (SP) − \$0001
⬇(IYH), SP ← (SP) − \$0001
⬇(IXL), SP ← (SP) − \$0001
⬇(IXH), SP ← (SP) − \$0001
⬇(ACCA), SP ← (SP) − \$0001
⬇(ACCB), SP ← (SP) − \$0001
⬇(CCR), SP ← (SP) − \$0001

Description: The program counter is incremented by one. The program counter, index registers Y and X, and accumulators A and B are pushed onto the stack. The CCR is then pushed onto the stack. The stack pointer is decremented by one after each byte of data is stored on the stack.

The MPU then enters a wait state for an integer number of MPU E-clock cycles. While in the wait state, the address/data bus repeatedly runs read bus cycles to the address where the CCR contents were stacked. The MPU leaves the wait state when it senses any interrupt that has not been masked.

Upon leaving the wait state, the MPU sets the I bit in the CCR, fetches the vector (address) corresponding to the interrupt sensed, and instruction execution is resumed at this location.

Condition Codes and Boolean Formulae:

S	X	H	I	N	Z	V	C
—	—	—	—	—	—	—	—

Although the WAI instruction itself does not alter the condition code bits, the interrupt which causes the MCU to resume processing causes the I bit (and the X bit if the interrupt was XIRQ) to be set as the interrupt vector is being fetched.

WAI

Wait for Interrupt

WAI

Source Form: WAI

Addressing Modes, Machine Code, and Cycle-by-Cycle Execution:

Cycle	WAI (INH)		
	Addr	Data	R/$\overline{W}$
1	OP	3E	1
2	OP+1	—	1
3	SP	Rtn lo	0
4	SP−1	Rtn hi	0
5	SP−2	(IYL)	0
6	SP−3	(IYH)	0
7	SP−4	(IXL)	0
8	SP−5	(IXH)	0
9	SP−6	(A)	0
10	SP−7	(B)	0
11	SP−8	(CCR)	0
12 to 12+n	SP−8	(CCR)	1
13+n	Vec hi	Svc hi	1
14+n	Vec lo	Svc lo	1

XGDX

**Exchange Double Accumulator and
Index Register X**

XGDX

Operation: (IX) ⬌ (ACCD)

Description: Exchanges the contents of double accumulator ACCD and the contents of index register X. A common use for XGDX is to move an index value into the double accumulator to allow 16-bit arithmetic calculations on the index value before exchanging the updated index value back into the X index register.

Condition Codes and Boolean Formulae:

S	X	H	I	N	Z	V	C
—	—	—	—	—	—	—	—

None affected

Source Form: XGDX

Addressing Modes, Machine Code, and Cycle-by-Cycle Execution:

Cycle	XGDX (INH)		
	Addr	Data	R/W̄
1	OP	8F	1
2	OP+1	—	1
3	FFFF	—	1

XGDY

**Exchange Double Accumulator and
Index Register Y**

XGDY

Operation: (IY) ◆▶ (ACCD)

Description: Exchanges the contents of double accumulator ACCD and the contents of index register Y. A common use for XGDY is to move an index value into the double accumulator to allow 16-bit arithmetic calculations on the index value before exchanging the updated index value back into the Y index register.

Condition Codes and Boolean Formulae:

S	X	H.	I	N	Z	V	C
—	—	—	—	—	—	—	—

None affected

Source Form: XGDY

Addressing Modes, Machine Code, and Cycle-by-Cycle Execution:

Cycle	XGDY (INH)		
	Addr	Data	R/$\overline{W}$
1	OP	18	1
2	OP+1	8F	1
3	OP+2	—	1
4	FFFF	—	1

Appendix I.2

The MC68HC68T1 Real-Time Clock with Serial Interface

Motorola Semiconductor Technical Data

MC68HC68T1

Real-Time Clock plus RAM with Serial Interface
CMOS

The MC68HC68T1 HCMOS Clock/RAM peripheral contains a real–time clock/calendar, a 32 x 8 static RAM, and a synchronous, serial, three–wire interface for communication with a microcontroller or processor. Operating in a burst mode, successive Clock/RAM locations can be read or written using only a single starting address. An on–chip oscillator allows acceptance of a selectable crystal frequency or the device can be programmed to accept a 50/60 Hz line input frequency.

The LINE and system voltage (V_{SYS}) pins give the MC68HC68T1 the capability for sensing power–up/power–down conditions, a capability useful for battery–backup systems. The device has an interrupt output capable of signaling a microcontroller or processor of an alarm, periodic interrupt, or power sense condition. An alarm can be set for comparison with the seconds, minutes, and hours registers. This alarm can be used in conjunction with the power supply enable (PSE) output to initiate a system power–up sequence if the V_{SYS} pin is powered to the proper level.

A software power–down sequence can be initiated by setting a bit in the interrupt control register. This applies a reset to the CPU via the CPUR pin, sets the clock out (CLKOUT) and PSE pins low, and disables the serial interface. This condition is held until a rising edge is sensed on the V_{SYS} input pin, signaling system power coming on, or by activation of a previously enabled interrupt if the V_{SYS} pin is powered up.

A watchdog circuit can be enabled that requires the microcontroller or processor to toggle the slave select (SS) pin of the MC68HC68T1 periodically without performing a serial transfer. If this condition is not met, the CPUR line resets the CPU.

- Full Clock Features — Seconds, Minutes, Hours (AM/PM), Day–of–Week, Date, Month, Year (0 – 99), Auto Leap Year
- 32–Byte General Purpose RAM
- Direct Interface to Motorola SPI and National MICROWIRE™ Serial Data Ports
- Minimum Timekeeping Voltage: 2.2 V
- Burst Mode for Reading/Writing Successive Addresses in Clock/RAM
- Selectable Crystal or 50/60 Hz Line Input Frequency
- Clock Registers Utilize BCD Data
- Buffered Clock Output for Driving CPU Clock, Timer, Colon, or LCD Backplane
- Power–On Reset with First Time–Up Bit
- Freeze Circuit Eliminates Software Overhead During a Clock Read
- Three Independent Interrupt Modes — Alarm, Periodic, or Power–Down
- CPU Reset Output — Provides Orderly Power–Up/Power–Down
- Watchdog Circuit
- Pin–for–Pin Replacement for CDP68HC68T1
- Chip Complexity: 8500 FETs or 2125 Equivalent Gates
- Also See Application Notes ANE425 "Use of the MC68HC68T1 RTC with M6805 Microprocessor", AN457 "Providing a Real–Time Clock for the MC68302", and AN1065 "Use of the MC68HC68T1 Real–Time Clock with Multiple Time Bases"

MICROWIRE is a trademark of National Semiconductor Inc.

REV 2
2/96

P SUFFIX
PLASTIC DIP
CASE 648

DW SUFFIX
SOG PACKAGE
CASE 751G

ORDERING INFORMATION

MC68HC68T1P	Plastic DIP
MC68HC68T1DW	SOG Package

PIN ASSIGNMENT

CLKOUT	1	16	V_{DD}
CPUR	2	15	$XTAL_{out}$
INT	3	14	$XTAL_{in}$
SCK	4	13	V_{BATT}
MOSI	5	12	V_{SYS}
MISO	6	11	LINE
SS	7	10	POR
V_{SS}	8	9	PSE

BLOCK DIAGRAM

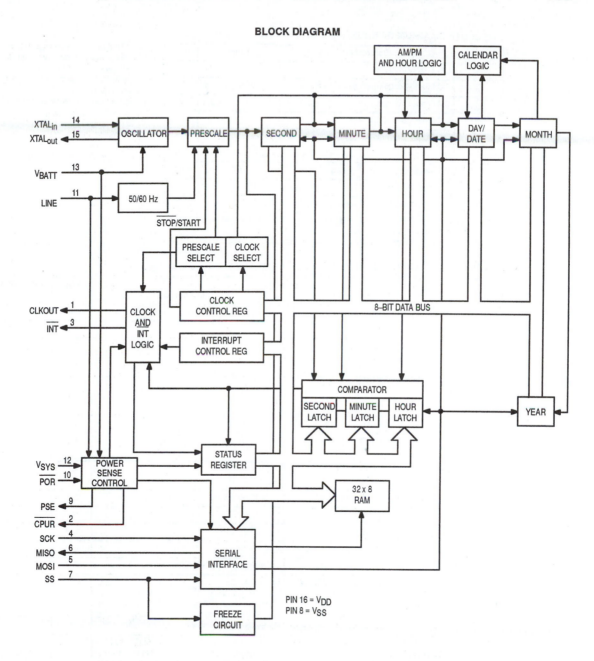

ABSOLUTE MAXIMUM RATINGS* (Voltages Referenced to V_{SS})

Symbol	Parameter	Value	Unit
V_{DD}	DC Supply Voltage	− 0.5 to + 7.0	V
V_{in}	DC Input Voltage (except Line Input**)	− 0.5 to V_{DD} + 0.5	V
V_{out}	DC Output Voltage	− 0.5 to V_{DD} + 0.5	V
I_{in}	DC Input Current, per Pin	± 10	mA
I_{out}	DC Output Current, per Pin	± 10	mA
I_{DD}	DC Supply Current, V_{DD} and V_{SS} Pins	± 30	mA
P_D	Power Dissipation, per Package***	500	mW
T_{stg}	Storage Temperature	− 65 to + 150	°C
T_L	Lead Temperature (10–Second Soldering)	260	°C

*Maximum Ratings are those values beyond which damage to the device may occur.
**See Electrical Characteristics Table.
***Power Dissipation Temperature Derating: — 12 mW/°C from 65 to 85°C.

This device contains circuitry to protect the inputs against damage due to high static voltages or electric fields. However, precautions must be taken to avoid applications of any voltage higher than maximum rated voltages to this high–impedance circuit. For proper operation, V_{in} and V_{out} should be constrained to the range $V_{SS} \leq (V_{in}$ or $V_{out}) \leq V_{DD}$.

Unused inputs must always be tied to an appropriate logic voltage level (e.g., either V_{SS} or V_{DD}). Unused outputs must be left open.

ELECTRICAL CHARACTERISTICS (T_A = − 40 to + 85°C, Voltages Referenced to V_{SS})

Symbol	Parameter	Test Condition	V_{DD} V	Guaranteed Limit	Unit
V_{DD}	Power Supply Voltage Range		—	3.0 to 6.0	V
$V_{(stdby)}$	Minimum Standby (Timekeeping) Voltage*		—	2.2	V
V_{IL}	Maximum Low–Level Input Voltage		3.0 4.5 6.0	0.9 1.35 1.8	V
V_{IH}	Minimum High–Level Input Voltage		3.0 4.5 6.0	2.1 3.15 4.2	V
V_{in}	Maximum Input Voltage, Line Input	Power Sense Mode	5.0	12	V p–p
V_{OL}	Maximum Low–Level Output Voltage	I_{out} = 0 µA I_{out} = 1.6 mA	4.5	0.1 0.4	V
V_{OH}	Minimum High–Level Output Voltage	I_{out} = 0 µA I_{out} = 1.6 mA	4.5	4.4 3.7	V
I_{in}	Maximum Input Current, Except SS	V_{in} = V_{DD} or V_{SS}	6.0	± 1	µA
I_{IL}	Maximum Low–Level Input Current, SS	V_{in} = V_{SS}	6.0	− 1.0	µA
I_{IH}	Maximum Pull–Down Current, SS	V_{in} = V_{DD}	6.0	100	µA
I_{OZ}	Maximum Three–State Leakage Current	V_{out} = V_{DD} or V_{SS}	6.0	± 10	µA
I_{DD}	Maximum Quiescent Supply Current	V_{in} = V_{DD} or V_{SS}, All Input; I_{out} = 0 µA	6.0	50	µA
I_{DD}	Maximum RMS Operating Supply Current Crystal Operation	I_{out} = 0 µA, V_{in} = V_{DD} or V_{SS}, all inputs except XTAL$_{in}$, Clock Out Disabled, No Serial Access Cycles	fXTAL$_{in}$ = 32 kHz fXTAL$_{in}$ = 1 MHz fXTAL$_{in}$ = 2 MHz fXTAL$_{in}$ = 4 MHz → 5.0	0.1 0.6 0.84 1.2	mA
	Maximum RMS Operating Supply Current External Frequency Source Driving XTAL$_{in}$, XTAL$_{out}$ Open	I_{out} = 0 µA, V_{in} = V_{DD} or V_{SS}, Clock Out Disabled, No Serial Access Cycles	fXTAL$_{in}$ = 32 kHz fXTAL$_{in}$ = 1 MHz fXTAL$_{in}$ = 2 MHz fXTAL$_{in}$ = 4 MHz → 5.0	0.024 0.12 0.24 0.5	
I_{batt}	Maximum RMS Standby Current Crystal Operation	V_{BATT} = 3.0 V, V_{SYS} = 0.0 V, V_{DD} = 0.0 V, I_{out} = 0 µA, V_{in} = Don't Care, all inputs except XTAL$_{in}$, Clock Out Disabled, No Serial Access Cycles	fXTAL$_{in}$ = 32 kHz fXTAL$_{in}$ = 1 MHz fXTAL$_{in}$ = 2 MHz fXTAL$_{in}$ = 4 MHz → 0.0	25 250 360 600	µA

*Timekeeping function only, no read/write accesses. Data in the registers and RAM retained.

AC ELECTRICAL CHARACTERISTICS ($T_A = -40$ to $+85°C$, $C_L = 200$ pF, Input $t_r = t_f = 6$ ns, Voltages Referenced to V_{SS})

Symbol	Parameter	Figure No.	V_{DD} V	Guaranteed Limit	Unit
f_{SCK}	Maximum Clock Frequency (Refer to SCK t_w, below)	1, 2, 3	3.0 4.5 6.0	— 2.1 2.1	MHz
t_{PLH}, t_{PHL}	Maximum Propagation Delay, SCK to MISO	2, 3	3.0 4.5 6.0	200 100 100	ns
t_{PLZ}, t_{PHZ}	Maximum Propagation Delay, SS to MISO	2, 4	3.0 4.5 6.0	200 100 100	ns
t_{PZL}, t_{PZH}	Maximum Propagation Delay, SCK to MISO	2, 4	3.0 4.5 6.0	200 100 100	ns
t_{TLH}, t_{THL}	Maximum Output Transition Time, Any Output (Measured Between 70% V_{DD} and 20% V_{DD})	2, 3	3.0 4.5 6.0	200 100 100	ns
C_{in}	Maximum Input Capacitance		—	10	pF

TIMING REQUIREMENTS ($T_A = -40$ to $+85°C$, Input $t_r = t_f = 6$ ns, Voltages Referenced to V_{SS})

Symbol	Parameter	Figure No.	V_{DD} V	Guaranteed Limit	Unit
t_{su}	Minimum Setup Time, SS to SCK	1, 2	3.0 4.5 6.0	200 100 100	ns
t_{su}	Minimum Setup Time, MOSI to SCK	1, 2	3.0 4.5 6.0	200 100 100	ns
t_h	Minimum Hold Time, SCK to SS	1, 2	3.0 4.5 6.0	250 125 125	ns
t_h	Minimum Hold Time, SCK to MOSI	1, 2	3.0 4.5 6.0	200 100 100	ns
t_{rec}	Minimum Recovery Time, SCK	1, 2	3.0 4.5 6.0	200 200 200	ns
$t_{w(H)}$, $t_{w(L)}$	Minimum Pulse Width, SCK	1, 2	3.0 4.5 6.0	400 200 200	ns
t_w	Minimum Pulse Width, POR		3.0 4.5 6.0	— 100 100	ns
t_r, t_f	Maximum Input Rise and Fall Times (Except $XTAL_{in}$ and POR) (Measured Between 70% V_{DD} and 20% V_{DD})	1, 2	3.0 4.5 6.0	— 2 2	µs

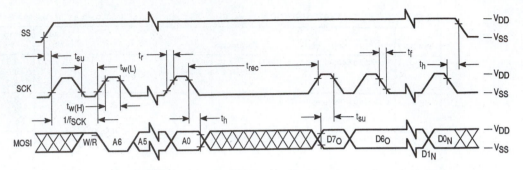

NOTE: Measurement points are V_{IL} and V_{IH} unless otherwise noted on the **AC Electrical Characteristics** table.

Figure 1. Write Cycle

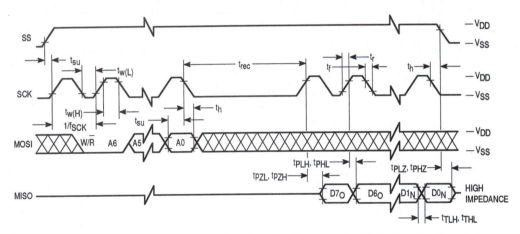

NOTE: Measurement points are V_{OL}, V_{OH}, V_{IL}, and V_{IH} unless otherwise noted on the **AC Electrical Characteristics** table.

Figure 2. Read Cycle

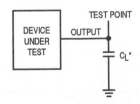

* Includes all probe and fixture capacitance.

Figure 3. Test Circuit

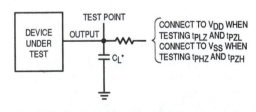

* Includes all probe and fixture capacitance.

Figure 4. Test Circuit

Reprinted with permission of Motorola

OPERATING CHARACTERISTICS

The real–time clock consists of a clock/calendar and a 32 x 8 RAM (see Figure 5). Communication with the device may be established via a serial peripheral interface (SPI) or MICROWIRE bus. In addition to the clock/calendar data from seconds to years, and systems flexibility provided by the 32–byte RAM, the clock features computer handshaking with an interrupt output and a separate square–wave clock output that can be one of seven different frequencies. An alarm circuit is available that compares the alarm latches with the seconds, minutes, and hours time counters and activates the interrupt output when they are equal. The clock is specifically designed to aid in power–up/power–down applications and offers several pins to aid the designer of battery–backup systems.

CLOCK/CALENDAR

The clock/calendar portion of this device consists of a long string of counters that is toggled by a 1 Hz input. The 1 Hz input is derived from the on–chip oscillator that utilizes one of four possible external crystals or that can be driven by an external frequency source. The 1 Hz trigger to the counters can also be supplied by a 50 or 60 Hz source that is connected to the LINE input pin.

The time counters offer seconds, minutes, and hours data in 12– or 24–hour format. An AM/PM indicator is available that once set, toggles at 12:00 AM and 12:00 PM. The calendar counters consist of day of week, date of month, month, and year information. Data in the counters is in BCD format. The hours counter utilizes BCD for hours data plus bits for 12/24 hour and AM/PM modes. The seven time counters are read serially at addresses $20 through $26. The time counters are written to at addresses $A0 through $A6. (See Figures 5 and 6 and Table 1.)

32 x 8 GENERAL–PURPOSE RAM

The real–time clock also has a static 32 x 8 RAM. The RAM is read at addresses $00 through $1F and written to at addresses $80 through $9F (see Figure 5).

ALARM

The alarm is set by accessing the three alarm latches and loading the desired data. (See **Serial Peripheral Interface**.) The alarm latches consist of seconds, minutes, and hours registers. When their outputs equal the values of the seconds, minutes, and hours time counters, an interrupt is generated. The interrupt output goes low if the alarm bit in the status register is set and the interrupt output is activated after an alarm time is sensed (see **Pin Descriptions, INT Pin**). To preclude a false interrupt when loading the time counters, the alarm interrupt bit in the interrupt control register should be reset. This procedure is not required when the alarm time is being loaded.

WATCHDOG FUNCTION

When Watchdog (bit 7) in the interrupt control register is set high, the clock's slave select pin must be toggled at regular intervals without a serial data transfer. If SS is not toggled at the rate shown in Table 2, the MC68HC68T1 supplies a

CPU reset pulse at Pin 2 and Watchdog (bit 6) in the status register is set (see Figure 7). Typical service and reset times are shown in Table 2.

CLOCK OUT

The value in the three least significant bits of the clock control register selects one of seven possible output frequencies. (See **Clock Control Register**.) This square–wave signal is available at the CLKOUT pin. When the power–down operation is initialized, the output is reset low.

CONTROL REGISTER AND STATUS REGISTER

The operation of the real–time clock is controlled by the clock control and interrupt control registers, which are read/write registers. Another register, the status register, is available to indicate the operating conditions. The status register is a read–only register, and a read operation resets status bits.

MODE SELECT

The voltage level that is present at the V_{SYS} input pin at the end of power–on reset selects the device to be in the single–supply mode or battery–backup mode.

Single–Supply Mode

If V_{SYS} is powered up when power–on reset is completed; CLKOUT, PSE, and CPUR are enabled high and the device is completely operational. CPUR is asserted low if the voltage level at the V_{SYS} pin subsequently falls below V_{BATT} + 0.7 V. If CLKOUT, PSE, and CPUR are reset low due to a power–down instruction, V_{SYS} brought low and then powered high re–enables these outputs.

An example of the single–supply mode is where only one supply is available and V_{DD}, V_{BATT}, and V_{SYS} are tied together to the supply.

Battery–Backup Mode

If V_{SYS} is not powered up (V_{SYS} = 0 V) at the end of power–on reset, CLKOUT, PSE, CPUR, and SS are disabled (CLKOUT, PSE, and CPUR low). This condition is held until V_{SYS} rises to a threshold (approximately 0.7 V) above V_{BATT}. CLKOUT, PSE, and CPUR are then enabled and the device is operational. If V_{SYS} falls below a threshold above V_{BATT}, the outputs CLKOUT, PSE, and CPUR are reset low.

An example of battery–backup operation occurs if V_{SYS} is tied to the 5 V supply and is not receiving voltage from a supply. A rechargeable battery is connected to the V_{BATT} pin, causing a POR while V_{SYS} = 0 V. The device retains data and keeps time down to a minimum V_{BATT} voltage of 2.2 V.

The power consumption may not settle to the specified limit until main power is cycled once.

POWER CONTROL

Power control is composed of two operations, power–sense and power–down/power–up. Two pins are involved in power sensing, the LINE input pin and the INT output pin. Two additional pins, PSE and V_{SYS}, are utilized during power–down/power–up operation.

FREEZE FUNCTION

The freeze function prevents an increment of the time counters, if any of the registers are being read. Also, alarm operation is delayed if the registers are being read. This causes the clock to lose time with increasing rates of acceleration.

POWER SENSING

When power sensing is enabled (Power Sense Bit in the interrupt control register), ac/dc transitions are sensed at the LINE input pin. Threshold detectors determine when transitions cease. After a delay of 2.68 to 4.64 ms plus the external input RC circuit time constant, an interrupt true bit is set high in the status register. This bit can then be sampled to see if system power has turned back on (see Figure 8).

The power–sense circuitry operates by sensing the level of the voltage present at the LINE input pin. This voltage is centered around V_{DD}, and as long as the voltage is either plus or minus a threshold (approximately 0.7 V) from V_{DD}, a power sense failure is not indicated. With an ac signal present, remaining in this V_{DD} window longer than a maximum of 4.64 ms activates the power–sense circuit. The larger the amplitude of the signal, the less likely a power failure would be detected. A 50 or 60 Hz, 10 V p–p sine–wave voltage is an acceptable signal to present at the LINE input pin to set up the power–sense function. When ac power fails, an internal circuit pulls the voltage at the line pin within the detection window.

Power–Down

Power–down is a processor–directed operation. The power–down bit is set in the interrupt control register to initiate power–down operation. During power–down, the power supply enable (PSE) output, normally high, is driven low. The CLKOUT pin is driven low. The CPUR output, connected to the processor reset input pin, is also driven low. In addition, the serial interface (MOSI and MISO) is disabled (see Figure 9).

Power–Up

There are four methods that can initiate the power–up mode. Two of the methods require an interrupt to the microcontroller or processor by programming the interrupt control register. The interrupts can be generated by the alarm circuit by setting the alarm bit and the appropriate alarm registers. Also, an interrupt can be generated by programming the periodic interrupt bits in the interrupt control register. V_{SYS} must be at 5 volts for this operation to occur.

The third method is by initiating the power sense circuit with the power sense bit in the interrupt control register set to sense power loss along with the V_{SYS} pin to sense subsequent power–up condition (see Figure 10). (Reference Figure 19 for application circuit for third method.)

The fourth method that initiates power–up occurs when the level on the V_{SYS} pin rises 0.7 V above the level of the V_{BATT} pin, after previously falling to the level of V_{BATT} while in the battery–backup mode. An interrupt is not generated when the fourth method is utilized.

While in the single–supply mode, power–up is initiated when the V_{SYS} pin loses power and then returns high. There is no interrupt generated when using this method (see Figure 11).

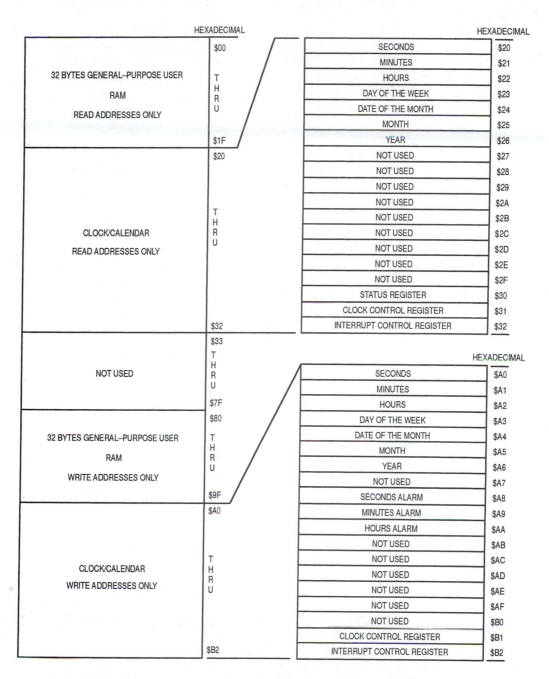

Figure 5. Address Map

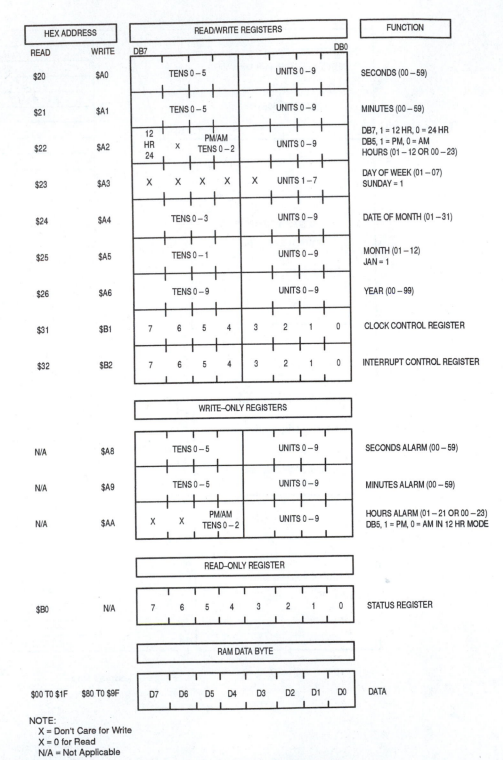

HEX ADDRESS		READ/WRITE REGISTERS		FUNCTION
READ	WRITE	DB7	DB0	
$20	$A0	TENS 0 – 5	UNITS 0 – 9	SECONDS (00 – 59)
$21	$A1	TENS 0 – 5	UNITS 0 – 9	MINUTES (00 – 59)
$22	$A2	12 HR 24 / X / PM/AM TENS 0 – 2	UNITS 0 – 9	DB7, 1 = 12 HR, 0 = 24 HR / DB5, 1 = PM, 0 = AM / HOURS (01 – 12 OR 00 – 23)
$23	$A3	X X X X X UNITS 1 – 7		DAY OF WEEK (01 – 07) / SUNDAY = 1
$24	$A4	TENS 0 – 3	UNITS 0 – 9	DATE OF MONTH (01 – 31)
$25	$A5	TENS 0 – 1	UNITS 0 – 9	MONTH (01 – 12) / JAN = 1
$26	$A6	TENS 0 – 9	UNITS 0 – 9	YEAR (00 – 99)
$31	$B1	7 6 5 4	3 2 1 0	CLOCK CONTROL REGISTER
$32	$B2	7 6 5 4	3 2 1 0	INTERRUPT CONTROL REGISTER

		WRITE–ONLY REGISTERS		
N/A	$A8	TENS 0 – 5	UNITS 0 – 9	SECONDS ALARM (00 – 59)
N/A	$A9	TENS 0 – 5	UNITS 0 – 9	MINUTES ALARM (00 – 59)
N/A	$AA	X X PM/AM TENS 0 – 2	UNITS 0 – 9	HOURS ALARM (01 – 21 OR 00 – 23) / DB5, 1 = PM, 0 = AM IN 12 HR MODE

		READ–ONLY REGISTER		
$B0	N/A	7 6 5 4 3 2 1 0		STATUS REGISTER

		RAM DATA BYTE		
$00 TO $1F	$80 TO $9F	D7 D6 D5 D4 D3 D2 D1 D0		DATA

NOTE:
X = Don't Care for Write
X = 0 for Read
N/A = Not Applicable

Figure 6. Clock/RAM Registers

Table 1. Clock/Calendar and Alarm Data Modes

Address Location		Function	Decimal Range	BCD Data Range	BCD Date* Example
Read	Write				
$20	$A0	Seconds	0 − 59	00 − 59	21
$21	$A1	Minutes	0 − 59	00 − 59	40
$22	$A2	Hours** (12 Hour Mode)	1 − 12	81 − 92 (AM) A1 − B2 (PM)	90
		Hours (24 Hour Mode)	0 − 23	00 − 23	10
$23	$A3	Day of Week (Sunday = 1)	1 − 7	01 − 07	03
$24	$A4	Date of Month	1 − 31	01 − 31	16
$25	$A5	Month (Jan = 1)	1 − 12	01 − 12	06
$26	$A6	Year	0 − 99	00 − 99	87
N/A	$A8	Seconds Alarm	0 − 59	00 − 59	21
N/A	$A9	Minutes Alarm	0 − 59	00 − 59	40
N/A	$AA	Hours Alarm*** (12 Hour Mode)	1 − 12	01 − 12 (AM) 21 − 32 (PM)	10
		Hours Alarm (24 Hour Mode)	0 − 23	00 − 23	10

N/A = Not Applicable
  * Example: 10:40:21 AM, Tuesday, June 16, 1987.
 ** Most significant data bit, D7, is "0" for 24–hour mode and "1" for 12–hour mode. Data bit D5 is "1" for PM and "0" for AM in 12–hour mode.
*** Data bit D5 is "1" for PM and "0" for AM in 12–hour mode. Data bits D7 and D6 are Don't Cares.

Table 2. Watchdog Service and Reset Times

	50 Hz		60 Hz		XTAL	
	Min	Max	Min	Max	Min	Max
Service Time	—	10 ms	—	8.3 ms	—	7.8 ms
Reset Time	20 ms	40 ms	16.7 ms	33.3 ms	15.6 ms	31.3 ms

NOTE: Reset does not occur immediately after slave select is toggled. Approximately two clock cycles later, reset initiates.

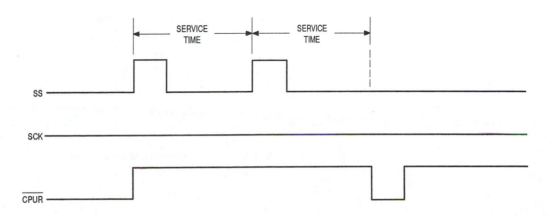

Figure 7. Watchdog Operation Waveforms

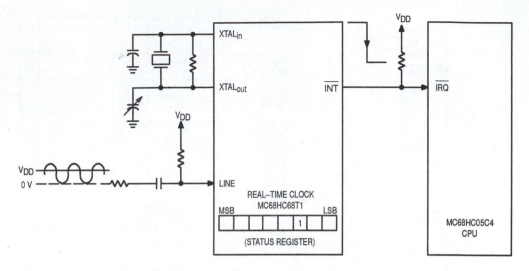

NOTE: A 60 Hz, 10 V p–p sine–wave voltage is an acceptable signal to present at the LINE input pin.

Figure 8. Power Sensing Functional Diagram

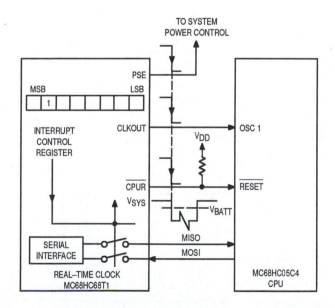

Figure 9. Software Power–Down Functional Diagram

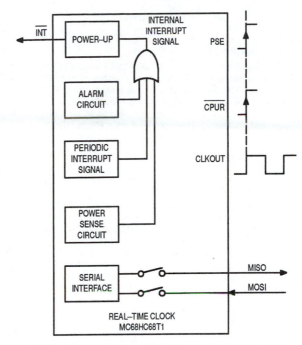

NOTE: The V$_{SYS}$ pin must be powered up.

Figure 10. Power–Up Functional Diagram (Initiated by Internal Interrupt Signal Generation)

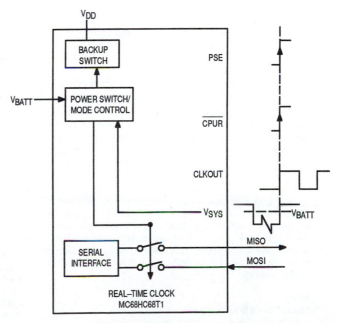

Figure 11. Power–Up Functional Diagram (Initiated by a Rise in Voltage on the V$_{SYS}$ Pin)

Reprinted with permission of Motorola

PIN DESCRIPTIONS

CLKOUT
Clock Output (Pin 1)

This signal is the buffered clock output which can provide one of the seven selectable frequencies (or this output can be reset low). The contents of the three least significant bit positions in the clock control register determine the output frequency (50% duty cycle, except 2 Hz in the 50 Hz time–base mode). During power–down operation (Power–Down bit in the interrupt control register set high), the CLKOUT pin is reset low.

$\overline{\text{CPUR}}$
CPU Reset (Pin 2)

This pin provides an N channel, open–drain output and requires an external pullup resistor. This active low output can be used to drive the reset pin of a microprocessor to permit orderly power–up/power–down. The CPUR output is low from 15 to 40 ms when the watchdog function detects a CPU failure (see Table 2). The low level time is determined by the input frequency source selected as the time standard. CPUR is reset low when power–down is initiated.

$\overline{\text{INT}}$
Interrupt (Pin 3)

This active–low output is driven from a single N channel transistor and must be tied to an external pullup resistor. Interrupt is activated to a low level when any one of the following takes place:

1. Power sense operation is selected (Power Sense Bit in the interrupt control register is set high) and a power failure occurs.

2. A previously set alarm time occurs. The alarm bit in the status register and the interrupt signal are delayed 30.5 ms when 32 kHz or 1 MHz operation is selected, 15.3 ms for 2 MHz operation, and 7.6 ms for 4 MHz operation.

3. A previously selected periodic interrupt signal activates.

The status register must be read to reset the interrupt output after the selected periodic interval occurs. This is also true when conditions 1 and 2 activate the interrupt. If power–down has been previously selected, the interrupt also sets the power–up function only if power is supplied to the V_{SYS} pin to the proper threshold level above V_{BATT}.

SCK
Serial Clock (Pin 4)

This serial clock input is used to shift data into and out of the on–chip interface logic. SCK retains its previous state if the line driving it goes into a high–impedance state. In other words, if the source driving SCK goes to the high–impedance state, the previous low or high level is retained by on–chip control circuitry.

MOSI
Master Out Slave In (Pin 5)

The serial data present at this port is latched into the interface logic by SCK if the logic is enabled. Data is shifted in, either on the rising or falling edges of SCK, with the most significant bit (MSB) first.

In Motorola's microcomputers with SPI, the state of the CPOL bit determines which is the active edge of SCK. If SCK is high when SS goes high, the state of the CPOL bit is high. Likewise, if a rising edge of SS occurs while SCK is low (see Figure 13), then the CPOL bit in the microcomputer is low.

MOSI retains its previous state if the line driving it goes into high–impedance state. In other words, if the source driving MOSI goes to the high–impedance state, the previous low or high level is retained by on–chip control circuitry.

MISO
Master In Slave Out (Pin 6)

The serial data present at this port is shifted out of the interface logic by SCK if the logic is enabled. Data is shifted out, either on the rising or falling edge of SCK, with the most significant bit (MSB) first. The state of the CPOL bit in the microcomputer determines which is the active edge of SCK (see Figure 13).

SS
Slave Select (Pin 7)

When high, the slave select input activates the interface logic; otherwise the logic is in a reset state and the MISO pin is in the high–impedance state. The watchdog circuit is toggled at this pin. SS has an internal pulldown device. Therefore, if SS is in a low state before going to high impedance, SS can be left in a high–impedance state. That is, if the source driving SS goes to the high–impedance state, the previous low level is retained by on–chip control circuitry.

V_{SS}
Ground (Pin 8)

This pin is connected to ground.

PSE
Power Supply Enable (Pin 9)

The power supply enable output is used to control system power and is enabled high under any one of the following conditions:

1. V_{SYS} rises above the V_{BATT} voltage after V_{SYS} is reset low by a system failure.

2. An interrupt occurs (if the V_{SYS} pin is powered up 0.7 V above V_{BATT}).

3. A power–on reset occurs (if the V_{SYS} pin is powered up 0.7 V above V_{BATT}).

PSE is reset low by writing a high into the power–down bit of the interrupt control register.

$\overline{\text{POR}}$
Power–On Reset (Pin 10)

This active–low Schmitt–trigger input generates an internal power–on reset signal using an external RC network (see Figures 18 through 21). Both control registers and frequency dividers for the oscillator and line inputs are reset. The status register is reset except for the first time–up bit (bit 4), which is set high. At the end of the power–on reset, single–supply or battery–backup mode is selected at this time, determined by the state of V_{SYS}.

This pin may be more aptly named first–time–up reset.

LINE
Line Sense (Pin 11)

The LINE sense input can be used to drive one of two functions. The first function utilizes the input signal as the frequency source for the timekeeping counters. This function is selected by setting the line/XTAL bit high in the clock control register. The second function enables the LINE input to detect a power failure. Threshold detectors operating above and below V_{DD} sense an ac voltage loss. The Power Sense bit in the interrupt control register must be set high, and crystal or external clock source operation is required. The line/XTAL bit in the clock control register must be low to select crystal operation. When Power Sense is enabled, this pin, left unconnected, floats to V_{DD}.

This output has no ESD protection diode tied to V_{DD} which allows this pin's voltage to rise above V_{DD}. Care must be taken in the handling of this device.

V_{SYS}
System Voltage (Pin 12)

This input is connected to system voltage. The level on this pin initiates power–up if it rises 0.7 V above the level at the V_{BATT} input pin after previously falling below 0.7 V below V_{BATT}. When power–up is initiated, the PSE pin returns high and the CLKOUT pin is enabled. The CPUR output pin is also set high. Conversely, if the level of the V_{SYS} pin falls below V_{BATT} + 0.7 V, the PSE, CLKOUT, and CPUR pins are placed low. The voltage level present at this pin at the end of POR determines the device's operating mode.

V_{BATT}
Battery Voltage (Pin 13)

This pin is the *only* oscillator power source and should be connected to the positive terminal of the battery. The V_{BATT} pin **always** supplies power to the MC68HC68T1, even when the device is not in the battery–backup mode. To maintain timekeeping, the V_{BATT} pin must be at least 2.2 V. When the level on the V_{SYS} pin falls below V_{BATT} + 0.7 V, **V_{BATT} is internally connected to the V_{DD} pin.**

When the LINE input is used as the frequency source, the unused V_{BATT} and XTAL pins may be tied to V_{SS}. Alternatively, if V_{BATT} is connected to V_{DD}, $XTAL_{in}$ can be tied to either V_{SS} or V_{DD}.

This output has no ESD protection diode tied to V_{DD} which allows this pin's voltage to rise above V_{DD}. Care must be taken in the handling of this device.

$XTAL_{in}$, $XTAL_{out}$
Crystal Input/Output (Pins 14, 15)

For crystal operation, these two pins are connected to a 32.768 kHz, 1.048576 MHz, 2.097152 MHz, or 4.194304 MHz crystal. If crystal operation is not desired and Line Sense is used as frequency source, connect $XTAL_{in}$ to V_{DD} or V_{SS} (caution: see V_{BATT} pin description) and leave XTA_{out} open. If an external clock is used, connect the external clock to $XTAL_{in}$ and leave $XTAL_{out}$ open. The external clock must swing from at least 30 to 70% of ($V_{DD} - V_{SS}$). Preferably, this input should swing from V_{SS} to V_{DD}.

V_{DD}
Positive Power Supply (Pin 16)

For full functionality, the positive power supply pin may range from 3.0 to 6.0 V with respect to V_{SS}. To maintain time-keeping, the minimum standby voltage is 2.2 V with respect to V_{SS}. For proper operation in battery–backup mode, a diode **must** be placed in series with V_{DD}.

CAUTION

Data transfer to/from the MC68HC68T1 must not be attempted if the supply voltage falls below 3.0 V.

REGISTERS

CLOCK CONTROL REGISTER (READ/WRITE) — READ ADDRESS $31/WRITE ADDRESS $B1

MSB D7	D6	D5	D4	D3	D2	D1	LSB D0
START — STOP	LINE — XTAL	XTAL SELECT 1	XTAL SELECT 0	50 Hz — 60 Hz	CLK OUT 2	CLK OUT 1	CLK OUT 0

All bits are reset low by a power–on reset.

Start–Stop

A high written into this bit enables the counter stages of clock circuitry. A low holds all bits reset in the divider chain from 32 Hz to 1 Hz. The clock out signal selected by bits D0, D1, and D2 is not affected by the stop function except the 1 and 2 Hz outputs.

Line/ XTAL

When this bit is high, clock operation uses the 50 or 60 cycle input present at the LINE input pin. When the bit is low, the $XTAL_{in}$ pin is the source of the time update.

XTAL Select

Accommodation of one of four possible crystals are selected by the value in bits D4 and D5.

0 = 4.194304 MHz   2 = 1.048576 MHz

1 = 2.097152 MHz   3 = 32.768 kHz

The MC68HC68T1 has an on–chip 150 kΩ resistor that is switched in series with the internal inverter when 32 kHz is selected via the clock control register. At power–up, the device sets up for a 4 MHz oscillator and the series resistor is not part of the oscillator circuit. Until this resistor is switched in, oscillations may be unstable with the 32 kHz crystal. (See Figure 12.)

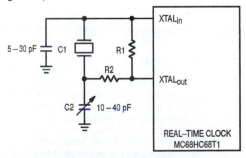

Figure 12. Recommended Oscillator Circuit (C1, C2 Values Depend Upon the Crystal Frequency)

Resistor R1 is recommended to be 10 MΩ for 32 kHz operation. Consult crystal manufacturer for R1 value for other frequencies. Resistor R2 must be used in 32 kHz operation only. Use a 200 to 300 kΩ range. This stabilizes the oscillator until the control register is set properly and reduces standby current.

50 Hz — 60 Hz

50 Hz may be used as the input frequency at the LINE input when this bit is set high; a low accommodates 60 Hz. The power sense bit in the interrupt control register must be reset low for line frequency operation.

Clock Out

Three bits specify one of the seven frequencies to be used as the square–wave clock output (CLKOUT).

0 = XTAL	4 = Disable (low output)
1 = XTAL/2	5 = 1 Hz
2 = XTAL/4	6 = 2 Hz
3 = XTAL/8	7 = 50/60 Hz for LINE operation
	7 = 64 Hz for XTAL operation

All bits in the clock control register are reset by a power–on reset. Therefore, XTAL is selected as the clock output at this time.

INTERRUPT CONTROL REGISTER (READ/WRITE) — READ ADDRESS $32/WRITE ADDRESS $B2

MSB D7	D6	D5	D4	D3	D2	D1	LSB D0
WATCH–DOG	POWER–DOWN	POWER SENSE	ALARM	PERIODIC SELECT			

All bits are reset low by power–on reset.

Watchdog

When this bit is set high, the watchdog operation is enabled. This function requires the CPU to toggle the SS pin periodically without a serial transfer requirement. In the event this does not occur, a CPU reset is issued at the CPUR pin. The status register must be read before re–enabling the watchdog function.

Power–Down

A high in this location initiates a power–down. A CPU reset occurs via the CPUR output, the CLKOUT and PSE output pins are reset low, and the serial interface is disabled.

Power Sense

When set high, this bit is used to enable the LINE input pin to sense a power failure. When power sense is selected, the input to the 50/60 Hz prescaler is disconnected; therefore, crystal operation is required. An interrupt is generated when a power failure is sensed and the power sense and interrupt true bit in the status register are set. When power sense is activated, a logic low must be written to this location followed by a high to re–enable power sense.

Alarm

The output of the alarm comparator is enabled when this bit is set high. When an equal comparison occurs between the seconds, minutes, and hours time counters and alarm latches, the interrupt output is activated. When loading the time counters, this bit should be reset low to avoid a false interrupt. This is not required when loading the alarm latches. See INT pin description for explanation of alarm delay.

Periodic Select

The value in these four bits (D0, D1, D2, and D3) selects the frequency of the periodic output (see Table 3).

Table 3. Periodic Interrupt Output Frequencies (at INT Pin)

D3 – D0 Value (Hex)	Periodic Interrupt Output Frequency	Frequency Timebase	
		XTAL	Line
0	Disable		
1	2048 Hz	X	
2	1024 Hz	X	
3	512 Hz	X	
4	256 Hz	X	
5	128 Hz	X	
6	64 Hz	X	
	50 or 60 Hz		X
7	32 Hz	X	
8	16 Hz	X	
9	8 Hz	X	
A	4 Hz	X	
B	2 Hz	X	X
C	1 Hz	X	X
D	1 Cycle per Minute	X	X
E	1 Cycle per Hour	X	X
F	1 Cycle per Day	X	X

STATUS REGISTER (READ ONLY) — ADDRESS $30

MSB D7	D6	D5	D4	D3	D2	D1	LSB D0
0	WATCH–DOG	0	FIRST TIME–UP	INTER–RUPT TRUE	POWER SENSE INT	ALARM INT	CLOCK INT

NOTE

All bits are reset low by a power–on reset except the first time–up bit which is set high. All bits except the power sense bit are reset after a read of the status register.

Watchdog

If this bit is set high, the watchdog circuit has detected a CPU failure.

First Time–Up

Power–on reset sets this bit high. This signifies the data in the RAM and Clock is not valid and should be initialized.

After the status register is read, the first time–up bit is set low if the POR pin is high. Conversely, if the POR pin is held low, the first time–up bit remains set high.

Interrupt True

A high in this bit signifies that one of the three interrupts (power sense, alarm, or clock) is valid.

Power–Sense Interrupt

This bit set high signifies that the power–sense circuit has generated an interrupt. This bit is not reset after a read of this register.

Alarm Interrupt

When the contents of the seconds, minutes, and hours time counters and alarm latches are equal, this bit is set high. The status register must be read before loading the interrupt control register for valid alarm indication after the alarm activates.

Clock Interrupt

A periodic interrupt sets this bit high (see Table 3).

SERIAL PERIPHERAL INTERFACE (SPI)

The serial peripheral interface (SPI) utilized by the MC68HC68T1 is a serial synchronous bus for address and data transfers. The shift clock (SCK), which is generated by the microcomputer, is active only during address and data transfer. In systems using the MC68HC05C4 or MC68HC11A8, the inactive clock polarity is determined by the clock polarity (CPOL) bit in the microcomputer's control register.

A unique feature of the MC68HC68T1 is that the level of the inactive clock is determined by sampling SCK when SS becomes active. Therefore, either SCK polarity is accommodated. Input data (MOSI) is latched internally on the internal strobe edge and output data (MISO) is shifted out on the shift edge (see Table 4 and Figure 13). There is one clock for each bit transferred. Address as well as data bits are transferred in groups of eight.

Table 4. Function Table

Mode	SS	SCK	MOSI	MISO
Disabled Reset	L	Input Disabled	Input Disabled	High–Z
Write	H	CPOL = 1 CPOL = 0	Data Bit Latch	High–Z
Read	H	CPOL = 1 CPOL = 0	X	Next Data Bit Shifted Out*

* MISO remains at a High–Z until eight bits of data are ready to be shifted out during a read. MISO remains at a High–Z during the entire write cycle.

ADDRESS AND DATA FORMAT

There are three types of serial transfers:

1. Read or write address
2. Read or write data
3. Watchdog reset (actually a non–transfer)

The address and data bytes are shifted MSB first, into the serial data input (MOSI) and out of the serial data output (MISO). Any transfer of data requires the address of the byte to specify a write or read Clock or RAM location, followed by one or more bytes of data. Data is transferred out of MISO for a read operation and into MOSI for a write operation (see Figures 14 and 15).

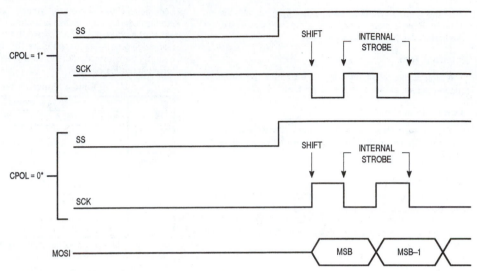

* CPOL is a bit that is set in the microcomputer's Control Register.

Figure 13. Serial Clock (SCK) as a Function of MCU Clock Polarity (CPOL)

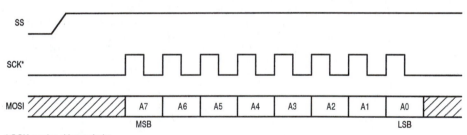

* SCK can be either polarity.

Figure 14. Address Byte Transfer Waveforms

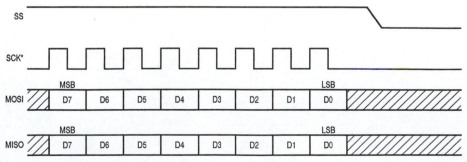

* SCK can be either polarity.

Figure 15. Read/Write Data Transfer Waveforms

Address Byte

The address byte is always the first byte entered after SS goes true. To transmit a new address, SS must first be brought low and then taken high again.

MSB							LSB
A7	A6	A5	A4	A3	A2	A1	A0

A7 — High initiates one or more write cycles.
Low initiates one or more read cycles.

A6 — Must be low (zero) for normal operation.

A5 — High signifies a clock/calendar location.
Low signifies a RAM location.

A0 – A4 — Remaining address bits (see Figure 5).

Address and Data

Data transfers can occur one byte at a time or in multi-byte burst mode (see Figures 16 and 17). After the MC68HC68T1 is enabled (SS = high), an address byte selects either a read or a write of the Clock/Calendar or RAM. For a single-byte read or write, one byte is transferred to or from the Clock/Calendar register or RAM location specified by an address. Additional reading or writing requires re-enabling the device and providing a new address byte. If the MC68HC68T1 is not disabled, additional bytes can be read or written in a burst mode. Each read or write cycle causes the Clock/Calendar register or RAM address to automatically increment. Incrementing continues after each byte transfer until the device is disabled. After incrementing to $1F or $9F, the address wraps to $00 and continues if the RAM is selected. When the Clock/Calendar is selected, the address wraps to $20 after incrementing to $32 to $B2.

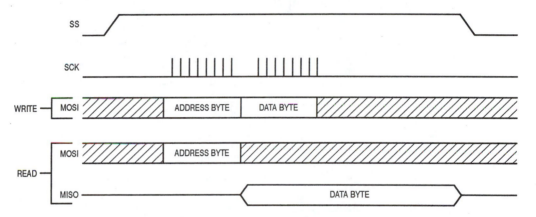

Figure 16. Single-Byte Transfer Waveforms

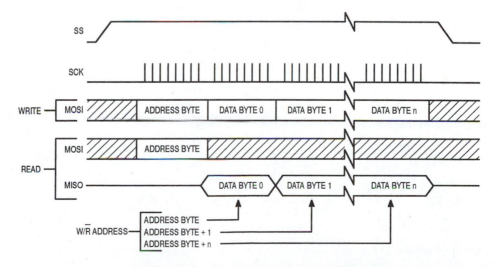

Figure 17. Multiple-Byte Transfer Waveforms

APPLICATION CIRCUITS

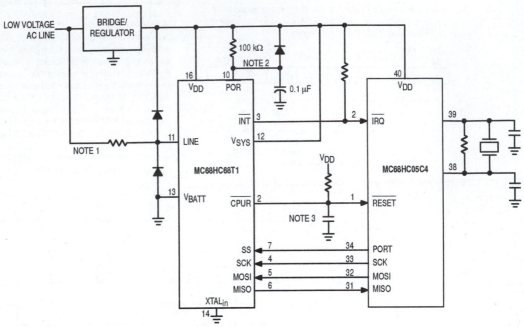

NOTES:
1. Clock circuit driven by line input frequency.
2. Power–on reset circuit included to detect power failure.
3. If an MC68HC11 MCU is used, delete the capacitor at the $\overline{\text{RESET}}$ pin.

Figure 18. Power–Always–On System

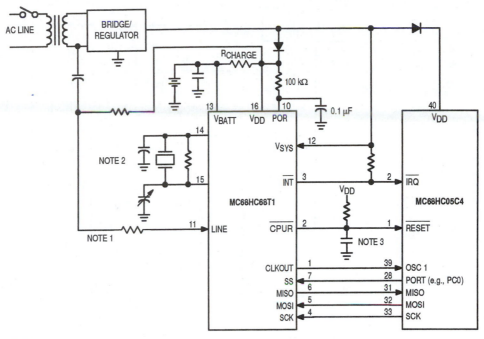

NOTES:
1. The LINE input pin can sense when the switch opens by use of the power sense interrupt. The MC68HC68T1 crystal drives the clock input to the CPU using the CLKOUT pin. On power–down when $V_{SYS} < V_{BATT} + 0.7$ V, V_{BATT} powers the clock. A threshold detect activates an on–chip P channel switch, connecting V_{BATT} to V_{DD}. V_{BATT} always supplies power to the oscillator, keeping voltage frequency variation to a minimum.
2. For 32.768 kHz oscillator, see Figure 12. This configuration, when the MC68HC68T1 supplies the MCU clock, usually requires a 1 to 4 MHz clock.
3. If an MC68HC11 MCU is used, delete the capacitor at the RESET pin.

Figure 19. Externally–Controlled Power System

POWER–SENSING POWER–DOWN PROCEDURE

A procedure for power–down operation consists of the following:

1. Set power sense operation by writing bit 5 high in the interrupt control register.
2. When an interrupt occurs, the CPU reads the status register to determine the interrupt source.
3. Sensing a power failure, the CPU does the necessary housekeeping to prepare for shutdown.

4. The CPU reads the status register again after several milliseconds to determine validity of power failure.
5. The CPU sets power–down (bit 6) and disables all interrupts in the interrupt control register when power–down is verified. This causes the CPU reset and Clock Out pins to be held low and disconnects the serial interface.
6. When power returns and V_{SYS} rises above $V_{BATT} + 0.7$ V, power–up is initiated. The CPU reset is released and serial communication is established.

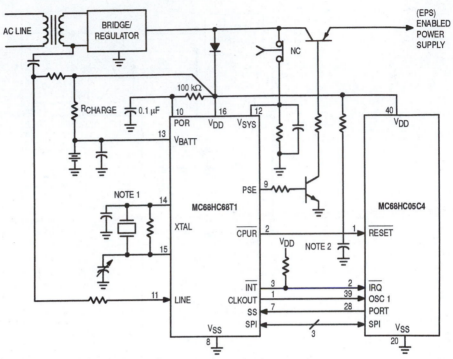

NOTES:
1. See Figure 12 for 32.768 kHz operation. This configuration, where the MC68HC68T1 supplies the MCU clock, usually requires a 1 to 4 MHz crystal.
2. If an MC68HC11 MCU is used, delete the capacitor at the $\overline{\text{RESET}}$ pin.

Figure 20. Rechargeable Battery–Backup System

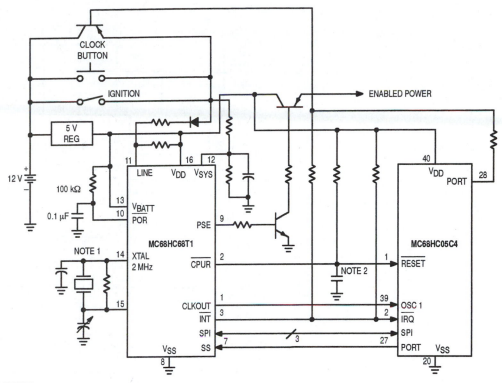

NOTES:
1. The V_{SYS} and Line inputs can be used to sense the ignition turning on and off. An external switch is included to activate the system without turning on the ignition. Also, the CMOS CPU is not powered down with the system V_{DD}, but is held in a low power reset mode during power–down. When restoring power, the MC68HC68T1 enables the CLKOUT pin and sets the PSE and CPUR pins high.
2. If an MC68HC11 MCU is used, delete the capacitor at the RESET pin.
3. Voltage at pin must not exceed absolute maximum V_{in} specification.

Figure 21. Automotive System

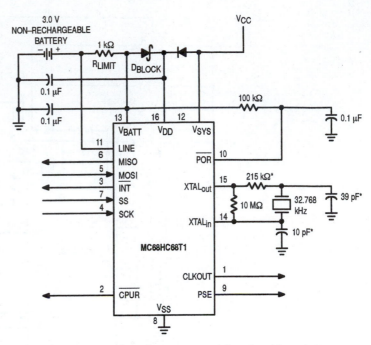

* Actual values may vary, depending on recommendations of crystal manufacturer.

Figure 22. Non–Rechargeable Battery–Backup System

TROUBLESHOOTING

1. *The circuit works, but the standby current is well above the spec. How can the standby current be reduced?*

   a. If using a 32.768 kHz crystal, include a series resistor in the circuit per Figure 12 of the data sheet. A good value to start with is 200 kΩ. The signals at $XTAL_{out}$ and $XTAL_{in}$ pins should look similar to Figure 23 when the correct value is selected. The sharp, clean edges on the $XTAL_{out}$ pin reduces current on the totem pole drivers internal to the device.

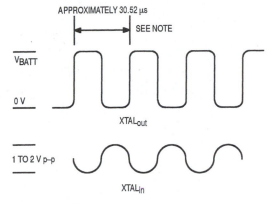

NOTE: Refer to item 8.

Figure 23. XTAL Waveforms

   b. Connect the LINE pin to something other than V_{DD} (e.g., V_{BATT}, V_{SS}, V_{SYS}).

   c. Ensure that the Power–On–Reset (POR) has a time constant of at least 100 ms.

   d. Ensure that there is a diode from V_{DD} to + 5 V of the system, in battery–backup applications. See **Application Circuits**.

2. *When power is applied, the clock does not start up nor does it hold data in the control registers.*

   Make sure the POR circuit is connected and working.

3. *The clock loses time, but the oscillator is tuned.*

   Do not make constant accesses to the clock. When a read or write cycle is started, the clock stops incrementing time.

4. *When the part is power cycled, the clock loses all time and data.*

   Check the battery installation and ensure that a diode is in the circuit from V_{DD} to + 5 V.

5. *Can a non–rechargeable lithium battery be used?*

   Yes, but the battery must have a large capacity. Careful attention MUST be given if the end unit needs to be UL approved. The circuit of Figure 22 is a good start.

6. *Able to read/write data to the RAM but not to the clock registers, or vice versa.*

   There is a software problem. There is no internal difference from reading/writing to the RAM or clock locations.

7. *How is the oscillator tuned?*

   The best way to tune the oscillator is to set the clock out bits of the Clock Control Register (bits 0, 1, and 2) to output the primary XTAL frequency (000). The frequency can then be more accurately measured from the CLKOUT pin. This prevents the measuring device from loading the oscillator circuit, which may shift the frequency.

8. *What is the accuracy of the oscillator?*

   The oscillator accuracy is dependent on the quality of the crystal used. For every 1 ppm variance in crystal frequency, the clock gains or loses 2.6 seconds per month. 25 ppm is a typical spec for a crystal, which translates to ± 65 seconds per month.

9. *Can the Line pin sense a dc failure?*

   Yes, the Line input is threshold triggered in a window from one diode drop above and below V_{DD}. If supply is removed in the low cycle of a sine wave, the internal network pulls the line pin to within the threshold in a few milliseconds. In the absence of a dc voltage outside the V_{DD}± 0.7 V window, the internal network pulls the signal to within the window and triggers the interrupt.

10. *Can the V_{SYS} line be more than 0.5 V above V_{DD}?*

    No. There is an ESD protection network that causes a supply problem with this application.

11. *The CLKOUT, CPUR, and PSE pins do not go inactive when V_{DD} and V_{SYS} are removed. The CLKOUT, CPUR, and PSE are not active immediately when V_{DD} and V_{SYS} is applied.*

    The problem is related to the power up procedure (battery–backup mode or single–supply mode). See these sections in the data sheet for more information.

PACKAGE DIMENSIONS

P SUFFIX
PLASTIC DIP (DUAL IN–LINE PACKAGE)
CASE 648–08

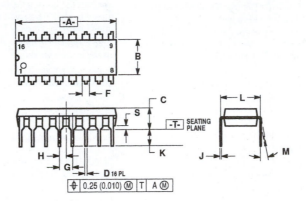

NOTES:
1. DIMENSIONING AND TOLERANCING PER ANSI Y14.5M, 1982.
2. CONTROLLING DIMENSION: INCH.
3. DIMENSION "L" TO CENTER OF LEADS WHEN FORMED PARALLEL.
4. DIMENSION "B" DOES NOT INCLUDE MOLD FLASH.
5. ROUNDED CORNERS OPTIONAL.
6. 648-01 THRU -07 OBSOLETE, NEW STANDARD 648-08.

DIM	MILLIMETERS		INCHES	
	MIN	MAX	MIN	MAX
A	18.80	19.55	0.740	0.770
B	6.35	6.85	0.250	0.270
C	3.69	4.44	0.145	0.175
D	0.39	0.53	0.015	0.021
F	1.02	1.77	0.040	0.070
G	2.54 BSC		0.100 BSC	
H	1.27 BSC		0.050 BSC	
J	0.21	0.38	0.008	0.015
K	2.80	3.30	0.110	0.130
L	7.50	7.74	0.295	0.305
M	0°	10°	0°	10°
S	0.51	1.01	0.020	0.040

DW SUFFIX
SOG (SMALL OUTLINE GULL–WING) PACKAGE
CASE 751G–01

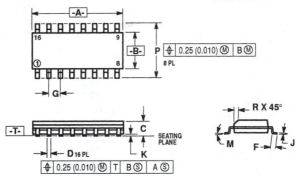

NOTES:
1. DIMENSIONING AND TOLERANCING PER ANSI Y14.5M, 1982.
2. CONTROLLING DIMENSION: MILLIMETER.
3. DIMENSION A AND B DO NOT INCLUDE MOLD PROTRUSION.
4. MAXIMUM MOLD PROTRUSION 0.15 (0.006) PER SIDE.

DIM	MILLIMETERS		INCHES	
	MIN	MAX	MIN	MAX
A	10.15	10.45	0.400	0.411
B	7.40	7.60	0.292	0.299
C	2.35	2.65	0.093	0.104
D	0.35	0.49	0.014	0.019
F	0.50	0.90	0.020	0.035
G	1.27 BSC		0.050 BSC	
J	0.25	0.32	0.010	0.012
K	0.10	0.25	0.004	0.009
M	0°	7°	0°	7°
P	10.05	10.55	0.395	0.415
R	0.25	0.75	0.010	0.029

Glossary

Accumulator   A register in a computer that contains an operand to be used in an arithmetic operation.

Activation record   Another term for stack frame.

Address access time   The amount of time it takes for a memory component to send out valid data to the external data pins after address signals have been applied (assuming that all other control signals have been asserted).

Addressing   The application of a unique combination of high and low logic levels to select a corresponding unique memory location.

Address multiplexing   A technique that allows the same address pin to carry different signals at different time; used mainly by DRAM technology. Address multiplexing can dramatically reduce the number of address pins required by DRAM chips and reduce the size of the memory chip package.

ALU (arithmetic logic unit)   The part of the processor in which all arithmetic and logical operations are performed.

Array   An ordered set of elements of the same type. The elements of the array are arranged so that there is a zeroth, first, second, third, and so forth. An array may be one-, two-, or multidimensional.

ASCII (American Standard Code for Information Interchange) code   A code that uses seven bits to encode all printable and control characters.

Assembler   A program that converts a program in assembly language into machine instructions so that it can be executed by a computer.

Assembler directive   A command to the assembler for defining data and symbols, setting assembler conditions, and specifying output format. Assembler directives do not produce machine code.

Assembly instruction   A mnemonic representation of a machine instruction.

Binary Coded Decimal (BCD)   A coding method that uses four binary digits to represent a decimal digit. The binary codes 0000_2 to 1001_2 correspond to the decimal digits 0 to 9.

Bootstrap mode   The operation mode in which the 68HC11 will start to execute the bootloader program residing at \$BF40-\$BFFF after it gets out of the reset state. The bootloader program will use the serial communication interface to read a 256-byte program into the on-chip RAM at locations \$0000-\$00FF. After the last byte is received, control is passed to that program at location \$0000.

Break   The transmission or reception of a low for at least one complete character time.

Breakpoint   A memory location in a program where the user program execution will be stopped and the monitor program will take over the CPU control and display the contents of CPU registers.

Bus   A set of signal lines through which the processor of a computer communicates with memory and I/O devices.

Bus cycle timing diagram   A diagram that describes the transitions of all the involved signals during a read or write operation.

Central processing unit (CPU)   The combination of the register file, the ALU, and the control unit.

Charge pump   A circuit technique that can raise a low voltage to a level above the power supply. A charge pump is often used in A/D converter, in EEPROM and EPROM programming, etc.

Column address strobe (CAS)   The signal used by DRAM chips to indicate that column address logic levels are applied to the address input pins.

Comment   A statement that explains the function of a single instruction or directive or a group of instructions or directives. Comments make a program more readable.

Computer   A computer consists of hardware and software. The hardware includes four major parts: the central processing unit, the memory unit, the input unit, and the output unit. Software is a sequence of instructions that control the operations of the hardware.

Computer operating properly (COP) watchdog timer   A system for detecting software processing errors. If software is written correctly, then it should complete all operations within some time limit. Software problems can be detected by enabling a watchdog timer so that the software resets the watchdog timer before it times out.

Control unit   The part of the processor that decodes and monitors the execution of instructions. It arbitrates the use of computer resources and makes sure that all computer operations are performed in proper order.

Cross assembler   An assembler that runs on one computer but generates machine instructions that will be executed by another computer with a different instruction set.

Cross compiler   A compiler that runs on one computer but generates machine instructions that will be executed by another computer with a different instruction set.

Data hold time   The length of time over which the data must remain stable after the edge of the control signal that latches the data arrives.

Data setup time   The amount of time over which the data must become valid before the edge of the control signal that latches the data arrives.

DCE   The acronym of data communication equipment. DCE usually refers to equipment such as a modem, concentrator, router, etc.

Dequeue   A data structure in which elements can be added and deleted at both ends.

DTE   The acronym for data terminal equipment. DTE usually refers to a computer or terminal.

Dynamic memories   Memory devices that require periodic refreshing of the stored information, even when power is on.

EBCDIC (Extended Binary Coded Decimal Interchange Code)   A code used mainly in IBM mainframe computers; it uses eight bits to represent each character.

EIA   The acronym of the Electronic Industry Association.

Electrically erasable programmable read-only memory (EEPROM) EEPROM can be erased and reprogrammed using electrical signals. EEPROM allows each individual location inside the chip to be erased and reprogrammed.

Erasable programmable read-only memory (EPROM) A type of read-only memory that can be erased by subjecting it to strong ultraviolet light. It can be reprogrammed using an EPROM programmer. A quartz window on top of the EPROM chip allows light to be shone directly on the silicon chip inside.

Exception A software interrupt, such as an illegal opcode, an overflow, division by zero, or an underflow.

Excess-3 A coding method for decimal digits; each excess-3 code is larger than its corresponding decimal digit by 3.

Expanded mode The operation mode in which the 68HC11 can access external memory components by sending out address signals. A 64 KB memory space is available in this mode.

Fall time The amount of time a digital signal takes to go from logic high to logic low.

Floating signal An undriven signal.

Framing error A data communication error in which a received character is not properly framed by the start and stop bits.

Frame pointer A pointer used to facilitate access to parameters in a stack frame.

Frame sync A signal that synchronizes the updating of the new display data between the LCD master (MC145000) and slave (MC145001) driver chips.

Framing error A data communication error in which a received character is not properly framed by the start and stop bits.

Full-duplex link A four-wire communication link that allows both transmission and reception to proceed simultaneously.

Global memory Memory that is available for all programs in a computer system.

Graph A data structure that consists of a set of nodes and arcs (or edges). Each arc in a graph is specified by a pair of nodes.

Gray code A coding method for decimal digits in which successive coded characters never differ in more than one bit.

Half duplex link A communication link that can be used for either transmission or reception, but only in one direction at a time.

Idle A continuous logic high on the RxD line for one complete character time.

Illegal opcode A binary bit pattern of the opcode byte for which an operation is not defined.

Input-capture The 68HC11 function that captures the value of the 16-bit free-running counter into a latch when the falling or rising edge of the signal connected to the input-capture pin arrives.

Interrupt An unusual event that requires the CPU to stop normal program execution and perform some service to the event.

Interrupt priority   The order in which the CPU will service interrupts when all of them occur at the same time.

Interrupt service   The service provided to a pending interrupt by CPU execution of a program called a service routine.

Interrupt vector   The starting address of an interrupt service routine.

Interrupt vector table   A table that stores all interrupt vectors.

ISO   The acronym for the International Standard Organization.

Linked list   A data structure that consists of linked nodes. Each node consists of two fields, an information field and a next address field. The information field holds the actual element on the list, and the next address field contains the address of the next node in the list.

Load cell   A transducer that can convert weight into a voltage.

Local variable   Temporary variables that exist only when a subroutine is called. They are used as loop indices, working buffers, etc. Local variables are often allocated in the system stack.

Low power mode   An operation mode in which less power is consumed. In CMOS technology, the low-power mode is implemented by either slowing down the clock frequency or turning off some circuit modules within a chip.

Machine instruction   A set of binary digits that tells the computer what operation to perform.

Mark   A term used to indicate a binary 1.

Maskable interrupts   Interrupts that can be ignored by the CPU. This type of interrupt can be disabled by setting a mask bit or by clearing an enable bit.

Masked ROM (MROM)   A type of ROM that is programmed when it is fabricated.

Matrix   A two-dimensional data structure that is organized into rows and columns. The elements of a matrix are of the same length and are accessed using their row and column numbers (i, j), where i is the row number and j is the column number.

Memory   Storage for software and information.

Memory capacity   The total amount of information that a memory device can store; also called memory density.

Memory organization   A description of the number of bits that can be read from or written into a memory chip during a read or write operation.

Microprocessor   A CPU packaged in a single integrated circuit.

Microcontroller   A computer system implemented on a single, very large-scale integrated circuit. A microcontroller contains everything that is in a microprocessor and may contain memories, an I/O device interface, a timer circuit, an A/D converter, and so on.

Mode fault   An SPI error that indicates that there may have been a multimaster conflict for system control. Mode fault is detected when the master SPI device has its SS pin pulled low.

Modem   A device that can accept digital bits and change them into a form suitable for analog transmission (modulation) and can also receive a modulated signal and transform it back to its original digital representation (demodulation).

Multidrop   A data communication scheme in which more than two stations share the same data link. One station is designated as the master, and the other stations are designated as slaves. Each station has its own unique address, with the primary station controlling all data transfers over the link.

Multitasking   A computing technique in which CPU time is divided into slots that are usually 10 to 20 ms in length. When multiple programs are resident in the main memory waiting for execution, the operating system assigns a program to be executed to one time slot. At the end of a time slot or when a program is waiting for completion of I/O, the operating system takes over and assigns another program to be executed.

Nibble   A group of four-bit information.

Nonmaskable interrupts   Interrupts that the CPU cannot ignore.

Nonvolatile memory   Memory that retains stored information even when power to the memory is removed.

Null modem   A circuit connection between two DTEs in which the leads are interconnected in such a way as to fool both DTEs into thinking that they are connected to modems. A null modem is only used for short distance interconnections.

Object code   The sequence of machine instructions that results from the process of assembling and/or compiling a source program.

Output-compare   A 68HC11 timer function that allows the user to make a copy of the value of the 16-bit free-running timer, add a delay to the copy, and then store the sum in a register. The output-compare function compares the sum with the free-running timer in each of the following E clock cycles. When these two values are equal, the circuit can trigger a signal change on an output-compare pin and may also generate an interrupt request to the 68HC11.

Overflow   A condition that occurs when the result of an arithmetic operation cannot be accommodated by the preset number of bits (say, 8 or 16 bits); it occurs fairly often when numbers are represented by fixed numbers of bits.

Parameter passing   The process and mechanism of sending parameters from a caller to a subroutine, where they are used in computations; parameters can be sent to a subroutine using CPU registers, the stack, program memory, or global memory.

Parity error   An error in which odd number of bits change value; it can be detected by a parity checking circuit.

Physical time   In the 68HC11 timer system, the time represented by the count in the 16-bit free-running timer counter.

Point-to-point   A data communication scheme in which two stations communicate as peers.

Power-down   An operation that responds to the detection of low power supply; normally it performs some housekeeping function and then resets the system.

Power sense   A function available in several power management chips for detecting power source problems (for example, power drops below normal level).

Power-up   An operation that occurs in response to the detection of the power supply returning to normal. A power-up operation enables the microcomputer to boot.

Program   A set of instructions that the computer hardware can execute.

Program counter (PC)   A register that keeps track of the address of the next instruction to be executed.

PROM (programmable read-only memory)   A type of ROM that allows the end user to program it once and only once using a device called PROM programmer.

Pull   The operation that removes the top element from a stack data structure.

Pulse accumulator   A 68HC11 timer function that uses an 8-bit counter to count the number of events that occur or to measure the duration of a single pulse.

Push   The operation that adds a new element to the top of a stack data structure.

Queue   A data structure to which elements can be added at only one end and removed only from the other end. The end to which new elements can be added is called the tail of the queue, and the end from which elements can be removed is called the head of the queue.

RAM (random-access memory)   RAM allows read and write access to every location inside the memory chip. Furthermore, read access and write access take the same amount of time for any location within the RAM chip.

Receiver overrun   A data communication error in which a character or a number of characters were received but not read from the buffer before subsequent characters being received.

Refresh   An operation performed on dynamic memories in order to retain the stored information during normal operation.

Refresh period   The time interval which each location of a DRAM chip must be refreshed at least once in order to retain its stored information.

Register   A storage location in the CPU. It is used to hold data and/or a memory address during the execution of an instruction.

Reset   A signal or operation that sets the flip-flops and registers of a chip or microprocessor to some predefined values or states so that the circuit or microprocessor can start from a known state.

Reset handling routine   The routine that will be executed when the microcontroller or microprocessor gets out of the reset state.

Reset state   The state in which the voltage level of the RESET pin of the 68HC11 is low. In this state, a default value is established for most on-chip registers, including the program counter. The operation mode is established when the 68HC11 exits the reset state.

Return address   The address of the instruction that immediately follows the subroutine call instruction (either JSR or BSR).

Rise time   The amount of time a digital signal takes to go from logic low to logic high.

ROM (read-only memory)   A type of memory that is nonvolatile in the sense that when power is removed from ROM and then reapplied, the original data are still there. ROM data can only be read—not written—during normal computer operation.

Row address strobe (RAS)   The signal used by DRAM chips to indicate that row address logic levels are applied to the address input pins.

RS232   An interface standard recommended for interfacing between a computer and a modem. This standard was established by EIA in 1960 and has since then been revised several times.

Simplex link   A line is dedicated either for transmission or reception, but not both.

Single-chip mode   The operation mode in which the 68HC11 functions without external address and data buses.

Source code   A program written in either assembly language or a high-level language; also called a source program.

Space   A term used to indicate a binary 0.

Special test mode   The 68HC11 operation mode used primarily during Motorola's internal production testing.

Stack   A last-in-first-out data structure whose elements can be accessed only from one end. A stack structure has a top and a bottom. A new item can be added only to the top, and the stack elements can be removed only from the top.

Stack frame   A region in the stack that holds incoming parameters, the sub-routine return address, local variables, saved registers, and so on.

Static memories   Memory devices that do not require periodic refreshing in order to retain the stored information as long as power is applied.

Status register   A register located in the CPU. It keeps track of the status of instruction execution by noting the presence of carries, zeros, negatives, overflows, and so on.

String   A sequence of characters.

Subroutine   A sequence of instructions that can be called from various places in the program and will return to the caller after its execution. When a subroutine is called, the return address will be saved on the stack.

Subroutine call   The process of invoking the subroutine to perform the desired operations. The 68HC11 has BSR and JSR instructions for making subroutine calls.

Temperature sensor   A transducer that can convert temperature into a voltage.

Thermocouple   A transducer that converts a high temperature into a voltage.

Time-of-day chip   An integrated circuit that can keep track of the current year, month, day, hour, minute, and seconds and may also provide the alarm function to the computer system. Like any other peripheral chip, the time-of-day chip must be initialized by the user before it can function properly.

Transducer   A device that can convert a nonelectric quantity into a voltage.

Transpose   An operation that converts the rows of a matrix into columns and vice versa.

Trap   A software interrupt; an exception.

Tree   A finite nonempty set of elements in which one element is called the root and the remaining elements are partitioned into m (≥ 0) disjoint subsets (branches), each of which is itself a tree. Each element in a tree is called a node of the tree. A node that does not have any branches is called a *leaf*.

UART   The acronym of universal asynchronous receiver and transmitter; an interface chip that allows the microprocessor to perform asynchronous serial data communication.

Vector   A vector is a unidimensional data structure in which each element is associated with an index i. The elements of a vector are of the same length.

Volatile memory   Semiconductor memory that loses its stored information when power is removed.

Write collision   The SPI error that occurs when an attempt is made to write to the SPDR register while data transfer is taking place.

Index